T0211514

The Mechanics of
Engineering Structures

The Mechanics of
Engineering Structures

David W. A. Rees D. Sc.

Brunel University, UK

Imperial College Press

ICP

Published by

Imperial College Press
57 Shelton Street
Covent Garden
London WC2H 9HE

Distributed by

World Scientific Publishing Co. Pte. Ltd.

5 Toh Tuck Link, Singapore 596224

USA office: 27 Warren Street, Suite 401-402, Hackensack, NJ 07601

UK office: 57 Shelton Street, Covent Garden, London WC2H 9HE

Library of Congress Cataloging-in-Publication Data
Rees, D. W. A. (David W. A.), 1947–
   The mechanics of engineering structures / by David W A Rees (Brunel University, UK).
    pages cm
   Includes bibliographical references and index.
   ISBN 978-1-78326-401-8 (hardback : alk. paper) -- ISBN 978-1-78326-408-7 (pbk. : alk. paper)
   1. Structural analysis (Engineering) 2. Mechanics, Applied. I. Title.
   TA645.R385 2014
   624.1'7--dc23
                               2013051297

British Library Cataloguing-in-Publication Data
A catalogue record for this book is available from the British Library.

Copyright © 2015 by Imperial College Press

All rights reserved. This book, or parts thereof, may not be reproduced in any form or by any means, electronic or mechanical, including photocopying, recording or any information storage and retrieval system now known or to be invented, without written permission from the Publisher.

For photocopying of material in this volume, please pay a copying fee through the Copyright Clearance Center, Inc., 222 Rosewood Drive, Danvers, MA 01923, USA. In this case permission to photocopy is not required from the publisher.

Printed in Singapore

CONTENTS

PREFACE

This book has been compiled from lecture notes and examples that I have used in my teaching of solid mechanics in various forms (including strength of materials, stress and structural analysis), over many years. It is intended for undergraduate and postgraduate engineering courses in which statics, solid mechanics and structures are taught from an intermediary to advanced level. The contents should serve most courses in mechanical, civil, aeronautical and materials engineering. The approach employed is to intersperse theory with many illustrative examples and exercises. As readers work through these it will become apparent what the engineer's practical interests in structural mechanics are. They will see that all calculations made are related to a safe load-carrying capacity and the deformation that materials used in structural design undergo. Amongst the specific design considerations are: the choice of material, its physical shape, the nature of imposed loading and its effect on the internal stress and strain. The loadings refer to: tension, compression, bending, torsion and shear. Typical structures upon which these loadings are applied in a multitude of applications include: bars, columns, struts, tubes, vessels, beams, springs and frames.

The chapters follow an orderly sequence, loosely connected to their degree of difficulty, in which the more fundamental material appears first. Thus, the properties of areas, the conditions for static equilibrium, definitions of stress and strain and linear elasticity theory underpin the structural analyses that follow. Therein lie those structures commonplace in many applications: beam bending, torsion of bars and tubes, buckling of struts and plates and tubes under pressure. The final four chapters examine more advanced analytical techniques, including the use of energy methods, plane stress and strain analyses, yield and failure criteria and finite elements. The analyses given of stress, strain, load and deflection employ various techniques with which the reader should soon become familiar. For example, amongst these are: Mohr's circle, the free-body diagram, Hooke's law, Macaulay's step-function method and Castigliano's theorems. The text illustrates where and how to employ each technique effectively within a logical presentation of the subject matter.

In general, a unique solution to the stress and strain borne by a loaded structure will satisfy three requirements: equilibrium, compatibility and the boundary conditions. Throughout this book these three conditions have been imposed upon many structures to provide closed solutions. However, it may not always be possible to achieve a closed-form solution as the loading and geometry become more complex. The final chapter shows how the known stiffness matrix for simpler types of finite elements can be embodied within a numerical solution to displacement, stress and strain. The three aforementioned conditions are satisfied but, because it is necessary to assume a displacement function, the solutions found remain approximations. Because finite elements cannot improve the accuracy of structural analyses that appear in closed-form the latter are often used to validate the numerical solutions as confidence measure. Finally, it must be mentioned that all that appears in a book of this kind will serve the basic need to design safe structures. The text revisits this basic objective throughout, particularly in its examination of safe stress levels through the use of safety factors. The point is often made that it is only through having a complete grasp of the subject can one exercise a proper control upon the degree of safety required from a structure, especially where the design imposes an economical use of material.

Worked examples and exercise sections have been devised and compiled by the author to support the topics within each chapter. Some have been derived, often with a conversion to SI units, from past examination papers set by institutions with which the author has been associated, namely: Brunel, Dublin, Kingston and Surrey Universities, and the Council of Engineering Institutions (CEI).

D. W. A. REES

HISTORICAL OVERVIEW

I Introduction

The subject of structural mechanics has had a long history within the role it plays in engineering design. Thus it has long been recognised that the engineer needs to design a safe structure, be it a bridge, a pressure vessel, a ship or an aeroplane. Not only is the choice of material important to this goal but also is the correct analysis of the manner in which the external loads are to be supported. It is via this route that a safety factor is decided upon, given other constraints such as minimising weight while retaining stiffness and resistance to corrosion. Solid mechanics is concerned with an understanding of what happens within a body when it is expected to carry loads. We identify the external loads within applied forces, moments and torques and their transmission into an internal stress and an accompanying strain. Often, the subject is employed with a re-design, say in a beam (see Chapters 5 and 6), where the area is to be increased and the length shortened to reduce the maximum stress and deflection to an acceptably safe level for a chosen material. Alternatively, we may select a stronger, stiffer material where an alteration to shape is not permitted if we are to increase the margin of safety. This book shows how mechanics plays an important role within the analysis stage of structures required to bear load often requiring many iterations before the synthesis stage can begin.

A wide variety of structures under different loading modes are presented within the twelve chapters of this book. Throughout the reader will see the names of those men whose contributions to engineering mechanics have shaped the subject into its present form. What follows here is not intended as a detailed biography of each of them, only to recognise that their associations with the following elements of this text has ensured their immortality.

II Units and Conventions

Firstly, within the SI system of units we acknowledge Sir Isaac Newton (1642–1727) [1] for our unit of force. A Newton (N) is the force required to give a mass of 1 kg an acceleration of 1 m/s^2. Our measures of energy and work recognise the pioneering work of James Prescott Joule (1818–1889). The Joule (J) refers to the work done when a force of 1 N moves its point of application through a distance of 1 m. Thus 1 J = 1 N × 1 m. The unit of energy is identical to that of work. This we should expect from the conservation law: that energy can neither be created nor destroyed but only converted from one form to another. For example, in loading a beam, work is done in deflecting the beam and this is stored internally as strain energy. James Watt (1736–1819) [2] is remembered for the unit Watt (W) now assigned to power. Power is the rate of doing work: 1 W = 1 J/s. Two further fundamental units that appear occasionally in solid mechanics (statics) are: the measurement of temperature on the Kelvin scale (K), after Lord Kelvin (1824–1907), and frequency in Hertz (Hz), where 1 Hz = 1 cycle/s, after Rudolph Heinrich Hertz (1857–1894).

Other important derived units, which the reader will meet, refer to our measures of stress and pressure. Both of these are measures of force intensity, referring to a unit of area lying either normal or parallel to that force. The chosen unit of stress and pressure in the SI system of units is the Pascal (Pa) where $1 \text{ Pa} = 1 \text{ N/m}^2$, after Blaise Pascal (1623–1662). In fact we have become more accustomed to working in multiples in this unit to make the numbers more manageable and meaningful. Thus the pressure unit 1 bar = 10^5 Pa, is just a little less than atmospheric pressure (1.01325 bar). The preferred stress unit is the Mega Pascal (MPa) where 1 MPa = 10^6 Pa. The use of the MPa has the advantage of providing the same numerical stress value when a force in Newton is referred to a unit area of 1 mm^2. That is, 1 MPa = 1 N/mm^2. Strength, in the context of resisting failure from tension, tearing, crushing and shear, refers to the limiting stress values of a material. Strength therefore carries the same unit (MPa) as stress. Usually, a three-figure number applies to the strengths of metals. Typically, the tensile yield strength of a low-carbon steel is 300 MPa and its ultimate strength is 450 MPa.

In our definition of stress we refer the force to the original area despite the small changes in dimensions brought about by strain. There are two types of stress: direct and shear. The former arises when the force lies normal to the area and the latter when it is tangential (see Chapter 3). The sense of each is given a sign so that direct tensile stress is positive and direct compressive stress is negative. Clockwise shear stress is positive and anticlockwise shear stress is negative. Strain is the measure of the change to the original dimensions which occurs when a material is stressed. It is defined as the non-dimensional ratio between the change in length and the original length, often expressed as a percentage. Within the elastic limit metallic materials remain stiff and dimensions will not have changed (i.e. strained) by much more than 0.2% at the yield point. However, when the stresses exceed the yield point a metal loses its spring-like behaviour and becomes plastic. The metal lattice distorts permanently through shear slippage along planes most closely packed with atoms. What we can see in the plastic range of a very ductile material like aluminium are dimensional changes (strains) of the order of 50%. Similar strains are reached when forming a steel sheet at a high speed in a press. The need to refer the applied loading to the current dimensions as a material suffers larger strain was recognised by Augustin-Louis Cauchy (1789–1857). It was he who first analysed stress where a force is applied obliquely to a plane. Resolving this force normal and parallel to the plane he identified the stress state for a unit area of that plane. In general, the stress state consists of three components: one normal and two shear stresses for any plane with similar loading. In a three-dimensional analysis of stress at a point within a body loaded in multiple directions we may take three such reference planes to form a cube and then let their areas tend to zero. This reveals that six stress components, from the total of nine acting over the three faces, are independent and sufficient to define the stress at a point uniquely. A particularly convenient representation of these components has been borrowed from relativity theory (Albert Einstein, 1879–1955 [3]). Thus stress is denoted simply as σ_{ij} where i and j may take any value between 1 and 3. Stress at a point in the body is said to have a tensorial character because it can only be defined completely when the magnitude of the components σ_{ij} are connected to its reference planes set in the orthogonal directions 1, 2 and 3. Einstein's tensor subscript notation appears in Chapters 10–13 alongside an equivalent matrix notation, the latter being more popular nowadays for finite element analyses.

III Elasticity

The spring-like behaviour of metals, in which stress and strain are proportional, is expressed within the law attributed to Robert Hooke (1635–1693) [4]. Hooke's law was initially concealed within a Latin anagram *ceiiinosssttuu*. The letters were arranged later in 1678 into the phrase *ut tensio sic vis*, meaning: 'as the extension, so the force.' Written in this form Hooke's law embraces all elastic structures given in this book including: a bar in tension, a spring under load and a beam under bending. All display the proportionality implied within Hooke's secretive discovery. Despite there being no surviving portraits of Robert Hooke his law has ensured his immortality though, evidently, this would not have been of his own choosing! Once its importance was recognised, there followed applications of Hooke's law to describe the elastic behaviour of many metallic materials under axial tension. So it was that Thomas Young (1773–1829) [5] identified a modulus of elasticity (symbol E) as the constant ratio of proportionality between axial stress and strain. That each material has a different value of Young's modulus provides a measure of its spring-like stiffness. For example, the ratio shows that steel is three times stiffer than aluminium with their respective moduli being 210 GPa and 70 GPa. Further elastic constants were identified in the corresponding ratio between stress and strain for a shear force and for a uniform hydrostatic pressure. These constants are called the shear and bulk moduli, respectively (symbols G and K). If we wish to extend Hooke's law to two- and three-dimensional stress states an account of the lateral strain induced by an axial stress is also required. Simon-Denis Poisson (1781–1840) recognised that a bar in tension will contract in its lateral dimensions as its length increases. Poisson's ratio (symbol v) is the constant ratio between the corresponding lateral and axial elastic strains. The three moduli mentioned above, together with

Poisson's ratio, account for the elastic response of a material under any loading combination. With restricted loading fewer constants are needed. Originally, Gabriel Lamé (1795–1870) identified two independent (mathematical) elastic constants even for the general loading condition. This is because of the relationships that exist between the four engineering constants. The derivation of Lamé's constants is given in the book by August Edward Hough Love (1863–1940) [6]. Love also persues the history of the subject, as has Timoshenko [7] and others [8–10], in greater detail than is given here. Among today's engineers it is more common to adopt the four engineering constants E, G, K and v that appear in this book. Typical values for ten materials are given in Table 4.4, (see p. 113).

The theory of elasticity is concerned with these relations between stress and strain and the two accompanying conditions: (i) that variations in stress remain in equilibrium and (ii) strains are compatible with displacements. In addition (i) and (ii) must be matched to the respective forces and displacements that are known to exist around the boundary. This is the approach adopted throughout this book to give the analytical solutions to stress, strain and displacements within loaded bodies. However, it is recognised that not all problems lend themselves to a closed solution. In Chapter 13 it is shown how the stated conditions can then be implemented within a numerical procedure centred upon the structural stiffness matrix [11], more commonly known as the finite element method [12].

IV Structures

Many have examined the manner in which particular structures deform elastically under point and distributed loadings. The distinction can be made between large displacements, in the case of an unstable structure and small displacements, that conform to Hooke's law, in a stable structure.

In the former category, Leonhard Euler (1707–1783) [13] considered how a long thin strut behaved when subjected to an axial compressive load. The solution to this problem revealed that here Hooke's law does not apply, in that the lateral deflection within the length is not directly proportional to the load. Euler was able to show mathematically that beyond a critical load a strut would become unstable and buckle. Clearly, this is an important design consideration in engineering construction whenever pillars and columns are used as supports. For shorter strut designs, William John Macquorn Rankine (1820–1872) predicted the critical buckling load empirically. He and Lewis Gordon showed that Euler's mathematics of buckling does not account for a limiting strength of the strut material. In fact, Chapter 8 shows an empirical basis of strut design, is often the preferred approach.

In the latter category it is often necessary, when designing structures, to allow for the smaller elastic deflections that occur beneath the loads that are applied to it. Otto Mohr (1835–1918) proposed two theorems that enable both the slope and deflection of a beam to be found when carrying lateral loading (see Chapter 6). Mohr's method employs the bending moment diagram for the applied loading and is particularly useful for cantilever beams. Later, Rudolf Freidrich Alfred Clebsch (1833–1872) and William Herrick Macaulay (1853–1936) overcame the problem of the discontinuities that arise in this diagram at concentrated load points. Their elegant step-function approach can be applied to find deflections for a beam whose loading is supported in any manner. One of the major advances to our understanding of the deformation behaviour of any structure was to link the displacement and the applied loading to its store of energy. Albert Castigliano (1847–1884) [14] set out to discover the link within an equilibrium structure for his degree thesis in 1873. He showed that both the load and the displacement beneath it were separate partial derivatives of the structure's complementary energy (see Chapter 10). This was a remarkable achievement for a young Italian railway engineer of 28 years whose interest in mechanics was largely self-taught. We shall see how his two theorems are applied to a Hookean structure where strain energy and complementary energy take on the same meaning. At that time it was realised that a *statically determinate* structure need only its equilibrium equations when finding the displacements, stresses and strains arising from the applied forces. On the other hand, a *statically indeterminate* structure is insoluble from applying equilibrium principles alone, where an additional compatibility condition is required. Examples of both these structures appear throughout this text, where a similar division in determinancy applies to any load-bearing device including frames, pressure vessels, beams, torsion bars and tubes.

V Yielding

Long has it been known that a metallic material will continue to support load levels beyond its elastic limit, a property that has ensured their survival in the face of many man-made materials. What the early engineers were less clear about was how a combination of loads would affect the yield point (see Chapter 12). We could present the problem in general terms by asking what magnitudes of the six independent components of the stress tensor are required to produce yielding? Richard von Mises (1883–1953) [15] proposed a criterion of yielding admitting all stress states. His criterion recognises that the stress tensor has invariants which do not depend upon the co-ordinate directions. Because yielding should not depend upon co-ordinates it becomes linked to critical values of the stress invariants. This general condition for yielding envelopes many of the earlier proposals for yielding under two- and three-dimensional stress states. For example, by omitting shear stresses, James Clerk Maxwell (1831–1879) [16] believed intuitively that yielding would commence when the root mean square of the principal stresses attains a critical value. This mean value, natural to an electrical engineer, has provided us with a dependable, simpler form of the von Mises yield criterion. There are alternative yield criteria however. One, in particular, is based upon the maximum shear stress in a simple tension test attaining a critical value at yield. Engineers have always been confident in extending this approach to multi-axial yielding because its predictions are known to be conservative. Thus, the original hypothesis of

Charles Augustin Coulomb (1736–1806), re-discovered in 1865 by H. Tresca, continues to this day to serve engineers with safe designs. Less use is made now of the Barre de Saint-Venant (1797–1886) criterion, based upon a critical strain at yield but, as is the case with many of these early engineers, his name will appear elsewhere. In particular, Chapter 7 refers to the St.Venant torsion constant for providing the angular twist of thin tubes and rectangular strips under torsion.

VI Concluding Remarks

The pioneering work of these early engineers, physicists and mathematicians sets the scene for the research conducted today in many areas of solid mechanics. Amongst these are: numerical techniques of stress analysis including the finite element and boundary element methods, and the development of the mechanics appropriate to plasticity, creep, fatigue and fracture. The continued interest in the subject arises from the development of new materials to support any manner of applied loading. We might require, for example, an enhanced strength from a structure with reduced weight [17]. Alternatively, a life prediction is imposed upon the design of a component that is expected to become damaged operating at a high temperature under fluctuating loading [18].

References

[1]   More, L. T. *Sir Isaac Newton*, Dover, New York, 1962.
[2]   Meyer, H. W. *A History of Electricity and Magnetism*, MIT Press, Cambridge, MA, USA, 1971.
[3]   Bell, E. T. *Men of Mathematics*, Vols 1 and 2, Penguin Books, London, 1953.
[4]   Andrade, E. N. Da. C. *Robert Hooke, Proc Roy Soc, London*, **201A**, 439–472, 1950.
[5]   Lenard, P. *The Great Men of Science*, Macmillan, London, 1933.
[6]   Love, A. E. H. *A Treatise on the Mathematical Theory of Elasticity*, 4th Edition, Dover, New York, 1944.
[7]   Timoshenko, S. P. *History of the Strength of Materials*, McGraw-Hill, New York, 1951.
[8]   Todhunter, I. and Pearson, K. *History of the Theory of Elasticity and the Strength of Materials*, Vol I, 1639–1850; Vol II, 1851–1886, Dover, New York, 1960.
[9]   Volterra, E. and Gaines, J. H. *Advanced Strength of Materials*, Prentice Hall, Englewood Cliffs, NJ, USA, 1971.
[10] Struik, D. J. *A Concise History of Mathematics*, 4th Edition, Chelsea Pub Co, NY, USA, 1987.
[11] Argyris, J. *Recent Advances in Matrix Methods of Structural Analysis*, (Eds Stanet, L. and Kucheman, D.) Pergamon Press, Oxford, 1964.
[12] Zienkiewic, O. C. and Cheung, Y. K. *The Finite Element Method in Structural and Continuum Mechanics*, McGraw-Hill, New York, USA, 1990.
[13] Johnston, B. G. *Column Buckling Theory: Historical Highlights, Jl Structural Engng, Am Soc Civ Engrs*, **109**, 2086–2096, 1983.
[14] Nascè, V. *Alberto Castigliano, Railway Engineer His Life and Times, Meccanica*, **19**, 5-14, 1984.
[15] Förste, J., Ludford, G. and Birkhoff, G. von, *Richard von Mises, ZAMM*, **63**, 277–284, 1983.
[16] Campbell, L. and Garnett, W. *The Life of James Clerk Maxwell*, Macmillan, London, 1882.
[17] Rees, D. W. A. *Mechanics of Optimal Structural Design (Minimum Weight Structures)*, Wiley, UK, 2010.
[18] Rees, D. W. A. *Mechanics of Deformable Solids*, Studium Press, Houston, USA, 2011.

CHAPTER 1

PROPERTIES OF AREAS

Summary: In this chapter the basic properties of areas that underpin many of the topics that appear throughout this book are introduced. It will be seen later how the present topic is applied to the cross-sections of beams in bending, shafts in torsion and long struts under compressive loading. Examples are selected to show what is meant by the terms 'centroid', the 'first and second moments' of their plane section areas. Two theorems are derived that enable the transfer of second moments of area between parallel and perpendicular axes within a given cross-section. Both analytical and graphical techniques are available for dealing with co-ordinate rotations. Exercises are given to enable the reader to gain familiarity with these terms and techniques.

1.1 Centroid and Moments of Area

The properties of a cross-section that resist loading applied externally loading are the area A of that section and its first and second moments of area i and I respectively, about the centroidal axes x and y shown in Figure 1.1.

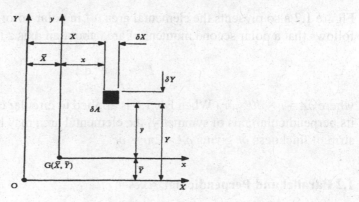

Figure 1.1 Elemental area δA set in Cartesian axes x, y

Firstly, consider axes X and Y that do not pass through the centroid, G. The following definitions apply to an element of area, $\delta A = \delta X \times \delta Y$, located at the position (X, Y) within the Cartesian co-ordinates frame X, Y, with an origin O, as shown. The area δA may also be located at the position (x, y) within a second frame x-y, with an origin at the centroid G of the area. When δA is a strip element of a given cross-section, whose area is A, the first moments of A about the axes X and Y are given by:

$$i_X = \int Y \, dA, \quad i_Y = \int X \, dA \qquad (1.1a,b)$$

The centroid position G($\overline{X}, \overline{Y}$) is found from Eqs 1.1a,b as follows:

$$\overline{X} = i_Y / A, \quad \overline{Y} = i_X / A \qquad (1.2a,b)$$

When the axes X, Y and x, y coincide then $\overline{X} = \overline{Y} = 0$ and Eqs 1.2a,b show that $i_X = i_Y = 0$. Hence, the first moments of area A is zero for any axis that passes through the centroid. This gives:

$$\int x \, dA = \int y \, dA = 0 \qquad (1.3)$$

The second moments of area A about the non-centroidal axes X and Y are defined as:

$$I_X = \int Y^2 \, dA, \quad I_Y = \int X^2 \, dA, \quad I_{XY} = \int XY \, dA \qquad (1.4a,b,c)$$

and about the centroidal axes x and y, the second moments become:

$$I_x = \int y^2 \, dA, \quad I_y = \int x^2 \, dA, \quad I_{xy} = \int xy \, dA \qquad (1.5a,b,c)$$

Figure 1.2 also presents the elemental area δA in polar co-ordinates (r, θ). Here it follows that a polar second moment of area about an axis z is defined as:

$$I_z = \int r^2 \, dA \qquad (1.6)$$

where $\delta A = r \times \delta\theta \times \delta r$. When Eq. 1.6 is applied to circular cross-section with z as its perpendicular axis of symmetry, the elemental area may be taken as an annular strip of thickness δr giving $\delta A = 2\pi r \times \delta r$.

1.2 Parallel and Perpendicular Axes

1.2.1 Parallel Axis Theorem

In Fig 1.1 the Cartesian axes x, y and X, Y are separated by $\overline{X}$ and $\overline{Y}$ but remain parallel as shown. Second moments of area may be transferred between the two pairs of parallel axes provided given that x, y are centroidal axes. Equations 1.4a,b,c give,

$$I_X = \int Y^2 dA = \int (y + \overline{Y})^2 dA = \int (y^2 + 2y\overline{Y} + \overline{Y}^2) dA \tag{1.7a}$$

$$I_Y = \int X^2 dA = \int (x + \overline{X})^2 dA = \int (x^2 + 2x\overline{X} + \overline{X}^2) dA \tag{1.7b}$$

$$I_{XY} = \int XY dA = \int (x + \overline{X})(y + \overline{Y}) dA = \int (xy + x\overline{Y} + y\overline{X} + \overline{X}\overline{Y}) dA \tag{1.7c}$$

Substituting Eqs 1.3 and 1.5a,b,c into Eqs 1.7a,b,c leads to three, *parallel axes theorems* that allow second moments to be found for any axes X, Y lying parallel to the centroidal axes:

$$I_X = I_x + A\overline{Y}^2 \tag{1.8a}$$

$$I_Y = I_y + A\overline{X}^2 \tag{1.8b}$$

$$I_{XY} = I_{xy} + A\overline{X}\overline{Y} \tag{1.8c}$$

where A is the cross-sectional area. The signs of $\overline{X}$ and $\overline{Y}$ are determined from the position of the centroid G. They are positive when G lies in the first quadrant of X, Y as shown in Fig. 1.1.

1.2.2 Perpendicular Axis Theorem

The *perpendicular axis theorem* enables the second moment of a plane area A to be found about its perpendicular centroidal axis z in terms of I_x and I_y. Let the elemental area δA lie in polar co-ordinates (see Figure 1.2).

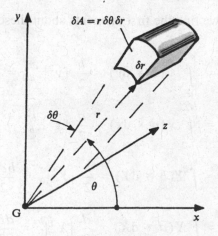

Figure 1.2 Perpendicular axes

Normally, the theorem provides the polar moment of area I_z for a bar of uniform cross-section where z is its centroidal, length axis. Representing δA as an element of A in polar co-ordinates it follows that:

$$I_z = \int r^2 dA = \int (x^2 + y^2) dA \qquad (1.9a)$$

Substituting I_x and I_y from Eqs 1.5a,b into Eq. 1.9a reveals the second theorem:

$$I_z = I_x + I_y \qquad (1.9b)$$

In Eq. 1.9b the subscripts x, y and z identify the Cartesian centroidal axes.

Example 1.1 Find all the area properties of a rectangle $b \times d$, for the axes (x, y) and (X, Y) shown in Figure 1.3.

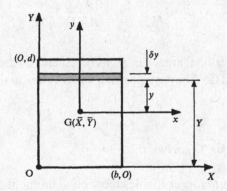

Figure 1.3 Plane rectangular area

Applying Eqs 1.1–1.9 we find the first moments about these axes:

$$i_x = \int_{-d/2}^{d/2} y(b \times dy) = \frac{b}{2} \, |y^2|_{-d/2}^{d/2} = 0$$

$$i_y = \int_{-b/2}^{b/2} x(d \times dx) = \frac{d}{2} \, |x^2|_{-b/2}^{b/2} = 0$$

$$i_X = \int_{0}^{d} Y(b \times dY) = \frac{b}{2} \, |Y^2|_{0}^{d} = \frac{bd^2}{2}$$

$$i_Y = \int_{0}^{b} X(d \times dX) = \frac{d}{2} \, |X^2|_{0}^{b} = \frac{db^2}{2}$$

The centroid position follows from Eqs 1.2a,b:

$$\bar{X} = i_Y/A = b/2$$

$$\bar{Y} = i_X/A = d/2$$

The 'centroidal' second moments are from Eqs 1.5a–c:

$$I_x = \int_{-d/2}^{d/2} y^2 (b \times dy) = \frac{b}{3} |y^3|_{-d/2}^{d/2} = \frac{bd^3}{12}$$

$$I_y = \int_{-d/2}^{b/2} x^2 (d \times dx) = \frac{d}{3} |x^3|_{-b/2}^{b/2} = \frac{db^3}{12}$$

and from their perpendicular axes, Eq. 1.9b gives

$$I_z = I_x + I_y = \frac{bd}{12}(b^2 + d^2)$$

$$I_{xy} = \int_{-b/2}^{b/2} \int_{-d/2}^{d/2} xy (dx\, dy) = \int_{-b/2}^{b/2} x i_y dx = 0$$

For the non-centroidal axes (X, Y), Eqs 1.4a–c give:

$$I_X = \int_0^d Y^2 (b \times dY) = \frac{b}{3} |Y^3|_0^d = \frac{bd^3}{3}$$

$$I_Y = \int_0^b X^2 (d \times dX) = \frac{d}{3} |X^3|_0^b = \frac{db^3}{3}$$

$$I_{XY} = \int_0^b \int_0^d XY (dX\, dY) = \frac{b^2}{2} \int_0^d y\, dy = \frac{b^2 d^2}{4}$$

or, from $I_{XY} = A\bar{X}\bar{Y}$. Also note that I_X, I_Y and I_{XY} can be checked from applying the parallel axes theorems (Eqs 1.8a–c):

$$I_X = I_x + A\bar{X}^2 = \frac{bd^3}{12} + bd \left(\frac{d}{2}\right)^2 = \frac{bd^3}{3}$$

$$I_Y = I_y + A\bar{Y}^2 = \frac{db^3}{12} + db \left(\frac{b}{2}\right)^2 = \frac{db^3}{3}$$

$$I_{XY} = I_{xy} + A\bar{X}\bar{Y} = 0 + bd \left(\frac{b}{2}\right)\left(\frac{d}{2}\right) = \frac{b^2 d^2}{4}$$

Example 1.2 Find I_x, I_y and I_z for the circular section in Figure 1.4.

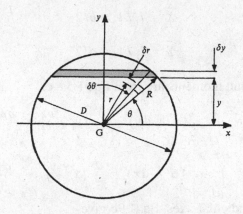

Figure 1.4 Solid circular plane area

Applying Eq. 1.5a

$$I_x = \int_{-R}^{R} y^2 \left[2(R^2 - y^2)^{1/2} dy \right]$$

Putting $y = R \sin \theta$, $dy = R\cos\theta \, d\theta$ and using trigonometric identities:

$$I_x = 4R^4 \int_{0}^{\pi/2} \sin^2\theta \, \cos^2\theta \, d\theta$$

$$= \frac{R^4}{2} \left| \theta - \frac{1}{4} \sin 4\theta \right|_{0}^{\pi/2}$$

$$= \frac{\pi R^4}{4} = \frac{\pi D^4}{64}$$

but $I_x = I_y$ and, therefore, from Eq. 1.9b

$$I_z = I_x + I_y = \frac{\pi R^4}{2} = \frac{\pi D^4}{32}$$

Alternatively, from Eq. 1.6, where $dA = r \times \delta\theta \times \delta r$ in Fig. 1.4:

$$I_z = \int_{0}^{R} \int_{0}^{2\pi} r^2 (dr \times r d\theta)$$

$$= 2\pi \int_{0}^{R} r^3 dr = \frac{\pi R^4}{2} = \frac{\pi D^4}{32}$$

Example 1.3 Find I_x, I_y and I_z for an ellipse with lengths of its major and minor axes $2a$ and $2b$ respectively, as shown in Figure 1.5.

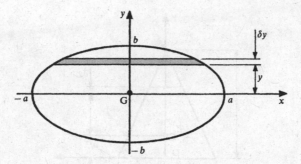

Figure 1.5 Plane elliptical area

The equation of an ellipse is

$$x^2/a^2 + y^2/b^2 = 1$$

Applying Eq. 1.5a, to the horizontal strip, the second moment of area about the centroidal x-axis is:

$$I_x = \int_{-b}^{b} y^2 (2x\,dy) = 2\int_{-b}^{b} a(1 - y^2/b^2)^{1/2}\, y^2\, dy$$

and applying standard integrals

$$I_x = \frac{2a}{b} \left| \frac{b^2 y}{8}(b^2 - y^2)^{1/2} - \frac{y}{4}(b^2 - y^2)^{3/2} + b^2 \sin^{-1}\left(\frac{y}{b}\right) \right|_{-b}^{b} = \frac{\pi a b^3}{4}$$

Similarly, taking a strip parallel to the centroidal y-axis and applying Eq.1.5b, the second moment of area about that axis is:

$$I_y = \int_{-a}^{a} x^2 (2y\,dx) = 2\int_{-a}^{a} b(1 - x^2/a^2)^{1/2}\, x^2\, dx = \frac{\pi b a^3}{4}$$

The perpendicular axis theorem (Eq. 1.9b) provides the second moment of area for the ellipse's centroidal z-axis:

$$I_z = I_x + I_y = \frac{\pi a b}{4}(a^2 + b^2)$$

Example 1.4 Find $\bar{Y}$, I_x and I_X for the area defined by an isosceles triangle, with base b and height a, as shown in Figure 1.6.

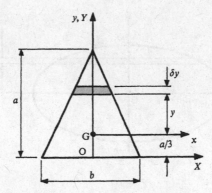

Figure 1.6 Isosceles triangular area

In the X, Y co-ordinate system the equation of the right sloping side is:

$$Y = a - \frac{2a}{b} X$$

The first moment and the centroid follow from Eqs 1.1 and 1.2 as

$$i_X = \int_0^a Y(2X\mathrm{d}Y) = \frac{b}{a} \int_0^a (aY - Y^2)\,\mathrm{d}Y = \frac{ba^2}{6}$$

$$\overline{Y} = \frac{i_X}{A} = \frac{ba^2/6}{ab/2} = \frac{a}{3}$$

In the x, y centroidal co-ordinate system the equation of the right sloping side provides expressions for y and x as:

$$y = -\frac{2a}{b} x + \frac{2a}{3} \quad \rightarrow \quad x = \frac{b}{2}\left(\frac{2}{3} - \frac{y}{a} \right)$$

Integrating the second moment of the shaded area about the x-axis, Eq. 1.5a gives:

$$I_x = \int y^2 \mathrm{d}A = \int y^2(2x\mathrm{d}y) = 2\int_{-a/3}^{2a/3} y^2 \left[b\left(\frac{2}{3} - \frac{y}{a} \right) \right] \mathrm{d}y$$

$$= b \int_{-a/3}^{2a/3} \left(\frac{2y^2}{3} - \frac{y^3}{a} \right) \mathrm{d}y = \frac{a^3 b}{36}$$

Now using the parallel axis Eq. 1.8a:

$$I_X = I_x + A\overline{Y}^2 = \frac{a^3 b}{36} + \left(\frac{ab}{2} \right)\left(\frac{a}{3} \right)^2 = \frac{a^3 b}{12}$$

Example 1.5 Find the centroidal position and $I_x (= I_X)$ for the wall area of a closed, equilateral tube having a mean side length a and thickness t, with one side lying parallel to the y-direction (see Figure 1.7).

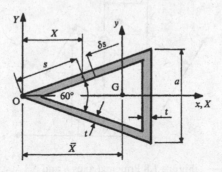

Figure 1.7 Closed equilateral tube

For simplicity a third variable s is introduced to define the distance along the sloping side as shown. Letting x and X coincide with the horizontal axis of symmetry it follows that $\overline{Y} = 0$. Equations 1.1b and 1.2a refer the first moment of the three side lengths to the centroid position G $(\overline{X})$ as:

$$i_Y = \int X\mathrm{d}A = (at)\left(\frac{\sqrt{3}a}{2}\right) + 2\int_0^a (s\cos 30°)(t\mathrm{d}s) = \sqrt{3}a^2 t$$

$$\overline{X} = \frac{i_Y}{A} = \frac{\sqrt{3}a^2 t}{3at} = \frac{a}{\sqrt{3}}$$

i.e. two thirds the length of the X-axis between the base and apex. Equation 1.4a gives the second moment of the tube wall area about the coincident axes x and X:

$$I_x = I_X = \int Y^2 \mathrm{d}A = \frac{ta^3}{12} + 2\int_0^a (s\sin 30°)^2 (t\mathrm{d}s) = \frac{a^3 t}{4}$$

in which the first term applies to the rectangular vertical wall area, i.e. the I_x expression taken from Example 1.1.

1.3 Principal Second Moments of Area

Principal second moments are by definition the centroidal values, I_x and I_y, when the product moment $I_{xy} = 0$. This will apply when x and y are axes of symmetry in a given cross section. When x and y are not axes of symmetry then $I_{xy} \neq 0$ and it becomes necessary to identify two principal I values (I_u and I_v) with another pair of orthogonal axes u and v for which $I_{uv} = 0$ (see Fig. 1.8).

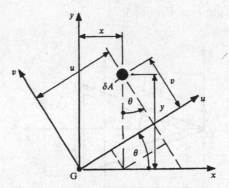

Figure 1.8 Principal axes u and v

The principal I values are, by definition:

$$I_u = \int v^2 \, dA, \quad I_v = \int u^2 \, dA, \quad I_{uv} = \int uv \, dA = 0 \qquad (1.10a,b,c)$$

Now, from the geometry given in Figure 1.8, the u, v co-ordinates are:

$$u = x\cos\theta + y\sin\theta, \quad v = y\cos\theta - x\sin\theta \qquad (1.11a,b)$$

Combining Eqs 1.10a,b,c and 1.11a,b and using Eqs 1.5a,b,c leads to:

$$I_u = I_x \cos^2\theta + I_y \sin^2\theta - I_{xy} \sin 2\theta \qquad (1.12a)$$

$$I_v = I_y \cos^2\theta + I_x \sin^2\theta + I_{xy} \sin 2\theta \qquad (1.12b)$$

$$I_{uv} = \tfrac{1}{2}(I_x - I_y)\sin 2\theta + I_{xy}\cos 2\theta = 0 \qquad (1.12c)$$

The inclination θ between the axes of u to x follows from Eq. 1.12c as:

$$\tan 2\theta = \frac{2I_{xy}}{I_y - I_x} \qquad (1.13)$$

Equation 1.13 allows θ to be eliminated between Eqs 1.12a,b and 1.13. This gives the principal values I_u, I_v in terms of I_x, I_y and I_{xy}:

$$I_{u,v} = \frac{1}{2}(I_x + I_y) \pm \frac{1}{2}\sqrt{(I_x - I_y)^2 + 4I_{xy}^2} \qquad (1.14a,b)$$

Note, from Eqs 1.14a,b that: $I_u + I_v = I_x + I_y$.

Example 1.6 Find I_x, I_y, I_{xy}, I_u, I_v and θ for a right-angled triangle in Fig. 1.9, having a base length b and height a, when lying in the positive quadrant of X, Y.

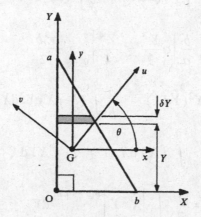

Figure 1.9 Right-angled triangle

The equation of the sloping side is expressed in two ways:

$$Y = a - \frac{a}{b}\, X, \quad \therefore X = \frac{b}{a}(a - Y)$$

Applying Eqs 1.1a and 1.2b, the first moment about the X-axis is non-zero:

$$i_X = \int Y(X\mathrm{d}Y) = \frac{b}{a} \int_0^a (a - Y)\, Y\mathrm{d}Y$$

$$= \frac{b}{a} \left| \frac{aY^2}{2} - \frac{Y^3}{3} \right|_0^a = \frac{a^2 b}{6}$$

from which the centroid co-ordinate is:

$$\overline{Y} = \frac{i_X}{A} = \frac{a^2 b/6}{ab/2} = a/3$$

Similarly, for the Y-axis the first moment of this area is

$$i_Y = \int X(Y\mathrm{d}X) = \frac{a}{b} \int_0^b (b - X)\, X\mathrm{d}X$$

$$= \frac{a}{b} \left| \frac{bX^2}{2} - \frac{X^3}{3} \right|_0^b = \frac{b^2 a}{6}$$

from which the centroid co-ordinate is:

$$\overline{X} = \frac{i_Y}{A} = \frac{b^2 a/6}{ab/2} = b/3$$

Equations 1.4a,b,c give the second moments of the area for the axes X, Y:

$$I_X = \int Y^2 (X\,dY) = \frac{b}{a} \int_0^a (a - Y) Y^2 \,dY$$

$$= \frac{b}{a} \left| a\frac{Y^3}{3} - \frac{Y^4}{4} \right|_0^a = \frac{a^3 b}{12}$$

$$I_Y = \int_0^b X^2 (Y\,dX) = \frac{a}{b} \int_0^b (b - X) X^2 \,dX = \frac{a^3 b}{12}$$

$$I_{XY} = \int\int XY\,dX\,dY = \int_0^a \left(\int_0^X XY\,dX \right) dY$$

$$= \frac{1}{2} \int_0^a \left| X^2 Y \right|_0^{b(a-Y)/a} \,dY$$

$$= \frac{b^2}{2a^2} \int_0^a (a^2 Y - 2aY^2 + Y^3)\,dY = \frac{a^2 b^2}{24}$$

I_{XY} may be checked from integrating the product moment of the elemental area $Y\,dA$ about its centroid $(X/2, Y)$. That is:

$$I_{XY} = \int XY\,dA = \frac{b}{2a} \int_0^a (a - Y) Y \left[\frac{b}{a}(a - Y)\,dY \right] = \frac{a^2 b^2}{24}$$

Then, by parallel axes, Eqs 1.8a,b,c, the following second moments apply to the centroidal axes x, y:

$$I_x = I_X - A\bar{Y}^2 = \frac{a^3 b}{12} - \frac{ab}{2} \left(\frac{a}{3} \right)^2 = \frac{a^3 b}{36}$$

$$I_y = I_Y - A\bar{X}^2 = \frac{b^3 a}{12} - \frac{ab}{2} \left(\frac{b}{3} \right)^2 = \frac{b^3 a}{36}$$

$$I_{xy} = I_{XY} - A\bar{X}\bar{Y} = \frac{a^2 b^2}{24} - \left(\frac{ab}{2} \right)\left(\frac{a}{3} \right)\left(\frac{b}{3} \right) = -\frac{a^2 b^2}{72}$$

and, finally, from Eqs 1.13 and 1.14:

$$\tan 2\theta = -\frac{ab}{b^2 - a^2}$$

$$I_{u,v} = \frac{ab}{72} \left[(a^2 + b^2) \pm \sqrt{a^4 + b^4 - a^2 b^2} \right]$$

Note from the above expressions that if $a = b$, $I_x = I_y = a^4/24$, $I_u = a^4/24$ and $I_v = a^4/72$. These show that I_u and I_v are, respectively, the greatest and least second moments for this area when $\theta = 45°$.

Example 1.7 Find the position of the centroid and I_x, I_y, I_{xy}, I_u and I_v for a quadrant of a circle, radius r, lying in the positive quadrant of X, Y in Figure 1.10.

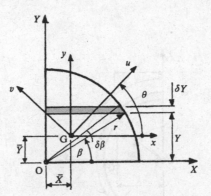

Figure 1.10 Quadrant

Because of symmetry it is only necessary to find the centroid position relative to one axis (say X). Equations 1.1a and 1.2b give

$$i_X = \int_0^r YX \, dY = \int_0^r Y(r^2 - Y^2)^{1/2} \, dY = \frac{r^3}{3} \; (= i_Y)$$

$$\bar{Y} = \frac{i_X}{A} = \frac{r^3/3}{\pi r^2/4} = \frac{4r}{3\pi} \; (= \bar{X})$$

when, from Eqs 1.4a,c:

$$I_X = \int_0^r Y^2(r^2 - Y^2)^{1/2} \, dY = \frac{r^4}{4} \int_0^{\pi/2} \sin^2 2\beta \, d\beta = \frac{\pi r^4}{16} \; (= I_Y)$$

$$I_{XY} = \int_0^r (X \, dY) \, YX/2 = \frac{1}{2} \int_0^r (r^2 - Y^2) Y \, dY = \frac{r^4}{8}$$

where $Y = r \sin\beta$ and $X^2 + Y^2 = r^2$. Using the parallel axes Eqs 1.8a,c, the centroidal, second moments of area become:

$$I_x = I_X - A\bar{Y}^2 = \frac{\pi r^4}{16} - \frac{\pi r^2}{4}\left(\frac{4r}{3\pi}\right)^2 = \left(\frac{\pi}{16} - \frac{4}{9\pi}\right) r^4 \; (= I_y)$$

$$I_{xy} = I_{XY} - A\bar{X}\bar{Y} = \frac{r^4}{8} - \frac{\pi r^2}{4}\left(\frac{4r}{3\pi}\right)^2 = r^4\left(\frac{1}{8} - \frac{4}{9\pi}\right)$$

Then, applying Eqs 1.13 and 1.14a,b, it is seen that the u, v axes are inclined at $\theta = 45°$ for which the principal second moments are

$$I_u = r^4\left(\frac{\pi}{16} + \frac{1}{8} - \frac{8}{9\pi}\right), \quad I_v = r^4\left(\frac{\pi}{16} - \frac{1}{8}\right)$$

Example 1.8 Find the position of the centroid and I_X, I_Y and I_{XY} for an open semi-circular channel section (see Figure 1.11) with mean radius a and thickness t, when the diameter coincides with the Y-axis. What are the principal values and the orientation of their axes?

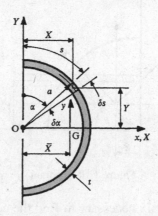

Figure 1.11 Open, semi-circular channel

Using the perimeter dimension, s, we see that $ds = a\,d\alpha$. Clearly, $Y = 0$ and the X position of the centroid is found from Eq. 1.2a as follows:

$$i_Y = \int X dA = 2a \int_0^{\pi/2} (ta)\sin\alpha\,d\alpha = 2a^2 t$$

$$\bar{X} = \frac{i_Y}{A} = \frac{2a^2 t}{\pi a t} = \frac{2a}{\pi}$$

From Eqs 1.4a,b the second moments about the X- and Y-axes are:

$$I_X = \int Y^2 dA = 2\int (a\cos\alpha)^2 (t\,ds)$$

$$= 2\int_0^{\pi/2} (a\cos\alpha)^2 (ta\,d\alpha) = \frac{\pi a^3 t}{2}$$

$$I_Y = \int X^2 dA = 2\int (a\sin\alpha)^2 (t\,ds)$$

$$= 2\int_0^{\pi/2} (a\sin\alpha)^2 (ta\,d\alpha) = \frac{\pi a^3 t}{2}$$

$$I_{XY} = \int XY dA = 2\int_0^{\pi/2} (a\sin\alpha)(a\cos\alpha)(ta\,d\alpha) = a^3 t$$

Substituting into Eqs 1.8a,c, since $\bar{Y} = 0$, it is obvious that:

$$I_x = I_X = \pi a^3 t/2, \quad I_{xy} = I_{XY} = a^3 t$$

and from Eq. 1.8b:

$$I_y = I_Y - A\bar{X}^2 = \frac{\pi a^3 t}{2} - (\pi a t)\left(\frac{2a}{\pi}\right)^2 = a^3 t\left(\frac{\pi}{2} - \frac{4}{\pi}\right)$$

Then, from Eq. 1.13:

$$\tan 2\theta = \frac{(2a^3 t)}{(-4a^3 t/\pi)} = -\pi/2$$

from which the expected result is that u and v are inclined at $\theta = -45°$ and $+45°$ to the x-axis respectively. Equations 1.14a,b give the principal second moments of area for the u- and v- directions:

$$I_u, I_v = \left(\frac{a^3 t}{2}\right)\left(\pi - \frac{4}{\pi}\right) \pm a^3 t\sqrt{1 + \left(\frac{2}{\pi}\right)^2}$$

1.4 Graphical Solution to I_u and I_v

In co-ordinates of $[(I_x, I_y), I_{xy}]$, Eqs 1.5a,b,c may be combined to describe the equation of a circle. The principal values I_u and I_v lie on opposite ends of the horizontal diameter, where $I_{xy} = I_{uv} = 0$, as shown in Figs 1.12a,b. In Eqs 1.14a,b the circle's centre co-ordinates are $\frac{1}{2}(I_x + I_y)$, 0 and its radius is: $\frac{1}{2}\sqrt{[(I_x - I_y)^2 + 4I_{xy}^2]}$. To construct the circle the points A and B are plotted initially with their co-ordinates derived from I_x, I_y and I_{xy}, as shown. There are four possible positions for A and B depending upon the relative magnitudes of I_x and I_y and the sign, as calculated, for I_{xy}. Consistently, $I_u > I_v$ when θ is calculated from Eq. 1.13 in each of the following:

(a) $I_x > I_y$ with I_{xy} positive gives θ negative
(b) $I_x > I_y$ with I_{xy} negative gives θ positive
(c) $I_y > I_x$ with I_{xy} positive gives θ positive
(d) $I_y > I_x$ with I_{xy} negative gives θ negative

In the graphical construction the calculated value of I_{xy}, together with its sign, is plotted with the larger of I_x or I_y. Thus in (a) and (b) the calculated I_{xy} is plotted with I_x as the two co-ordinates for point A. The sign of I_{xy} is changed when accompanying I_y as the two co-ordinates for point B. The circle is then drawn with AB as its diameter. Figures 1.12a,b show the constructions for (a) and (b). In Fig. 1.12a, where I_{xy} is positive and $I_x > I_y$, then θ is negative for the u-axis to make an acute angle θ with the x-axis so giving $I_u > I_v$. Figure 1.12b gives the construction corresponding to $I_x > I_y$ and a negative I_{xy}. Here $I_u > I_v$ gives a positive acute angle θ between the axes u and x. The u- and v-directions are established from the focus point F: a singular point on the circumference through which *any* pair of perpendicular axes may be projected. Where the projections cut the circle will determine the associated I_x, I_y and I_{xy} for the chosen pair of centroidal axes x, y, including the principal axes u and v. It follows that F is located by projecting the chosen x-direction through point A or the perpendicular y-direction through point B. All such projections meet at F.

(a)

(b)

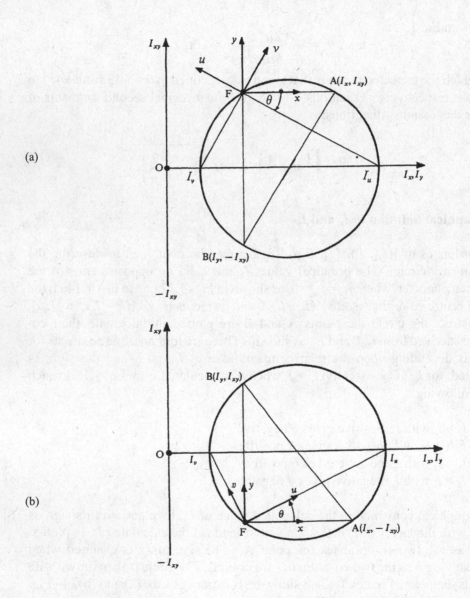

Figure 1.12 Circle construction showing focus and principal second moments of area

It is left as an exercise for the reader to show that $I_u > I_v$ for conditions (c) and (d) in two circle constructions where the calculated I_{xy} (with sign) accompanies I_y to become the co-ordinates for point B. Here the change in sign of I_{xy} accompanies I_x to provide the co-ordinates for point A. The following example shows how the graphical method can be applied to a channel-section.

Example 1.9 Determine, for the section given in Figure 1.13a, the position of the centroid G($\overline{X}, \overline{Y}$) and I_x, I_y, I_z for the centroidal axes x, y and z. Using both analytical and graphical methods, determine I_u, I_v and the inclination θ, of the principal axes u and v, relative to the x- and y-directions.

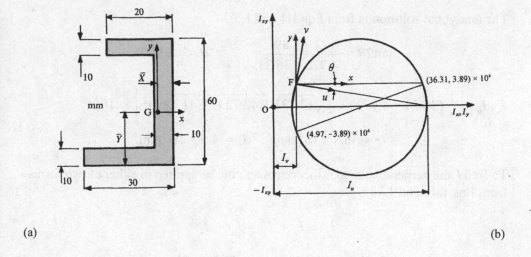

(a)   (b)

Figure 1.13 Construction showing principal I's for a channel section

Taking first moments about axes X, Y in Figure 1.13a gives the centroid position G($\overline{X}, \overline{Y}$):

$$(30 \times 10 \times 5) + (40 \times 10 \times 30) + (20 \times 10 \times 55) = 900\,\overline{Y}$$

$$\therefore \ \overline{Y} = 27.22 \text{ mm}$$

$$(60 \times 10 \times 5) + (10 \times 10 \times 15) + (20 \times 10 \times 20) = 900\,\overline{X}$$

$$\therefore \ \overline{X} = 9.45 \text{ mm}$$

The second moments of area are found from taking the sum and difference of rectangular components of the section area all 'based' on the centroidal axes x and y. For this $I = \sum bd^3/3$ (from Example 1.1) is applied as follows:

$$I_x = \frac{20}{3} \times (60 - 27.22)^3 - \frac{10}{3}(60 - 37.22)^3 + \frac{30}{3}(27.22)^3 - \frac{20}{3}(17.22)^3$$

$$= 36.31 \times 10^4 \ mm^4$$

$$I_y = \frac{60}{3}(9.45)^3 + \frac{10}{3}(20 - 9.45)^3 + \frac{10}{3}(30 - 9.45)^3 + \frac{40}{3}(10 - 9.45)^3$$

$$= 4.972 \times 10^4 \ mm^4$$

$$I_{xy} = (60 \times 10)(30 - 27.22)(9.45 - 5) - (10 \times 10)(60 - 32.22)(15 - 9.45)$$

$$+ (20 \times 10)(27.22 - 5)(20 - 9.45) = 3.89 \times 10^4 \ mm$$

The analytical solution is from Eqs 1.4 and 1.5:

$$\tan 2\theta = \frac{2 \times 3.89}{4.972 - 36.31} \qquad \therefore \qquad \theta = -6.97°$$

$$I_u, I_v = \left[\frac{1}{2}(36.31 + 4.972) \pm \frac{1}{2}\sqrt{(36.31 - 4.972)^2 + 4(3.89)^2} \right] \times 10^4$$

$$\therefore I_u = 36.8 \times 10^4 \ mm^4, \quad I_v = 4.5 \times 10^4 \ mm^4$$

To find I_z the perpendicular axis theorem may now be applied in either of two forms from Eqs 1.4b and 1.9b:

$$I_z = I_x + I_y = I_u + I_v$$

and substituting the I-values from above: $I_z = 41.29 \times 10^4 \ mm^4$.

The corresponding graphical solution is given in Figure 1.13b. This corresponds to condition (a) above. It is seen that since $I_x > I_y$ and I_{xy} is positive, then θ is negative and $I_u > I_v$ as shown in Figure 1.13b.

1.5 Matrix Method

The fact that I depends upon the orientation of the in-plane, centroidal Cartesian co-ordinates (x, y) and (u, v) in the analytical and graphical analyses given above, shows that I is a quantity that can be transformed. It will be seen that I has similar transformation properties to those of stress and strain, these being known as *tensor* quantities in mechanics. It is convenient to assemble I_x, I_y and I_{xy} as the components of a symmetrical 2×2 matrix $\mathbf{I}$ when applying Eqs 1.12a,b to find I_u and I_v by the matrix method. More generally, the reader should note that Eqs 1.5a,b,c will conform to the following matrix transformation equation for a rotation of centroidal axes from x, y to x', y':

$$\mathbf{I}' = \mathbf{LIL}^{\mathrm{T}} \tag{1.15a}$$

where $\mathbf{I}$ is the matrix containing I_x, I_y, I_{xy} and $\mathbf{I}'$ is the matrix of I values for the rotated axes x' and y', which include u and v. In Eq. 1.15a, $\mathbf{L}$ is another 2×2 matrix of direction cosines l, constructed from a given inclination θ between x and x', as follows:

$$\mathbf{L} = \begin{bmatrix} l_{xx} & l_{xy} \\ l_{yx} & l_{yy} \end{bmatrix} = \begin{bmatrix} \cos\theta & \cos(90-\theta) \\ \cos(90+\theta) & \cos\theta \end{bmatrix} = \begin{bmatrix} \cos\theta & \sin\theta \\ -\sin\theta & \cos\theta \end{bmatrix} \tag{1.15b}$$

where, referring to Fig. 1.14, the components of $\mathbf{L}$ are the cosines between the primed and un-primed axes as follows:

$l_{xx} = \cos x'x$ is the cosine of the angle x' makes with x
$l_{xy} = \cos x'y$ is the cosine of the angle x' makes with y
$l_{yx} = \cos y'x$ is the cosine of the angle y' makes with x
$l_{yy} = \cos y'y$ is the cosine of the angle y' makes with y

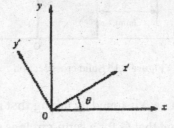

Figure 1.14 Rotation in axes x, y to x', y'

Since $\cos(90 - \theta) = \sin\theta$ and $\cos(90 + \theta) = -\sin\theta$, Eq. 1.15a appears in full when written as:

$$\begin{bmatrix} I_x' & I_{xy}' \\ I_{yx}' & I_y' \end{bmatrix} = \begin{bmatrix} \cos\theta & \sin\theta \\ -\sin\theta & \cos\theta \end{bmatrix} \begin{bmatrix} I_x & I_{xy} \\ I_{yx} & I_y \end{bmatrix} \begin{bmatrix} \cos\theta & -\sin\theta \\ \sin\theta & \cos\theta \end{bmatrix} \tag{1.15c}$$

When x' and y' coincide with the principal axes, I_u and I_v replace I_x' and I_y', respectively, in the left-hand matrix. Here the off-diagonal terms are zero ($I_{xy}' = 0$) and Eq. 1.13 specifies the inclination θ of u to x (see Figure 1.8). Equation 1.15c will then reduce to finding the roots of a quadratic equation that results from expanding the following determinant:

$$\det \begin{bmatrix} (I_x - I) & I_{xy} \\ I_{xy} & (I_y - I) \end{bmatrix} = 0 \qquad (1.16)$$

The expansion of Eq. 1.16 confirms that the roots to the following *characteristic equation* are I_u and I_v, as given in Eqs 1.14a,b.

$$(I_x - I)(I_y - I) = (I_{xy})^2 \quad \rightarrow \quad I^2 - I(I_x + I_y) + [I_x I_y - (I_{xy})^2] = 0$$

Example 1.10 Determine, from the matrix method, I_u, I_v and θ for the solid cross section shown given in Figure 1.15.

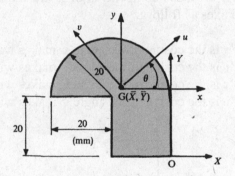

Figure 1.15 Solid cross section

The centroid position G($\overline{X}, \overline{Y}$) is found by taking first moments of area about the corner axes X and Y. Noting that G for a semi-circle is $4r/3\pi$ above its horizontal diameter, $\overline{X}$ and $\overline{Y}$ follow from

$$(20 \times 20 \times 10) + \frac{\pi}{2}(20)^2 \times 20 = \left[(20 \times 20) + \frac{\pi}{2}(20)^2 \right] \overline{X}$$

$$\therefore \overline{X} = -16.11 \text{ mm}$$

$$(20 \times 20 \times 10) + \frac{\pi}{2}(20)^2 \left[20 + \frac{4(20)}{3\pi} \right] = \left[(20 \times 20) + \frac{\pi}{2}(20)^2 \right] \overline{Y}$$

$$\therefore \overline{Y} = 21.32 \text{ mm}$$

As a check on these centroidal co-ordinates we should now expect that $i_x = i_y = 0$ for the centroidal axes x, y. Applying the parallel axes theorem the second moments are referred to the centroidal axes x and y as:

$$I_x = \frac{20(20)^3}{12} + 20^2 (21.32 - 10)^2 + (20)^4 \left(\frac{\pi}{8} - \frac{8}{9\pi} \right) + \frac{\pi(20)^2}{2} \left(20 + \frac{80}{3\pi} - 21.32 \right)^2$$

$$\therefore I_x = 11.44 \times 10^4 \text{ mm}^4$$

The first two terms in this calculation refer to the square. The remaining two terms refer to the semi-circle. The first in each pair of terms applies to individual centroidal axes and the second term to the centroidal axis for the whole section.

$$I_y = \frac{20(20)^3}{12} + (16.11 - 10)^2(20)^2 + \frac{\pi(20)^4}{8} + \frac{\pi}{2}(20)^2(20 - 16.11)^2$$

$$\therefore I_y = 10.06 \times 10^4 \text{ mm}^4$$

$$I_{xy} = \frac{\pi}{8}(40)^2(16.11 - 20)\left(\frac{80}{3\pi} - 1.3\right) + (20)^2(16.11 - 10)(10 - 21.3)$$

$$\therefore I_{xy} = -4.52 \times 10^4 \text{ mm}^4$$

Substitution into Eq. 1.13 gives $\theta = 41°$. The principal values are found from the expansion to the determinant in Eq. 1.16 as follows:

$$\begin{bmatrix} (11.44 - I) & -4.52 \\ -4.52 & (10.06 - I) \end{bmatrix} \times 10^4 = 0$$

which means

$$(11.44 - I)(10.06 - I) + (4.52)^2 = 0$$

for which the roots are:

$$I_u = 15.32 \times 10^4 \text{ mm}^4 \quad \text{and} \quad I_v = 6.18 \times 10^4 \text{ mm}^4$$

which are re-confirmed directly from substituting I_x, I_y and I_{xy} into Eqs 1.14a,b.

Exercises

1.1 Write down the expression for the polar second moment of area (I_z) for the ellipse shown in Fig. 1.5.

1.2 Find I_X by integration for the triangular section in Figure 1.6, then, given the position of the centroid, find I_x and I_z from the appropriate theorems.

1.3 Find I_Y, I_y and I_z for the equilateral triangular tube shown in Figure 1.7.

1.4 Show, from first principles, that I_X for the thin-walled trapezoidal section in Figure 1.16 is given by:

$$I_X = t\left[\frac{a^2c}{2} + ac^2\sin\theta + \frac{2c^2}{3}\sin^2\theta + \frac{b^3}{12} + \frac{a^3}{12}\right]$$

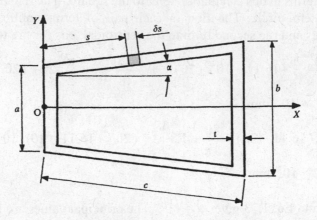

Figure 1.16

Hint: Integrate the δI_X for the element shown within the sloping sides, then add standard expressions for the vertical walls.

1.5 Find the position of the centroid and the first moments of an area enclosed between the X-axis, the parabola $Y^2 = 4aX$ and $X = b$, as shown in Figure 1.17.

$$\left[\text{Ans: } \bar{X} = \frac{3b}{5}, \ \bar{Y} = \frac{3}{4}\sqrt{(ab)}, \ i_X = ab^2, \ i_Y = \frac{4}{5}\sqrt{(ab^5)}\right]$$

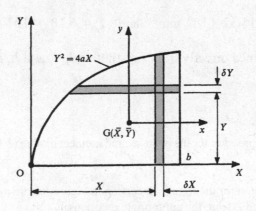

Figure 1.17

1.6 Show that I_X, I_Y, I_{XY} and I_x, I_y, I_{xy} for the parabola in Figure 1.17 are given by:

$$I_X = \frac{16}{15}\sqrt{a^3b^5}, \quad I_Y = \frac{4}{7}\sqrt{ab^7}, \quad I_{XY} = \frac{2}{3}ab^3$$

$$I_x = \frac{19}{60}\sqrt{a^3b^5}, \quad I_y = \frac{16}{175}\sqrt{ab^7}, \quad I_{xy} = \frac{ab^3}{15}$$

1.7 Examine the effect on I_{XY}, I_{xy}, I_u, I_v and θ when the areas in Figures 1.9 and 1.10 lie in the second quadrant of X, Y, i.e. they are mirrored about Y, shifting the position of the centroidal x, y axes with G but leaving the original positions of the X, Y axes unchanged.

1.8 Show, for a thin-walled circular tube of mean radius a and wall thickness t, that the second moment of the wall area is $I_x = \pi a^3 t$, where the x-axis is coincident with the horizontal diameter. Determine I_z from the perpendicular axis theorem and check I_z from integration within using polar co-ordinates.

1.9 Find I_x and I_y for each of the four symmetrical sections shown in Figures 1.18a–d. Check that $I_{xy} = 0$ in each case.

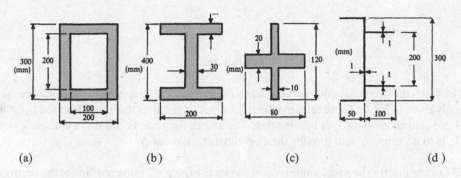

<table>
<tr><td>(a)</td><td>(b)</td><td>(c)</td><td>(d)</td></tr>
</table>

Figure 1.18

1.10 Find the position of the centroid and I_x for each of the asymmetric sections shown in Figures 1.19a–c.
(Ans: 200 mm, 208×10^6 mm^4; 30.1 mm, 31.6×10^4 mm^4; 54.7 mm, 11.87×10^6 mm^4)

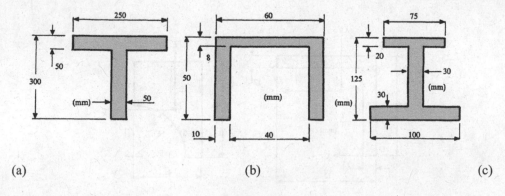

(a)   (b)   (c)

Figure 1.19

1.11 Use the parallel axis theorem to find the second moment of area about a horizontal axis x passing through the centroid for each of the three sections shown in Figs 1.20a–c.
Hint: Locate, firstly, each position of the centroid $\bar{y}$ with respect to a datum base axis (X).
(Ans to I_x and $\bar{y}$: (a) 67.6×10^4 mm^4, 28.8 mm; (b) 59.6×10^4 mm^4, 28.75 mm; (c) 87.1×10^4 mm^4, 22.86 mm)

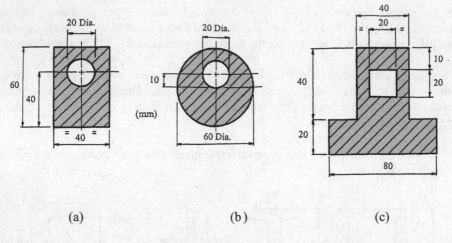

(a)   (b)   (c)

Figure 1.20

1.12 Sketch the two circle constructions when finding I_u and I_v graphically for conditions (c) and (d) given in § 1.4. These refer to where $I_y > I_x$ which requires that the calculated I_{xy} (with its sign) is to accompany I_y to become the co-ordinates for point B and that a change in sign of I_{xy} is to accompany I_x to provide the co-ordinates for point A.

1.13 Determine, for the equal-angle section shown in Figure 1.21, the position of the centroid G $(\overline{X}, \overline{Y})$ and I_x, I_y and I_{xy}. Find, by both analytical and graphical methods, the principal second moments of area I_u and I_v and the inclination θ of the principal axes u and v to the axes x and y, in which all axes originate from the centroid.

(Ans: $\overline{X}$ = -22.4 mm, $\overline{Y}$ = 49.6 mm, $I_x = I_y = 73.5 \times 10^4$ mm⁴, I_{xy} = 42.4 × 10⁴ mm⁴, I_u = 31.1 × 10⁴ mm⁴, I_v = 115.9 × 10⁴ mm⁴, θ = 45°).

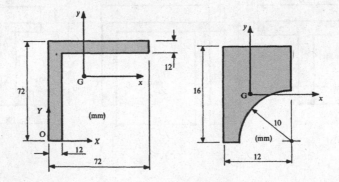

Figure 1.21   **Figure 1.22**

1.14 Find the principal second moments I_u, I_v and the inclination θ of u to x, for the solid cross sectional area, machined as shown in Figure 1.22.

(Ans: I_u = 791.2 mm⁴, I_v = 2241.8 mm⁴ and θ = 51.35°)

CHAPTER 2

STATIC EQUILIBRIUM

Summary: The following analyses are restricted to systems of two-dimensional forces all lying in the same plane. The distinction is made between forces which pass through a common point and those that do not. The conditions given refer, firstly, to when each co-planar force system is in equilibrium. Their non-equilibrium conditions follow in which a resultant force and a moment is identified. Applications to free-body diagrams include frames, plates and beams.

2.1 Co-Planar, Concurrent Forces

2.1.1 Equilibrium Conditions

A system of force vectors **A**, **B**, **C**, **D** and **E**, acting through a common point O, in the same plane (see Fig. 2.1a), is in equilibrium when the algebraic sum of their scalar magnitude components, aligned with the x- and y-directions, is zero.

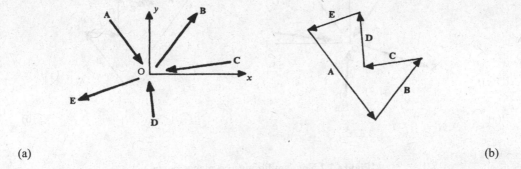

(a)   (b)

Figure 2.1 Co-planar, concurrent forces in equilibrium

Taking the sense of these components to be as positive for the positive x- and y-directions, it follows, for the directions of forces given in Fig. 2.1a, that their scalar components A_x, B_x ... and A_y, B_y ... etc, obey two *equilibrium equations*:

$$A_x + B_x - C_x - D_x - E_x = 0 \qquad (2.1a)$$

$$-A_y + B_y - C_y + D_y - E_y = 0 \qquad (2.1b)$$

Alternatively, it follows from Eqs 2.1a,b that the polygon of these force vectors will close (see Fig. 2.1b). Here vectors representing the *magnitude, direction* and *sense* of each force **A**, **B**, **C** etc. are laid 'tail to head' starting with **A** in the manner shown. The magnitude is the length of each vector, direction is its orientation and sense refers to where the arrow at its tip points.

In general, for any system of n co-planar, concurrent forces F_i, $(i = 1, 2, ... n)$, each with components F_{ix} and F_{iy}, Eqs 2.1a,b may be generalised:

$$\sum_{i=1}^{n} F_{ix} = 0 \tag{2.2a}$$

$$\sum_{i=1}^{n} F_{iy} = 0 \tag{2.2b}$$

The simplified notation, adopted within Eqs 2.2a,b, embodies any number n of force components: $F_{1x}, F_{2x} ... F_{nx}$ and $F_{1y}, F_{2y}, ... F_{ny}$, with each 'summation' accounting for either a positive or a negative sense in each component.

2.1.2 Non-Equilibrium Condition

When the system of co-planar, concurrent forces **A**, **B**, **C** and **D** in Figure 2.2a is not in equilibrium the polygon of these forces will not close. Figure 2.2b indicates how the vector **R** required to close the polygon may be interpreted in two ways.

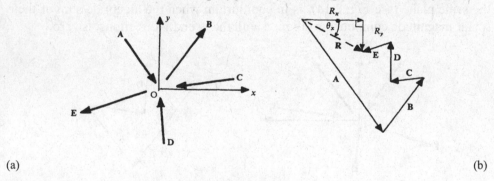

(a)   (b)

Figure 2.2 Non-equilibrium, co-planar forces

Firstly, with its sense shown, the vector **R** represents the single resultant force for the five forces. Correspondingly, the polygon identifies components of **R** in the x- and y-directions as R_x and R_y. These two components become the algebraic sum of the five force components: A_x, B_x, C_x, D_x, E_x and A_y, B_y, C_y, D_y, E_y, accounting for their signs, as follows:

$$R_x = A_x + B_x - C_x - D_x - E_x \tag{2.3a}$$

$$R_y = -A_y + B_y - C_y + D_y - E_y \tag{2.3b}$$

Figure 2.2b shows that the magnitude and direction of the resultant vector **R** is:

$$|\mathbf{R}| = \sqrt{R_x^2 + R_y^2}, \quad \theta_x = \tan^{-1}\left(R_y/R_x\right) \tag{2.4a,b}$$

where θ_x is referred to the x-direction. Equations 2.4a,b apply to any system of co-planar, concurrent forces $\mathbf{F}_1, \mathbf{F}_2, \mathbf{F}_3$, etc. For this non-equilibrium system the x- and y-components of its resultant are written as the sums: $R_x = \sum F_x$ and $R_y = \sum F_y$, in which positive signs identify with the positive sense of x and y. The signs of R_x and R_y will determine the direction and sense of $|\mathbf{R}|$ as Example 2.1 will show.

Secondly, the magnitude of the resultant vector $|\mathbf{R}|$ must equal the magnitude of the opposing *equilibrant* vector required to close the force polygon in Figure 2.2b. Clearly, the equilibrant identifies the magnitude and direction of an additional force that would need to be added to the system of forces to restore equilibrium. It follows that the magnitude and direction of the equilibrant are provided again by the resultant's Eqs 2.4a,b, with only the required change in sense to distinguish them.

Example 2.1 Four concurrent forces **A**, **B**, **C**, and **D** with respective magnitudes of 20, 30, 40 and 25 N act at a point in a plane as shown in Figure 2.3a. Find the magnitude, direction and sense of their resultant force.

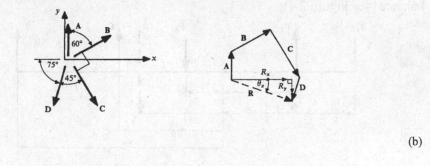

(a)   (b)

Figure 2.3 Non-equilibrium co-planar, concurrent forces

Applying Eqs 2.3a,b, with the appropriate sense signs, the components of the resultant force are:

$$R_y = A_y + B_y - C_y - D_y$$

$$= 20 + 30\sin 30° - 40\sin 60° - 25\sin 75° = -23.79 \text{ N}$$

$$R_x = B_x + C_x - D_x$$

$$= 30\cos 30° + 40\cos 60° - 25\cos 75° = 39.51 \text{ N}$$

Equations 2.4a,b provide the magnitude and direction of the resultant:

$$|\mathbf{R}| = \sqrt{[(-23.79)^2 + (39.51)^2]} = 46.12 \text{ N}$$

$$\theta_x = \tan^{-1}(-23.79/39.51) = -31.05° \text{ or } 328.95° \text{ (ACW from } x)$$

The sense of the resultant force vector **R** follows from the signs of R_x and R_y appearing within a triangle of forces. This triangle is shown in Fig. 2.3b overlaid with the alternative force polygon solution. Note that positive signs for force components identify with the co-ordinates x and y shown in Fig. 2.3a.

2.2 Co-Planar, Non-Concurrent Forces

More often than not forces lying in the same plane do not pass through a common point. Two systems of such non-concurrent, co-planar forces may be identified: *parallel* forces are often found among beam loadings while *non-parallel* forces are found more commonly within free-body diagram components of load-bearing structures. The equilibrium and non-equilibrium conditions for the two non-concurrent force systems involve an additional moment equation as now follows.

2.2.1 Parallel Forces in Equilibrium

When a parallel force system, F_i with magnitudes $|F_i| \equiv F_1, F_2, F_3, ...$, is in equilibrium both these forces and their moments about any arbitrary point, A, will balance (see Figure 2.4).

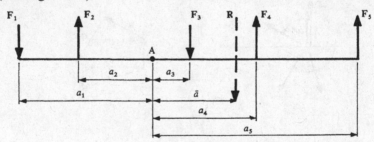

Figure 2.4 Co-planar, non-concurrent parallel forces

Thus, accounting for the sense of vertical forces shown in Figure 2.4:

$$-F_1 + F_2 - F_3 + F_4 + F_5 = 0 \qquad (2.5a)$$

and with positive, clockwise moments taken about A:

$$-F_1 a_1 + F_2 a_2 + F_3 a_3 - F_4 a_4 - F_5 a_5 = 0 \qquad (2.5b)$$

When any number of parallel forces, with magnitudes, $|F_i| \equiv F_i$ $(i = 1, 2, ... n)$, are in equilibrium, Eqs 2.5a,b are generalised to:

$$\sum_{i=1}^{n} F_i = 0 \qquad (2.6a)$$

$$M_A = \sum_{i=1}^{n} F_i a_i = 0 \qquad (2.6b)$$

where each summation is 'aligned' with the vertical forces so that a_i become their perpendicular distances from the reference point A.

2.2.2 Parallel Forces in Non-Equilibrium

If the parallel force system (Fig. 2.4) is not in equilibrium it will exert both a force and a moment that are its resultant values. The force polygon becomes a line within which the vector joining the starting point to the finishing point is the resultant force for this system:

$$|\mathbf{R}| = \sum_{i=1}^{5} F_i = -F_1 + F_2 - F_3 + F_4 + F_5 \tag{2.7a}$$

Taking the resultant force $\mathbf{R}$ to act vertically downward at a distance $\bar{a}$ to the right of point A, the resultant moment for this system becomes:

$$M_A = |\mathbf{R}|\bar{a} = -F_1 a_1 + F_2 a_2 + F_3 a_3 - F_4 a_4 - F_5 a_5 \tag{2.7b}$$

It follows from Eqs 2.7a,b that the resultant force and moment for any parallel, non-equilibrium force system, F_i, are:

$$|\mathbf{R}| = \sum_{i=1}^{n} F_i \tag{2.8a}$$

$$M_A = |\mathbf{R}|\bar{a} = \sum_{i=1}^{n} F_i a_i \tag{2.8b}$$

The following example shows how Eqs 2.8a,b require an account of the sense of each force and of the moment it produces in relation to the reference point A.

Example 2.2 Determine the resultant of the three forces shown in Fig. 2.5 and the position $\bar{a}$ at which this acts from the point A. If these forces are to be equilibrated by (i) a single balancing force and (ii) two reactive forces acting at A and B, find their magnitudes.

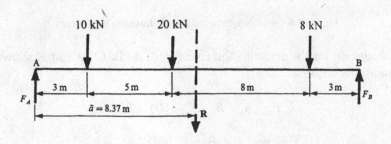

Figure 2.5 Resultant of three parallel forces

Equations 2.8a,b provide the resultant force and moment:

$$|\mathbf{R}| = \sum F_i = 10 + 20 + 8 = 38 \text{ kN } (\downarrow)$$

$$M_A = 38\bar{a} = (3 \times 10) + (8 \times 20) + (16 \times 8)$$

from which $\mathbf{R}$ lies at its perpendicular distance $\bar{a} = 8.37$ m from point A. It follows that the single balancing force of 38 kN with opposite sense must be placed at this position. The force would act downwards upon a roller or knife edge when acting as the fulcrum of this balanced lever. Equations 2.4a,b provide the magnitudes of upward end forces to be applied at A and B required to restore equilibrium:

$$\sum M_A = 0 = (8.37 \times 38) - 19F_B, \; \therefore F_B = 16.74 \text{ kN } (\uparrow)$$

$$\sum F_i = 0 = -38 + 16.74 + F_A, \; \therefore F_A = 21.26 \text{ kN } (\uparrow)$$

Again, the balanced beam requires suitable supports to be placed at its ends, upon which these reactive forces act.

2.2.3 Non-Parallel Forces in Equilibrium

The conditions for equilibrium for this most general 2D force system are: the balance of force components in the x and y directions and a balance of moments for any arbitrary point P (see Figure 2.6).

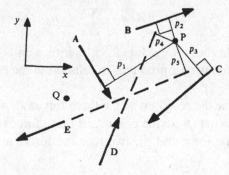

Figure 2.6 Co-planar, non-concurrent forces in equilibrium

Thus, accounting for the sense of the five forces **A, B, C, D** and **E** shown, these three conditions become

$$\sum F_x = A_x + B_x - C_x + D_x - E_x = 0 \tag{2.9a}$$

$$\sum F_y = -A_y + B_y - C_y + D_y - E_y = 0 \tag{2.9b}$$

$$\sum M_P = -|\mathbf{A}|\, p_1 + |\mathbf{B}|\, p_2 + |\mathbf{C}|\, p_3 + |\mathbf{D}|\, p_4 + |\mathbf{E}|\, p_5 = 0 \tag{2.9c}$$

Alternatively, or in addition to Eq. 2.9c, $\sum M_Q = 0$ for any another point Q, where perpendicular distances are identified in a similar manner with the usual convention that clockwise moments are positive. It follows from Eqs 2.9a–c, that for a number of forces $\mathbf{F}_i$ $(i = 1, 2 \dots n)$, with magnitudes $|\mathbf{F}_i| \equiv F_i$ and components F_{ix} and F_{iy}, two force and one moment condition apply:

$$\sum_{i=1}^{n} F_{ix} = 0 \tag{2.10a}$$

$$\sum_{i=1}^{n} F_{iy} = 0 \tag{2.10b}$$

$$M_P = \sum_{i=1}^{n} F_i \, p_i = 0 \tag{2.10c}$$

Additional moment equilibrium conditions, similar to Eq. 2.10c, may be applied repeatedly to other points as required. For example, at point Q:

$$M_Q = \sum_{i=1}^{n} F_i \, q_i = 0 \tag{2.10d}$$

2.2.4 Non-Parallel Forces in Non-Equilibrium

When any system of co-planar, non-concurrent forces is not in equilibrium the resultant force $\mathbf{R}$ and its inclination to x are again given by Eqs 2.4a,b. In summary, the resultant properties for the forces within this system are found from:

$$R_x = \sum_{i=1}^{n} F_{ix} \tag{2.11a}$$

$$R_y = \sum_{i=1}^{n} F_{iy} \tag{2.11b}$$

$$|\mathbf{R}| = \sqrt{R_x^2 + R_y^2}, \quad \theta_x = \tan^{-1}\left(\frac{R_y}{R_x}\right) \tag{2.11c,d}$$

In addition, there will be a resultant moment which is entirely dependent upon the reference position. Thus at point P the moment from the resultant is statically equivalent to the moment sum from the applied forces:

$$M_P = |\mathbf{R}|\bar{p} = \sum_{i=1}^{n} |\mathbf{F}_i| \, p_i = F_i p_i \tag{2.12a}$$

where $\bar{p}$ is the perpendicular distance of $\mathbf{R}$ from P. Alternatively, for point Q the resultant moment and its position are found from:

$$M_Q = |\mathbf{R}|\bar{q} = \sum_{i=1}^{n} |\mathbf{F}_i| \, q_i = F_i q_i \tag{2.12b}$$

The following examples show how these resultants may be found analytically.

Example 2.3 A horizontal beam is loaded with four forces as shown in Fig. 2.7. Find the magnitude, direction and position of the resultant force. Also find the magnitude and direction of the supporting reactions when the beam shown rests upon rollers at A and is hinged at B.

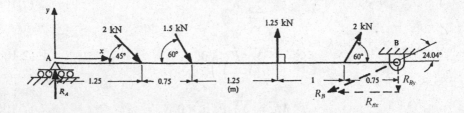

Figure 2.7 Beam in equilibrium under four non-parallel forces

Equations 2.11a,b supply the components of the resultant force:

$$R_x = 2\cos 45° + 1.5\cos 60° + 2\cos 60° = 3.164\,\text{kN}$$

$$R_y = -2\sin 45° - 1.5\sin 60° + 1.25 + 2\sin 60° = 0.269\,\text{kN}$$

and Eqs 2.11c,d give the magnitude and direction of the resultant as:

$$|\mathbf{R}| = \sqrt{(3.164)^2 + (0.269)^2} = 3.176\,\text{kN}$$

$$\theta_x = \tan^{-1}\left(\frac{0.269}{3.164}\right) = 4.86°$$

The inclination θ_x shows that $\mathbf{R}$ exerts an anticlockwise moment about A. Hence, the position of $\mathbf{R}$ may be referred to its perpendicular distance $\bar{a}$ from A when, from Eq. 2.12a:

$$M_A = |\mathbf{R}|\,\bar{a} = \sum_{i=1}^{n} |\mathbf{F}_i|\,a_i = a_i F_i$$

$$-3.176\,\bar{a} = 1.25\,(2\sin 45°) + 2\,(1.5\sin 60°) - (3.25 \times 1.25) - 4.25\,(2\sin 60°)$$

from which $\bar{a} = 2.222$ m Hence the four forces would be balanced by a single equilibrant force that has the magnitude and orientation of resultant force $\mathbf{R}$ but must act with an opposing sense.

In the case of an end-supported, balanced beam we apply the four equilibrium conditions in Eqs 2.10a–d. Firstly, the moment Eq. 2.10c is applied to B to provide the roller's vertical support reaction R_A:

$$\sum M_B = (0.75 \times 2\sin 60°) + (1.75 \times 1.25) - (3 \times 1.5\sin 60°)$$
$$- (3.75 \times 2\sin 45°) + 5R_A = 0$$

from which $R_A = 1.143$ N. Let the components of the hinge reaction be R_{Bx} and R_{By} (assumed positive). Equations 2.10a,b express the force equilibrium conitions:

$$(+ \rightarrow) \sum R_x = 0 = 2 \cos 45° + 1.5 \cos 60° + 2 \cos 60° + R_{Bx}$$
$$\therefore R_{Bx} = -3.164 \text{ N}$$

$$(+ \uparrow) \sum R_y = 0 = 1.143 - 2 \sin 45° - 1.5 \sin 60° + 1.25 + 2\sin 60° + R_{By}$$
$$\therefore R_{By} = -1.412 \text{ N}$$

In fact, the negative signs show that both components act in the negative x- and y-directions, as shown. Equations 2.11c,d provide the magnitude and direction of the resultant:

$$R_B = \sqrt{(-3.164)^2 + (1.412)^2} = 3.465 \text{ N}$$

$$\theta_x = \tan^{-1} \frac{R_{By}}{R_{Bx}} = \tan^{-1} \left(\frac{-1.412}{-3.164} \right) = 24.04°$$

Using the second moment Eq. 2.10d (i.e. $\sum M_A = 0$) as a check upon R_B:

$$\sum M_A = (1.25 \times 2 \sin 45°) + (2 \times 1.5 \sin 60°) - (1.25 \times 3.25)$$
$$- (4.25 \times 2 \sin 60°) + (5 \times 3.465 \sin 24.04°) \approx 0$$

Note that the final term equals the clockwise moment produced by R_{By} from A, i.e. (1.412×5). Taking moments of a force using its x- and y-components often simplifies a solution, as the following example will show.

Example 2.4 A 12 m square plate ABCD weighs 6 kN. If it is to remain in equilibrium under the action of the six forces shown, including self-weight, find the magnitude and direction of the force **V** and the perpendicular distance at which this force should act from A.

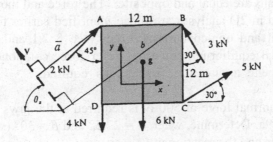

Figure 2.8 Plate in equilibrium

Let the unknown force **V** with magnitude V act at a perpendicular distance $\bar{a}$ from A in the direction θ_x shown. The three unknowns, V, θ_x and $\bar{a}$ are found from the three Eqs 2.10a–c as follows:

$$\sum F_x = 0 = 2\sin 45° - V\cos\theta_x - 3\sin 30° + 5\cos 30°$$
$$\therefore \ V\cos\theta_x = 4.2442$$

$$\sum F_y = 0 = 2\cos 45° + V\sin\theta_x + 3\cos 30° + 5\sin 30° - 6 - 4 = 0$$
$$\therefore \ V\sin\theta_x = 3.4887$$

$$V = \sqrt{[(4.2442)^2 + (3.4887)^2]} = 5.4934 \text{ kN}$$

$$\theta_x = \tan^{-1}(3.4887/4.2442) = 39.41°$$

$$\sum M_A = 5.4934\,\bar{a} + (6 \times 6) - (12 \times 3\cos 30°) - (12 \times 5\cos 30°) - (12 \times 5\sin 30°) = 0$$

This gives $\bar{a} = 14.04$ m. To check this answer take moments about any other point, say B, where we should expect the moment sum to be zero:

$$\sum M_B = (12 \times 2\cos 45°) - (4 \times 12) - (6 \times 6) - (12 \times 5\cos 30°) + 5.4934(14.04 + 12\sin 39.41°) \approx 0$$

The analysis shows that the direction and sense assumed for **V** was correct. Here it is helpful to identify the equilibrant force **V** with the closing vector in the force polygon. Alternatively, if the magnitude and direction of the resultant of the five known forces is found, these also apply to the equilibrant vector but with a change in sense.

2.3 The Free-Body Diagram

To examine the equilibrium of a mechanism consisting of many load-bearing components it is necessary to isolate each component as a free-body diagram (FBD). The diagram shows each component with all the external actions imposed upon it. The force directions are provided by Newton's third law which states that actions and reactions are equal and opposite. The force and moment equilibrium equations required in 2D analyses are those identified earlier for systems of co-planar, concurrent and non-concurrent forces (see § 2.1 and § 2.2). For 3D analyses, additional equilibrium equations refer force components to a third z-direction and moments to planes x, z and y, z, as required.

Example 2.5 A normal force of 500 N is required at the jaws of the adjustable pliers in Figure 2.9a. Determine, when $\alpha = 27.5°$ and $\beta = 30°$ (see Fig. 2.9c), the force **Q** at the grips and the reaction at O_2.

The FBD for the jaw and lower grip (Figs 2.9b,c, respectively) are examples of co-planar forces in equilibrium. The three forces **A**, **B** and **Q** (Fig. 2.9c) upon the grip are concurrent at O_1 and therefore the directions of **A** and **B** are defined by α and β, respectively. Applying the moment equilibrium Eq. 2.10c to the non-concurrent forces upon the jaw in Figure 2.10b gives:

$$\sum M_{O_2} = 0 = - (500 \times p) + (|\mathbf{A}| \times q),$$

$$(|\mathbf{A}| \times 45 \cos 27.5°) - (500 \times 50) = 0$$

giving $|\mathbf{A}| = 626.3$ N. With the lower jaw FBD in Fig. 2.9b being in equilibrium, the components of the reaction $\mathbf{R}$ vector at O_2, are: $R_x = 626.3 \cos 27.5° = 555.54$ N and $R_y = 500 + 626.3 \sin 27.5° = 789.2$ N. These give: $|\mathbf{R}| = 965.1$ N and $\theta = 54.86°$.

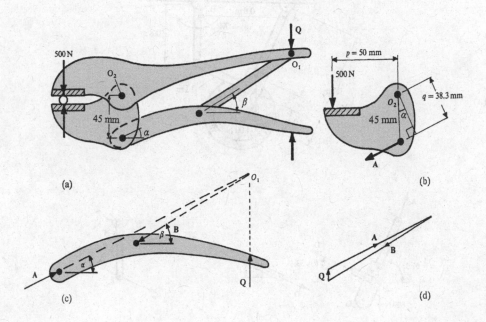

Figure 2.9 Forces applied to pliers' grip and jaw

Resolving forces in the x- and y-directions (Fig. 2.9c), we have from Eqs 2.10a,b:

$$\sum F_y = A_y - B_y + |\mathbf{Q}| = 0$$

$$|\mathbf{A}| \sin \alpha - |\mathbf{B}| \sin \beta + |\mathbf{Q}| = 0$$

$$626.3 \sin 27.5° - |\mathbf{B}| \sin 30° + |\mathbf{Q}| = 0 \qquad\qquad \text{(i)}$$

$$\sum F_x = A_x - B_x = 0$$

$$|\mathbf{A}| \cos \alpha - |\mathbf{B}| \cos \beta = 0$$

$$626.3 \cos 27.5° - |\mathbf{B}| \cos 30° = 0 \qquad\qquad \text{(ii)}$$

The solution to Eqs i and ii gives $|\mathbf{B}| = 641.5$ N and $|\mathbf{Q}| = 31.56$ N. Alternatively, a force polygon (see Fig. 2.9d) supplies $\mathbf{Q}$ directly once $\mathbf{A}$ is known. Note that α and β, which must be known to apply either method, can be found from the geometry of the pliers in Fig. 2.9a when gripping.

Example 2.6 Find the pin forces at A and B and the moment M required to lift the nose wheel assembly in Fig. 2.10a when the link BC is vertical. The fork link AO weighs 800 N acting vertically downwards at its centroid g as shown. By comparison the weights of connecting links BC and DC may be neglected.

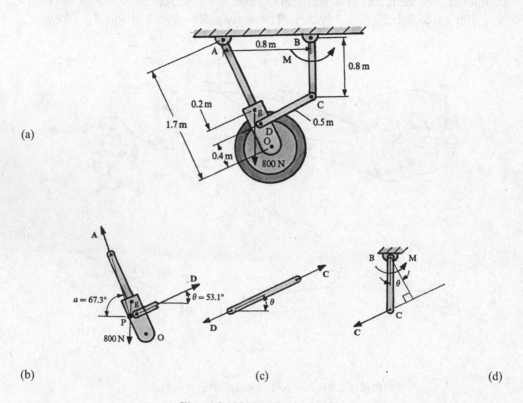

(a)

(b)   (c)   (d)

Figure 2.10 Nose wheel assembly

From the geometry of the link in Fig. 2.10a we have: $\theta = 53.1°$ and $\alpha = 67.3°$. These could also be found with sufficient accuracy from a scaled line diagram in which D lies at the intersection between arcs of radii 0.5 m and 1.3 m with their respective centres at C and A.

The FBD for each separated link is given in Figure 2.10b–d. Recall that each diagram shows the external forces and moments acting upon each link from their connections within the assembly. With no moment applied at A in (b), the three forces must be concurrent at point P and in equilibrium. Thus, from Eqs 2.1a,b:

$$\sum F_x = |\mathbf{D}| \cos \theta - |\mathbf{A}| \cos \alpha = 0$$

$$-0.386 |\mathbf{A}| + 0.600 |\mathbf{D}| = 0 \tag{i}$$

$$\sum F_y = 0 = -800 + |\mathbf{D}| \sin \theta + |\mathbf{A}| \sin \alpha$$

$$-0.923 |\mathbf{A}| + 0.800 |\mathbf{D}| = 800 \tag{ii}$$

The solutions to the simultaneous Eqs i and ii are: $|\mathbf{A}| = 556.5$ N and $|\mathbf{D}| = 358$ N. Force $|\mathbf{A}|$ acts upon the shear pin at A whose diameter is selected appropriately (e.g. see Ex. 3.4, p. 69). Now $|\mathbf{D}|=|\mathbf{C}|$, indicating in Figure 2.10c, that force $\mathbf{D}$ is transmitted to C in the direction of link DC. Referring this to Figure 2.10d, the moment M is found from Eq. 2.10c as follows:

$$\sum M_B = - M + |\mathbf{C}| \times (BC \cos\theta) = 0$$

$$\therefore M = 358 \times 0.8 \cos 53.1° = 172 \text{ Nm}$$

Finally, the transverse shear force upon the pin at B is $|\mathbf{B}| = |\mathbf{C}| = 358$ N, acting in the direction of C. Hence, a pin B of lesser diameter than that of pin A may be chosen but the link BC requires that it be designed as a beam supporting both shear force and bending moment (as in § 2.5).

2.4 Bar Forces in Plane Frames

In two-dimensional plane frames any loading applied externally to a pinned joint forms a concurrent force system with the internal bar forces. Hence, the bar forces may be determined from solving the two simultaneous equations that arise from applying Eqs 2.2a,b to each joint. This method of *joint tension* assumes that each bar pulls at its joint as the analysis proceeds from joint to joint. This can be very tedious and when only a single bar force is required, the *method of sections* is preferred. The section must cut the frame including the bar of interest so that each half of the frame appears as a free-body diagram. The forces exposed within the cut bars together with the external nodal forces and support reactions form a co-planar, non-concurrent equilibrium force system to which Eqs 2.10a–d are applied. The following example illustrates how the joint tension and section methods are applied.

Example 2.7 Determine the forces in kN within each bar of the Warren frame shown in Figure 2.11. The girder is loaded at E and D when suspended upon pins at A and C.

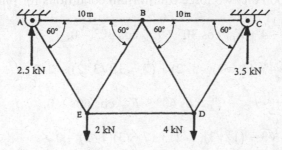

Figure 2.11 Warren frame under load

The first step is to find the support reactions from the external forces. These are found conveniently from applying $\sum M = 0$ and then $\sum F = 0$ in the following manner. Equating clockwise to anticlockwise moments about point A:

$$(2 \times 5) + (15 \times 4) = 20R_C, \quad \therefore \quad R_C = 3.5 \text{ kN}$$

Then equating upward forces to downward forces:

$$R_A + 3.5 = 2 + 4, \quad \therefore \quad R_A = 2.5 \text{ kN}$$

Joint Tension. Assuming tension at each joint in Figure 2.11a we apply vertical and horizontal force equilibrium conditions $\sum F_x = 0$ and $\sum F_y = 0$, to give the required simultaneous equations. Bar forces so found, whether they be positive or negative, are carried between the ends so that $F_{AB} = F_{BA}$, $F_{ED} = F_{DE}$, etc. in the manner now shown. The analysis must start at a joint where there are two bars and proceed from joint to joint where there are never more than two unknown bar forces:

joint A: $\uparrow$ $2.5 - F_{AE} \cos 30° = 0,$ $\hspace{3cm}$ $\therefore F_{AE} = 5/\sqrt{3}$

$\rightarrow$ $F_{AB} + F_{AE} \cos 60° = 0,$ $\hspace{3cm}$ $\therefore F_{AB} = -5/(2\sqrt{3})$

joint E: $\uparrow$ $- 2 + F_{EA} \cos 30° + F_{EB} \cos 30° = 0,$ $\hspace{1cm}$ $\therefore F_{EB} = -1/\sqrt{3}$

$\rightarrow$ $- F_{EA} \cos 60° + F_{EB} \cos 60° + F_{ED} = 0,$ $\hspace{1cm}$ $\therefore F_{ED} = \sqrt{3}$

joint B: $\uparrow$ $- F_{BE} \cos 30° - F_{BD} \cos 30° = 0,$ $\hspace{2cm}$ $\therefore F_{BD} = 1/\sqrt{3}$

$\rightarrow$ $- F_{BE} \cos 60° - F_{BA} + F_{BD} \cos 60° + F_{BC} = 0,$ $\hspace{0.3cm}$ $\therefore F_{BC} = -7/(2\sqrt{3})$

joint C: $\uparrow$ $3.5 - F_{CD} \cos 30° = 0,$ $\hspace{3cm}$ $\therefore F_{CD} = 7/\sqrt{3}$

$\rightarrow$ $- F_{CB} - F_{CD} \cos 60° = 0,$ $\hspace{3cm}$ $\therefore F_{BC} = -7/(2\sqrt{3})$

and, as a check upon the two force equilibrium conditions for joint D:

$$\uparrow - 4 + F_{DB} \cos 30° + F_{DC} \cos 30° = 0,$$

$$- 4 + (1/\sqrt{3})(\sqrt{3}/2) + (7/\sqrt{3})(\sqrt{3}/2) = 0 \checkmark$$

$$\rightarrow - F_{DE} - F_{DB} \cos 60° - F_{DC} \cos 60° = 0,$$

$$- \sqrt{3} - (1/\sqrt{3})(1/2) + (7/\sqrt{3})(1/2) = 0 \checkmark$$

Method of Sections. When this method is used to find all the bar forces, these forces are exposed firstly within the sections shown in Figs 2.12a–d.

(a)   (b)

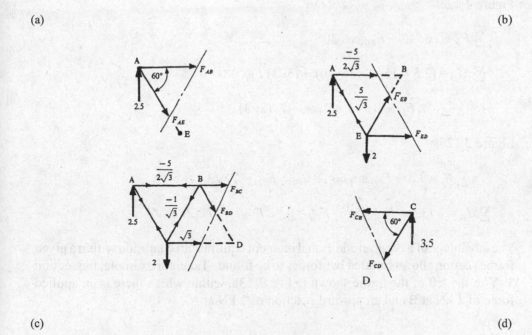

(c)   (d)

Figure 2.12 Bar forces by method of sections

The force and moment conditions $\sum F_x = 0$, $\sum F_y = 0$ and $\sum M = 0$, as required, are then applied to each section, i.e. to the left of the sections in Figs 2.12a,b and c and to the right in Fig. 2.12d:

Figure 2.12a:

$$\sum F_y = 2.5 - F_{AE} \sin 60° = 0$$

$$\sum M_E = (5 \times 2.5) + (5\sqrt3) F_{AB} = 0$$

$$\therefore F_{AE} = 5/\sqrt3, \ F_{AB} = -5/(2\sqrt3)$$

Figure 2.12b:

$$\sum F_x = -5/(2\sqrt3) + (5/\sqrt3) \cos 60° - (5/\sqrt3) \cos 60° + F_{ED} + F_{EB} \cos 60° = 0$$

$$\sum M_B = (2.5 \times 10) - (5 \times 2) - (5\sqrt3) F_{ED} - (5\sqrt3) \, 10 \cos 30°$$
$$+ (5\sqrt3) \, 10 \cos 30° = 0$$

$$\therefore F_{ED} = \sqrt3, \ F_{EB} = -1/\sqrt3$$

Notice here that the bar forces and the moments from AE cancel since this bar has not been cut. Consequently, the analysis may proceed to Figs 2.12c and d more simply by omitting any uncut bar forces.

Figure 2.12c:

$$\sum F_y = 2.5 - 2 - F_{BD} \cos 30°$$

$$\sum M_D = (2.5 \times 15) - (2 \times 10) + (5\sqrt{3}) F_{BC} = 0$$

$$\therefore F_{BD} = 1/\sqrt{3}, \ F_{BC} = -7/(2\sqrt{3})$$

Figure 2.12d:

$$\sum F_y = 3.5 - F_{CD} \sin 60° = 0, \ \ \therefore F_{CD} = 7/\sqrt{3}$$

$$\sum M_D = -(3.5 \times 5) - (5\sqrt{3}) F_{CB} = 0, \ \ \therefore F_{CB} = -7/(2\sqrt{3}) \checkmark$$

The advantage of having an additional moment equilibrium condition within a given frame section allows isolated bar forces to be found. Take, for example, the section Y-Y to the left of the frame shown in Fig. 2.13a, within which there is an applied force of 1 kN at B and an upward reaction of 2 kN at A.

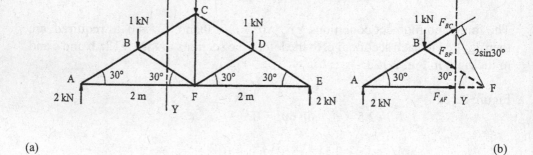

(a)   (b)

Figure 2.13 Bar forces from method of sections

The three bar forces F_{BF}, F_{AF} and F_{BC}, exposed by this section in Fig. 2.13b, are found from applying the moment equilibrium condition to points A, B and F, respectively:

$$\sum M_A = 0 = (1 \times 1) + (2 \sin 30°) F_{BF}, \ \ \ \therefore \ F_{BF} = -1 \text{ kN}$$

$$\sum M_B = 0 = (1 \times 2) - (1 \tan 30°) F_{AF}, \ \ \ \therefore \ F_{AF} = 2\sqrt{3} \text{ kN}$$

$$\sum M_F = 0 = (2 \times 2) - (1 \times 1) + (2 \sin 30°) F_{BC}, \ \ \ \therefore \ F_{BC} = -3 \text{ kN}$$

in which the frame geometry provides the perpendicular distances to each force as indicated: (2 sin 30°) and (1 tan 30°).

2.5 Forces and Moments in Beams

When applying the theory of bending to find stresses in beams (see Chapter 5) it is necessary to know the manner in which the transverse shear force, F and the bending moment, M vary over the length. For the cantilever and a simply supported beam resting upon two supports, the application of equilibrium principles alone is sufficient to establish the F and M distributions. The beam is said to be *statically determinate*. Both F and M variations appear in diagrams whose construction adopts the following procedure:

(i)   Find the support reactions from applying the external force and moment equilibrium equations across the whole beam (see Eqs 2.6a,b).

(ii)   To find F at any section, calculate the resultant force to the left of that section taking downward forces as positive. This is equivalent to plotting F (↑ +ve) from right to left along the length of the beam.

(iii)   To find M at any section, take moments either to the right or left of that section, depending upon which is easier. Employ the convention that hogging moments (see Fig. 2.14) are positive. Sagging moments which produce the opposite curvature, are negative.

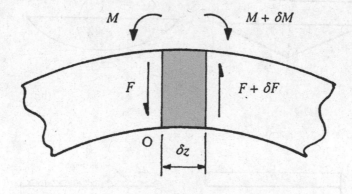

Figure 2.14 Convention for positive shear forces and bending moments

Calculations (ii) and (iii) apply Eqs 2.8a,b to find the resultant force and moment that is carried by the beam at each section. Some texts make use of the alternative convention in which upward forces and sagging moments are positive. This produces F- and M-diagrams that are inversions of those that are given here. It is important to note that the sign of the moment is now given by the *curvature* it produces and not whether it is clockwise or anticlockwise. This ensures that the net moment at any section carries a similar sign for both its left and right sides. The present convention takes hogging as positive to avoid the need to introduce a minus sign later when calculating stress and deflection in beams and struts arising from the bending moments (see Chapters 5 and 8).

2.5.1 Relationship Between F and M

Let the force and moment increase by amounts δF and δM over a length δz, of the beam (see Fig. 2.14). Taking moments about O:

$$(M + \delta M) = M + (F + \delta F)\delta z \qquad (2.13a)$$

Neglecting the product $\delta F \times \delta z$ in Eq. 2.13a, we find:

$$F = \delta M / \delta z \qquad (2.13b)$$

Equation 2.13b may be interpreted in two ways: (i) F is the derivative of M and (ii) M is the area beneath the F-z diagram. Note also from Eq. 2.13b that a maximum or a minimum M will occur where F is zero.

In Figures 2.15a–d the F and M diagrams are given for standard simply supported beams and cantilevers with concentrated and distributed loading.

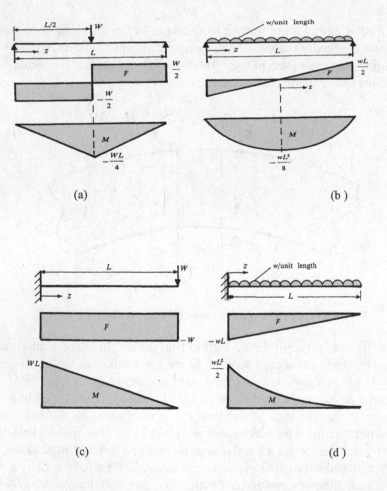

Figure 2.15 F- and M-diagrams for simply loaded standard beams

At the position defined by z the following expressions for F and M are seen to conform to Eq. 2.13b:

(a) $F = -W/2$; $M = -Wz/2$; $M_{max} = -WL/4$

(b) $F = -wL/2 + wz$; $M = -wLz/2 + wz^2/2$; $M_{max} = -wL^2/8$

(c) $F = -W$; $M = W(L - z)$; $M_{max} = wL$

(d) $F = -w(L - z)$; $M = w(L - z)^2/2$; $M_{max} = wL^2/2$

For simply supported beams (a) and (b) the maximum, central moments appear where $F = 0$ (or F is crossed). Their sagging curvatures (negative) produce compression along the top surface. In cantilevers (c) and (d), the fixed end lies at the origin where the maximum fixing moments appear. Their hogging curvatures (positive) produce tension along the top surface. Expressions given for maximum F and M should be remembered. In each case the area beneath between F-diagram from its origin to the position defined by z is the ordinate in the M-diagram.

Example 2.8 Draw the shear force and bending moment diagrams for the beam with mixed loading shown in Figure 2.16.

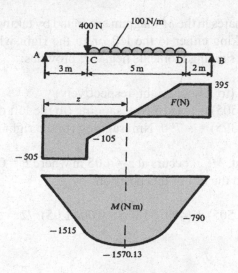

Figure 2.16 F- and M-diagrams

Reactions. Using Eqs 2.6a,b:

$$(3 \times 400) + (5.5 \times 5 \times 100) = 10 R_B, \quad \therefore R_B = 395 \text{ N}$$

$$R_A + R_B = 900, \quad \therefore R_A = 505 \text{ N}$$

F-Diagram. Calculating the resultant forces to the left of sections at A, C and D:

$$F_A = -R_A = -505 \text{ N}$$
$$F_C = -505 + 400 = -105 \text{ N}$$
$$F_D = -505 + 400 + (100 \times 5) = 395 \text{ N} = F_B$$

Alternatively, if we work to the right of these positions we may take upward forces to be positive:

$$F_A = 395 - (100 \times 5) - 400 = -505 \text{ N}$$
$$F_C = 395 - (100 \times 5) = -105 \text{ N}$$
$$F_D = 395 \text{ N}$$

Applying $\sum F = 0$ to the left of the position of zero shear force gives the position where M is a minimum:

$$F = 0 = -505 + 400 + 100 (z - 3)$$

from which $z = 4.05$ m. Note that the distributed loading, $w = 100$ N/m, is the slope of F-diagram for the 5 m length over which it is applied (i.e. $w = \delta F/\delta z$).

M-diagram. The ordinates in the M-diagram are found by taking moments at points A, B, C and D. Working either to the left or to the right when taking hogging moments positive and sagging moments negative provides:

$$M_A = M_B = 0 \text{ (to left and right, respectively)}$$
$$M_C = - (3 \times 505) = - 1515 \text{ Nm sagging (to the left of C)}$$
$$M_D = - (2 \times 395) = - 790 \text{ Nm sagging (to the right of D)}$$

The maximum moment, M_{max}, occurs at $z = 4.05$ m where $F = 0$. The magnitude is found from working to the left of this position:

$$M_{max} = - (4.05 \times 505) + (400 \times 1.05) + 100 (1.05)^2/2 = - 1570.13 \text{ Nm}$$

2.5.2 Contraflexure

In the previous examples the bending moment diagrams were all of the same sign. With other forms of loading, the bending moment diagrams may contain both positive and negative portions. The points of contraflexure (or inflection) are points of zero bending moment, i.e. where the beam changes its curvature from hogging to sagging. Calculating the position of the point of contraflexure is illustrated in the following three examples.

Example 2.9 Construct the F- and M-diagrams for the loaded beam in Fig. 2.17. Find the position and magnitude of the maximum bending moment and the points of inflection.

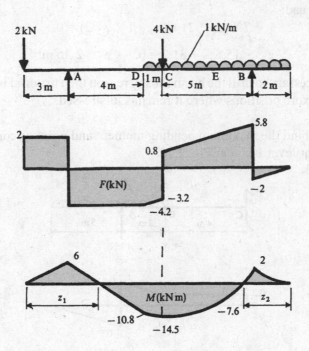

Figure 2.17 F- and M-diagrams

Reactions. Take moments about A:

$$(5 \times 4) + (8 \times 8) - (2 \times 3) = 10R_B$$
$$R_B = 7.8 \text{ kN}, \quad R_A = 2 + 4 + 8 - 7.8 = 6.2 \text{ kN}$$

F-diagram. The construction has worked from the right to left taking downward forces -ve and upward forces +ve. In this way the forces appear in the diagram with the same sense as they are applied.

M-diagram. Work to left or right of points A, B, C, D and E as indicated:

$$M_A = +(2 \times 3) = 6 \text{ kNm (to left)},$$
$$M_B = +(2 \times 1 \times 1) = 2 \text{ kNm (to right)},$$
$$M_C = (1 \times 1 \times 0.5) + (8 \times 2) - (5 \times 6.2) = -14.5 \text{ kNm (to left)},$$
$$M_D = (7 \times 2) - (4 \times 6.2) = -10.8 \text{ kNm (to left)},$$
$$M_E = (4 \times 1 \times 2) - (2 \times 7.8) = 7.6 \text{ kNm (to right)}.$$

Figure 2.17 shows that the greatest moment $M_{max} = 14.5$ kNm (sagging) occurs at C where F is crossed. Two *points of contraflexure* appear in the M-diagram. Their positions, z_1 and z_2, are found from equating the moment to zero at each point:

(i) z_1 from LH end:

$$(2 z_1) - (z_1 - 3) 6.2 = 0, \quad \therefore z_1 = 4.45 \text{ m}$$

(ii) z_2 from RH end:

$$-7.8 (z_2 - 2) + (z_2 \times 1 \times z_2/2) = 0$$

$$z_2{}^2 - 15.6z_2 + 31.2 = 0, \quad \therefore z_2 = 2.36 \text{ m}$$

Should it be necessary to drill the beam for any reason then material is best removed at the contraflexure positions where it remains unstressed.

Example 2.10 Find the maximum bending moment and point of contraflexure for the propped cantilever beam in Fig. 2.18.

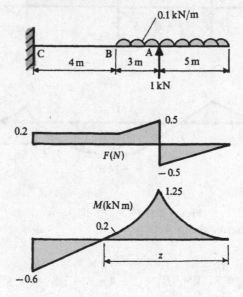

Figure 2.18 *F*- and *M*-diagrams for a cantilever beam

F-diagram. With cantilevers, it is more convenient to arrange that the free end lies at the RH end. The diagram may be constructed as before working from right to left taking ↑ $F(+\text{ve})$. This avoids the need to calculate the fixing reaction at the LH end, which appears as the 0.2 kN force required to close the *F*-diagram.

M-diagram. Working to the right of points A, B and C:

$$M_A = 5 \times 0.1 \times 2.5 = 1.25 \text{ kNm} = M_{max}$$
$$M_B = (8 \times 0.1 \times 4) - (3 \times 1) = 0.2 \text{ kNm}$$
$$M_C = (8 \times 0.1 \times 8) - (7 \times 1) = -0.6 \text{ kNm}$$

The maximum moment occurs at A, where the *F*-diagram crosses its axis. Let the *point of contraflexure* lie at a distance z from RH end. Equating $M = 0$ at this point:

$$M = [8 \times 0.1 \, (z - 8 + 4)] - 1 \, (z - 5) = 0$$

from which $z = 9$ m. Zero moment at contraflexure means that the beam cross section remains unstressed at this position. Consequently, this is the best position to offset a potential weakening of the beam where there is the requirement for material removal and welding attachments.

Exercises

Non-Concurrent Forces

2.1 Calculate the force F required to balance the lever ABCD in Figure 2.19 and find the magnitude and direction of the support reaction at B. Why would a knife-edge support be unsatisfactory in this instance and with what should it be replaced?
(Ans: $F = 7.81$ kN, $R_B = 1.64$ kN at $9.76°$ clockwise from the horizontal)

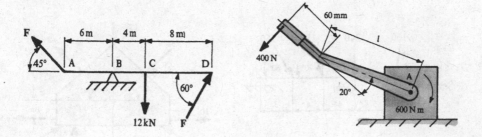

Figure 2.19   **Figure 2.20**

2.2 The action of the control lever in Figure 2.20 is such that a force of 400 N, when applied at the lever end, is resisted by a clockwise couple of 600 Nm at A. Determine the necessary dimension l for the cranked lever.
(Ans: $l = 1.53$ m)

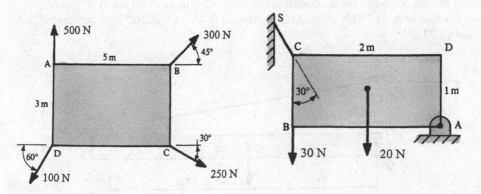

Figure 2.21   **Figure 2.22**

2.3 Four forces act on the rectangular plate ABCD in Figure 2.21. Find the magnitude, direction and position from B of the resultant force.
(Ans: 0.627 kN, 52.8° ACW from horizontal, 2.5 m ⊥ to B)

2.4 The rectangular lamina ABCD in Figure 2.22 is hinged at A and supported by the string SC. Determine the tension in the string and the reaction components at the hinge A when the lamina supports a vertical load of 30 N at B in addition to its self-weight of 20 N. What is the magnitude and orientation of the shear force acting upon the pin at A?
(Ans: 64.9 N, A_x = 32.5 N, A_y = − 6.2 N, |**A**| = 33.1, θ_x = − 10.6°)

2.5 The light, horizontal, 20 m length beam in Figure 2.23 is hinged at the left end and is simply supported at the right end. Determine the resultant of the four applied forces shown and each support reaction in magnitude and direction.
(Ans: resultant 705.1 N at 81.1° to the horizontal, 10.21 m ⊥ to LH end; RH support reaction 360 N↑; LH support reaction 353.8 N at 72.1 to the horizontal)

2.6 Show that the gusset plate in Figure 2.24 satisfies both force and moment equilibrium conditions under the action of the eight forces (in kN) as shown.

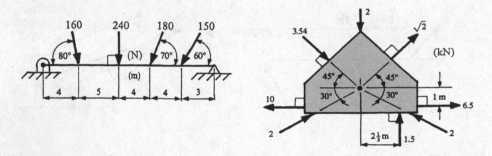

Figure 2.23   **Figure 2.24**

Free-Body Diagrams

2.7 Determine, for the jib crane shown in Figure 2.25, the wall reactions at A and B and the force in the rod AC when the horizontal beam supports a force of 50 kN at mid-span.
(Ans: reaction at A 43.73 kN along AC; reaction at B 46.5 kN at 36.25° to horizontal; rod tension 43.73 kN)

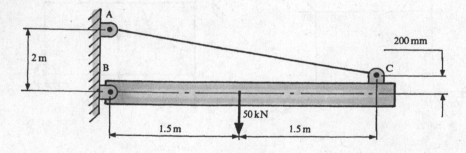

Figure 2.25

2.8 The jib crane shown in Figure 2.26 supports a vertical load of 5 kN at 2 m from B. Determine the forces acting: (i) at joint A, (ii) at the hinge support B, (iii) at the contact support E and (iv) along the rod DC.
(Ans: 12.2 kN at 65.75°, 11.11 kN at 3.1°; 11.1 kN at 0°; 12.4 kN at 26.6°)

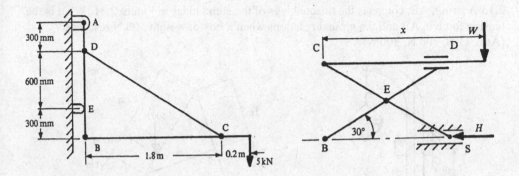

Figure 2.26   **Figure 2.27**

2.9 The principle of the lifting mechanism shown in Figure 2.27 is that a horizontal force H, applied at the slider S, will raise a weight W through the pin joints at B, C and E and a further slider at D. Find the force, H, necessary to support a weight $W = 150$ kN and the total vertical shear force at the central pin joint E. Take $CS = BD = x = 1$ m.
(Ans: 260 kN, 103.6 kN)

2.10 Find the initial force exerted by the jack in Figure 2.28 as it raises a triangular block of 5 kN while the apex remains in contact with the ground. What is the corresponding tension in the vertical chain BC and the shear force in the pin at A?

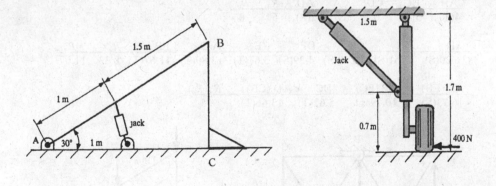

Figure 2.28   **Figure 2.29**

2.11 During taxiing a lateral force of 400 N is developed at the wheel of the retractable aircraft undercarriage shown in Figure 2.29. The internal diameter of the hydraulic jack is 50 mm. Find the internal pressure within the jack necessary to react to the lateral force.
(Ans: 4.12 bar)

2.12 The boat in Figure 2.30 is supported at its ends by two identical davits ABC. Find the reactions at the hinge A and the guide B when a vertical load of 3.84 kN is transmitted to each davit at C.

(Ans: R_A = 6.4 kN at 36.9° ACW from horizontal; R_B = 5.12 kN horizontal)

2.13 A string, AB, connects the unequal legs of the step ladder in Figure 2.31. What is the tension force in AB and the ground reactions when a boy of weight 500 N rests at C?
(Ans: 73 N, 200 N, 300 N)

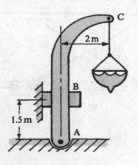

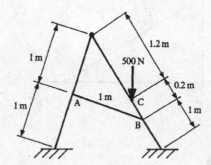

Figure 2.30   **Figure 2.31**

Plane Frames

2.14 The simply-supported frames in Figures 2.32–2.35 are subjected to the vertical loadings shown. Determine the forces in the bars using the joint tension method.
(Respective answers are below where: T - tie, S - strut, 1t = 10^3 kg)

AB	BC	AE	EB	ED	BD	DC	R_A	R_B
23.09(S)	2.89(T)	11.55(T)	57.73(T)	23.09(S)	40.41(T)	5.77(S)	10	85

AB	BC	CD	CA	AD	R_A	R_B
4(T)	5(S)	0	5(S)	2(S)	5	6

AD	DC	CB	DE	CE	A E	EB	R_A	R_B
16.20(S)	13.61(S)	16.50(S)	4.29(S)	3.67(T)	13.89(T)	11.67(T)	8.33	11.67

AE (EB)	DE (EC)	DC	AD (CB)	$R_A (R_B)$
16.73(S)	10.79(T)	5.62(T)	13.66(T)	5

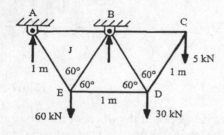

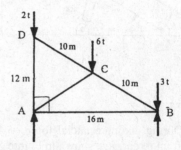

Figure 2.32   **Figure 2.33**

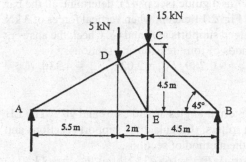

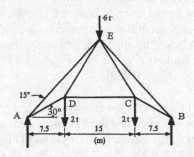

Figure 2.34   **Figure 2.35**

2.15 Find the remaining bar forces for the frame in Fig. 2.13a using the method of sections. Take one further vertical section cutting bars AB and AF and one horizontal section cutting bars BC, CF and CD.   Confirm all the bar forces using the joint tension method.
(Ans: F_{AB} = – 4 kN, F_{CF} = 1 kN, other forces follow from frame symmetry)

2.16 Using the method of sections, determine the forces in the lettered bars for the frames in Figures 2.36 to 2.38 when each frame is hinged at A and rests on rollers at B.
[Ans (T-tie, S-strut): XY = 0.901 kN (T), YA = 0.161 kN (T), R_B = 1.194 kN; XY = 2.15 t (S), ZB = 2.08 t (T), YZ = 2.73 t (T), YB = 1.57 t (S), R_A = 5.17 t at 41.5°, R_B = 0.61 t; BX = 10.04 t (S), BY = 6.98 t (T), XY = 9.07 t (T), XC = 13.22 t (S), R_A = 10.8 t at 20° ACW from vertical, R_B = 8.46 t]

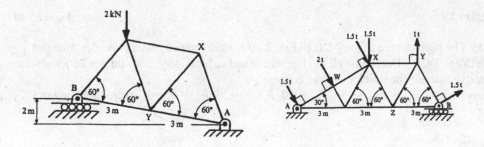

Figure 2.36   **Figure 2.37**

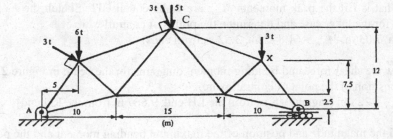

Figure 2.38

2.17 Using the worked solution to Example 2.7 as a guide (see p. 43), determine all the bar forces for the inverted Warren frame shown in Fig. 2.11 except when vertical forces of 3 kN and 4 kN act at E and D, respectively, with hinged supports at A and C. Check the answers given below by applying, separately, the methods of joint tension and sections.
[Ans (in kN): F_{AE} = 3.753, F_{AB} = -1.877, F_{EB} = -0.289, F_{ED} = 2.021, F_{CD} = 4.330, F_{CB} = -2.165, F_{DB} = 0.289]

2.18 In the roof truss shown in Fig. 2.39 the internal angles are 30°, 60° and 90°. The LH end A is hinged and the RH end E rests upon rollers. Calculate the support reactions and forces in the members BC, FG and FD using the method of sections.
[Ans: R_A = 8 kN, R_E = 7 kN; F_{BC} = - 15 kN (strut), F_{FG} = 16/√3 kN (tie), F_{FD} = 3/2 kN]

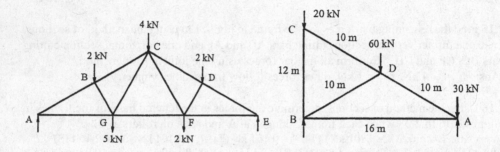

Figure 2.39   **Figure 2.40**

2.19 The right-angled frame ADC in Fig. 2.40 consists of two, inner isosceles triangles ABD and BCD. Determine all the bar forces and explain why it would not be feasible to remove that bar for which the force is zero.
[Ans (in kN): F_{AB} = F_{BD} = -50, F_{AD} = 40, F_{CD} = - 20, F_{BC} = 0; risk of strut buckling in DC]

Beams

2.20 A beam of 12 m length carries a uniformly distributed load of 200 kg/m and rests on two simple supports distance l apart with an equal overhang at each end. If M = 0 at mid-span, find l. What is l if the peak moments M_{max} are to be equalised? Sketch the F- and M-diagrams for the latter case and determine the points of contraflexure.
(Ans: 6 m, 7.03 m, M_{max} = 61.8 kNm, 3.52 m from ends)

2.21 Draw the shear force and bending moment diagrams for the beam in Figure 2.41 and find the position of the point of contraflexure.
[Ans: M_{max} = 22 t m (sagging), 8 m from the LH end, 9.667 m from the RH end]

2.22 Find the magnitude and position of the maximum bending moment and the points of contraflexure for the beam in Figure 2.42.
[Ans: M_{max} = 7.26 tm (sagging), 5.15 m from RH end; contraflexure positions: 4.34 m from RH end, 3.71 m from LH end]

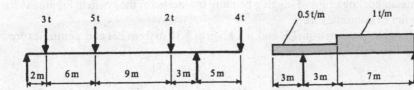

Figure 2.41   Figure 2.42

2.23 Sketch the F- and M-diagrams for the beam shown in Figure 2.43. Determine the position and magnitude of the maximum bending moment and the point of contraflexure. [Ans: M_{max} = 20.1 tm (sagging), 6 m from LH-end: 7.82 m from RH-end]

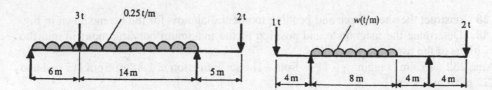

Figure 2.43   Figure 2.44

2.24 Find w in tonne/m and each reaction for the simply supported beam in Figure 2.44 in order that $F = 0$ at 5 m from the LH end. Draw the F- and M-diagrams showing the maximum and minimum moments.
[Ans: w = 1/13 t/m, 1.077 t, 2.539 t, 8 tm (hogging), 3.96 tm (hogging)]

2.25 Calculate, for the beam in Figure 2.45, the position and magnitude of the maximum hogging and sagging bending moments. Locate the point of contraflexure.
[Ans: 120 kNm (hogging) at 4 m from RH end, 15.3 kNm (sagging) at 1.75 m from LH end, 3.6 m from LH end]

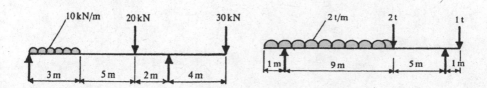

Figure 2.45   Figure 2.46

2.26 Find the points of contraflexure and the magnitude and position of F_{max} and M_{max} for the beam shown in Figure 2.46.
[Ans: 1.141 m from RH end, 1.078 m from LH end; F_{max} = 12.93 t at left support; M_{max} = 40.79 t m (sagging) at 7.46 m from LH end]

2.27 Find the maximum hogging and sagging bending moments for the beam in Figure 2.47. Where is the point of contraflexure?
[Ans: 10 kNm (hogging) at 1 m from LH end, 4.1 kNm at 3.38 m from LH end, contraflexure at 2.094 m from LH end]

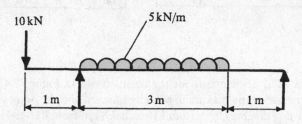

Figure 2.47

2.28 Construct the shear force and bending moment diagrams for the beam shown in Fig. 2.48. Determine the magnitude and position of the maximum bending moment and the positions of the two inflection points.
[Ans: 232.64 kNm (sagging), 6.41 m from LH end; inflection at 2.48 m from LH end and 5.22 m from RH end]

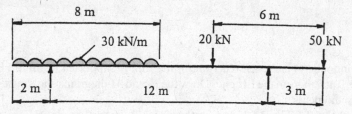

Figure 2.48

CHAPTER 3

BASIC STRESS AND STRAIN

Summary: There are a limited number of forces which, when applied to structural materials, cause their fibres to suffer stress and become strained. These 'forces' refer to tension, compression, shear, bending and torsion. This chapter's title refers to the basic stress and strain arising from the simplest applications of an axial tension (or compression) to a bar and of a tangential shear force to a block. These two contrasting loadings will be examined here in terms of the manner in which both their stress and strain are defined and to the relationship that exists between stress and strain up to the yield point and beyond. The method of 'designing against' these forces through the use of safety factors is given within typical practical applications including a similar loading of composite bars. The parallel analyses of bending and torsion appear later in Chapters 5 and 7 as these are more complicated involving stress and strain gradients. In practice, these four sources of stress and strain often act in combination. Various accounts when dealing with their more common combinations are considered here and in Chapters 11 and 12, where both the analytical and graphical methods are employed. An additional source arising from constrained thermal expansion is also given here.

3.1 Direct Stress and Strain

These arise from a simple uniaxial loading of bars, rods, cylinders, etc. in which the load is applied to their ends with no eccentricity. Two such loading modes arise in practice. The stress and strain are regarded as positive with the application of an axial tensile force. Negative stress and strain apply to axial compression over relatively short lengths where there is no influence of buckling. Hence, the uni-axial compression analysis given here is restricted to columns with comparable dimensions in length and cross section. Struts, on the other hand, have far greater length dimension where a different analysis (see Chapter 8) provides the critical elastic buckling load

3.1.1 Uniform Cross Section

Consider a uniform rod or bar of parallel length l with uniform cross-sectional area A subjected to an applied axial force, W (see Fig. 3.1). Direct stress refers to the internal force a material exerts per unit of its normal area A, in resisting W.

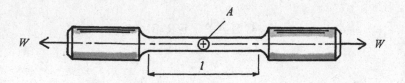

Figure 3.1 Bar under an axial force

The bar is shown with enlarged ends for gripping in laboratory test machines where the parallel 'gauge' dimensions are used in stress and strain calculations. Since the external force may be either tensile (+ as shown) or compressive (−) the direct stress is written as:

$$\sigma = \pm\, W/A \qquad (3.1)$$

The corresponding direct strain, ε, is the amount by which the material will extend or compress per unit of its length. When the extension or compression of the length l is $\pm x$, the direct strain is:

$$\varepsilon = \pm\, x/l \qquad (3.2)$$

Provided the loading is elastic the displacement x will recover fully upon unloading. Hooke (1635–1693) showed that for loading most metallic materials within their elastic limit the plot between W and x is linear. Later, Young (1773–1829) recognised that within this limit the ratio between σ and ε will be constant for a particular material. This constant is the *modulus of elasticity* (Young's modulus), defined from Eqs 3.1 and 3.2 as:

$$E = \sigma/\varepsilon = Wl/Ax \qquad (3.3)$$

When solving problems based upon Eqs 3.1-3.3 the elastic displacements are, of course, very small. Hence it may be more meaningful to work in units of mm and N, rather than the standard SI units of m and N. The following example illustrates this but, firstly, note the conversions from MPa to N/mm^2 and GPa to MN/m^2, where $1\ Pa = 1\ N/m^2$:

$$1\ MPa = 1\ MN/m^2 = 10^6\ N/m^2 = 1\ N/mm^2$$

$$1\ GPa = 10^9\ N/m^2 = 10^3\ MPa = 10^3\ N/mm^2$$

Example 3.1 A steel tie bar on a pressing machine is 2 m long and 40 mm in diameter. What is the elastic stress and extension under a tensile load of 100 kN? Take $E = 205$ GPa.

Substitution into Eqs 3.1 and 3.3 in N and m gives:

$\sigma = W/A = (100 \times 10^3)/[\pi(0.04)^2/4] = 79.6 \times 10^6 \text{ Pa} = 79.6 \text{ MPa}$

$x = Wl/AE = (100 \times 10^3 \times 2)/[(\pi \times 0.04^2/4) \times (205 \times 10^9)]$
$= 0.78 \times 10^{-3} \text{ m} = 0.78 \text{ mm}$

Alternatively, knowing that 1 MPa = 1 N/mm^2 using units of N and mm will lead to the stress and displacement more directly:

$\sigma = W/A = (100 \times 10^3)/(\pi \times 40^2/4) = 79.6 \text{ MPa}$

$x = Wl/AE = (100 \times 10^3 \times 2 \times 10^3)/[(\pi \times 40^2 \times 205 \times 10^3] = 0.78 \text{ mm}$

3.1.2 Variable Cross Section

When the section area A varies with the length we must write A as a function of a length variable z, i.e. $A = A(z)$. It is then possible to rearrange Eq. 3.3 to provide the incremental displacement δx that occurs over an increment of length δz as:

$$\delta x = \frac{W \times \delta z}{E \times A(z)} \qquad (3.4)$$

Equation 3.4 may be integrated over the full length l to find the displacement x between the ends as the following example shows.

Example 3.2 The diameter of the tapered, solid alloy column in Fig. 3.2 varies uniformly from 25 mm to 75 mm over a length of 500 mm. Find the change in length and the maximum and minimum stresses in the column under an axial compressive force of 50 kN. Take $E = 100$ GPa.

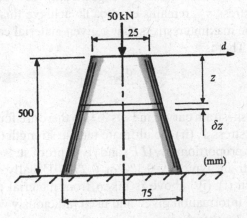

Figure 3.2 Tapered column under axial compression

Within the chosen d versus z axes shown, the equation of the sloping side is: $d/2 = 12.5 + (0.05)z$. Hence the required area expression in mm^2, within Eq. 3.4, becomes:

$$A(z) = \pi(d/2)^2 = \pi(12.5 + 0.05z)^2$$

The reduction in length follows from the integral:

$$x = \frac{W}{E} \int_0^l \frac{dz}{A(z)} = \frac{W}{\pi E} \int_0^{500} \frac{dz}{(12.5 + 0.05z)^2}$$

$$= \frac{-W}{0.05\pi E} \left| (12.5 + 0.05z)^{-1} \right|_0^{500}$$

$$= \frac{-50 \times 10^3}{0.05\pi \times 100 \times 10^3} \left(\frac{1}{12.5} - \frac{1}{37.5} \right) = -0.17 \text{ mm}$$

Ignoring contact friction, the maximum, axial compressive stress occurs in the minimum cross section:

$$\therefore \sigma_{max} = W/A_{min} = -(50 \times 10^3)/(\pi \times 25^2/4) = -101.86 \text{ MPa}$$

and the minimum compressive stress occurs in the maximum section:

$$\therefore \sigma_{min} = W/A_{max} = -(50 \times 10^3)/(\pi \times 75^2/4) = -11.32 \text{ MPa}$$

3.1.3 Safety Factor

Tensile tests conducted upon metals reveal regions of elastic and plastic behaviour within the plot of stress versus strain (see Figs 3.3a,b). A sound design is ensured when the working stress σ_w, remains elastic. To achieve this, a *safety factor* S is applied to reduce the maximum stress that a given material can sustain (σ_{max}), to a safe working level. That is:

$$S = \frac{\sigma_{max}}{\sigma_w} > 1 \tag{3.5}$$

Referring to the stress–strain curves in Figs 3.3a,b there are four common measures of σ_{max}: (i) the yield stress Y, (ii) the ultimate tensile strength (UTS), (iii) the stress at the limit of elastic proportionality (LP) and (iv) a proof stress (PS) corresponding to a given offset strain, typically 0.1% or 0.2%. Usually, one or other of the strength measures in (i)–(iv) above is taken from material property handbooks, depending upon the information given and used judiciously with a safety factor S. It follows from Eq. 3.5 that the respective working stress may become one of the following ratios: Y/S, UTS/S, LP/S, and $(0.1\% \ PS)/S$.

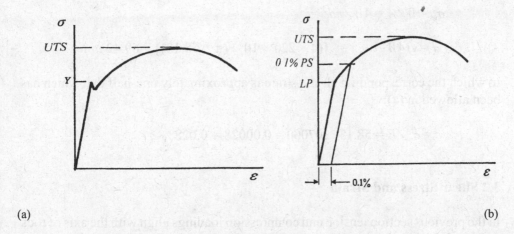

(a)   (b)

Figure 3.3 Typical stress–strain behaviour for metallic materials

The *LP* strength measure is used for materials that do not display a sharp yield point (see Fig. 3.3b). The proof stress is applied where the transition from linear elasticity to non-linear, elastic-plastic behaviour is very gradual. For example, to find a 0.1% proof stress value a line is drawn parallel to the initial elastic line offset by 0.1% strain as shown. If this proof stress should be reached the measure accepts that a permanent strain of 0.1% would set in the material, but the use of a safety factor obviates this. Another, lesser-used maximum stress (strength) used within Eq. 3.5 is taken from the tangent to the elastic-plastic curve where the gradient is 50% that of the elastic modulus. Again, as this stress lies within the elastic-plastic region it would involve an (unspecified) amount of permanent set.

Example 3.3 A solid circular bar is to carry a force of 220 kN. Find a suitable bar diameter based upon two designs: (a) the elastic strain in the bar is not to exceed 0.05% and (b) applying a safety factor of safety of 8 to its ultimate tensile strength $UTS = 465$ MPa. What is the strain in (b)? Take $E = 207$ GPa.

(a) From Eq. 3.3 that for the strain-based diameter follows as:

$$E = W/A\varepsilon = 4W/\pi d^2 \varepsilon$$

$$\therefore d = \sqrt{(4W/\pi E\varepsilon)}$$

$$= \sqrt{[(4 \times 220 \times 10^3 \times 10^2)/(\pi \times 207000 \times 0.05)]} = 52 \text{ mm}$$

(b) From Eq. 3.5, the safety factor limits the working stress to:

$$\sigma_w = \sigma_{max}/S = 465/8 = 58.13 \text{ MPa}$$

from which the stress-based diameter follows:

$$\sigma_w = W/A = 4W/\pi d^2$$

$$\therefore d = \sqrt{(4W/\pi\sigma_w)} = \sqrt{[(4 \times 220 \times 10^3)/(\pi \times 58.13)]} = 69.42 \text{ mm}$$

in which the corresponding elastic strain is approximately one-half that which has been allowed in (a):

$$\varepsilon = \sigma_w/E = 58.13/207000 = 0.00028 = 0.028\%$$

3.2 Shear Stress and Strain

In the previous section tension and compression loadings align with the axis of rods and shafts, etc. acting normal to the cross-sectional area. Shear refers to a tangential force lying parallel to the section area. The shear mode arises typically in the bolts and rivets used to join plates bearing axial loading. In addition, we saw in § 2.5, p. 47, with the construction of the shear force diagram, how shear is distributed in long beams supporting transverse loading that is either concentrated or distributed.

3.2.1 Shear Stress

An internal shear stress τ, is sustained by a material when it is subjected to an external shear force F. The accompanying deformation mode is one of an angular distortion, shown within the block of Fig. 3.4.

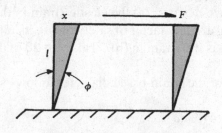

Figure 3.4 Block in shear

In this case, where the top surface area A_s supports the tangential force F, the shear stress is simply:

$$\tau = F/A_s \tag{3.6a}$$

Note, however, that the shear area A_s of the top surface will depend upon the mode of shear. For example, in the riveted joint of Fig. 3.5a the rivet is in single shear and therefore Eq. 3.6a applies.

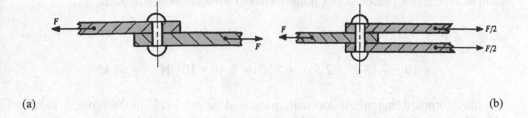

F F F $F/2$ $F/2$

Figure 3.5 Single- and double-lapped riveted joints

The strength of a single spot welded joint also follows from Eq. 3.6a when τ becomes the ultimate shear strength (*USS*) for the plate material. For example, consider the strength of a 7 mm spot welded joint between two mild steel plates, 2 mm thick, 20 mm wide, with *USS* = 265 MPa. The shear force required to separate the joint is found from Eq. 3.6a:

$$F = A_s \times USS = \left(\frac{\pi}{4} \times 7^2 \right) \times 265 = 10.2 \times 10^3 \text{ N} = 10.2 \text{ kN}$$

We may compare this prediction with the author's measured failure loads in the range 9–11.5 kN in which each weld was formed for different times with various amounts of power consumed. Where the weld power index was high, plate failures occurred in the greater load range 12.9–14.3 kN. Here the plate's tensile failure load is predicted as the product of its ultimate tensile strength *UTS* = 350 MPa and cross-sectional area A:

$$F = A \times UTS = (2 \times 20) \times 350 = 14 \times 10^3 \text{ N} = 14 \text{ kN}$$

It is seen that the simple prediction to each mode of failure is reasonably accurate despite neither providing an account of the machine settings nor of any loss in strength within the heat-affected zone.

In a double lap joint (see Fig. 3.5b) the rivet is placed in double shear so that with two rivet section areas resisting F, the rivet's average shear stress becomes

$$\tau = F / 2A_s \tag{3.6b}$$

Comparing Eqs 3.6a,b suggests that the lapped joint is twice as strong but usually the joint is less than 100% efficient with slight imperfections arising from its manufacture (see § 3.2.5). To take a further illustration from the author's archive, consider the strength of a double-lapped joint formed from three aluminium alloy plates 3.2 mm thick and 58 mm wide. Five bifurcated steel rivets, with inner and outer diameters 2.5 and 3.2 mm, were placed symmetrically within a 58 mm plate overlap in a 38 mm square arrangement, with one rivet placed at each corner and another at its centre. The breaking force F, when applied to the plates in the manner of Fig. 3.5b, will have overcome the ultimate shear strength ($USS = 270$ MPa) in each of five rivets placed under double shear. That is, from Eq. 3.6b:

$$F = 5 \times 2 \times A_s \times USS$$

$$= 10 \times \frac{\pi}{4}(3.2^2 - 2.5^2) \times 270 = 8.46 \times 10^3 \text{ N} = 8.46 \text{ kN}$$

We may compare this prediction with measured failure loads in the range 8.7–9.8 kN. The greatest load was found for a joint manufactured with tighter tolerances where all rivets failed together. Tensile failure within the single plate side of the joint would occur when the stress under F attains the plate's ultimate tensile strength ($UTS = 385$ MPa). This gives a breaking force for the plate section area

$$F = A \times UTS$$

$$= 3.2 \times 58 \times 385 = 71.46 \times 10^3 = 71.46 \text{ kN}$$

showing that the rivet failure is far more likely but this requires one final check: namely, the possibility of a shear failure in this plate. As with the tensile failure a shear failure would occur perpendicular to the width but with a 45° orientation through the thickness (a 90° through-thickness orientation applies to the tensile failure). Taking USS = 220 MPa the force F required to produce a shear failure in the single aluminium plate side becomes:

$$F = A_s \times USS$$

$$= (3.2 \times \sqrt{2} \times 58) \times 220 = 57.75 \times 10^3 = 57.75 \text{ kN}$$

where here A_s is the sloping shear area (in brackets). Again this failure mode is unlikely given the far lower force calculated above for rivet shear failure. An optimum design would ensure that all the possible failure modes occur simultaneously under a single critical load. When the plate geometry is adjusted to achieve this there results a worthwhile saving in material for each joint given above.

3.2.2 Shear Strain

The shear stress produces angular distortion from which a non-dimensional measure of shear strain is derived. To refer the shear strain γ to the distorted block in Fig. 3.4, where φ is the change to the right angle in radians:

$$\gamma = \tan \varphi \tag{3.7a}$$

While Eq. 3.7a will apply generally to all materials we note that for metallic materials the elastic shear displacement x in Fig. 3.4 will be small when a good approximation to the shear strain (in radians) holds:

$$\gamma \approx \varphi \approx x/l \tag{3.7b}$$

Conversely, both x and φ can attain high values with the elastic distortion of rubbers and elastomers for which this approximation is invalid.

3.2.3 Shear Modulus

Within the elastic region x will increase linearly with an increasing shear force F. Given that A_s and l are constants it can be seen from Eqs 3.6 and 3.7b that F and x are proportional to τ and γ. Consequently, an elastic *modulus of rigidity* (shear modulus) G is the constant ratio between shear stress and shear strain:

$$G = \tau / \gamma \tag{3.8a}$$

It follows from Eqs 3.6a and 3.7b that the shear modulus is:

$$G = \tau / \gamma = Fl / Ax \tag{3.8b}$$

Again, Eq. 3.8b is restricted to small elastic displacements, typical of metals and alloys, where the approximation in Eq. 3.7b applies.

3.2.4 Safety Factor

A safety factor, S, in shear is defined as the ratio between the maximum allowable shear stress for a material and its working shear-stress level:

$$S = \frac{\tau_{max}}{\tau_w} > 1 \tag{3.9}$$

In Eq. 3.9 τ_{max} may be equated to the shear yield stress k, a suitable proof stress or the ultimate shear strength (USS). There is a simple relationship between these shear strengths and the corresponding tensile strengths used previously within the safety factor S in Eq. 3.5. The yield an ultimate strength relationships are:

$$k = Y/Q \text{ and } USS = UTS/Q$$

where Q lies between $\sqrt{3}$ and 2, these conversion factors provided by the respective strength theories of von Mises and Tresca in Chapter 12 (see § 12.1, p. 487).

Example 3.4 The maximum allowable shear stress in a steel axle is 150 MPa. Using a safety factor of 4, find the smallest diameter of axle that is permissible when it is to carry a shear force of 20 kN (a) in double shear and (b) in single shear.

The working stress for the axle is, from Eq. 3.9:

$$\tau_w = \tau_{max} / S = 150/4 = 37.5 \text{ MPa}$$

(a) The double shearing action is that shown in Fig. 3.5b. Two section areas resist the force F and therefore the shear stress in the axle becomes:

$$\tau_w = F / 2A_s = F / (2\pi d^2/4)$$

Rearranging this equation provides the axle diameter d, in mm directly, when we recall that 1 MPa = 1 N/mm^2:

$$d = \sqrt{(2F/\pi\tau_w)} = \sqrt{[(2 \times 20 \times 10^3)/(\pi \times 37.5)]} = 18.43 \text{ mm}$$

Alternatively, d is found in m from knowing 1 MPa = 10^6 N/m^2:

$$d = \sqrt{(2F/\pi\tau_w)} = \sqrt{[(2 \times 20 \times 10^3)/(\pi \times 37.5 \times 10^6)]} = 18.43 \times 10^{-3} \text{ m}$$

(b) In single shear $\tau_w = F/A_s$, modifying the above calculations to give a larger, safe diameter required under this mode. For the units chosen similarly:

$$d = \sqrt{(4F/\pi\,\tau_w)} = \sqrt{[(4 \times 20 \times 10^3)/(\pi \times 37.5)]} = 26.06 \text{ mm}$$

or,

$$d = \sqrt{(4F/\pi\,\tau_w)} = \sqrt{[(4 \times 20 \times 10^3)/(\pi \times 37.5 \times 10^6)]} = 26.06 \times 10^{-3} \text{ m}$$

3.2.5 Pinned and Riveted Joints

Joints and connections formed with pins and rivets may appear in single or multiple rows. The safe design of pins and their manner of supporting must be based upon both the ultimate tensile and shear strengths of the materials used. In a riveted joint, the pitch, the rivet diameter and the efficiency of the joint are determined from considering all the failure modes possible: tearing of the plate, shearing and crushing of the rivets. The following examples illustrate how to design each type of joint safely.

Example 3.5 A rocker arm is to carry an axial force F of 35.5 kN when connected to a fork-end as shown in Figs 3.6a,b. Using a safety factor of 2 throughout, determine: (a) the diameter, d, of the connecting pin given that its ultimate shear strength $USS = 386$ MPa; (b) the thickness, t, of the mild steel rocker arm required to avoid its bearing failure under an ultimate compressive stress $UCS = 620$ MPa; (c) dimension A of the rocker to prevent the pull-out shown, given its ultimate shear strength $USS = 250$ MPa; and (d) dimension B of the rocker to a prevent tensile failure across each ligament shown with an ultimate strength $UTS = 310$ MPa.

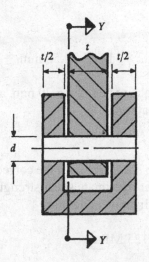
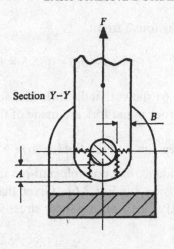

Figure 3.6 Loaded fork-end assembly

(a) Diameter d of Connecting Pin

Using Eq. 3.9, provides the working shear stress for the pin:

$$\tau_w = USS/S = 386/2 = 193 \text{ MPa}$$

Let a be the pin area. When under double shear, Eq. 3.6b applies

$$\tau_w = F/2a = 4F/2\pi d^2$$

which provides the pin diameter (in units of $\underline{N}$ and $\underline{mm}$):

$$d = \sqrt{(2F/\pi\tau_w)} = \sqrt{[(2 \times 35.5 \times 10^3)/(\pi \times 193)]} = 10.82 \text{ mm}$$

(b) Thickness t of the Mild Steel Rocker Arm

Bearing failure is concerned with permanent deformation arising in the material in contact with the pin. Failure would occur when its ultimate compressive stress *UCS* = 620 GPa is reached in the semi-circular bearing area. The use of a safety factor avoids a bearing failure giving the working stress as:

$$\sigma_w = UCS/S = 620/2 = 310 \text{ MPa}$$

This stress is to be carried by the lower semi-circular area in contact with the pin. Given the pre-determined diameter d in (a), the thickness t must be chosen to bear this stress. The area taken for this calculation is $d \times t$, the contact area projected perpendicular to the force F. This gives simply:

$$\sigma_w = F/dt$$

Hence t is found from:

$$t = F/(d \times \sigma_w) = (35.5 \times 10^3)/(10.82 \times 310) = 10.58 \text{ mm}$$

Figure 3.6a shows that the fork thicknesses are each to be $t/2 = 5.29$ mm, assuming that the rocker and fork are made of the same material.

(c) Dimension A of the Rocker Arm

The avoidance of a possible pull-out failure from shear lies with the correct choice of dimension A in Fig. 3.6b. Given that the rocker's ultimate shear strength is $USS = 250$ MPa, the working shear stress within the ligament becomes

$$\tau_w = USS/S = 250/2 = 125 \text{ MPa}$$

This shear stress is to lie within the two shear areas each $[(A + d/2) \times t]$ so that:

$$\tau_w = F/[2t(A + d/2)]$$

Hence, to avoid the pull-out failure shown in Fig. 3.6b, A follows as

$$A = F/(2t\tau_w) - d/2$$
$$= \frac{35.5 \times 10^3}{2 \times 10.58 \times 125} - \frac{10.82}{2} = 8.01 \text{ mm}$$

(d) Dimension B of the Rocker Arm

The avoidance of a possible tensile failure shown within the two ligament areas $B \times t$ lies with the correct choice of dimension B (Fig. 3.6b). Given the rocker's tensile strength: $UTS = 310$ MPa its working stress is to be

$$\sigma_w = UTS/S = 310/2 = 155 \text{ MPa}$$

This tensile stress is borne by two ligament areas, each $B \times t$, so that:

$$\sigma_w = F/(2Bt)$$

Hence B follows as

$$B = F/(2t\sigma_w) = (35.5 \times 10^3)/(2 \times 10.58 \times 155) = 10.82 \text{ mm}$$

Notice the sequence of dependent calculations that starts with the pin diameter leading to the dimensions of the connecting rocker. It can be seen that between the calculations for d, A and B that the radius of the rocker end should be the greater of $A + d/2$ and $B + d/2$, i.e. the latter, which gives 16.23 mm. In this example, due to the manner of loading, the ratio between the ultimate tensile and compressive strengths for the rocker is 1/2. It is known from conducting a simple compression test upon a short, solid cylinder that barrelling occurs in the testpiece as the end-material in contact with compression plates remains elastic while the central region begins to deform plastically (see § 8.3, p. 306). As the load is raised plasticity spreads to the ends. Similarly, the spread in the load over the pin's contact area requires a greater compressive stress for plasticity to cause permanent damage in the material around the hole. In material well away from the contact zone, where frictional shear does not play a role, the two strengths may be assumed equal.

A similar series of calculations would need to be made for the larger radius required of a fork-end with each fork having one half the rocker's thickness. We can estimate from the above that this radius should be a minimum of twice that calculated for the rocker (see Exercise 3.12).

Example 3.6 A double butt-riveted joint is shown in Fig. 3.7 in which two long plates 25 mm thick are connected with cover plates and 30 mm diameter rivets. Determine the necessary pitch p of the rivets given that their ultimate shear and compressive strengths are $USS = 370$ MPa and $UCS = 695$ MPa respectively. What is the efficiency of the joint when the ultimate tensile strength for the plate material is 465 MPa?

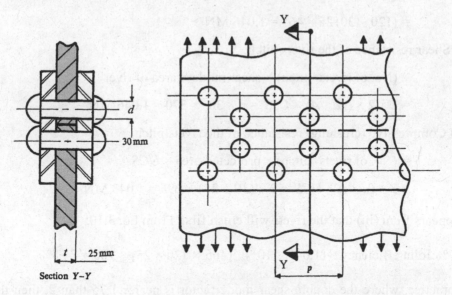

Figure 3.7 Double, butt-riveted joint

There is one rivet per pitch p within their inner and outer lines. Each rivet, of diameter d, is subjected to double shear when the plates carry a uniform tension shown. The joint can be made equally strong in tension and shear when the tensile force carried by plate material along each rivet line equals the shear force supported by all the rivets. Within the pitch length, where there are two rivets per plate, each in double shear, this gives the force balance equation required:

$$(p - d)t \times UTS = 2 \times 2 \times (\pi \times d^2/4) \times USS$$

$$(p - 30)\, 25 \times 465 = \pi \times 30^2 \times 370$$

$$\therefore \; p = 120 \text{ mm}$$

Note that through the use of different safety factors in tension and shear the mode of failure can be controlled. The joint efficiency is defined for one pitch length as:

$$Joint\ efficiency \; = \; \frac{Least\ value\ of\ failure\ load}{Ultimate\ tensile\ force\ of\ plate\ material} \qquad (3.10)$$

where the denominator is found from the ultimate tensile strength of the plate as: $UTS \times p \times t$. The numerator in Eq. 3.10 normally refers to the least of the following three calculations (i)–(iii) but here (i) and (ii) have been equalised above for determining p:

(i) Tensile (tearing) force of the plate/pitch:

$$= (p - d)t \times UTS$$

$$= (120 - 30)\, 25 \times 465 = 1.046 \text{ MN}$$

(ii) Shear resistance of the rivets/pitch:

$$= (\text{No. of rivets}/\text{pitch}) \times \text{shear mode} \times \text{area of rivet}$$

$$= 2 \times 2 \times (\pi d^2/4) \times USS = \pi \times 30^2 \times 370 = 1.046 \text{ MN}$$

(iii) Compressive (crushing) resistance of the rivets/pitch:

$$= (\text{No. of rivets}/\text{pitch}) \times \text{projected area} \times UCS$$

$$= 2 \times (d \times t) \times UCS = 2 \times (30 \times 25) \times 695 = 1.043 \text{ MN}$$

It appears from (iii) that the rivets will crush first. From Eq. 3.10:

$$\% \text{ Joint efficiency} = (1.043 \times 10^6)/(465 \times 120 \times 25) \times 100 = 74.8\%$$

In practice, where the double shear mode factor is nearer 1.75 than 2, then the rivets' failure load would dictate joint efficiency for this design. The failure load calculation in (ii) above would become: $2 \times 1.75 \times (\pi d^2/4) \times USS = 7/8 \times \pi \times 30^2 \times 370 = 0.9154$ MN, giving an efficiency of 65.6%. This efficiency refers to a pitch of 120 mm, where the plate tearing force in (i) remains unaltered at 1.046 MN, this giving a greater efficiency of 75%. For a similar efficiency (65.6%) between rivet and plate tearing failure the calculation in (i) above requires a revised pitch $p = 108.74$ mm. At this pitch, the crushing resistance of the rivets found in calculation (iii) remains unaltered whilst the efficiency on this basis is raised from 74.8% to 82.5%. Hence, a joint efficiency of 65.6% applies to each pitch length.

3.3 Compound Bars

Compounding refers to where an assembly of bars and plates, often in differing materials, is used to support axial loading. Mechanical strain may be induced in compound bars by the direct application of axial tension or compression and by bolt tightening across clamped ends. In each case the principles of force equilibrium and strain compatibility enable the stress and strain in the bar to be found.

3.3.1 Composite Section Under Axial Force

Figures 3.8a and b show, respectively, layered and reinforced bar section designs in two different materials A and B of similar length l, used for supporting tension and compression.

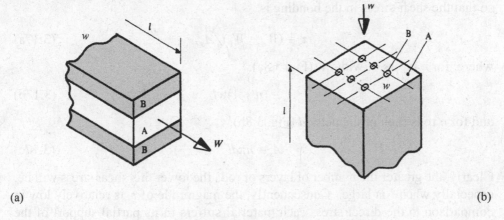

(a)   (b)

Figure 3.8 Composite bars

The materials are bonded securely so that when either bar design is to support an axial force, W, each component material contributes to the force while suffering the same amount of axial strain, ε. The respective *equilibrium* and *compatibility* conditions are expressed as:

$$W = W_A + W_B \tag{3.11}$$

$$\varepsilon = (\sigma/E)_A = (\sigma/E)_B \tag{3.12a,b}$$

The axial stresses in each material are:

$$\sigma_A = (W/A)_A \text{ and } \sigma_B = (W/A)_B \tag{3.13a,b}$$

Combining Eqs 3.12a,b and 3.13a,b gives:

$$\frac{W_A}{A_A E_A} = \frac{W_B}{A_B E_B} \tag{3.14}$$

Combining Eqs 3.11 and 3.14 shows how the *load sharing* is achieved:

$$W_A = \frac{WA_A E_A}{A_A E_A + A_B E_B}$$

(3.15a)

$$W_B = \frac{WA_B E_B}{A_A E_A + A_B E_B}$$

(3.15b)

The common axial displacement x over a length l, for each material is:

$$x_A = \frac{W_A l}{A_A E_A} \quad \text{and} \quad x_B = \frac{W_B l}{A_B E_B}$$

(3.16a,b)

in which $x_A = x_B$ when bonding ensures that the loads are shared in this manner. An interfacial shear stress τ_i arises from the difference between the loads supported by each material. Thus $W_A - W_B$ is the load carried by the total interface shear area A_s so that the shear stress in the bonding is

$$\tau_s = (W_A - W_B)/A_s$$

(3.17a)

where, for n layers of width w (Fig. 3.8a):

$$A_s = (n - 1)wl$$

(3.17b)

and for n rods each of diameter d (Fig. 3.8b):

$$A_s = n\pi dl$$

(3.17c)

Clearly, the greater the number of layers or rods the lower this shear stress will be, especially when l is large. Consequently, the magnitude of τ_s is relatively low in comparison to the direct stress each material suffers in its partial support of the load, as the following example shows.

Example 3.7 The composite bar in Fig. 3.8a consists of a 25 mm × 6 mm strip of steel bonded between two 25 mm × 10 mm strips of brass for a common length of 130 mm. Calculate the stresses in the strips and the bonding and the extension of the bar under a tensile force of 100 kN. Take the elastic constants as: $E = 84.8$ GPa for brass and $E = 207$ GPa for steel.

Using subscript A for steel and B for brass, their respective cross-sectional areas are:

$$A_A = 25 \times 6 = 150 \text{ mm}^2; \quad A_B = 2 \times 25 \times 10 = 500 \text{ mm}^2$$

Equations 3.13a,b and 3.15a,b provide the stresses in each material:

$$\sigma_A = WE_A/(A_A E_A + A_B E_B)$$

$$= (100 \times 10^3 \times 207 \times 10^3)/[(150 \times 207000) + (500 \times 84800)]$$

$$= 281.82 \text{ MPa}$$

$$\sigma_B = E_B \sigma_A / E_A$$

$$= (84800 \times 281.82)/207000 = 115.45 \text{ MPa}$$

The loads supported are:

$$W_A = \sigma_A A_A = 281.82 \times 150 = 42.273 \times 10^3 \text{ N}$$

$$W_B = \sigma_B A_B = 115.45 \times 500 = 57.725 \times 10^3 \text{ N}$$

Their difference results in a shear stress within the bonding, found from combining Eqs 3.17a,b:

$$\tau_s = (W_A - W_B)/[(n-1)wl]$$

$$= (42.273 - 57.725)10^3/(2 \times 25 \times 130) = -2.38 \text{ MPa}$$

The sign refers to the direction of shear which here is aligned with the axial stress in material B. The axial displacements follow from Eqs 3.13a,b and 3.16a,b

$$x_A = (\sigma l / E)_A$$

$$= (281.82 \times 130)/207000 = 0.177 \text{ mm}$$

$$x_B = (\sigma l / E)_B$$

$$= (115.45 \times 130)/84800 = 0.177 \text{ mm} \checkmark$$

Using units of N and mm have provided the required measures more expediently but if N and m are preferred the σ_A stress calculation should appear as:

$$\sigma_A = WE_A/(A_A E_A + A_B E_B)$$

$$= \frac{(100 \times 10^3 \times 207 \times 10^9)}{(150 \times 10^{-6} \times 207 \times 10^9) + (150 \times 10^{-6} \times 84.8 \times 10^9)}$$

$$= 281.82 \times 10^6 \text{ N/m}^2 = 281.82 \text{ MPa}$$

and the x_A displacement calculation becomes

$$x_A = \left(\frac{\sigma l}{E}\right)_A = \frac{281.82 \times 10^6 \times 130 \times 10^{-3}}{207 \times 10^9} = 0.177 \times 10^{-3} \text{ m}$$

Example 3.8 The reinforced concrete column in Fig. 3.8b is 375 mm square with eight steel rods, each of area 625 mm², embedded over a length of 1.8 m. If the compressive stress for concrete is not to exceed 4.1 MPa, compare the total compressive loads that the column can carry with and without reinforcement. Find the axial stress in the steel, the shear stress at the steel-concrete interface and the axial compression. Take $E = 207$ GPa for steel and $E = 13.75$ GPa for concrete.

Without a reinforcement the maximum load in the concrete is restricted to:

$$W_{max} = A_c \sigma_c = (375)^2 \times 4.1 = 576.56 \times 10^3 = 576.56 \text{ kN}$$

Now refer to Fig. 3.8b where the areas of the concrete (material A) and the steel reinforcement (material B) are:

$$A_B = 8 \times 625 = 5000 \text{ mm}^2$$

$$A_A = (375)^2 - 5000 = 135625 \text{ mm}^2$$

From Eqs 3.11 and 3.13a,b, the total load that can now be carried is:

$$W = W_A + W_B = \sigma_A A_A + \sigma_B A_B = \sigma_A (A_A + E_B A_B / E_A)$$

$$= 4.1 \left[135625 + \frac{(207 \times 10^3) \times 5000}{(13.75 \times 10^3)} \right] = 864.68 \text{ kN}$$

for which the steel stress follows from Eqs 3.12a,b:

$$\sigma_B = E_B \sigma_A / E_A = 207 \times 4.1 / 13.75 = 61.72 \text{ MPa}$$

This steel stress, σ_B may be checked directly from Eq. 3.15b:

$$\sigma_B = W E_B / (A_B E_B + A_A E_A)$$

$$= \frac{(864.68 \times 10^3) \times (207 \times 10^3)}{5000 (207 \times 10^3) + 135625 (13.75 \times 10^3)}$$

$$= 61.72 \text{ MPa } \checkmark$$

Hence the loads supported by the steel and concrete are:

$$W_A = \sigma_A A_A = 4.1 \times 135625 = 556.06 \times 10^3 \text{ N} = 556.06 \text{ kN}$$

$$W_B = \sigma_B A_B = 61.72 \times 5000 = 308.6 \times 10^3 \text{ N} = 308.6 \text{ kN}$$

$$W_A = W - W_B = 864.68 - 308.6 = 556.08 \text{ kN } \checkmark$$

The difference $W_A - W_B$ is the load carried by the total interface area. The interfacial shear stress τ_i follows from Eqs 3.17a,c:

$$\tau_i = (W_A - W_B)/n\pi dl$$

Here $n = 8$ and $d = \sqrt{(625 \times 4/\pi)} = 28.21$ mm, giving

$$\tau_i = (556.06 - 308.62)10^3 / (8\pi \times 28.21 \times 1800) = 0.194 \text{ MPa}$$

Assuming that the bonding is capable of withstanding this interfacial shear, the compression in the column is found from either of Eqs 13.16a or b:

$$x_A = (\sigma l/E)_A = (4.1 \times 1800)/(13.75 \times 10^3) = 0.537 \text{ mm}$$

$$x_B = (\sigma l/E)_B = (61.72 \times 1800)/(207 \times 10^3) = 0.537 \text{ mm} \checkmark$$

3.3.2 Bolted Assembly

On pre-tightening the bolted assembly in Fig. 3.9, the bolt and cylinder (respective subscripts b and c) will carry equal but opposing loads P, under the same strain.

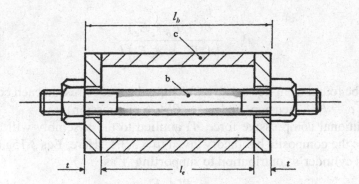

Figure 3.9 Bolt-loaded compound assembly

Because P exists in the absence of an external force it is said to be *self-equilibrating* within the assembly. When the contraction x_c of the original cylinder length l_c is measured after tightening nuts on rigid end plates, the load P may be calculated directly:

$$x_c = \frac{Pl_c}{A_c E_c} \qquad \therefore \quad P = \frac{A_c E_c x_c}{l_c} \qquad (3.18a,b)$$

Alternatively, the travel of the nuts when tightened is $\delta = pN$ where p is the thread pitch for the bolt and N its number of turns. The travel may be connected to the corresponding displacements of the bolt and cylinder x_b and x_c, respectively. Thus, with the bolt in tension and the cylinder in compression, the final length of bolt must equal the final length of the tube. This gives:

$$l_b + x_b - \delta = (l_c + 2t) - x_c \tag{3.19a}$$

Now the initial bolt length l_b is greater than the initial cylinder length l_c by two end-plate thicknesses. That is: $l_b = l_c + 2t$, so that Eq. 3.19a gives

$$x_b + x_c = \delta \tag{3.19b}$$

in which the two displacements are given by:

$$x_b = \frac{Pl_b}{A_b E_b} \quad \text{and} \quad x_c = \frac{Pl_c}{A_c E_c} \tag{3.20a,b}$$

P is found from substituting Eqs 3.20a,b into Eq. 3.19b. This gives:

$$\frac{P(l_c + 2t)}{A_b E_b} + \frac{Pl_c}{A_c E_c} = \delta \tag{3.21a}$$

When t is small then $l = l_c \approx l_b$ and Eq. 3.21a yields the bolt/cylinder load as:

$$P = \frac{\delta(A_b E_b A_c E_c)}{(A_b E_b + A_c E_c)l} \tag{3.21b}$$

Whether P be found from Eqs 3.18b or 3.21a,b, the axial stress in each component remains as: $\sigma_b = P/A_b$ (tension) and $\sigma_c = -P/A_c$ (compression).

An additional compressive force W, applied to the assembly will be shared according to the composite bar theory given in § 3.3.1. Here, Eqs 3.15a,b provide the bolt and cylinder's contribution to supporting W as:

$$W_b = \frac{WA_b E_b}{A_b E_b + A_c E_c} \tag{3.22a}$$

$$W_c = \frac{WA_c E_c}{A_b E_b + A_c E_c} \tag{3.22b}$$

which 'add' to the pre-tightening force P, as shown in the following example.

Example 3.9 Find, for the assembly shown in Fig. 3.9, the stress in the steel bolt b and the copper cylinder c when the nuts are tightened by one revolution on a thread pitch of ½ mm. Find the net stresses when the pre-stressed assembly is then subjected to an axial compressive force of 5 kN. Take, for the cylinder: $l_c = 2.5$ m, i.d.= 38 mm, o.d.= 50 mm and $E_c = 104$ GPa. For the bolt: diameter = 25 mm and $E_b = 207$ GPa with the thickness of each steel end-plate at 30 mm.

The bolt and cylinder section areas are:

$$A_b = (\pi/4)(12.5)^2 = 122.72 \text{ mm}^2$$

$$A_c = (\pi/4)[(50)^2 - (38)^2] = 829.38 \text{ mm}^2$$

Re-arranging Eq. 3.21a:

$$P = \frac{\delta(A_b E_b A_c E_c)}{[(1 + 2t/l_c)A_c E_c + A_b E_b]l_c}$$

$$= \frac{0.5\,(122.72 \times 207 \times 829.38 \times 104) \times 10^3}{[(1 + 2 \times 30/2500)\,829.38 \times 104 + 122.72 \times 207]2.5 \times 10^3} = 3853.3 \text{ N}$$

under which the bolt and cylinder pre-tightening stresses become:

$$\sigma_b = P/A_b = 3853.3/122.72 = 31.40 \text{ MPa}$$

$$\sigma_c = -P/A_c = 3853.3/829.38 = -4.646 \text{ MPa}$$

Equations 3.22a,b provide the bolt and cylinder stresses due to the axial force W:

$$\sigma_b = -\frac{W_b}{A_b} = \frac{-WE_b}{A_b E_b + A_c E_c}$$

$$= \frac{-5 \times 10^3 \times 207 \times 10^3}{[(122.72 \times 207) + (829.38 \times 104)]10^3} = -9.27 \text{ MPa}$$

$$\sigma_c = \frac{W_c}{A_c} = \frac{E_c \sigma_b}{E_b} = -\frac{104 \times 9.27}{207} = -4.657 \text{ MPa}$$

The net stresses become:

$$\sigma_b = 31.40 - 9.27 = 22.13 \text{ MPa}$$

$$\sigma_c = -4.646 - 4.657 = -9.303 \text{ MPa}$$

When the end-plate thickness is neglected, Eq. 3.21b applies:

$$P = \frac{\delta(A_b E_b A_c E_c)}{(A_b E_b + A_c E_c)l}$$

$$= \frac{0.5\,(122.72 \times 207 \times 829.38 \times 104) \times 10^3}{[(122.72 \times 207) + (829.38 \times 104)]2.5 \times 10^3} = 3924.74 \text{ N}$$

This increased bolt load modifies the pre-tightening stresses slightly:

$$\sigma_b = 3924.74 / 122.72 = 31.98 \text{ MPa}$$

$$\sigma_c = -3924.74 / 829.38 = -4.732 \text{ MPa}$$

so that the net stresses become:

$$\sigma_b = 31.98 - 9.27 = 22.71 \text{ MPa}$$

$$\sigma_c = -4.732 - 4.657 = -9.389 \text{ MPa}$$

Clearly, when $t \ll l_c$ the end-plate thickness does not alter the net stresses significantly.

It can be seen in the design of Fig. 3.9 that the bolt will carry the whole of an applied tension along its axis. This will add to the pre-tightening stress in the bolt by the amount W/A_b whilst relieving the cylinder stress by an amount corresponding to the displacement recovered. The latter equals that in the bolt under W where the end-plates remain in contact with the cylinder. The net stresses become:

$$\sum \sigma_b = \frac{\delta(A_c E_b E_c)}{(A_b E_b + A_c E_c)l} + \frac{W}{A_b}$$

$$\sum \sigma_c = -\frac{\delta(A_b E_b E_c)}{(A_b E_b + A_c E_c)l} + \frac{W E_c}{A_b E_b}$$

However, if the end plates are welded to the cylinder and are located upon a stepped shaft (see Fig. 3.10a) then the load sharing Eqs 3.22a,b will again apply.

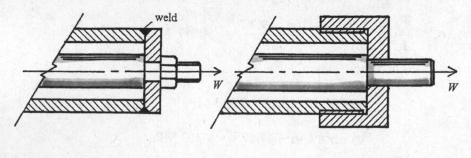

(a)   (b)

Figure 3.10 Tensile load sharing between cylinder and end fittings

A tensile load sharing between bolt and cylinder also applies where screwed plugs or collars engage with a stepped shaft and cylinder ends (see Figs 3.10b). Thus, in the last of the calculations made in Example 3.9, if the 5 kN load were to be applied in tension to a cylinder with either of the two designs shown in Figs 3.10a,b, the net stresses become:

$$\sigma_b = 31.98 + 9.27 = 41.25 \text{ MPa}$$

$$\sigma_c = -4.732 + 4.657 = -0.075 \text{ MPa}$$

This calculation shows that the cylinder becomes almost unstressed under the external force. In fact, here it can be arranged that the cylinder is free of stress under any external load from the degree of initial tightening (see Exercise 3.25).

3.4 Temperature Effects

So-called 'temperature stresses' arise where bars are prevented from attaining their natural expansions through various connections and fittings. Consider a bar of length, l, in a material with a coefficient of linear expansion, α. When the temperature of the bar is raised by Δt, the bar would normally extend by an amount $x = \alpha(\Delta t)l$. If this extension is prevented by constraining its ends the bar becomes strained by the amount $\varepsilon = x/l$, for which the compressive internal stress is

$$\sigma = E\varepsilon = Ex/l = E\alpha\Delta t \tag{3.23a}$$

By the constraining the expansion in this way, the end plates must sustain the force F exerted upon them from the stress within the bar's cross-sectional area A:

$$F = \sigma A = AE\alpha\Delta t \tag{3.23b}$$

In a similar manner, constraining the free expansion of a composite bars in Figs 3.8a,b will result in thermally induced stress and strain for which Eqs 3.23a,b apply. The following cases arise when length changes are restrained within compound bar assemblies.

3.4.1 Bars between Fixed Ends

Figure 3.11 shows two bars 1 and 2, of dissimilar materials, lengths and sectional areas, connected end to end between rigid walls.

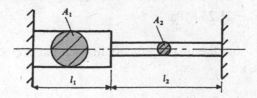

Figure 3.11 Composite bar with expansion prevented

With a temperature rise Δt, the free displacements in each bar sum to give a total unconstrained displacement:

$$x = x_1 + x_2 = \Delta t (\alpha_1 l_1 + \alpha_2 l_2) \tag{3.24a}$$

When this displacement is prevented, each bar is placed in compression, ·thereby exerting a similar axial force upon the restraining walls:

$$\sigma_1 A_1 = \sigma_2 A_2 \tag{3.24b}$$

Here the compressive stresses σ_1 and σ_2 within Eq. 3.24b must reappear within Eq. 3.24a to ensure zero axial displacement from end to end:

$$\sigma_1 l_1 / E_1 + \sigma_2 l_2 / E_2 + \Delta t (\alpha_1 l_1 + \alpha_2 l_2) = 0 \tag{3.25}$$

in which each individual bar may extend due to stress and temperature. Combining the respective *equilibrium* and *compatibility*, Eqs 3.24b and 3.25 provide the axial stress within each bar:

$$\sigma_1 = - \frac{\Delta t (\alpha_1 l_1 + \alpha_2 l_2)}{(l_1 / E_1 + A_1 l_2 / A_2 E_2)} \tag{3.26a}$$

$$\sigma_2 = \frac{\sigma_1 A_1}{A_2} \tag{3.26b}$$

Example 3.10 Copper and steel bars, each with the same length and diameter are assembled as in Fig. 3.11. Determine the stress and strain in each material following a temperature rise of 10°C. Take, for steel: $E = 207$ GPa, $\alpha = 12.5 \times 10^{-6}/$°C and for copper: $E = 105$ GPa, $\alpha = 18 \times 10^{-6}/$°C.

Since $A_1 = A_2$ and $l_1 = l_2$, substituting into Eqs 3.26a,b, with units of N and mm, gives the bar stresses directly:

$$\sigma_1 = - \Delta t (\alpha_1 + \alpha_2) E_1 E_2 / (E_1 + E_2) = \sigma_2$$

$$= - \frac{10 (12.5 + 18) 10^{-6} \times (207 \times 105) 10^3}{(207 + 105) 10^3}$$

$$= - 21.25 \text{ MPa}$$

Equation 3.25 shows that the displacements in each bar are equal but they must have opposing signs, i.e. $x_1 = -x_2$:

$$\sigma_1 l_1 / E_1 + \Delta t \, \alpha_1 l_1 = - (\sigma_2 l_2 / E_2 + \Delta t \, \alpha_2 l_2)$$

With $l_1 = l_2$, it follows that the bar strains are also equal but opposite (allowing for rounding error):

$$\varepsilon_1 = x_1/l_1 = \sigma_1/E_1 + \Delta t\, \alpha_1\, l_1$$

$$= -21.25/(207 \times 10^3) + (10 \times 12.5 \times 10^{-6})$$

$$= (-102.66 + 125)\, 10^{-6} = 22.34\ \mu\varepsilon$$

$$\varepsilon_2 = x_2/l_2 = \sigma_2/E_2 + \Delta t\, \alpha_2\, l_2$$

$$= -21.25/(105 \times 10^3) + (10 \times 18 \times 10^{-6})$$

$$= (-202.38 + 180)\, 10^{-6} = -22.38\ \mu\varepsilon$$

3.4.2 Tubes or Composite Bar with Constrained Ends

Figures 3.12a,b show, respectively, concentric tubes and balanced rectangular strips of different materials A and B. The manner in which each assembly is held together will equalise the displacements within A and B. Displacements are zero only in the special case of (b) where the end plates become rigid walls, but, in general, the length l will change in the absence of this limiting constraint.

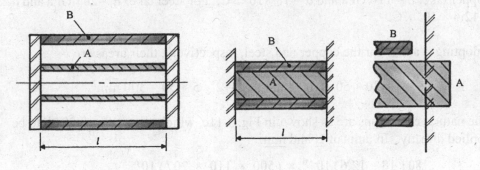

(a)   (b)   (c)

Figure 3.12 Effect of temperature upon expansion of composite bars

Assuming $\alpha_A > \alpha_B$, the free expansion caused by Δt in each material is shown in Fig. 3.12c. When the ends are joined, the displacements in A and B equalise to a value intermediate to their free expansions as shown. Consequently, axial stresses are induced in each material placing B in tension and A in compression. Here the stress magnitudes are found from the equilibrium and compatibility conditions to be satisfied. These are:

(i) With no external force and no end constraint, the *equilibrium* condition becomes:

$$\sigma_A A_A + \sigma_B A_B = 0 \qquad\qquad (3.27)$$

(ii) Equal displacements (and strains) arise from stress and temperature effects. For equal displacements, the *compatibility* condition applies:

$$\alpha_A l \, \Delta t + \sigma_A l / E_A = \alpha_B l \Delta t + \sigma_B l / E_B \tag{3.28}$$

Combining Eqs 3.27 and 3.28 provides each stress:

$$\sigma_A = - \frac{\Delta t \, (\alpha_A - \alpha_B)(A_B \, E_A \, E_B)}{A_A \, E_A + A_B \, E_B} \tag{3.29a}$$

$$\sigma_B = - \frac{\sigma_A \, A_A}{A_B} \tag{3.29b}$$

Example 3.11 A rectangular composite section is 50 mm wide by 30 mm deep. The depth is made up from a central 20 mm layer of copper sandwiched between two 5 mm layers of steel. The 50 mm width is common to all three strips, as are their 900 mm initial lengths. Determine the axial stresses induced in each material by an 80°C temperature rise when the strips are bonded together over their contact areas. Examine the effect of interchanging the materials upon their stress levels. For copper take: $E = 110$ GPa and $\alpha = 18 \times 10^{-6}/°C$. For steel take: $E = 207$ GPa and $\alpha = 12.6 \times 10^{-6}/°C$.

Adopting A and B for the copper and steel, respectively, their areas are

$$A_A = 20 \times 50 = 1000 \text{ mm}^2: A_B = 2 \times 5 \times 50 = 500 \text{ mm}^2$$

The natural extensions are as shown in Fig. 3.11c, which allows Eqs 3.29a,b to be applied directly. In units of N and mm:

$$\sigma_A = - \frac{80 \, (18 - 12.6) \, 10^{-6} \times (500 \times 110 \times 207) \, 10^6}{(1000 \times 110) \, 10^3 + (500 \times 207) \, 10^3} = -23.04 \text{ MPa}$$

$$\sigma_B = 23.04 \times 1000/500 = 46.08 \text{ MPa}$$

These show compression in the central copper strip and tension in the outer steel strips. Equation 3.28 ensures that their lengths change by an equal amount:

$$x_A = \alpha_A l \Delta t + \sigma_A l / E_A$$

$$= (18 \times 10^{-6} \times 900 \times 80) - (23.04 \times 900) / (110 \times 10^3) = 1.108 \text{ mm}$$

$$x_B = \alpha_B l \Delta t + \sigma_B l / E_B$$

$$= (12.6 \times 10^{-6} \times 900 \times 80) + (46.08 \times 900) / (207 \times 10^3) = 1.108 \text{ mm}$$

When the materials are interchanged $(A$ - steel, B - copper), this alters the substitution into Eqs 3.29a,b as follows:

$$\sigma_A = -\frac{80\,(12.6 - 18)\,10^{-6} \times (500 \times 110 \times 207)\,10^6}{(1000 \times 207)\,10^3 + (500 \times 110)\,10^3} = 18.77 \text{ MPa}$$

$$\sigma_B = -18.77 \times 1000/500 = -37.54 \text{ MPa}$$

Conversely, these show tension in the central steel strip and compression in the outer copper strips. Correspondingly, their lengths change by an equal amount:

$$x_A = \alpha_A l\Delta t + \sigma_A l/E_A$$

$$= (12.6 \times 10^{-6} \times 900 \times 80) + (18.77 \times 900)/(207 \times 10^3) = 0.989 \text{ mm}$$

$$x_B = \alpha_B l\Delta t + \sigma_B l/E_B$$

$$= (18 \times 10^{-6} \times 900 \times 80) - (37.54 \times 900)/(110 \times 10^3) = 0.989 \text{ mm}$$

Note that should there be an initial difference in the lengths of A and B, say where the end plates are stepped, then l in Eq. 3.28 cannot be cancelled.

3.5 Combined Mechanical and Thermal Effects

The solutions to problems of this nature are found from superimposing the separate effects of mechanical and thermal straining considered previously. The following examples will illustrate how this is done for the composite section and the bolted assembly given in § 3.3.1 and 3.3.2, respectively.

3.5.1 Composite Section

Consider an axial force W applied to either of the composite sections shown in Figs 3.8a,b at a raised temperature. The net stress for the composite section, given in § 3.3.1, is found from adding the thermal stresses Eqs 3.29a,b, arising from the change in temperature, to the mechanical stresses under the applied load. The latter, as provided by Eqs 3.13a,b and 3.15a,b, are applied in the following example:

$$\sigma_A = \frac{WE_A}{E_A A_A + E_B A_B} \tag{3.30a}$$

$$\sigma_B = \frac{WE_B}{E_A A_A + E_B A_B} \tag{3.30b}$$

Example 3.12 In a composite section copper is sandwiched between two steel strips with similar dimensions and properties given in Example 3.11. The section is first raised in temperature by 80°C and then subjected to an axial tensile force of 25 kN. Find the net stresses in the copper and steel. Examine how the net stresses are altered when the materials are interchanged.

In Example 3.11, with a central copper strip, the temperature stresses were found from Eqs 3.29a,b to be:

$$\sigma_A = -23.04 \text{ MPa for copper and } \sigma_B = 46.08 \text{ MPa for steel.}$$

The mechanical stresses from the applied force are provided by Eqs 3.30a,b:

$$\sigma_A = \frac{25 \times 10^3 \times 110 \times 10^3}{(110 \times 10^3 \times 1000) + (207 \times 10^3 \times 500)} = 12.88 \text{ MPa}$$

$$\sigma_B = \sigma_A E_B / E_A = 12.88 \times 110/207 = 6.85 \text{ MPa}$$

Hence the net stresses become:

$$\sigma_A = -23.04 + 12.88 = -10.16 \text{ MPa}$$

$$\sigma_B = 46.08 + 6.85 = 52.93 \text{ MPa}$$

Here the thermal stresses in the copper and steel are respectfully partially canceled and reinforced by the application of a force. This is a sensible arrangement for higher load applications knowing that steel has the greater yield stress: 300 MPa, compared to 70 MPa for copper.

When the materials are interchanged the temperature stresses were found from Eqs 3.29a,b to be:

$$\sigma_A = 18.77 \text{ MPa for the steel core steel}$$

$$\sigma_B = -37.54 \text{ MPa for the outer copper strips}$$

The mechanical stresses from the applied force are provided by Eqs 3.30a,b:

$$\sigma_A = \frac{25 \times 10^3 \times 207 \times 10^3}{(207 \times 10^3 \times 1000) + (110 \times 10^3 \times 500)} = 19.75 \text{ MPa}$$

$$\sigma_B = \sigma_A E_B / E_A = 19.75 \times 110/207 = 10.50 \text{ MPa}$$

Hence the net stresses become:

$$\sigma_A = 18.77 + 19.75 = 38.52 \text{ MPa}$$

$$\sigma_B = -37.54 + 10.50 = -27.24 \text{ MPa}$$

As both net stresses have been reduced from 52.93 MPa and −10.16 MPa interchanging materials has provided a greater tensile load carrying capacity.

3.5.2 Bolted Assembly

Equations 3.27–3.29a,b will also provide the amount by which the stresses are altered by raising the temperature in a pre-tensioned, bolted-flange assembly. For example, in the case of Fig. 3.9, the pre-tightening stress for the bolt and cylinder follow from Eq. 3.21b as:

$$\sigma_b = \frac{P}{A_b} = \frac{\delta E_b A_c E_c}{l(A_b E_b + A_c Ec)} \qquad (3.31\text{a})$$

$$\sigma_c = -\frac{P}{A_c} = \frac{-\delta E_b A_b E_c}{l(A_b E_b + A_c E_c)} \qquad (3.31\text{b})$$

Next, consider raising the temperature of the assembly for a long steel bolt and a copper cylinder (say), of approximately equal lengths. With b and c redefining the bolt A and cylinder B, respectively, then $\alpha_b < \alpha_c$ in Eqs 3.29a,b and the stresses become:

$$\sigma_b = \frac{\Delta t (\alpha_c - \alpha_b)(A_c E_b E_c)}{A_b E_b + A_c E_c} \qquad (3.32\text{a})$$

$$\sigma_c = -\frac{\sigma_b A_b}{A_c} \qquad (3.32\text{b})$$

Hence the net tensile stress in the bolt is the sum of Eqs 3.31a and 3.32a. The net compressive stress in the cylinder becomes the sum of Eqs 3.31b and 3.32b. It appears the tightening stress will only 'relax' when $\alpha_b > \alpha_c$.

Example 3.13 Find the net stresses for the pre-tensioned bolted assembly in Example 3.9 when its temperature is raised by 100°C. Take $\alpha = 12.2 \times 10^{-6}/°C$ for the steel bolt and $\alpha = 18 \times 10^{-6}/°C$ for the copper cylinder. Neglect the thickness of the end plates.

The pre-tightening force was shown in Example 3.9 to be: $P = 3294.74$ N. Hence the bolt and cylinder pre-tightening stresses become:

$$\sigma_b = P/A_b = 3294.74/122.72 = 31.98 \text{ MPa}$$

$$\sigma_c = -P/A_c = -3294.74/829.38 = -4.732 \text{ MPa}$$

The temperature stresses follow directly from Eqs 3.32a,b:

$$\sigma_b = \frac{100\,(18 - 12.2)\,10^{-6} \times 829.38 \times 207 \times 10^3 \times 104 \times 10^3}{(122.72 \times 207 \times 10^3) + (829.38 \times 104 \times 10^3)} = 92.75\ \text{MPa}$$

$$\sigma_c = -\frac{92.75 \times 122.72}{828.38} = -13.72\ \text{MPa}$$

Hence the net stresses become

$$\sigma_b = 31.98 + 92.75 = 124.73\ \text{MPa}$$

$$\sigma_c = -4.73 - 13.72 = -18.45\ \text{MPa}$$

which have been increased considerably from a moderate temperature rise. Here the safety factor for steel is $300/124.7 \approx 2.5$ in tension and for copper $65/18.5 \approx 3$ in compression. Given the relative difference between their yield stresses but with their comparable safety factors, shows that the materials have been chosen well.

Finally, consider a pre-tightened bolted assembly in Fig. 3.9 carrying an axial load at elevated temperature, as it may well be subjected to in service. We have seen that the net stress in the bolt b and cylinder c arise from three sources: pre-tightening, raising the temperature and applying the force. Correspondingly, the respective stresses are those given in Eqs 3.31a,b, 3.32a,b and 3.22a,b. The latter applies when the plates are screwed or welded to the cylinder as in Figs 3.10a,b, so that they each bear a proportion of the load. Hence, for the bolt, the net stress is:

$$\sum \sigma_b = \frac{\delta(A_c E_b E_c)}{(A_b E_b + A_c E_c)\,l} + \frac{\Delta t\,(\alpha_c - \alpha_b)(A_c E_b E_c)}{A_b E_b + A_c E_c} + \frac{W E_b}{A_b E_b + A_c E_c}$$

and for the cylinder, the net stress is

$$\sum \sigma_c = -\frac{\delta(A_b E_b E_c)}{(A_b E_b + A_c E_c)\,l} - \frac{\Delta t\,(\alpha_c - \alpha_b)(A_b E_b E_c)}{A_b E_b + A_c E_c} + \frac{W E_c}{A_b E_b + A_c E_c}$$

The reader will recognise here that the order of these summations is unimportant to the final net stress values. Thus, the *principal of superposition* accepts a summation of the separate effects in any order provided they are each elastic. Should plasticity arise within one or more of these contributions then the order in which they are applied does matter as the deformation becomes *path dependent*. Moreover, there are likely to be time-dependent influences from creep and relaxation effects at moderately high temperature. The pre-tightening stress may relax under the raised temperature and a creep strain develop in the cylinder under the constant applied load. As far as an elastic response of this assembly is concerned the above equations may be used to good effect in various ways. For example, we may wish to ensure that the net stress in the cylinder is zero ($\sum \sigma_c / E_c = 0$) through the control of the variables δ, Δt and W. Alternatively, we may wish to maintain a seal between the cylinder and its end plates, say, where the cylinder is a container for high-temperature fluid. Here it can be arranged that their final displacements will remain equal, as with their strains, when: $\sum \sigma_b / E_b = \sum \sigma_c / E_c$.

Exercises

Direct Stress and Strain

3.1 A 180 m length of pipe run is subjected to a longitudinal tensile force of 50 kN. If the outside and inside diameters are 110 and 100 mm, respectively, calculate the increase in the pipe length given its modulus of elasticity: $E = 193$ GPa.
(Ans: 28.3 mm)

3.2 A steel bar 300 mm long is 6.5 mm square for 100 mm of its length, the remaining length being 50 mm diameter. Determine the extension of the bar when it is subjected to a tensile load of 220 kN. Take $E = 207$ GPa.
(Ans: 0.13 mm)

3.3 A metal tube 76 mm o.d. and 1.65 m long is to carry a compressive load of 60 kN. If the allowable tensile stress is 77 MPa, calculate the internal diameter of the tube. Given its modulus $E = 92.6$ MPa find by how much the tube will shorten.
(Ans: 69.3 mm, 1.4 mm)

3.4 A 25 mm diameter steel bolt, with an effective length of 250 mm and a thread pitch of 2.5 mm, passes through a rigid block. Calculate the load imparted to the bolt when the clamping nut is initially tightened 1/4 turn with a spanner. Given that the ultimate strength for the bolt material is: $UTS = 925$ MPa, what safety factor has been applied to the bolt? Take: $E = 208$ GPa.
(Ans: 265 kN, 1.8)

3.5 A tapered alloy column 300 mm long with end diameters of 25 and 50 mm, is subjected to an axial tensile load of 100 kN. Calculate the axial displacement and the maximum stress. Take $E = 80$ GPa.
(Ans: 0.38 mm, 203.7 MPa)

3.6 A solid steel bar has a square cross section with side length increasing linearly from 20 mm at one end to 30 mm at the other over a length of 0.4 m. What is the greatest axial tensile load that this bar can carry when the maximum permissible stress is 300 MPa and by how much does the bar extend under this load if $E = 200$ GPa?
(Ans: 120 kN, 0.4 mm)

3.7 The following load and extension readings were obtained from conducting a tensile test upon a 12.5 mm diameter testpiece, with a 50 mm gauge length. Determine modulus of elasticity, the 0.1% proof stress and the stress at the limit of elastic proportionality.

Load (kN)	0	5.56	11.13	16.45	22.25	27.80	33.38	38.94	44.50	53.40
Ext. (mm)	0	0.01	0.020	0.031	0.041	0.051	0.061	0.071	0.081	0.102

Load (kN)	58.30	64.97	70.09	76.10	81.44	85.00	88.33	91.67	93.90
Ext. (mm)	0.112	0.127	0.142	0.163	0.183	0.203	0.224	0.254	0.279

Shear Force, Stress and Strain

3.8 A rivet 5 mm diameter joins three plates, as in Fig. 3.5b, so that the rivet is placed under double shear when a force of 5 kN is applied to the plates. Find the shear stress in the rivet.
(Ans: 127.3 MPa)

3.9 Estimate the forces required to punch: (i) a 75 mm diameter hole in a 5 mm-thick aluminium plate with an ultimate shear strength, $USS = 95$ MPa and (ii) a 20 mm diameter hole in a 2.5 mm-thick boiler plate with $USS = 415$ MPa. What are the compressive stresses in the punch in each case?
(Ans: 112 kN, 65.2 kN, 25.3 MPa, 207.5 MPa)

3.10 A tie bar in a steel structure (see Fig. 3.13) carries a load of 100 kN through a 25 mm diameter pin in double shear. Calculate, using a safety factor of 3, suitable dimensions t, B and D of the bar end if tensile failure of the bar and bursting of the lug are to be avoided. The respective ultimate strengths are 260 MPa and 320 MPa. What safety factor has been used for the pin given that its ultimate strength in single shear is 400 MPa?
(Ans: 4)

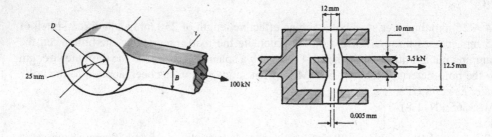

Figure 3.13   Figure 3.14

3.11 The central portion of a circular pin, 12 mm diameter, is displaced 0.005 mm from its unstrained centre line as shown in Fig.3.14, due to a 3.5 kN force. The ends of the pin pass through fork-ends that maintain the original centre line in position. Find the shear stress and strain in the pin and the value of its shear modulus.
(Ans: 154.7 MPa, 0.004, 38.75 GPa)

3.12 Find the radius of the fork-end in Figs 3.6a,b when it is made from similar material to the rocker at half its thickness as shown. The design should consider bearing failure around the pin and failures from tearing and pull-out, similar to those shown for the rocker in Fig. 3.6b. Take the same ultimate strengths for the rocker and fork-end, as given in Example 3.5.

3.13 A press tool produces 100 washers/stroke in a 1.5 mm thick material whose ultimate shear strength is 110 MPa. If the dimensions of the washers are to be 12 mm o.d. and 6 mm i.d. calculate the power required to drive a crankshaft when rotating at 30 rev/min given that the rod, which connects the crankshaft to the press, has a 25 mm stroke. Hint: The power is found from Eqs 7.7a,b. (Ans: 2.443 kW)

3.14 A lap joint in two plates 12.5 mm thick is formed with a single row of 20 mm diameter rivets. Determine the maximum possible pitch of the rivets and the efficiency of the joint with possible failure from tearing of the plate and shearing and crushing of the rivets. The respective ultimate strengths are: 430 MPa, 355 MPa and 710 MPa.
(Ans: 48 mm, 53.3%, 52.8%, 76.5%)

3.15 Three rows of 20-mm diameter rivets are used to form a lap joint between two 15 mm thick plates. Find the maximum allowable pitch of the rivets and the efficiency of the joint when designing to avoid tearing of the plates and shearing of the rivets. Take the respective ultimate strengths as: 90 MPa and 70 MPa. What is the bearing stress?
(Ans: 80 mm, 72%, 74%)

3.16 A simple lap joint consists of a single row of rivets in a plate 20 mm thick. Using a safety factor of 3 for ultimate strengths of 380 MPa and 700 MPa in shear and compression respectively for the rivets and 350 MPa for the plate in tension, determine a suitable rivet diameter and pitch. (Ans 46.9 mm, 140.7 mm)

Compound Bars: Mechanical Effects

3.17 Strips of brass 10 mm thick are bonded to a central 10 mm thick steel strip for a common depth of 25 mm and common length of 150 mm (see Fig. 3.8a). If the compound bar is to carry an axial compressive force of 50 kN find the compression, the stress in the strips and in the bonding. Take, for steel: $E = 210$ GPa and for brass: $E = 85$ GPa.
[Ans: - 0.102 mm, 142.37 MPa (steel), 57.63 MPa (brass), 2.825 MPa (bonding)]

3.18 Two bronze strips 8 mm thick and 250 mm wide are fixed securely to the sides of a rectangular steel bar of the same width and length. Find the thickness of the steel required to limit its stress to 155 MPa when the compound bar is loaded to 1000 kN in tension. Find also the stress in the bronze and the extension of the bar over 1 m length. Take the moduli as: $E = 110$ GPa for bronze and $E = 207$ GPa for steel.
(Ans: 17.3 mm, 82.4 MPa, 0.75 mm)

3.19 A circular mild steel bar 37.5 mm diameter is firmly enclosed by a bronze tube of outer diameter 62.5 mm for a common length of 150 mm. Find the stress in each material and the change in length under an axial compressive force of 200 kN. Take $E = 200$ GPa for steel and $E = 100$ GPa for bronze.
(Ans: 95.9 MPa, 47.9 MPa, 0.07 mm)

3.20 A bar of steel and two bars of brass each with a 25 x 15 mm cross section are to carry jointly a load of 20 kN over a common length of 180 mm. If the bars are bonded in parallel so that each extends by the same amount, determine the load carried by each bar. Take $E = 207$ GPa and $E = 83$ GPa for steel and brass, respectively.
(Ans: 8.9 kN brass, 11.1 kN steel)

3.21 A 25 mm composite square section is formed from a central brass strip 5 mm thick bonded between two 10 mm strips of epoxy resin for a length of 750 mm. Calculate the maximum axial load and the extension given that the respective allowable stresses for brass and epoxy are 155 MPa and 60 MPa, respectively. Take for brass: $E = 86$ GPa and for epoxy resin: $E = 3.1$ GPa.
(Ans: 27.5 kN, 1.4 mm)

3.22 A short steel tube 100 mm o.d. and 10 mm thick is firmly surrounded by a brass tube of the same thickness for a common length of 100 mm. If the compound tube carries a total compressive load of 5 kN, calculate by how much the tube will shorten. What is the load carried by each material? Take, for steel: $E = 201$ GPa, for brass: $E = 85$ GPa.
(Ans: 3.2 kN, 1.8 kN, 0.00046 mm)

3.23 Determine the maximum axial tensile load that may be applied to a 250 mm long bonded composite tube without causing yielding in either of its components. The brass inner ring is 50 mm i.d., 75 mm o.d., with a yield stress of 160 MPa and $E = 105$ GPa. The steel outer ring is 75 mm i.d., 100 mm o.d with a yield stress of 210 GPa and $E = 210$ GPa.

3.24 A concrete column, 3.05 m high with 380 mm square section, is reinforced by four steel rods, 25 mm diameter, evenly spaced within the cross section. If the concrete carries an axial load of 600 kN determine the stresses in the steel and concrete and the change in length of the column. Take, for steel: $E = 207$ GPa, for concrete: $E_c = 13.8$ GPa.
(Ans: 52.4 MPa, 3.49 MPa, 0.772 mm)

3.25 Show that the number of bolt rotations N required for pre-tightening the bolt, of thread pitch p, is given by: $N = Wl/(A_b E_b p)$ if the cylinder in Fig. 3.9 is to becomes free of stress when an axial load W is applied to the assembly. Neglect the thickness of the end plates.

3.26 A 12.5 mm diameter steel rod passes centrally through a 2.5 m length of copper tube: 50 mm o.d. and 38 mm i.d. The tube is closed at each end by 25 mm thick rigid steel plates and secured by nuts on the threaded ends of the rod (see Fig. 3.9). Calculate the stresses set up in the steel and the copper (a) when the tube contracts 0.5 mm after the nuts are tightened and (b) when the nuts, of 0.5 mm thread pitch, are tightened by two revolutions. Note: the rod and cylinder are of unequal lengths. For steel: $E_s = 207$ GPa; for copper, $E_c = 105$ GPa.
(Ans: (a) 140.6 MPa, – 20.8 MPa)

Compound Bars: Temperature Effects

3.27 A tie bar 50 mm diameter and 6 m long realigns a bulging vessel by being alternately heated and cooled to a maximum of 300°C. The vessel offers a maximum resistance to the bar of 200 kN. Determine the greatest alignment for each heating cycle when the nuts at the end of the bar are finger tight at 300°C. Take $E = 207$ GPa, $\alpha = 12 \times 10^{-6}/°C$ for the tie.
(Ans: 17.2 mm)

3.28 Solid bronze and steel bars of equal lengths with diameters of 50 mm and 30 mm, respectively, are placed together end to end with axes horizontal between close-fitting rigid walls. Calculate the stresses induced in each material when the temperature is raised by 37.78°C (100°F). Take $E = 208$ GPa, $\alpha = 11.7 \times 10^{-6}/°C$ ($6.5 \times 10^{-6}/°F$) for steel and $E = 108$ GPa, $\alpha = 18 \times 10^{-6}/°C$ ($10 \times 10^{-6}/°F$) for bronze. (Ans: 72.9 MPa, 202.6 MPa)

3.29 A 40 mm long phosphor bronze tube, 40 mm i.d. and 50 mm o.d., forms part of a pipeline with a second tube of the same i.d. but with a 55 mm i.d. and a 60 mm length. The assembly is held together by rigid end plates that prevent axial expansion of the tubes but allow free fluid flow. Calculate the stresses induced in each tube when the fluid temperature is raised by 75°C from the unstressed state at room temperature. Take $E = 80$ GPa, $\alpha = 20 \times 10^{-6}/°C$ for the smaller tube and $E = 12$ GPa, $\alpha = 18 \times 10^{-6}/°C$ for the larger tube. (Ans: 179 MPa, 112 MPa)

3.30 A compound tube is formed from an outer stainless steel tube, 50 mm o.d., 47.5 mm i.d. and an inner mild steel tube of 6 mm wall thickness. The two tubes are welded together at their ends to give a 1.5 mm radial clearance. Assuming that the tubes are free to expand between the ends when heated, calculate the tube stresses when the temperature is raised by 56°C. For stainless steel: $E = 172$ GPa and $\alpha = 18 \times 10^{-6}/°C$; for mild steel: $E = 207$ GPa and $\alpha = 12 \times 10^{-6}/°C$. (Ans: 44.8 MPa, 14.5 MPa)

3.31 A compound bar is formed from a stainless steel rod placed inside a mild steel tube, welded together at their ends, over a length of 600 mm. The cross-sectional area of the rod is 750 mm^2 and that of the tube is 1625 mm^2. If an extension of the compound bar of 0.375 mm results from a temperature rise of 38.88°C, determine the load induced in the rod and the stress in the tube. What is α for the tube if $E = 207$ GPa? Take: $E = 165.4$ GPa and $\alpha = 18 \times 10^{-6}/°C$ for stainless steel. (Ans: 9.6 kN, 5.72 MPa, $\alpha = 15.35 \times 10^{-6}/°C$)

Compound Bars: Mechanical and Temperature Effects

3.32 Solid copper and steel bars of respective lengths 100 mm and 150 mm and diameters 30 mm and 15 mm, are placed together end to end, horizontally between rigid walls. Calculate the stresses induced in each material when the temperature is raised by 20°C and the force exerted by each bar upon the wall. If at the raised temperature the walls are moved together by 0.1 mm, what then are the net stresses and the external force that needs to be applied? For steel take: $E = 207$ GPa and $\alpha = 12.5 \times 10^{-6}/°C$. For copper, $E = 105$ GPa, and $\alpha = 18 \times 10^{-6}/°C$. [Ans: -19.1 MPa (copper), -76.4 MPa, -13.5 kN; -45.1 MPa (copper), -180.2 MPa, -18.36 kN]

3.33 Find, for the welded assembly in Fig. 3.10a, the stresses in the steel bolt (s) and copper tube (c) when the nut is tightened by one revolution on a thread pitch of ½ mm. Find the net stresses when the pre-stressed assembly is subjected to: (i) an axial tensile force $W = 5$ kN and (ii) a temperature rise $\Delta t = 10°C$. Neglect the thickness of the end plates. For the cylinder: $l_c = 2.5$ m, i.d.= 38 mm, o.d.= 50 mm, $E_c = 104$ GPa and $\alpha_c = 18 \times 10^{-6}°C^{-1}$. For the 25 mm bolt: $E_s = 207$ GPa and $\alpha_s = 12.2 \times 10^{-6}°C^{-1}$. [Ans: (i) $\sigma_c = -0.075$ MPa (ii), $\sigma_s = 41.25$ MPa]

3.34 A rectangular compound bar is 50 mm wide × 30 mm deep. The depth is made up from a 20 mm layer of steel sandwiched between 5 mm thick copper strips. The strips are held firmly together at the ends of their common 900 mm lengths at 20°C. Determine the stresses in the steel and the copper when the temperature is raised by 80°C. What are the stresses and the final extension of the assembly when a tensile force of 5 kN is applied at the increased temperature? For steel: $E = 207$ GPa and $\alpha = 12.6 \times 10^{-6}/°C$. For copper: $E = 110$ GPa and $\alpha = 18 \times 10^{-6}/°C$. [Ans: 18.77 MPa (steel), -37.55 MPa; 22.72 MPa (steel), -35.45 MPa, 1.01 mm]

3.35 A flat steel bar 25 mm wide and 5 mm thick is placed between two aluminium alloy bars each 25 mm wide × 10 mm thick to form a composite bar with 25 mm square section. The bars are fixed together at their ends at a temperature of 10°C. Find the stress in each bar when the temperature is then raised to 49°C and, also, when an additional force of 20 kN is applied at a temperature of 49°C. Take $E = 207$ GPa and $\alpha = 11.5 \times 10^{-6}/°C$ for steel. Take $E = 69$ GPa, $\alpha = 23 \times 10^{-6}/°C$ for aluminium alloy.
[Ans: 46.3 MPa (steel), 15.44 MPa (Al alloy); 108.4 MPa (steel), 5.24 MPa (Al alloy)]

3.36 A horizontal brass tube is clamped between rigid end covers by six equally spaced, 6 mm diameter steel bolts as shown in Fig. 3.15. The tube is pre-compressed at 15°C by tensioning the bolts to 40 MPa. Determine the stresses in both the bolts and the tube when the temperature of the bolts is 0°C and the temperature of the tube is 60°C. Assume the following material constants for steel: $E = 200$ GPa and $\alpha = 12 \times 10^{-6}/°C$ and for brass take: $E = 100$ GPa and $\alpha = 18 \times 10^{-6}/°C$.

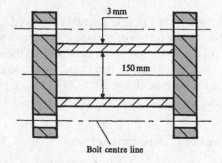

3 mm

150 mm

Bolt centre line

Figure 3.15

3.37 The steel bolt shown in Fig. 3.16 is pre-tensioned over a fixed length of 150 mm. A hole drilled through the centre of the bolt contains a loose-fitting aluminium alloy rod which is attached to the left-hand bolt head. In this way tightening the right-hand nut strains the bolt but not the pin. Determine the design load for the bolt when the right-hand ends of the rod and bolt shank are level. By how much is this load and the end alignment altered when the introduction of fluid to the bolt hole raises the temperature of the whole bolt assembly by 60°C? Assume that the 150 mm length is unchanged. Take, for steel: $E = 207$ GPa and $\alpha = 11.5 \times 10^{-6}/°C$. Take for aluminium alloy: $E = 70$ GPa and $\alpha = 24 \times 10^{-6}/°C$.

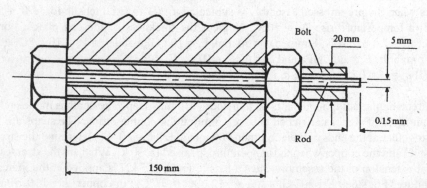

Bolt

20 mm

5 mm

0.15 mm

Rod

150 mm

Figure 3.16

CHAPTER 4

LINEAR ELASTICITY

Summary: A stress state may exist in one, two or three dimensions, which is more commonly referred to as being, respectively, a uniaxial, biaxial or triaxial state. The elastic constants provide constitutive relations, which connect each stress state to strain. Strain is usually three-dimensional for each stress state. For example, under uniaxial tension a rectangular bar suffers contractions to its width and thickness as it extends. A plate thins as it is stretched biaxially in perpendicular directions within its plane. In contrast, a pure shear distortion (see Fig. 3.4) would not produce thinning unless it is combined with an axial loading. The engineering strains, both normal and shear, are defined as the ratio between each distortion and the original length, using the formulae in Eqs 3.2 and 3.7b. At its most general, a constitutive relation can provide the three-dimensional strains for a stress state arising from any combination of loading: tension, shear, bending and torsion. The relation requires that the normal and shear stress components be identified either within a Cartesian or polar co-ordinate system that is more suited to the geometry of the solid subjected to these loads. Finally, it is in the meaning of the term *linear elasticity* that all distortion is recoverable when the solid is relieved of its loading. Here the loading must lie within the elastic limit within which the analyses of stress and strain are simplified with the solid responding in a linearly elastic manner to its loading and unloading. While most metals and their alloys behave in this way, non-metals may not. We shall examine here and later in Chapter 12 how ductile metals behave when loaded beyond the elastic limit and where brittle, elastic failure applies to solids and fibrous composites incapable of sustaining large extensions.

4.1 Elastic Constants

The *modulus of elasticity* for uniaxial tension and a *modulus of rigidity* for shear were defined in Chapter 3 as the ratio between stress and strain. Firstly, we shall define these two elastic constants again and compare their values across a range of metals and alloys used in engineering structures. This amounts to a comparison in their stiffness. The coupling between a required strength and stiffness is often the basis for design in structures. The full range of elastic behaviour requires two further elastic constants: *Poisson's ratio* and the *bulk modulus*. All four constants can be defined for any material when placed under specific elastic states of stress.

4.1.1 Modulus of Elasticity

For simple tension and compression of a bar, the *modulus of elasticity E* (Young's modulus) identifies directly with the gradient to the linear plot between axial stress σ and axial strain ε. This gives:

$$E = \sigma/\varepsilon \qquad (4.1)$$

The engineering stress and strain, that determine the ratio in Eq. 4.1, are said to be nominal quantities since they refer to the original geometry of the bar in the manner of Eqs 3.1–3.3. Equation 4.1 applies to both loading and unloading for stress levels within the limit of proportionality (elastic limit). It is important to attach the word 'axial' to the strain in Eq. 4.1 because lateral strain is also produced from a uniaxial stress. The distinction between lateral and axial strains appears in the definition of Poisson's ratio (see § 4.1.3).

4.1.2 Tension Beyond the Elastic Limit

Loading beyond the elastic limit is non-linear when Eq. 4.1 no longer applies. However, for metals the deformation is said to be elastic-plastic, as the total strain is the sum of elastic and plastic components. Consequently, Eq. 4.1 may be re-applied to the elastic strain that recovers upon unloading. The plastic component of strain remains in the form of a permanent set, that is associated with the prior work hardening and an increase in the proportional limit for the plastically pre-strained material. Thus, upon re-loading this material, Eq. 4.1 applies again to an increased elastic range. Such an application of Eq. 4.1 ignores the slight hysteresis that often appears between unloading and reloading from within the plastic range. In metals the separation between each slightly non-linear plot is negligible. Equation 4.1 cannot be applied in a similar manner to polymers and rubbers where hysteresis is known to be significant.

Example 4.1 The results given in Table 4.1 apply to tensile loading, unloading and reloading cycles for a half-hard grade of C101 copper sheet in the form of a testpiece having a width 15.03 mm and thickness 3.32 mm over a parallel length of 75 mm. An extensometer provided extensions x (mm) over a 50 mm gauge length corresponding to each cycle for the loads W (kN) indicated. The reloading cycle was continued until the testpiece failed. Determine: (a) the initial elastic modulus; (b) the modulus for unloading; (c) the modulus for re-loading; (d) the limit of proportionality; (e) the 0.1% proof stress; (f) the ultimate tensile strength; (g) the % range of uniform strain; and (h) the % strain to fracture.

The plot (see Fig. 4.1) between load and extension for the three cycles is easily converted to one of stress σ in MPa and percentage strain since the following relationships hold:

$$\sigma = W/A = (W \times 10^3)/(15.32 \times 3.32) = 20W \text{ (MPa)} \qquad (4.2a)$$

$$\varepsilon = x/l = (x \times 100)/50 = 2x \text{ (\%)} \qquad (4.2b)$$

Table 4.1 Tensile loading and unloading cycles for C101 copper

Load	Displacements x, mm		
W, kN	Loading	Unloading	Re-Loading
	↓		↓
0	0	0.068 →	0.068
1	0.0085	0.078	0.073
2	0.017	0.085	0.082
3	0.025	0.094	0.090
4	0.033	0.104	0.098
5	0.041	0.109	0.105
6	0.049	0.118	0.114
7	0.058	0.125	0.122
8	0.066	0.134	0.131
9	0.074	0.142	0.138
10	0.090	0.149	0.148
10.5	0.113	0.153	0.158
10.7	0.155 →	↑	0.175
10.8			0.200
10.9			0.245
11.0			0.487
11.25			2.300
11.40			3.330
11.53			4.390
11.69			6.340
11.75			7.760
11.75			10.85
11.75			13.74
11.68			14.24
11.63			14.61 fracture

Consequently, rather than convert every pair of readings in Table 4.1 to stress and strain, Eqs 4.2a and b show that it is simpler to convert the scales of a W versus x plot to σ versus ε by applying multiplication factors of 20 to W and 2 to x as shown in Fig. 4.1. Then taking the axes of stress versus strain, the gradients of the loading, unloading and re-loading lines identify with Eq. 4.1, which provides the moduli required in parts (a)–(c):

(a) $E = 122.2$ GPa, (b) $E = 125.7$ GPa and (c) $E = 122$ GPa

Note that the moduli for (b) and (c) are based upon the central linear regions across the narrow hysteresis loop shown where minor variations in gradient occur. Far greater variations occur in polymers with wide hysteresis loops.

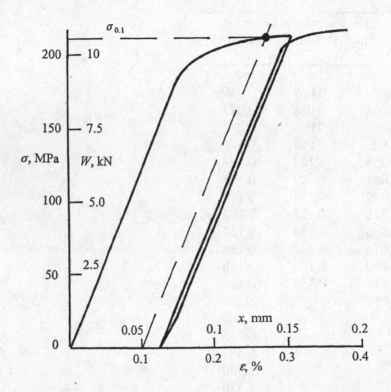

Figure 4.1 Tensile loading and unloading cycles for C101 copper

(d) Figure 4.1 shows that initial elasticity ceases at the limit of proportionality at which the stress is estimated as 175 MPa.

(e) The 0.1% proof stress is found by drawing a line parallel to the elastic loading line offset by 0.1% strain as shown. This line intersects the non-linear region of the loading plot at the required proof stress $\sigma_{0.1} = 214$ MPa. Had the testpiece been unloaded from this stress level there would remain 0.1% permanent set. The proof stress is adopted for materials that do not show a clear yield point within a very gradual transition from elasticity to plasticity. This amount of set is tolerable for design applications with a material capable of sustaining large strain to fracture.

(f) The re-loading cycle was continued to fracture for which only a selected number of data are given. These are sufficient to show that the stress level becomes asymptotic at its ultimate level, i.e. *UTS* = 235.5 MPa.

(g) The ultimate tensile strength (*UTS*) marks the point of instability beyond which the nominal stress level falls with the development of a local neck. The instability strain ε_I corresponds to the ultimate stress at peak load and defines the limit of uniform tensile straining. From Table 4.1:

$$\varepsilon_I = (13.74/50) = 0.275 \text{ or } 27.5\%$$

Note also that this instability strain equals the hardening exponent n in Hollomon's parabolic hardening law, given that his law connects the true stress to natural strain within the plastic range:

$$\sigma = A\varepsilon^n \tag{i}$$

where A is the strength coefficient. To show this the Considéré condition for tensile instability applies:

$$\frac{d\sigma}{d\varepsilon} = \sigma \tag{ii}$$

and combining Eqs i and ii:

$$\frac{d\sigma}{d\varepsilon} = nA\varepsilon^{n-1} = \frac{n\sigma}{\varepsilon} = \sigma \tag{iii}$$

Equation iii limits the uniform strain in reaching the point of instability at the ultimate stress to $\varepsilon_I = n$. Correspondingly, A follows from Eqs i and iii as:

$$A = \frac{\sigma}{n^n} = \frac{235.5}{0.275^{0.275}} = 335.9 \text{ MPa}$$

Hollomon's Eq. i describes parabolic hardening within the plastic range of metals fairly well at moderately high strain. However, it should be noted as the law does not account for an initial linear elastic region it is valid only for post-yield stress levels at moderate strains (beyond 1%).

(h) A greater % strain to fracture applies to the final extension reading given in Table 4.1 as:

$$\varepsilon_f = (14.61/50) \times 100 = 29.2\%$$

Normally, the fracture strain equates to the % elongation at fracture as it is estimated from the permanent extension within two broken halves of the testpiece.

4.1.3 Modulus of Rigidity

Similarly, for shear loading and unloading within the elastic limit, the *modulus of rigidity* G identifies with the gradient to the linear plot between shear stress and shear strain.

It can be seen how a pure shear mode arises within the distortion of the rectangular block shown in Fig. 3.4. This block may be thought of as an elemental length of a beam carrying lateral loads. Moreover, we saw from the examples considered in the previous chapter that transverse shear arises in connecting pins, rivets and bolts, these having circular cross sections. Because elasticity for metals and alloys involves small strain it is usually sufficient to adopt Eqs 3.6–3.8 to provide G in these cases. In fact, $\tau = F/A$ is the average of a distribution in vertical shear stress that arises from lateral loading applied to a beam (see § 5.4, p. 171).

The shear distortion produced in a circular bar under torsion is another important example where Eq. 3.8a applies (see Chapter 7). Note that here the shear stress τ and shear strain γ are calculated from the torque T and angular twist θ and therefore Eqs 3.6 and 3.7 do not apply. Torsion theory (see § 7.1, p. 243) provides the appropriate definitions of τ and γ as

$$\tau = \frac{Tr}{J}, \quad \gamma = \frac{r\theta}{L} \qquad (4.3\text{a,b})$$

Hence the ratio between Eqs 4.3a and b provide G as:

$$G = \frac{\tau}{\gamma} = \frac{TL}{J\theta} \qquad (4.4\text{a,b})$$

where θ is the angular twist in radians between the ends of a shaft of length L and $J = \pi d^4/32$ is its polar second moment of area for a solid circular cross section of diameter d. In the case of a hollow circular shaft, with outer and inner diameters d_o and d_i respectively, the polar second moment of area J within Eq. 4.4b is reduced to: $J = \pi d_o^4/32 - \pi d_i^4/32$ but the twist increases under a given torque to provide the consistent material constant G. Thin tubes are the better choice for modulus determination, eliminating the stress gradient to a near uniform state within the wall. On the other hand, tubes are less satisfactory for plasticity studies where the range of strain is limited by buckling. A solid circular section in carbon steel, especially, is capable of many full rotations before final failure. However, an account of the effect of the shear stress gradient upon its torque versus twist plot is required for a conversion into a useful equivalent flow curve for the material.

As an alternative to angular twist measurement, strain gauges are often bonded to the shaft's outer surface to find γ directly in the torsion test. A pair of perpendicular gauges 1 and 2 overlaid in a 'rosette', with their axes aligned at $\pm 45°$ to the shaft axis, provides $\gamma = \varepsilon_1 - \varepsilon_2$ where ε_1 and ε_2 are the direct gauge strains. Here ε_1 is the major principal (tensile) strain and ε_2 is the minor principal strain (compressive), these lying in perpendicular directions with inclinations at $+45°$ and $-45°$, respectively, to the axis of torsion (see Table 11.1 and § 11.4, pp.450–453). As the principal strains have equal magnitude they may be connected, conveniently, as a half-bridge within a strain meter to display γ directly, usually in 'micro-strain' (symbol $\mu\varepsilon$). This means that the actual strain is a factor of 10^6 less than its meter display, e.g. a display of 1000 $\mu\varepsilon$ = 1000 $\times$ 10^{-6} = 0.001. Note that if this strain is quoted or displayed as a percentage it is 100 times greater than its actual value, i.e. 0.1 % or 1000 $\mu\varepsilon$ represents an absolute strain of 0.001. When this technique is used to determine γ, then within the G-definition in Eq. 4.4a, the shear stress τ provided by Eq. 4.3a, applies to the surface of the shaft.

4.1.4 Torsion Beyond the Elastic Limit

Loading beyond the limit of proportionality involves both elastic and plastic strain, similar to elastic-plastic tension (see § 4.1.2). Only the elastic component of strain recovers allowing G to be re-determined with permanent set in the testpiece. The following example illustrates a suitable method for obtaining a single G-value from a table of results for the torsional loading and unloading of a hollow shaft.

Example 4.2 The results in Table 4.2 apply to torsional loading, unloading and reloading cycles for a mild steel tube having inner and outer radii of 6.29 mm and 7.97 mm respectively over its parallel length of 120 mm. Torque T (Nm) was transmitted axially through flats machined upon enlarged ends. The shear strain γ (micro-strain) was measured at the testpiece centre using strain gauges aligned at $\pm 45°$ to the shaft axis. Determine: (a) the initial elastic modulus; (b) the modulus for unloading; (c) the modulus for re-loading; (d) the limit of proportionality; and (e) the 0.05% proof stress.

Table 4.2 Torsional loading and unloading cycles for mild steel

Torque	Shear strain γ (με)		
T, Nm	Loading	Unloading	Re-Loading
	↓		↓
0	0	1344 →	1344
10	296	1696	1540
20	552	1988	1900
30	814	2240	2130
40	1084	2545	2400
50	1344	2760	2650
60	1594	3070	2950
70	1866	3300	3200
80	2140	3615	3500
90	2406	3840	3750
100	2695	4080	4020
110	2985	4740	4300
120	3236	4695	4550
130	3526	4900	4850
140	3800	5030	5110
150	4120	5472	5500
160	4752	5776	5850
162	5832 →	↑	6500

Firstly, Eq. 4.3a is used to convert the torque T (Nm) to an outer diameter shear stress τ (MPa) as follows:

$$\tau = \frac{Tr}{J} = \frac{Tr_o}{\pi(r_o - r_i)^4/2} = \frac{T \times 10^3 \times 7.97}{\pi(7.97^4 - 6.29^4)/2} = 2.055\, T \text{ (MPa)}$$

The plot between τ and γ for the three cycles is shown in Fig. 4.2.

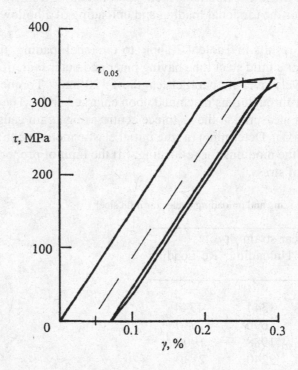

Figure 4.2 Torsional loading, unloading and re-loading cycles for a mild steel tube

Within these axes the gradients of the loading, unloading and re-loading lines provide the three moduli required: (a) $G = 76.8$ GPa, (b) $G = 76.6$ GPa and (c) $G = 75.0$ GPa. The shear moduli in (b) and (c) are identified with the central linear gradients within the narrow hysteresis loop for which a slight variation is found.

(d) Elasticity ceases at the limit of proportionality at which the stress is estimated from Fig. 4.2 as 275 MPa.

(e) The 0.05% proof stress is found by drawing a line parallel to the elastic loading line offset by 0.1% strain as shown. This line intersects the non-linear region of the loading plot at the required proof stress $\tau_{0.05} = 330$ MPa. This means that had the testpiece been unloaded from this stress level there would remain a 0.05% permanent shear strain, which lies well within the shear strain to fracture expected for a mild steel tube ($\approx 20\%$). When reduced to a working stress level with a safety factor, the 0.05% proof stress provides a suitable basis for an elastic design.

4.1.3 Poisson's Ratio, v

Simon-Denis Poisson (1781–1840) discovered that the lateral contraction which occurs in a tension test is proportional to the axial extension. Poisson's ratio refers to the constant ratio between the corresponding elastic strains for a given material. The ratio is an non-dimensional elastic constant, which lies in the range ¼–⅓ for most metals. In order to provide a positive ratio, the conventional minus sign, which accompanies the lateral contraction, is annulled by introducing a further minus sign in the definition. Thus, if the lateral strain is ε_w and the axial strain ε_l, Poisson's ratio v is given as:

$$v = \frac{- \; Lateral \; strain}{Axial \; strain} = \frac{- \; \varepsilon_w}{\varepsilon_l} \tag{4.5a}$$

Though the two uniaxial strains in Eq. 4.5a are principal strains, the use of their subscripts 1 and 2 will not appear until the stress state is two- or three-dimensional.

(a) Tension and Compression
Figure 4.3 gives the elevation of a rectangular bar under a tensile stress σ in which Δl denotes the increase in the original length l and Δw the negative contraction in the original width w.

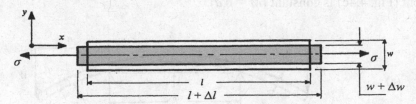

Figure 4.3 Extension and lateral contraction under tension

The lateral and axial strains in Eq. 4.5a are:

$$\varepsilon_w = \frac{\Delta w}{w}, \qquad \varepsilon_l = \frac{\Delta l}{l} \tag{4.5b,c}$$

Poisson's ratio, Eq. 4.5a, becomes:

$$v = \frac{- \; \varepsilon_w}{\varepsilon_l} = - \; \frac{l \times \Delta w}{w \times \Delta l} \tag{4.6a}$$

For a testpiece with a solid circular cross section the diameter d contracts by an amount Δd when the length l and extends by Δl. Poisson's ratio is found from:

$$v = \frac{- \; \varepsilon_w}{\varepsilon_l} = - \; \frac{l \times \Delta d}{d \times \Delta l} \tag{4.6b}$$

The mechanical measurement of small, elastic length and width/diameter changes is likely to be inaccurate unless the tensile test can be conducted upon a large testpiece. Consequently, strain gauges are often employed to find v in which ε_l and

ε_w in Eq. 4.5b are measured directly with gauges bonded to a flat surface in alignment with the axial and lateral directions. The quarter bridge connections provide the strains in each gauge separately when v identifies with the gradient of the plot between the two strains (see Example 4.3).

When the elastic constants E and v are known, the lateral strain may be found at a given stress σ from Eq. 4.6b as:

$$\varepsilon_w = -v\varepsilon_l = -v\sigma/E \tag{4.6c}$$

Equations 4.5a–c and 4.6a–c also apply to compression where Δl is negative and Δw (or Δd) is positive. The positive ratio should agree with that found under tension but, as discussed in Chapter 3 (see § 3.1, p. 61), it is difficult to achieve a purely uniaxial compression with load eccentricity and the risk of buckling.

(b) Four-Point Beam Bending
Beam bending may be used to good effect when finding v under tension and compression. In fact, strain gauges are calibrated using the analytical axial surface strain expression which derives from the central deflection δ of a beam under four-point loading, shown in Fig. 4.4a. The loading places the support length L under pure bending within which the shear force is absent (Fig. 4.4b) and the bending moment (Fig. 4.4c) is constant ($M = Wa$).

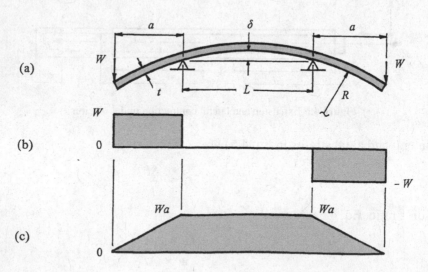

Figure 4.4 F- and M-diagrams for a beam in four-point loading

Hence, in the length L between the supports, the beam deflects into a circular arc with a radius R that is found from Pythagoras's theorem:

$$R^2 = (R - \delta)^2 + (L/2)^2 \tag{4.7a}$$

Eq. 4.7a gives $R \approx L^2/(8\delta)$ provided $L \gg \delta$. The tensile surface strain becomes:

$$\varepsilon_l = \frac{t/2}{R} = \frac{t/2}{L^2/(8\delta)} = \frac{4 \times \delta \times t}{L^2} \qquad (4.7b)$$

where t is the beam thickness and L is the length between the two supports, beyond which equal end-weights are hung at similar overhang distances a (see Fig. 4.4a). The calibration requires that the strain gauge factor, which converts the change in gauge resistance to strain, be set to match Eq. 4.7b. In practice, when this factor is set on the strain meter dial and with the ¼ bridge connections made, a channel selector allows the meter to display micro-strain directly in each gauge as the loading is increased. Further details are given in the following example.

Example 4.3 Longitudinal and width micro-strains, $\mu\varepsilon_l$ and $\mu\varepsilon_w$, given in Table 4.3 are found from bonding strain gauges in these directions to the top (tensile) surface of a steel beam subjected to increasing four-point loading. The central support length is 400 mm. Beyond each support the weight is applied at an overhang distance of 200 mm. The rectangular beam section is 25.4 mm width and 4.763 mm deep. Confirm the $\mu\varepsilon_l$ strains from the beam's central, vertical deflection reading δ and find the value of Poisson's ratio and Young's modulus.

Table 4.3 Beam's surface strains and central deflection

W, N	$\mu\varepsilon_l$	$\mu\varepsilon_w$	δ, mm	$\mu\varepsilon_l$	σ, MPa
5	52	−13	0.47	56	10.42
10	120	−26	0.95	113	20.84
15	176	−40	1.43	170	31.25
20	226	−53	1.90	225	41.68
25	272	−65	2.37	281	52.10
30	325	−76	2.85	338	62.52
35	373	−89	3.30	392	72.93

The fifth column gives the length strains calculated from Eq. 4.7b. The strain gauge factor has ensured that calculated strains compare well with the measured strains in the second column. The gradient of the plot between the measured strains in columns two and three (Fig. 4.5a) reveals $v = -\varepsilon_w/\varepsilon_l = 0.24$, which agrees with the accepted value for steel.

The results may be contracted to find a single elastic modulus from the plot between the bending stress and the length strain that apply to each load. The stress is calculated from the constant bending moment $M = Wa$ that exists between the supports (see Fig. 4.5a) and the second moment of area for the beam's rectangular section $I = bt^3/12$, as follows

$$\sigma = \frac{M \times t/2}{I} = \frac{Wa \times t/2}{bt^3/12} = \frac{6Wa}{bt^2} \qquad (4.8)$$

Taking the W as the weight applied to each hanger, Eq. 4.8 provides the surface stresses given in the final column of Table 4.3. We may plot these stresses against length strain recorded from the strain gauge (column two) or as calculated from the deflection (column five). The gradient of the former plot (see Fig. 4.5b) provides Young's modulus as $E = 203$ GPa, which lies within its accepted range for steel: $E = 200$–220 GPa.

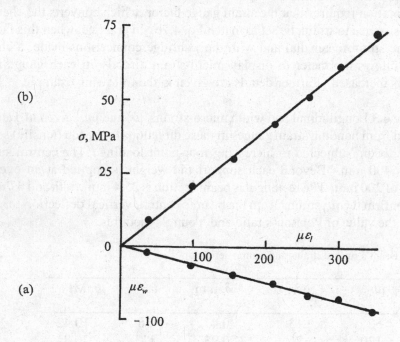

Figure 4.5 Poisson's ratio and modulus plots for steel beam in bending

Conveniently, when the elastic constants E and v are required for compression it is only necessary to turn the beam over so that the two gauges required are attached to the compressive surface. Thereby, the axial gauge is placed in compression and the lateral gauge in tension.

(c) Elastic-Plastic Range
An interesting study is the change produced in Poisson's ratio as the tensile deformation becomes elastic-plastic. In what follows, it is shown, theoretically, how the ratio between the corresponding plastic strains becomes ½. Consequently, the ratio between the total strains equals Poisson's elastic value at the proportionality limit but thereafter increases with elastic plus plastic straining asymptotically to ½, when the plastic strain becomes much greater than the elastic strain. It follows that the ratio between lateral (width) and axial (length) strains changes continuously during the transition from elasticity and plasticity. The elasticity that recovers from an unloading will conform to Poisson's elastic ratio. Figure 4.6 shows the transition behaviour and elastic recovery for an aluminium-copper alloy (AC 100) in which the gradient becomes the negative of the said ratio.

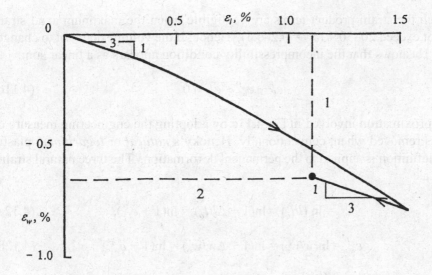

Figure 4.6 Total strain plot for aluminium alloy for its elastic-plastic transition

For example, the lateral to axial strain ratio: $\varepsilon_w/\varepsilon_l = -\frac{1}{3}$ ($\nu = \frac{1}{3}$), is seen to apply to both loading and unloading. The ratio between the permanent strains that remain, however, will conform to $\varepsilon_w/\varepsilon_l = -\frac{1}{2}$. This ratio is predicted from an incompressible plasticity in which the deformation occurs by shear along close-packed atomic planes. To show this, let the original length, width and thickness of the testpiece be l_o, w_o and t_o, respectively. Under load, these lengths will change permanently by amounts Δl, Δw and Δt, from which it follows that the change in volume is:

$$\Delta V = (l_o + \Delta l)(w_o + \Delta w)(t_o + \Delta t) - l_o w_o t_o \qquad (4.9a)$$

Dividing Eq. 4.9a throughout by the original volume $V_o = l_o w_o t_o$

$$\Delta V/V_o = (1 + \Delta l/l_o)(1 + \Delta w/w_o)(1 + \Delta t/t_o) - 1 \qquad (4.9b)$$

in which the three *engineering* plastic strains appear:

$$e_l^P = \Delta l/l_o, \; e_w^P = \Delta w/w_o \text{ and } e_t^P = \Delta t/t_o \qquad (4.10a,b,c)$$

It follows from Eqs 4.9b and 4.10a,b,c that

$$\Delta V/V_o = (1 + e_l^P)(1 + e_w^P)(1 + e_t^P) - 1 \qquad (4.11a)$$

Expanding Eq. 4.11a gives:

$$\Delta V/V_o = e_l^P + e_w^P + e_t^P + e_l^P e_w^P + e_w^P e_t^P + e_l^P e_t^P + e_l^P e_w^P e_t^P \qquad (4.11b)$$

in which the strain product terms are negligible when the maximum axial strain does not exceed 20–30% ($\varepsilon_1{}^P = 0.2$–0.3). Hence, if the volume does not to change, Eq. 4.11b shows that the incompressibility condition appears as a linear sum:

$$e_l{}^P + e_w{}^P + e_t{}^P = 0 \qquad (4.11c)$$

The approximation involved in Eq. 4.11c by adopting the engineering measure of strain is removed when, conventionally, Hencky's *natural* or *logarithmic* plastic strain definition is applied to the permanent deformation. The three natural strains appear as:

$$\varepsilon_l{}^P = \ln (l/l_o) = \ln(1 + \Delta l/l_o) = \ln(1 + e_l{}^P), \qquad (4.12a)$$

$$\varepsilon_w{}^P = \ln(w/w_o) = \ln(1 + \Delta w/w_o) = \ln(1 + e_w{}^P) \qquad (4.12b)$$

$$\varepsilon_t{}^P = \ln(t/t_o) = \ln(1 + \Delta t/t_o) = \ln(1 + e_t{}^P) \qquad (4.12c)$$

when from Eq. 4.11a it follows:

$$\Delta V/V_o = \exp(\varepsilon_l{}^P + \varepsilon_w{}^P + \varepsilon_t{}^P) - 1 \qquad (4.13a)$$

and, therefore, when $\Delta V = 0$, Eq. 4.13a becomes:

$$\varepsilon_l{}^P + \varepsilon_w{}^P + \varepsilon_t{}^P = 0 \qquad (4.13b)$$

A further advantage of the natural strain is that it allows the final plastic strain to be summed meaningfully from a series of increments. Say we wish to find the net plastic strain arising from a series of four loading-unloading cycles, where the initial length l_0, following each unloading, has become l_1, l_2, l_3, and l_4. The natural strain increment is referred to the current length at the start of each cycle so that the net axial plastic strain becomes:

$$\Sigma \varepsilon^P = \ln (l_1/l_0) + \ln (l_2/l_1) + \ln (l_3/l_2) + \ln (l_4/l_3) = \ln (l_4/l_0) \qquad (4.14a)$$

If the engineering strains are used to express the deformation within each cycle as e_1, e_2, e_3 and e_4, the summation in Eq. 4.14a becomes:

$$\Sigma \varepsilon^P = \ln(1 + e_1{}^P) + \ln(1 + e_2{}^P) + \ln(1 + e_3{}^P) + \ln(1 + e_4{}^P)$$

$$= \ln[(1 + e_1{}^P)(1 + e_2{}^P)(1 + e_3{}^P)(1 + e_4{}^P)] \qquad (4.14b)$$

However, as these engineering strains are not all referred to the original length it cannot be said that the total deformation is $e_1 + e_2 + e_3 + e_5$.

In a tension test the two lateral natural strains $\varepsilon_w{}^P$ and $\varepsilon_t{}^P$ in Eq. 4.13b may be assumed equal for an initially isotropic material, as found within the equi-axed grain structure provided by an annealing treatment. Here $\varepsilon_w^P = \varepsilon_t^P$ in Eq. 4.13b, from which it follows that:

$$\varepsilon_w^P/\varepsilon_t^P = 1 \quad \text{and} \quad \varepsilon_w^P/\varepsilon_l^P = \varepsilon_t^P/\varepsilon_l^P = -\tfrac{1}{2} \qquad (4.15a,b)$$

In heavily as-rolled material the grains are elongated in alignment with the rolling direction. Equations 4.15a,b are unlikely to apply to a structure with this preferred orientation. The degree of anisotropy is quantified by the amount the width to thickness strain ratio in Eq. 4.15a deviates from unity. In sheet-metal forming this strain ratio becomes an important parameter (the r-value), as the indicator for assessing the sheet's capacity for thinning. For example, rolled sheet steels with $r > 1$ have a greater capacity for thinning than aluminium ($r < 1$) when being formed under various combinations of biaxial, in-plane stress states.

4.1.4 Bulk Modulus, K

The cubic and hexagonal cells within polycrystalline metals change their volumes, elastically, as the van der Waal forces that maintain their atomic positions are influenced by external forces. The bulk modulus is the ratio between the mean or hydrostatic stress σ_m and volumetric strain, $\Delta V/V$. This is:

$$K = \sigma_m / (\Delta V/V) \qquad (4.16a)$$

A solid does change its volume under elastic straining and therefore K may be regarded a comparative measure of compressibility where high K values signify low volume change and vice versa.

(a) Uniform Pressure

A hydrostatic stress state arises directly when a solid is subjected to fluid pressure p (see Fig. 4.7).

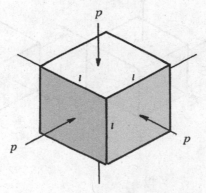

Figure 4.7 Hydrostatic pressure

Alternatively, a hydrostatic tension can arise from applying three equal, perpendicular loads to a cube but this is difficult to achieve in practice. The ratio in Eq. 4.16a remains positive irrespective of whether σ_m is tensile or compressive because ΔV will change sign accordingly. For example, if a strain ε exists within the sides of a unit cube under fluid pressure, then ε becomes the reduction in length of each side. The change in volume is:

$$\Delta V = Strained\ Volume - Initial\ Volume$$

from which the volumetric strain becomes

$$\Delta V/V = (1 - \varepsilon)^3 - 1 \approx -3\varepsilon \qquad (4.16b)$$

where terms in ε^2 and ε^3 have been ignored, justifiably, as these can never be greater than the strain at the proportional limit. The latter is found from dividing the limiting stress by the modulus, e.g. for steel the ratio becomes $300/210000 = 1/700$ and for aluminium alloy $150/75000 = 1/500$. Hence, setting $\sigma_m = -p$ and substituting from Eq. 4.16b, the bulk modulus Eq. 4.16a appears in the form that would be applied in the experimental determination of K:

$$K = p/3\varepsilon \qquad (4.16c)$$

where, here, as both p and ε are negative under compression their ratio is positive.

(b) Principal Stress State

A principal, tri-axial stress state refers to three normal stresses that act upon the sides of an element in the absence of shear stress (see Fig. 4.8a). When the principal stresses are unequal, conforming to the convention: $\sigma_1 > \sigma_2 > \sigma_3$, a hydrostatic stress exists indirectly. Thus, Fig. 4.8a may be decomposed into the *hydrostatic* and *deviatoric* stress components shown in Figs 4.8b and c.

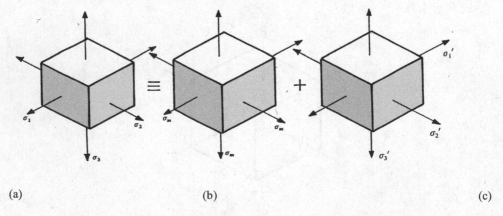

(a)   (b)   (c)

Figure 4.8 Hydrostatic and deviatoric components of a tri-axial stress

The hydrostatic stress σ_m, which is responsible for elastic compressibility, is the mean of the three normal stresses:

$$\sigma_m = \tfrac{1}{3}(\sigma_1 + \sigma_2 + \sigma_3) \tag{4.17a}$$

The deviatoric stress components σ_1', σ_2' and σ_3' are what remains of the normal stress when the mean stress has been removed. The *stress deviator* components are:

$$\sigma_1' = \sigma_1 - \sigma_m, \quad \sigma_2' = \sigma_2 - \sigma_m \text{ and } \sigma_3' = \sigma_3 - \sigma_m \tag{4.17b–d}$$

which are responsible for distortion without volume change. Again, taking a unit cube in Fig. 4.8a, the length changes to its three sides become the total elastic strains ε_1, ε_2 and ε_3 associated with the normal stress. It follows that the volumetric strain is:

$$\Delta V/V = (1 + \varepsilon_1)(1 + \varepsilon_2)(1 + \varepsilon_3) - 1 \approx \varepsilon_1 + \varepsilon_2 + \varepsilon_3 \tag{4.18}$$

Substituting Eqs 4.17a and 4.18 into Eq. 4.16a provides the bulk modulus for this stress state:

$$K = \frac{\sigma_1 + \sigma_2 + \sigma_3}{3(\varepsilon_1 + \varepsilon_2 + \varepsilon_3)} = \frac{\sigma_{kk}}{3\varepsilon_{kk}} \tag{4.19}$$

Table 4.4 Elastic constants for metals and metal alloys

Material	E (GPa)	$G\,(=\mu)$ (GPa)	K (GPa)	v	λ
Al and alloys	70-75	28	66	0.31	61.5
Antimony	78	29	87	0.35	67.5
Bismuth	32	12	31	0.33	23
Carbon & alloy steels	207	81	157	0.28	103
Cast iron	97	41	51	0.18	49
Copper	110	41.3	107.5	0.33	91.5
Brass	103.5	39	101.5	0.33	76
Bronze	117	44.8	102.6	0.31	73
Cobalt	206	79	172	0.30	119
Copper	95-120	40	88	0.31	91.5
Gold	79	27	66	0.30	48
Invar	145	56	121	0.30	83.5
Lead	17	6	28	0.40	24
Magnesium	44	17	35	0.29	23.5
Manganese	120	45	118	0.33	88
Platinum	150	54	208	0.38	172
Ni and alloys	200	80	184	0.31	124
Silver	78	29	87	0.35	67.5
Stainless steel (18/8)	200	77	167	0.30	115.5
Tin	40	15	39	0.33	29
Titanium and alloys	110-115	45	104	0.31	89.5
Tungsten	400	157	290	0.27	184.5
Zinc and alloys	90-100	38	63	0.25	37.5

Values of E, G, K and v for metals and some common engineering alloys are given in Table 4.4. Lamé's elastic constants λ and μ also appear in Table 4.4 (see § 4.3).

4.2 Analysis of Pure Shear

Figure 4.9a shows the state of stress produced in a material element under a shear force. This mode refers to a point in a block under direct shear force, as in Fig. 3.4, a beam cross section, or at a point on the surface of a shaft under torsion.

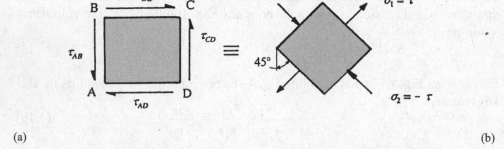

(a)   (b)

Figure 4.9 Analysis of stress in pure shear

The shear stresses τ_{AD} and τ_{BC}, acting on opposite parallel faces AD and BC, result from the shear force. The uniform shear stress acting upon these faces is simply the shear force divided by the surface area beneath it. For a beam subjected to a transverse force on its top surface a similar calculation provides the average of a vertical shear stress distribution through the depth. Here the shear stress varies from zero at the edges to a maximum at the central (neutral) axis. For a solid shaft under torsion, Eq. 4.3a shows that the shear stress varies from its maximum all around the outer surface to zero at the shaft axis. Given that the shear stress arises with a variation in each case it is customary to refer the following analyses to a point within the respective distributions.

4.2.1 Complementary Shear

The shear stresses cannot exist without the equilibrating actions occurring on adjacent faces AB and CD, known as *complementary shear*. The complementary shear stresses τ_{AB} and τ_{CD} are found from applying the moment equilibrium condition to point A. For an element of thickness t this condition gives:

$$\tau_{BC} \times (BC \times t) \times AB = \tau_{CD} \times (CD \times t) \times AD$$

Since $\tau_{AB} = \tau_{CD}$ and $\tau_{AD} = \tau_{BC}$ it follows that

$$\tau_{AB} = \tau_{AD} \text{ and } \tau_{BC} = \tau_{CD}$$

This shows that the shear stresses must all be equal (simply labeled as τ hereafter).

4.2.2 Principal Stresses and Planes

Within the element shown in Fig. 4.9a there exist two perpendicular directions, with a particular inclination, upon which shear stress disappears completely. To show this we need to consider, firstly, the state of stress induced along any oblique direction BT, inclined at θ to AB in Fig. 4.10a. The stress state for BT consists of a normal stress σ_θ combined with a shear stress τ_θ as shown. The conversion from stress to force and their resolution into components parallel and perpendicular to BT is given in Fig. 4.10b.

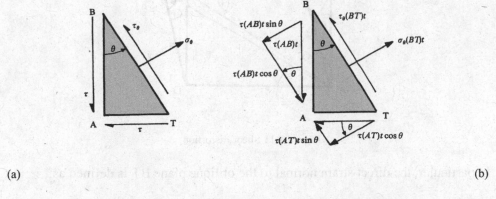

(a)   (b)

Figure 4.10 State of stress on an oblique plane in simple shear

For example, the shear force along side AB is $\tau \times$ AB $\times t$, which resolves into two force components: $(\tau \times$ AB $\times t)\cos\theta$ and $(\tau \times$ AB $\times t)\sin\theta$, as shown. The following two force equilibrium equations apply, parallel and perpendicular to BT, noting that: AB = BT$\cos\theta$ and AT = BT$\sin\theta$:

$$\sigma_\theta \,(\text{BT})\, t = \tau\,(\text{AB})\, t \sin\theta + \tau\,(\text{AT})\, t \cos\theta$$
$$= \tau\, t \,(\text{BT}\cos\theta)\sin\theta + \tau\, t \,(\text{BT}\sin\theta)\cos\theta$$
$$= 2\tau\, t \,(\text{BT}\sin\theta \cos\theta)$$
$$\therefore\ \sigma_\theta = \tau \sin 2\theta \qquad (4.20a)$$

$$\tau_\theta \,(\text{BT})\, t = \tau\,(\text{AB})\, t \cos\theta - \tau\,(\text{AT})\, t \sin\theta$$
$$= \tau\, t \,(\text{BT}\cos\theta)\cos\theta - \tau\, t \,(\text{BT}\sin\theta)\sin\theta$$
$$= \tau\, t \,(\text{BT})(\cos^2\theta - \sin^2\theta)$$
$$\therefore\ \tau_\theta = \tau \cos 2\theta \qquad (4.20b)$$

Equations 4.20a,b show that upon the plane for which $\theta = 45°$:

$$\sigma_{45°} = \tau \text{ and } \tau_{45°} = 0$$

and upon the plane for which $\theta = 135°$:

$$\sigma_{135°} = -\tau \text{ and } \tau_{135°} = 0$$

It follows that pure shear is equivalent to an element with a 45° orientation (see Fig. 4.9b) where direct tensile and compressive stresses, $\sigma_1 = \tau$ and $\sigma_2 = -\tau$ respectively, act upon its sides in the absence of shear stress. By definition, these are called *principal stresses* and *principal planes*.

The combined stress state that exists upon any non-principal plane BT is responsible for associated normal and shear strains, ε_θ and γ_θ respectively. Figure 4.11 shows the shear distortion produced in the element ABCD.

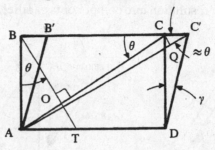

Figure 4.11 Shear distortion

In particular, the direct strain normal to the oblique plane BT is defined as:

$$\varepsilon_\theta = (AC' - AC)/AC \approx (CC'\cos\theta)/AC \tag{4.21a}$$

The approximation $\theta \approx \angle AC'B$ holds for small elastic distortions. The 'applied' shear strain referred to the planes AB and CD is, by definition: $\gamma = \tan CDC' = CC'/CD$. Hence, for small elastic distortion the shear strain equals the distortion angle in radians, i.e. $\gamma \approx \angle CDC'$ (rad). Now $CC' = CD \times \gamma = (AC\sin\theta)\,\gamma$, when Eq. 4.21a becomes:

$$\varepsilon_\theta = (AC\sin\theta)(\gamma\cos\theta)/AC$$

$$\varepsilon_\theta = (\gamma/2)\sin2\theta \tag{4.21b}$$

The shear strain γ_θ for BT is the change in the right angle BOC, components of which are the rotations to the plane BT and to its normal AC. The latter identifies with the rotation angle CAC', expressed in radians, as follows:

$$\angle CAC' = CQ/AC \approx (CC'\sin\theta)/AC \approx \gamma\sin^2\theta \tag{4.22a}$$

Replacing θ in Eq. 4.22a with $\theta + \pi/2$, the rotation angle for BT becomes $\gamma\sin^2(\theta + \pi/2) = -\gamma\cos^2\theta$. The sum of these two rotations is associated with the shear strain γ_θ for the oblique plane BT as:

$$\gamma_\theta = \gamma(\sin^2\theta - \cos^2\theta) = -\gamma\cos2\theta \tag{4.22b}$$

in which the minus sign implies that BT has rotated in a sense opposite to the rotation from AC to AC' shown in Fig. 4.11. Equation 4.22b shows that $\gamma_\theta = 0$ when

$\theta = 45°$ and $135°$, this confirming that the principal planes in Fig. 4.9b remain perpendicular. Correspondingly, Eq. 4.21b provides the principal strains for these directions, i.e. when $\theta = 45°$: $\varepsilon_{45°} = \gamma/2$ and when $\theta = 135°$: $\varepsilon_{135°} = -\gamma/2$.

The equivalence between the two elements in Figs 4.9a,b that the pure shear analysis reveals is important when deriving the following dependencies between the elastic constants.

4.3 Relationships Between Elastic Constants

The historical overview that began this book mentioned that the engineering elastic constants E, G, K and v are not independent of one another because their number is not minimal. Lamé's mathematical theory of elasticity does provide a minimum of just two independent elastic constants, λ and μ, to which the engineering constants are connected as follows:

$$E = \frac{\mu(2\mu + 3\lambda)}{(\mu + \lambda)}, \quad v = \frac{\lambda}{2(\mu + \lambda)} \qquad (4.23a,b)$$

$$G = \mu, \quad K = \frac{2}{3}\mu + \lambda \qquad (4.23c,d)$$

In fact, as Lamé's constant μ identifies with G directly, it can be seen that his second constant λ connects to all the remaining engineering constants E, v and K. The inverse relationships provide a triple choice for calculating λ when it is to be found from the known engineering constants:

$$\lambda = \frac{1}{3}(3K - 2G) = \frac{2vG}{1 - 2v} = \frac{G(E - 2G)}{3G - E} \qquad (4.24a,b,c)$$

It is worth noting that λ and μ do still appear in elasticity theory presented with a mathematical bias. Of the various relationships between the two sets of constants given above, those in Eqs 4.23c and 4.24a, especially, allow for the simplest interchange between them, as shown in Table 4.4 for common metals.

4.3.1 E, G and v

The pure shear analysis given in § 4.2 above shows that an equivalent, principal element exists with a 45° orientation (see Figs 4.9a,b). Recall that because there is no angular change between adjacent sides of Fig. 4.9b, they are coincident with the *principal directions* for pure shear. Using subscript 1 for the major principal direction (tensile) and 2 for the minor principal direction (compressive) the principal stresses are written as:

$$\sigma_1 = \tau \quad \text{and} \quad \sigma_2 = -\tau \qquad (4.25a,b)$$

Equations 4.25a,b show that the major and minor principal stresses are, respectively, tensile and compressive, both equal in magnitude to applied shear stress. The corresponding principal strains are:

$$\varepsilon_1 = \frac{\gamma}{2} \quad \text{and} \quad \varepsilon_2 = -\frac{\gamma}{2} \qquad (4.26\text{a,b})$$

Equations 4.26a,b reveal that the major and minor principal strains are, respectively, tensile and compressive, both equal in magnitude to one half the applied shear strain. The principal stress and strain for the $45°$ and $135°$ directions may now be related. The major principal strain ε_1 for the 1-direction (tensile) is composed of the sum of two parts: (i) a direct tensile strain σ_1/E from Eq. 4.1 and (ii) a lateral strain under σ_2, which is found from Eq. 4.6c as $-v\sigma_2/E$. Adding (i) and (ii):

$$\varepsilon_1 = \frac{1}{E}\left(\sigma_1 - v\sigma_2\right) \qquad (4.27\text{a})$$

Substituting from Eqs 4.25a,b and Eq. 4.26a into Eq. 4.27a will relate the principal stress and strain arising from pure shear:

$$\frac{\gamma}{2} = \frac{1}{E}\left[\tau - v(-\tau)\right] = (1+v)\frac{\tau}{E} \qquad (4.27\text{b})$$

Since $G = \tau/\gamma$ (see Eq. 3.8a) it follows that Eq. 4.27b will relate the three elastic constants E, G and v:

$$E = 2G(1+v) \qquad (4.28)$$

Here the reader should confirm Eq. 4.28 from applying the stress–strain relation to the 2-direction, i.e. $\varepsilon_2 = (\sigma_2 - v\sigma_1)/E$ and then examine here just how well the elastic constants given in Table 4.4 conform to Eq. 4.28.

4.3.2 E, K and v

Consider the linear (normal) strain ε, for one edge of the cubic element shown in Fig. 4.7. This strain is composed of three contributions from the uniform, tri-axial compression that prevails under the hydrostatic pressure: (i) a direct compressive strain $\varepsilon_{(i)} = p/E$ and (ii) two lateral tensile strains: $\varepsilon_{(ii)} = -vp/E$ (in which p is negative within Eq. 4.6c). The total strain $\varepsilon = \varepsilon_{(i)} + \varepsilon_{(ii)}$ becomes:

$$\varepsilon = \frac{p}{E} - \frac{vp}{E} - \frac{vp}{E} = \frac{p}{E}(1 - 2v) \qquad (4.29\text{a})$$

Substituting Eq. 4.29a into Eq. 4.16c:

$$K = \frac{p}{(3p/E)(1 - 2v)} \qquad (4.29\text{b})$$

Equation 4.29b shows:

$$E = 3K(1 - 2v) \qquad (4.30)$$

Given that K is generally expressed by Eq. 4.19, the section that follows this (see § 4.4) shows that Eq. 4.30 can also be proven from a principal tensile stress state (Fig. 4.8a) but this is more difficult to achieve in practice.

4.3.3 E G and K

Further relationships may be found between G, K and v by eliminating E and v in turn between Eqs 4.28 and 4.30. This gives:

$$G = \frac{3K(1 - 2v)}{2(1 + v)} = \frac{3KE}{(9K - E)} \qquad (4.31a,b)$$

Example 4.4 A metal bar, 50 mm diameter and 150 mm long, is subjected to a tensile force of 75 kN. If the change in the bar's length and diameter are 0.08 and -0.0065 mm, respectively, find its four elastic constants.

Let Δd and Δl be the change in diameter and length. Applying Eqs 4.1, 4.5a, 4.6b, 4.28 and 4.30:

$$E = \sigma/\varepsilon_1 = F\,l/A\,\Delta l$$

$$= (75 \times 10^3 \times 150) / [(\pi/4)(50)^2 \times 0.08]$$

$$= 71619.72 \text{ N/mm}^2 = 71.62 \text{ GPa}$$

$$v = -\varepsilon_2/\varepsilon_1 = -(\Delta d \times l)/(d \times \Delta l)$$

$$= -(-0.0065 \times 150) / (50 \times 0.08) = 0.244$$

$$G = E/2(1 + v)$$

$$= 71.62 / [2(1 + 0.244)] = 28.8 \text{ GPa}$$

$$K = E/3(1 - 2v)$$

$$= 71.62 / [3(1 - 0.488)] = 46.63 \text{ GPa}$$

Example 4.5 A cylindrical steel bar, 50 mm diameter and 300 mm long, is subjected to a fluid pressure of 800 bar. Find the change in volume and the modulus of rigidity, given that $E = 207$ GPa and $v = 0.28$.

Using Eq. 4.30, provides the bulk modulus for steel:

$$K = E/[3(1 - 2v)]$$

$$= 207/[3(1 - 0.56)] = 156.82 \text{ GPa}$$

Since the pressure is compressive Eq. 4.16a gives a reduction in volume :

$$\therefore \Delta V = - pV/K$$

$$= - 80 \times (\pi/4)(50)^2 \times 300/(156.82 \times 10^3)$$

$$= - 300.5 \text{ mm}^3$$

4.4 Extended Hooke's Law

4.4.1 Three Dimensions

The stress–strain relations given in Eqs 4.27a and 4.29a are particular reductions to the general 3D relation that applies to unequal, principal stresses: σ_1, σ_2 and σ_3. Each accompanying, principal strain is composed of one direct strain and two lateral strains. Thus, in Fig. 4.8a the principal strain ε_1 is the sum of the direct strain σ_1/E and two lateral strains: $-v\sigma_2/E$ and $-v\sigma_3/E$ aligned with the 1-direction. Similarly, the normal strains for the 2- and 3-directions are sums of their direct and lateral strain components. The three principal stress–strain relations become:

$$\varepsilon_1 = \frac{1}{E}[\sigma_1 - v(\sigma_2 + \sigma_3)] \qquad (4.32a)$$

$$\varepsilon_2 = \frac{1}{E}[\sigma_2 - v(\sigma_1 + \sigma_3)] \qquad (4.32b)$$

$$\varepsilon_3 = \frac{1}{E}[\sigma_3 - v(\sigma_1 + \sigma_2)] \qquad (4.32c)$$

The element may be expanded to a rectangular block where the principal stresses are uniform throughout its volume: $l_1 \times l_2 \times l_3$. Equations 4.32a–c provide the changes in length of its sides: $x_1 = \varepsilon_1 l_1$, $x_2 = \varepsilon_2 l_2$ and $x_3 = \varepsilon_3 l_3$. As an aid to memory, it can seen that Eqs 4.32b,c follow from rotating subscripts in Eq. 4.32a.

The volumetric strain within the denominator in Eq. 4.19 requires the addition of the three linear principal strains in Eqs 4.32a–c. This gives the their sum as:

$$\varepsilon_1 + \varepsilon_2 + \varepsilon_3 = \frac{(1 - 2v)}{E}(\sigma_1 + \sigma_2 + \sigma_3) \qquad (4.32d)$$

Substituting Eq. 4.32d into Eq. 4.19 it is seen how Eq. 4.30, connecting E, K and v, also applies to a tri-axial stress state. Note from Eq. 4.32d that when $v \rightarrow \frac{1}{2}$, then $\varepsilon_1 + \varepsilon_2 + \varepsilon_3 = 0$. This relationship applies to metals deforming plastically without volume change. A zero principal strain sum also applies to certain elastically incompressible materials, e.g. rubber and other elastomers, for which $K \rightarrow \infty$.

4.4.2 Two Dimensions

Figure 4.12 shows the reduction to a plane principal stress state arising from two-dimensional stretching of a thin plate.

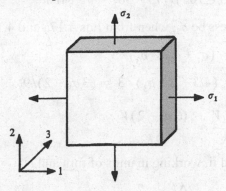

Figure 4.12 Plate under biaxial stress

Putting $\sigma_3 = 0$ in Eqs 4.32a–c gives the biaxial stress–strain reductions:

$$\varepsilon_1 = \frac{1}{E}(\sigma_1 - v\sigma_2) \tag{4.33a}$$

$$\varepsilon_2 = \frac{1}{E}(\sigma_2 - v\sigma_1) \tag{4.33b}$$

$$\varepsilon_3 = -\frac{v}{E}(\sigma_1 + \sigma_2) \tag{4.33c}$$

Equation 4.33c shows that a *through-thickness strain* is produced from the sum of lateral strains that accompany the in-plane stresses: σ_1 and σ_2. This is the *Poisson effect*, which appears in the thinning of the plate shown in Fig. 4.12. Solving Eqs 4.33a,b simultaneously, leads to the inverted forms:

$$\sigma_1 = \frac{E}{1-v^2}(\varepsilon_1 + v\varepsilon_2) \tag{4.34a}$$

$$\sigma_2 = \frac{E}{1-v^2}(\varepsilon_2 + v\varepsilon_1) \tag{4.34b}$$

Equations 4.34a,b are useful for calculating stress from strain measured by bonded gauges in alignment with the 1- and 2-directions (see Example 11.7, p. 456).

Example 4.6 The reduction in volume of a solid, rectangular block 50 mm × 75 mm × 100 mm is to be restricted to 1 mm³ under the action of three mutually perpendicular forces. Two of these forces are to be: 5 kN in tension normal to the 75 × 50 mm faces and 10 kN in compression normal to the 100 × 50 mm faces. Find the magnitude and sense of the maximum force that may act on the 100 × 75 mm faces and the change in length of each side. Take $K = 150$ GPa and $v = 0.27$.

The direct stresses are:

$$\sigma_2 = -(10 \times 10^3)/(50 \times 100) = -2 \text{ MPa}$$

$$\sigma_1 = (5 \times 10^3)/(75 \times 50) = 4/3 \text{ MPa}$$

Let the unknown stress be σ_3, when from Eqs 4.17a and 4.16a, respectively:

$$\sigma_m = (\sigma_1 + \sigma_2 + \sigma_3)/3$$
$$= (4/3 - 2 + \sigma_3)/3 = (3\sigma_3 - 2)/9 \qquad \text{(i)}$$

$$\therefore \Delta V = \frac{\sigma_m V}{K} = \frac{(3\sigma_3 - 2)V}{9K} \qquad \text{(ii)}$$

Then, from Eqs i and ii, working in units of mm and N:

$$\sigma_3 = 3K \times \frac{\Delta V}{V} + \frac{2}{3}$$

$$= 3 \times \frac{(150 \times 10^3) \times (-1)}{(100 \times 50 \times 75)} + \frac{2}{3}$$

$$= -1.2 + 0.667 = -0.533 \text{ MPa}$$

from which the third force is:

$$F_3 = \sigma_3 \times A$$

$$= -0.533(100 \times 75) = -4 \text{ kN (compression)}$$

Young's modulus follows from Eq. 4.30:

$$E = 3K(1 - 2v)$$

$$= 3 \times 150 \times 10^3(1 - 2 \times 0.27) = 207000 \text{ MPa}$$

From Eqs 4.32a–c, the strains are:

$$\varepsilon_1 = (1/E)[\sigma_1 - v(\sigma_2 + \sigma_3)]$$

$$= (10^{-3}/207)[1.333 - 0.27(-2 - 0.533)] = +9.75 \ \mu\varepsilon$$

$$\varepsilon_2 = (1/E)[\sigma_2 - v(\sigma_1 + \sigma_3)]$$

$$= (10^{-3}/207)[-2 - 0.27(1.333 - 0.533)] = -10.71 \ \mu\varepsilon$$

$$\varepsilon_3 = (1/E)[\sigma_3 - v(\sigma_1 + \sigma_2)]$$

$$= (10^{-3}/207)[-0.533 - 0.27(1.333 - 2)] = -1.71 \ \mu\varepsilon$$

where $\mu\varepsilon$ are *micro-strain* ($1\mu\varepsilon = 1 \times 10^{-6}$). Finally, the displacements in micro-metres follow as:

$$x_1 = \varepsilon_1 l_1 = (9.75 \times 0.10) = 0.975 \ \mu m$$

$$x_2 = \varepsilon_2 l_2 = -(10.71 \times 0.075) = -0.803 \ \mu m$$

$$x_3 = \varepsilon_3 l_3 = -(1.71 \times 0.05) = -0.0855 \ \mu m$$

Note that normally when principal stress magnitudes are known, their magnitudes are taken to obey: $\sigma_1 > \sigma_2 > \sigma_3$.

4.5 Thin-Walled Pressure Vessels

Cylindrical and spherical pressure vessels are said to be thin-walled when the ratio between the mean wall diameter and thickness is 20 or more. At these ratios it may be assumed that the wall is stressed biaxially in its own plane. Across the wall a radial stress variation does exist but the magnitude of this variation is negligibly small compared to the in-plane stresses. In a cylinder these stresses are referred to its axial and circumferential directions. In a sphere they refer to its meridional and circumferential directions. Thus, with the appropriate change in co-ordinates, Eqs 4.33a–c can be employed to convert each two-dimensional principal stress state into the corresponding strain state.

4.5.1 Cylinder

The dimensions of a cylindrical vessel will change under internal fluid pressure. Let the inner diameter d and the length l change by Δd and Δl, respectively. Axial and circumferential strains are defined in terms of polar co-ordinates (z and θ) as:

$$\varepsilon_z = \frac{\Delta l}{l} \tag{4.35a}$$

$$\varepsilon_\theta = \frac{\pi(d + \Delta d) - \pi d}{\pi d} = \frac{\Delta d}{d} \tag{4.35b}$$

The volumetric strain $\Delta V/V$ expresses the ratio between the change in internal volume to the original volume:

$$\Delta V/V = (\pi/4)[(d + \Delta d)^2 (l + \Delta l) - d^2 l]/(\pi/4)d^2 l$$

$$= [d^2 l + 2dl(\Delta d) + l(\Delta d)^2 + d^2(\Delta l) + 2d(\Delta d)(\Delta l) + (\Delta d)^2(\Delta l) - d^2 l]/(d^2 l)$$

$$\approx [2dl(\Delta d) + d^2(\Delta l)]/(d^2 l)$$

$$= 2(\Delta d/d) + (\Delta l/l) \tag{4.36a}$$

Substituting from Eqs 4.35a,b within Eq 4.36a gives:

$$\frac{\Delta V}{V} = 2\varepsilon_\theta + \varepsilon_z \tag{4.36b}$$

The strains ε_θ and ε_z depend upon the biaxial stress state in the wall of the vessel. These principal strains follow from Eqs 4.33a,b in polar co-ordinates as follows

$$\varepsilon_\theta = \frac{1}{E}(\sigma_\theta - v\sigma_z) \tag{4.37a}$$

$$\varepsilon_z = \frac{1}{E}(\sigma_z - v\sigma_\theta) \tag{4.37b}$$

A closed, thin-walled cylindrical vessel containing fluid under pressure is a *statically determinate* problem. This means that the stresses in the wall may be found directly from force equilibrium. Figure 4.13a shows that the stress state in the plane of the wall consists of an axial tensile stress σ_z and a circumferential (or hoop) stress σ_θ.

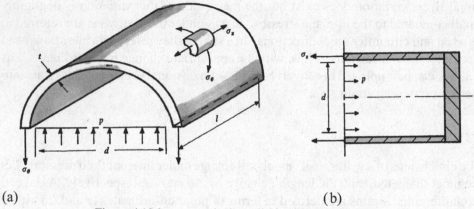

(a)   (b)

Figure 4.13 Stress state in the wall of a closed thin cylinder

Firstly, vertical force equilibrium must be satisfied upon the diameter plane $d \times l$, in Fig. 4.13a. Its area $d \times l$ is that projected from the semi-cylindrical area upon which p acts radially. We may then take p to act normally to this plane as shown to give a net force equivalent to that which would be found from resolving vertically the radial forces around the inner circumference. This force is balanced by the two vertical, circumferential wall forces exposed by the diameter plane:

$$2\sigma_\theta tl = pdl$$

$$\sigma_\theta = \frac{pd}{2t} \tag{4.38a}$$

The longitudinal section shown in Fig. 4.13b exposes the axial stress σ_z and the normal pressure as shown. The horizontal equilibrium condition is:

$$\pi dt\sigma_z = \frac{\pi}{4}d^2 p$$

$$\sigma_z = \frac{pd}{4t} \tag{4.38b}$$

Equations 4.38a,b provide the uniform, biaxial stress state (σ_θ, σ_z) everywhere in the wall from the known pressure, inner diameter and wall thickness. Note that the wall becomes stressed axially by the action of the pressure upon the end plates. There is no axial stress but the circumferential stress is present when the vessel is sealed by bore pistons that react the end-pressure.

Substituting Eqs 4.37a,b and 4.38a,b into Eq. 4.36b leads to a convenient expression for the increase in the volume capacity of a thin cylinder under internal pressure:

$$\frac{\Delta V}{V} = \frac{2}{E}(\sigma_\theta - v\sigma_z) + \frac{1}{E}(\sigma_z - v\sigma_\theta)$$

$$= \frac{1}{E}\left[\sigma_\theta(2 - v) + \sigma_z(1 - 2v)\right]$$

$$\therefore \frac{\Delta V}{V} = \frac{pd}{4tE}(5 - 4v) \qquad\qquad (4.39)$$

Example 4.7 A thin-walled closed cylinder with inner diameter 500 mm, thickness 10 mm and internal volume 0.5 m^3 is filled with a fluid at a gauge pressure of 20 bar (2 MPa). What volume of fluid would be ejected from the cylinder when the pressure is reduced to atmospheric assuming that the fluid is (a) incompressible and (b) compressible? Take $E = 210$ GPa, $v = 0.25$ for the cylinder and $K = 2500$ MPa for the compressible fluid. (1 bar = 10^5 Pa)

(a) Equation 4.39 gives the amount ΔV_c by which the cylinder recovers in volume following release of the internal pressure. This is equal to the amount of incompressible fluid ejected (in units of N and mm):

$$\Delta V_c = \frac{pdV}{4tE}(5 - 4v)$$

$$= \frac{2 \times 500 \times (0.5 \times 10^9) \times (5 - 4 \times 0.25)}{4 \times 10 \times (210 \times 10^3)} = 238.1 \times 10^3 \text{ mm}^3$$

(b)The amount ΔV_f, by which a compressible fluid is compressed under pressure, is found from setting $\sigma_m = -p$ in Eq. 4.16a:

$$\Delta V_f = -\frac{pV}{K}$$

$$= -\frac{2 \times (0.5 \times 10^9)}{2500} = -400 \times 10^3 \text{ mm}^3$$

On release of the pressure this additional volume is ejected to give a total volume spilled:

$$\Delta V = \Delta V_c + \Delta V_f = 638.1 \times 10^3 \text{ mm}^3$$

If units of $\underline{N}$ and $\underline{m}$ are preferred:

$$\Delta V_c = \frac{(20 \times 10^5) \times (500 \times 10^{-3}) \times 0.5 \times (5 - 4 \times 0.25)}{4 \times 0.010 \times (210 \times 10^9)} = 238.1 \times 10^{-6} \text{ m}^3$$

$$\Delta V_f = -\frac{(20 \times 10^5) \times 0.5}{(2500 \times 10^6)} = -400 \times 10^{-6} \text{ m}^3$$

4.5.2 Sphere

When the internal diameter d of a pressurised sphere increases by the amount Δd the volumetric strain is:

$$\frac{\Delta V}{V} = \frac{\frac{\pi}{6}\left[(d + \Delta d)^3 - d^3\right]}{\frac{\pi}{6}d^3} = \frac{d^3 + 3d^2\Delta d + 3d(\Delta d)^2 + (\Delta d)^3 - d^3}{d^3}$$

$$\approx 3d^2\left(\frac{\Delta d}{d^3}\right) = 3 \times \frac{\Delta d}{d} = 3\varepsilon_\theta \tag{4.40}$$

where the circumferential (hoop) strain $\varepsilon_\theta = \Delta d/d$ depends upon the stress state within the wall. Provided the ratio of inner diameter to thickness $d/t > 20$, then the radial stress may be neglected and a uniform, equi-biaxial stress state may be assumed, as shown in Fig. 4.14.

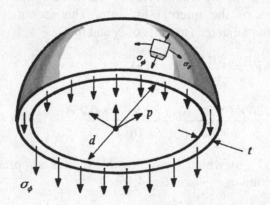

Figure 4.14 Thin-walled sphere

The thin-walled sphere is *statically determinate*. The meridional stress σ_φ exposed within its wall at the horizontal plane must remain in vertical equilibrium with the radial pressure acting upon the inner hemisphere. As with the cylinder, the resolution of the force vertically due to radial pressure is equivalent to a normal pressure upon the surface area projected upon the horizontal plane, i.e. a circle with diameter d. The equilibrium equation for vertical forces is thus simplified:

$$\pi d t \sigma_\varphi = \frac{\pi}{4} d^2 p$$

$$\sigma_\varphi = \frac{pd}{4t} \tag{4.41a}$$

In fact, Eq. 4.41a applies to the stress exposed by all horizontal sections taken above or below the diameter plane. Moreover, taking any vertical section to expose the hoop stress σ_θ in the plane of the wall, a similar horizontal force balance equilibrium applies, showing that the stress state within the wall is uniformly equi-biaxial, conforming to:

$$\sigma_\theta = \sigma_\varphi \tag{4.41b}$$

The hoop strain state follows from Eq. 4.33a as:

$$\varepsilon_\theta = \frac{1}{E}(\sigma_\theta - v\sigma_\varphi) \tag{4.42a}$$

Combining Eqs 4.40, 4.41a,b and 4.42a, the volumetric strain for a thin sphere is:

$$\frac{\Delta V}{V} = 3\varepsilon_\theta = \frac{3}{E}(\sigma_\theta - v\sigma_\theta)$$

$$\therefore \frac{\Delta V}{V} = \frac{3pd}{4tE}(1 - v) \tag{4.42b}$$

Example 4.8 A thin-walled cylinder has two hemispherical caps of the same material welded to its ends. Determine the sphere to cylinder thickness ratio that will ensure the same radial displacement at the weld when the vessel is pressurized. The cylinder is made from steel with dimensions: inner diameter 760 mm, wall thickness 37.5 mm and length 2 m. What additional volume of water must be pumped in to fill the vessel when the greatest stress in the wall is to be limited to 125 MPa? Take $E = 207$ GPa, $v = 0.25$ for steel and $K = 690$ MPa for water.

Equality of radial displacement requires equal hoop strains at the join between cylinder and sphere. With subscripts s and c for the sphere and cylinder, respectively, Eq. 4.33a gives:

$$\varepsilon_\theta = \frac{1}{E}(\sigma_\theta - v\sigma_z)_c = \frac{1}{E}(\sigma_\theta - v\sigma_\varphi)_s$$

Substituting from Eqs 4.38a,b and 4.41a,b

$$\left(\frac{pd}{4tE}\right)_c (2 - v)_c = \left(\frac{pd}{4tE}\right)_s (1 - v)_s$$

and given that p, d and E are common to both sphere and cylinder, their thickness ratio becomes:

$$\frac{t_s}{t_c} = \frac{1 - v}{2 - v} = \frac{1 - 0.25}{2 - 0.25} = \frac{3}{7}$$

It follows that $t_s = 16.07$ mm for $t_c = 37.5$ mm. By design the greatest (hoop) stress $\sigma_\theta = 125$ MPa is reached under the lesser of the two water pressures:

$$p = \left(\frac{2t\sigma_\theta}{d}\right)_c = \frac{2 \times 37.5 \times 125}{760} = 12.34 \text{ MPa}$$

$$p = \left(\frac{4t\sigma_\theta}{d}\right)_s = \frac{4 \times 16.07 \times 125}{760} = 10.57 \text{ MPa}$$

The total additional volume δV_t to be pumped in is given by the sum of the volume changes for the cylinder, sphere and fluid:

$$\Delta V_t = \Delta V_c + \Delta V_s + \Delta V_f \tag{i}$$

Substituting from Eqs 4.16a, 4.39 and 4.42b into Eq. i:

$$\Delta V_t = \frac{p d V_c}{4 t_c E}(5 - 4v) + \frac{3 p d V_s}{4 t_s E}(1 - v) + \frac{p V_t}{K}$$

where:
$$= \frac{pd}{4 t_c E}\left[(5 - 4v)V_c + 7(1 - v)V_s\right] + \frac{p V_f}{K} \tag{ii}$$

$$V_c = \frac{\pi}{4} \times 760^2 \times 2000 = 907.29 \times 10^6 \text{ mm}^3$$

$$V_s = \frac{\pi}{6} d^3 = \frac{\pi}{6} \times 760^3 = 229.85 \times 10^6 \text{ mm}^3$$

$$V_f = V_c + V_s = 1137.14 \times 10^6 \text{ mm}^3$$

Substituting these values into Eq. ii, for the additional volume of water required:

$$\Delta V_t = \left(\frac{10.57 \times 760 \times 10^6}{4 \times 37.5 \times 207 \times 10^3}\right)\left[907.29 + (7 \times 0.75 \times 229.85)\right] + \left(\frac{10.57 \times 1137.14 \times 10^6}{690}\right)$$

$$= 258.72\,(907.29 + 1206.71) + (17.42 \times 10^6)$$

$$= (0.547 + 17.42) \times 10^6 \text{ mm}^3 = 17.97 \times 10^6 \text{ mm}^3$$

Clearly, the final term shows that water compressibility is the dominant contributor. Note that the cylinder design pressure is altered when based upon a von Mises yield criterion (see § 12.1.5, p. 492). Given a safe working stress σ_w, the Mises criterion (Eq. 12.11b) accounts for the cylinder wall's biaxial stress state (σ_θ, σ_z) in the following manner:

$$\sigma_\theta^2 - \sigma_\theta \sigma_z + \sigma_z^2 = \sigma_w^2$$

and substituting from Eqs 4.38a,b:

$$\left(\frac{pd}{t}\right)^2\left(\frac{1}{4} - \frac{1}{8} + \frac{1}{16}\right) = \sigma_w^2$$

giving a greater operating pressure

$$p = \frac{4t\sigma_w}{\sqrt{3}d} = \frac{4 \times 37.5 \times 125}{\sqrt{3} \times 760} = 14.24 \text{ MPa} = 142.4 \text{ bar}$$

The more conservative cylinder pressure $p = 123.4$ bar, given above, corresponds to the Tresca yield criterion in which Eq. 12.5a gives $\sigma_\theta - \sigma_r = Y$. In taking $\sigma_r = 0$, Tresca's pressure prediction becomes based upon the limiting hoop stress.

The sphere pressure $p = 105.7$ bar would not be influenced by applications of these yield criteria since they all require stress differences. Thus, as the lower limiting sphere pressure dictates this design, the above solution remains valid.

4.5.3 Jointed Vessels

When a cylindrical or spherical pressure vessel is fabricated from riveted or welded plate the design of the joint is based upon the working stresses as they would be calculated from Eqs 4.38a,b and 4.41a,b. Safety factors and joint efficiencies, defined in Eqs 3.5, 3.9 and 3.10 of the previous chapter, are employed to avoid tearing of the plate and shearing of the rivets. Note, in applying Eq. 3.10, the least value of the failure load is determined from tearing of the plate along the joint:

$$\textit{Joint Efficiency} = \textit{Tearing Strength of Joint}/\textit{UTS of Plate} \qquad (4.43)$$

Example 4.9 A cylindrical boiler, welded from plates 15 mm thick, is to withstand a gauge pressure of 15 bar. If the efficiencies of the longitudinal and circumferential welded joints are 85% and 50% respectively, calculate the boiler diameter using a factor of safety $S = 4$, when for the plate material $UTS = 340$ MPa.

Equation 4.43 gives the tearing strength (TS) of a welded joint as:

$$TS = \textit{Joint Efficiency} \times UTS \qquad (i)$$

Now, from Eq. 3.5, the working stress becomes:

$$\sigma_w = \frac{TS}{S} \qquad (ii)$$

Combining Eqs i and ii:

$$\sigma_w = \textit{Joint Efficiency} \times \frac{UTS}{S} \qquad (iii)$$

Equation iii gives the working stress level for circumferential tension which lies normal to the longitudinal joint:

$$(\sigma_\theta)_w = 0.85 \times \frac{340}{4} = 72.25 \text{ MPa}$$

when from Eq. 4.38a:

$$d = \frac{2t(\sigma_\theta)_w}{p} = \frac{2 \times 15 \times 72.25}{1.5} = 1445 \text{ mm}$$

The axial stress acts normally to the circumferential joint, when Eq. iii gives its working level as:

$$(\sigma_z)_w = 0.5 \times \frac{340}{4} = 42.5 \text{ MPa}$$

and from Eq. 4.38b:

$$d = \frac{4t(\sigma_z)_w}{p} = \frac{4 \times 15 \times 42.5}{1.5} = 1700 \text{ mm}$$

The lesser diameter of 1.445 m governs the vessel's design capacity. This means that the circumferential stress attains its allowable working level but the axial stress does not. The latter is re-calculated at $(\sigma_z)_w = 36.12$ MPa. In fact, the overall safety factor for the cylinder exceeds 4 when based upon a *von Mises effective stress* (see Example 4.8):

$$\sigma_w = \sqrt{(\sigma_\theta)_w^2 - (\sigma_\theta)_w(\sigma_z)_w + (\sigma_z)_w^2} = 62.57 \text{ MPa}$$

$$S = \frac{UTS}{\sigma_w} = \frac{340}{62.57} = 5.43$$

4.6 Thick-Walled Cylinder

When $d/t < 20$ a pressure vessel is deemed to be thick walled. The full stress analysis of the thicker vessel under pressure must account for the variation in radial stress, σ_r, through its wall. Consequently, it will now be shown that the stress state for a point within the wall of both cylindrical and spherical vessels is 3D (triaxial). This state can arise from internal or external pressures acting separately or in combination.

4.6.1 Stress State

The triaxial stress within the wall of a thick-walled vessel cylinder consists of a radial stress together with hoop and axial stresses. This state is given in polar co-ordinates r, θ and z, so that the three stresses become σ_r, σ_θ and σ_z, respectively. The triaxial stress state at a point in the wall of thick cylinder under an internal pressure p, is shown in Fig. 4.15a. These are principal stresses since no shear stress is present on each face of the cylindrical element used to represent the point. The radial stress σ_r and the hoop stress σ_θ vary with radius through the wall but the axial stress σ_z is a constant.

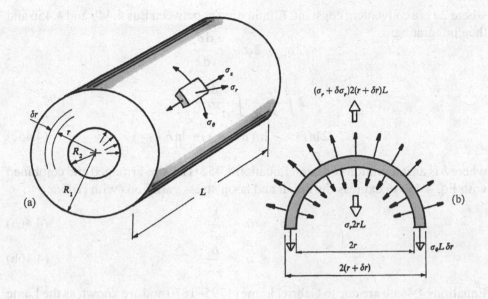

Figure 4.15 Thick-walled pressurised cylinder

4.6.2 Equilibrium and Compatibility

Consider, firstly, the vertical equilibrium requirement about a horizontal diameter for a cylindrical shell element of radial thickness δr in Fig. 4.15b. While the radial stress σ_r varies with r as shown the hoop stress σ_θ does not vary with θ. This gives the required force balance:

$$2\sigma_\theta \delta r L + \sigma_r(2rL) = (\sigma_r + \delta\sigma_r) \times 2(r + \delta r)L \qquad (4.44a)$$

in which σ_r (assumed tensile) acts on the inner and outer projected areas. The latter become rectangular areas $2rL$ and $(2r + \delta r)L$, respectively, upon a section plane through the horizontal diameter. Equation 4.44a reduces to a *radial equilibrium equation*:

$$\sigma_\theta - \sigma_r = r\,\frac{d\sigma_r}{dr} \qquad (4.44b)$$

The two stresses appearing in Eq. 4.44b may be separated knowing that the longitudinal stress, σ_z, is uniform in the length and independent of r. This means that every plane cross section remains plane when the cylinder is under pressure. Consequently the longitudinal (axial) strain, ε_z is also uniform in the length and cross section. Equation 4.32c expresses this in polar co-ordinates as:

$$\varepsilon_z = \frac{1}{E}\left[\sigma_z - v(\sigma_\theta + \sigma_r)\right] = constant \qquad (4.45a)$$

$$\therefore\quad \sigma_\theta + \sigma_r = \frac{(\sigma_z - \varepsilon_z E)}{v} = 2a \qquad (4.45b)$$

where $2a$ is a convenient constant. Eliminating σ_θ between Eqs 4.44b and 4.45b and then integrating:

$$2\sigma_r = 2a - r\frac{d\sigma_r}{dr}$$

$$2\int \frac{dr}{r} = \int \frac{d\sigma_r}{a - \sigma_r}$$

$$2\ln r = -\ln(a - \sigma_r) + \ln b \qquad (4.45c)$$

where b is an integration constant. Equation 4.45c is now re-arranged and combined with Eq. 4.45b to express the radial and hoop stress variations with radius:

$$\sigma_r = a - \frac{b}{r^2} \qquad (4.46a)$$

$$\sigma_\theta = a + \frac{b}{r^2} \qquad (4.46b)$$

Equations 4.46a,b are due to Gabriel Lamé (1795–1870 and are known as the Lamé equations. Unlike Lamé's elastic constants (see § 4.3) his thick-walled cylinder equations can never be superseded. Their uniqueness is ensured with their matching to the cylinder's equilibrium and compatibility conditions correctly.

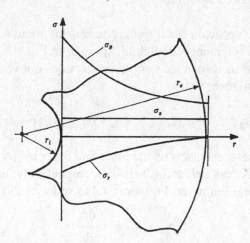

Figure 4.16 Stress variations through a cylinder's thick wall

The constants a and b in Eqs 4.46a,b are found from identifying the radial stress at the extreme radii with the internal and external pressure. In the case of internal pressure only, i.e. where only atmospheric pressure is applied to the outer radius, the cylinder's *boundary conditions* are: (i) $\sigma_r = -p$ for $r = r_i$ and (ii) $\sigma_r = 0$ for $r = r_o$. Substituting (i) and (ii) into Eq. 4.46a provides for a and b:

$$a = \frac{pr_i^2}{r_o^2 - r_i^2}, \qquad b = \frac{pr_i^2 r_o^2}{r_o^2 - r_i^2} \qquad (4.47a,b)$$

Substituting Eqs 4.47a,b into Eqs 4.46a,b leads to:

$$\sigma_r = \frac{p}{R^2 - 1}\left(1 - \frac{r_o^2}{r^2}\right), \qquad \sigma_\theta = \frac{p}{R^2 - 1}\left(1 + \frac{r_o^2}{r^2}\right) \qquad \text{(4.48a,b)}$$

where $R = r_o/r_i$. Equations 4.48a,b are the most common form of the Lamé equations, providing radial and hoop stress distributions shown in Fig. 4.16. It is seen that the hoop stress attains its maximum in tension at the bore. The radial stress attains its maximum in compression at the bore and falls to zero at the outer radius, these extreme values matching the gauge pressures that apply to the two radii (the boundary conditions). Here we should note that as the radial stress was taken to be tensile for the derivation but it must be equated to the negative of each pressure as the following example shows.

Example 4.10 Determine the maximum allowable stress for pressure piping, 150 mm i.d. and 200 mm o.d. rated for an internal pressure of 500 bar. Determine, for a safety factor of 4, the safe internal pressure for a second pipe of the same material and internal diameter but with a 40 mm wall thickness.

For the first pipe, the boundary conditions are: (a) $\sigma_r = -p = -50$ MPa for $r = 75$ mm and (b) $\sigma_r = 0$ for $r = 100$ mm. Substituting (a) and (b) into Eq. 4.46a:

$$-50 = a - \frac{b}{75^2} \qquad \text{(i)}$$

$$0 = a - \frac{b}{100^2} \qquad \text{(ii)}$$

Solving Eqs i and ii gives: $a = 64.286$ and $b = 642.86 \times 10^3$. The maximum allowable stress is the hoop stress in the bore, as found from setting $r = 75$ mm in Eq. 4.46b:

$$\sigma_\theta = a + \frac{b}{r^2} = 64.29 + \frac{642.86 \times 10^3}{75^2} = 178.6 \text{ MPa}$$

Hence, the safe working hoop stress for the *both pipes* becomes $\sigma_\theta = 178.6/4 = 44.65$ MPa. To find the corresponding internal pressure we must reapply Eqs 4.46a,b with new constants a' and b':

$$44.65 = a' + \frac{b'}{75^2} \qquad \text{(iii)}$$

$$0 = a' - \frac{b'}{100^2} \qquad \text{(iv)}$$

Solving Eqs iii and iv leads to: $a' = 16.074$ and $b' = 160.74 \times 10^3$. Equation 4.46a yields the corresponding internal pressure $(p = -\sigma_r)$ for $r = 75$ mm, as:

$$-p = a' - \frac{b'}{r^2} = 16.074 - \frac{160.74 \times 10^3}{75^2} = -12.5 \text{ MPa}$$

$$\therefore p = 125 \text{ bar}$$

For the second pipe Eqs iii and iv are modified to:

$$44.65 = a'' + \frac{b''}{75^2} \tag{v}$$

$$0 = a'' - \frac{b''}{115^2} \tag{vi}$$

Solving Eqs v and vi leads to: $a'' = 13.32$ and $b'' = 176.16 \times 10^3$. Equation 4.46a provides the corresponding internal pressure ($p = -\sigma_r$) for $r = 75$ mm:

$$-p = a'' - \frac{b''}{r^2} = 13.32 - \frac{176.16 \times 10^3}{75^2} = -18 \text{ MPa}$$

$$\therefore p = 180 \text{ bar}$$

4.6.3 End Conditions

Finally, the third, axial stress σ_z will depend upon the cylinder's *end condition*. For a *closed-end cylinder* the axial stress is found from the equilibrium equation between the opposing horizontal forces within the wall and acting upon the end plate (similar to Fig. 4.13b):

$$\sigma_z \pi (r_o^2 - r_i^2) = p \pi r_i^2$$

which gives:

$$\sigma_z = \frac{p r_i^2}{(r_o^2 - r_i^2)} = \frac{p}{R^2 - 1} \tag{4.49a}$$

This constant axial stress, shown in Fig. 4.16, may also be written from Eqs 4.46a,b and 4.48a,b as

$$\sigma_z = a = \frac{1}{2} \left(\sigma_\theta + \sigma_z \right) \tag{4.49b}$$

If the cylinder length is prevented from expanding by rigid end closures then σ_z must be found from setting $\varepsilon_z = 0$ in Eq. 4.45a. The cylinder is said to be under *plane strain* for which the axial stress in Eqs 4.45a,b becomes:

$$\sigma_z = v(\sigma_\theta + \sigma_r) = 2av \tag{4.50a}$$

An *open-end cylinder* arises when the axial force is reacted by pistons that compress the fluid within the cylinder bore.

The cylinder wall is then unstressed axially:

$$\sigma_z = 0 \qquad (4.50b)$$

but the radial and circumferential stresses are again given by the Lamé Eqs 4.46a,b. The latter also apply to a cylinder in which internal pressure p is combined with a separate axial force, F. Equations 4.46a,b continue to provide σ_θ and σ_r but the net axial stress becomes a sum due to p and F. For example, when a tensile force is applied to closed-end pressurised cylinder:

$$\sigma_z = \frac{pr_i^2}{(r_o^2 - r_i^2)} + \frac{F}{\pi(r_o^2 - r_i^2)} \qquad (4.50c)$$

4.6.4 Strain-Displacement Relationships

To derive the cylinder's strain-displacement relations consider an element in its cross section in Fig. 4.17, whose original side lengths are δr and $r\delta\theta$ within polar co-ordinates.

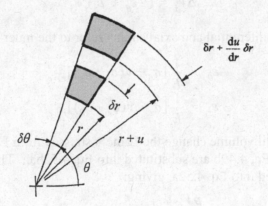

Figure 4.17 Displacements for a cylindrical element

An element of the radius δr changes its length to $\delta r + (du/dr)\delta r$. Hence its radial strain ε_r is the ratio between its radial extension and δr. This is:

$$\varepsilon_r = \frac{[\delta r + (du/dr)\delta r] - \delta r}{\delta r} = \frac{du}{dr} \qquad (4.51a)$$

Because the current radius has increased to $r + u$, then an element of the original circumference $r\delta\theta$ will have increased its length to $(r + u)\delta\theta$. The circumferential strain ε_θ is the ratio between the increase in this length and the original circumferential length:

$$\varepsilon_\theta = \frac{(r + u)\delta\theta - r\delta\theta}{r\delta\theta} = \frac{u}{r} \qquad (4.51b)$$

Because these two strains in Eqs 4.51a,b are seen to depend upon a single displacement a *compatibility relation* must exist between them:

$$\varepsilon_r = \frac{d}{dr}(r\varepsilon_\theta) = r\frac{d\varepsilon_\theta}{dr} + \varepsilon_\theta \qquad (4.51c)$$

The axial strain, which occurs independently of radial expansion, is the ratio between the cylinder's length change (when it occurs) and its original length:

$$\varepsilon_z = \Delta l / l \qquad (4.51d)$$

The length change is influenced by the cylinder's end condition that prevails (see § 4.6.3). Axial strain may not appear at all under plane strain though axial stress does, where Eq. 4.50a applies.

4.6.5 Change in Interval Volume

Equations 4.36a,b provide a close approximation to the change in the cylinder's closed, internal volume V_i when it contains a fluid at pressure:

$$\Delta V_i = (2\varepsilon_\theta + \varepsilon_z)_i V_i \qquad (4.52a)$$

in which the circumferential and axial strains refer to the inner radius:

$$\varepsilon_{\theta_i} = \frac{1}{E}[\sigma_\theta - v(\sigma_r + \sigma_z)]_{r=r_i} \qquad (4.52b)$$

$$\varepsilon_{z_i} = \frac{1}{E}[\sigma_z - v(\sigma_r + \sigma_\theta)]_{r=r_i} \qquad (4.52c)$$

To find the internal volume change the Lamé's stresses from Eqs 4.48a,b and the axial stress from Eq, 4.49b are substituted into Eq. 4.52b,c. The resulting strains are then substituted into Eq. 4.52a, giving:

$$\Delta V_i = \frac{pV_i}{E(R^2 - 1)}\left[3(1 - 2v) + 2R^2(1 - v)\right] \qquad (4.52d)$$

4.6.6 Changes to the Inner and Outer Radii

The hoop strain in Eq. 4.51b supplies directly the dimensional change to any radius (or diameter). For these u is interpreted as the radial displacement giving: $\varepsilon_\theta = \Delta r/r = \Delta d/d$. Applying this equation to the inner radius:

$$\varepsilon_{\theta_i} = \frac{\Delta r_i}{r_i} = \frac{1}{E}[\sigma_\theta - v(\sigma_r + \sigma_z)]_{r=r_i} \qquad (4.53a)$$

when from Eqs 4.48a,b, with $r = r_i$:

$$\Delta r_i = \frac{pr_i}{E(R^2 - 1)}\left[(1 - 2v) + (1 + v)R^2\right] \qquad (4.53b)$$

For the outer radius:

$$\varepsilon_{\theta_o} = \frac{\Delta r_o}{r_o} = \frac{1}{E}[\sigma_\theta - v(\sigma_r + \sigma_z)]_{r=r_o} \tag{4.53c}$$

Substituting again from Eqs 4.48a,b and 4.49b, with $r = r_o$:

$$\Delta r_o = \frac{pr_o(2-v)}{E(R^2-1)} \tag{4.53d}$$

4.6.7 Thickness Change

The change to the wall thickness produced by pressure within a closed-end, cylinder follows as:

$$\Delta t = [(r_o + \Delta r_o) - (r_i + \Delta r_i)] - (r_o - r_i) \tag{4.54a}$$

Equation 4.54a is the difference between Eqs 4.53d and 4.53b:

$$\Delta t = \frac{pr_i}{E(R^2-1)}\left[(2-v)R - (1-2v) - (1+v)R^2\right] \tag{4.54b}$$

Alternatively, Δt follows from integrating the radial strain between the inner and outer radii. The radial strain in Eq. 4.51a, depends upon the triaxial stress state:

$$\varepsilon_r = \frac{du}{dr} = \frac{1}{E}[\sigma_r - v(\sigma_\theta + \sigma_z)] \tag{4.55a}$$

Substituting stress expressions from Eqs 4.48a,b and 4.49b:

$$\frac{du}{dr} = \frac{p}{E(R^2-1)}\left[(1-2v) - (1+v)\frac{r_o^2}{r^2}\right] \tag{4.55b}$$

Integrating Eq. 4.55b with the appropriate limits gives the thickness change:

$$\Delta t = \int_{r_i}^{r_o} du = \frac{p}{E(R^2-1)}\int_{r_i}^{r_o}\left[(1-2v) - (1+v)\frac{r_o^2}{r^2}\right]dr$$

$$= \frac{p}{E(R^2-1)}\left|(1-2v)r + (1+v)\frac{r_o^2}{r}\right|_{r_i}^{r_o}$$

$$= \frac{p}{E(R^2-1)}\left[(1-2v)(r_o - r_i) + (1+v)r_o(1 - r_o/r_i)\right]$$

and substituting $R = r_o/r_i$ leads again to Eq. 4.54b.

Example 4.11 Find the necessary wall thickness for a pressure vessel with an internal diameter of 130 mm and length 500 mm that is to contain an internal (gauge) pressure of 100 bar while the maximum hoop stress is limited to 15 MPa. Find the changes in internal volume, inner and outer diameter and the wall thickness at this pressure.

Figure 4.16 shows that σ_θ is a maximum in tension at the inner diameter. Also, the pressure equals the radial compressive stress at the inner radius. Therefore, the boundary conditions are: (a) $\sigma_\theta = +15$ MPa for $r = 65$ mm, (b) $\sigma_r = -10$ MPa for $r = 65$ mm. Substituting these into Eqs 4.46a,b gives:

$$15 = a + \frac{b}{65^2} \tag{i}$$

$$-10 = a - \frac{b}{65^2} \tag{ii}$$

The solution to Eqs i and ii are: $a = 2.5$ and $b = 52812.5$. At the outer diameter the pressure is atmospheric, i.e. zero gauge pressure. Equation 4.46a becomes:

$$0 = 2.5 - \frac{52812.5}{r_o^2} \tag{iii}$$

Equation iii provides an outer radius $r_o = 145.4$ mm and wall thickness $t = 80.4$ mm. Equation 4.52d provides the volumetric strain:

$$\frac{\Delta V}{V} = \frac{10\left[3(1 - 0.54) + 2 \times 1.1185^2(1 - 0.27)\right]}{207 \times 10^3 (1.1185^2 - 1)} = 0.617 \times 10^{-3}$$

from which the change to the internal volume is

$$\Delta V = 0.617 \times 10^{-3} \times \left(\frac{\pi}{4} \times 130^2 \times 500\right) = 4094.9 \text{ mm}^3$$

Equation 4.53b proves the inner circumferential strain:

$$\frac{\Delta r_i}{r_i} = 0.617 \times 10^{-3}\left[(1 - 0.54) + 1.1185^2(1 + 0.27)\right] = 1.2642 \times 10^{-3} \text{ mm}$$

from which the change to the inner diameter follows as

$$\Delta d_i = 130 \times 1.2642 \times 10^{-3} = 0.164 \text{ mm}$$

Equation 4.53d proves the outer circumferential strain:

$$\frac{\Delta r_o}{r_o} = 0.617 \times 10^{-3}(2 - 0.27) = 1.0675 \times 10^{-3}$$

from which the outer diameter change is

$$\Delta d_o = 2 \times 145.4 \times 1.0675 \times 10^{-3} = 0.310 \text{ mm}$$

The thickness change is half the difference between Δd_o and Δd_i:

$$\Delta t = (0.310 - 0.164)/2 = 0.073 \text{ mm}$$

which may be confirmed from Eq. 4.54b.

4.7 Thick-Walled Sphere

The distinction between a thin-walled and thick-walled sphere is set at a maximum inner diameter to thickness ratio of 20, similar to that for thick cylinders. While there is no axial stress in this case the stress state is again triaxial with two orthogonal stresses acting around the wall and the third stress acting perpendicular to the wall. All three (principal) stresses vary through the thickness.

4.7.1 Equilibrium and Compatibility

Figure 4.18 shows a hemispherical shell element for a thick-walled spherical vessel. The element is taken from within the vessel wall at radius r and with radial thickness δr as shown. It is required to find the vertical equilibrium equation upon the horizontal plane of symmetry.

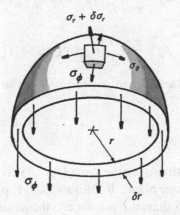

Figure 4.18 Shell element within thick-walled sphere

This plane exposes the meridional stress σ_φ acting vertically upon the element's annular rim area $(2\pi r)\delta r$. Allowing for a variation in radial stress across thickness δr we have σ_r for r and $\sigma_r + \delta\sigma_r$ for $r + \delta r$. Their force resultants may be projected to act normally to the symmetry plane in a similar manner to the analysis of a thin-walled sphere given in § 4.5.2. When both the radial and meridional stress are assumed tensile, they satisfy the vertical force equilibrium condition:

$$\sigma_\varphi (2\pi r)\delta r + \sigma_r \pi r^2 = (\sigma_r + \delta\sigma_r)\pi(r + \delta r)^2 \qquad (4.56a)$$

Expanding Eq. 4.56a leads to:

$$\sigma_\varphi - \sigma_r = \frac{1}{2}\left(\sigma_r \frac{\delta r}{r} + r\frac{\delta\sigma_r}{\delta r} + 2\,\delta\sigma_r + \delta\sigma_r \frac{\delta r}{r} \right) \qquad (4.56b)$$

In the limit, as $\delta r \to 0$ small quantities and their products are neglected. This provides this element's *differential equilibrium equation*:

$$\sigma_\varphi - \sigma_r = \frac{r}{2}\left(\frac{d\sigma_r}{dr} \right) \qquad (4.56c)$$

Had a vertical diameter section been taken instead, this would expose the hoop stress σ_θ, thereby modifying the equilibrium Eq. 4.56c to:

$$\sigma_\theta - \sigma_r = \frac{r}{2}\left(\frac{d\sigma_r}{dr} \right) \qquad (4.56d)$$

The strain-displacement relations for the radial, hoop and meridional strains are similar to those derived for the cylinder in Eqs 4.51a,b. These are combined with the sphere's triaxial stress–strain relations as follows:

$$\varepsilon_r = \frac{du}{dr} = \frac{1}{E}\left[\sigma_r - v(\sigma_\theta + \sigma_\varphi)\right] \qquad (4.57a)$$

$$\varepsilon_\theta = \frac{u}{r} = \frac{1}{E}\left[\sigma_\theta - v(\sigma_r + \sigma_\varphi)\right] \qquad (4.57b)$$

$$\varepsilon_\varphi = \frac{u}{r} = \frac{1}{E}\left[\sigma_\varphi - v(\sigma_r + \sigma_\theta)\right] \qquad (4.57c)$$

where u is the radial displacement at any radius r in the wall.

4.7.2 Stress Distributions

The hoop and meridional strains are the same ($\varepsilon_\theta = \varepsilon_\varphi$) given the spherical symmetry that exists upon any diameter plane. It follows from Eqs 4.57b,c that the hoop and meridional stresses for all diameter planes are the same, i.e. $\sigma_\theta = \sigma_\varphi$. This equibiaxial stress state exists in the presence of a variation in radial stress throughout the wall. Now, from Eqs 4.57a,b, as the two strains depend upon a single displacement u they may be combined within a single *compatibility condition*, Eq. 4.51c, as follows:

$$(\sigma_r - 2v\sigma_\theta) = \frac{d}{dr}\left[r\sigma_\theta(1-v) - vr\sigma_r \right]$$

$$(1+v)(\sigma_\theta - \sigma_r) = vr\frac{d\sigma_r}{dr} - (1-v)r\frac{d\sigma_\theta}{dr} \qquad (4.58a)$$

Substituting Eq. 4.56d into Eq. 4.58a and integrating leads to:

$$\frac{r}{2}(1-v)\frac{d\sigma_r}{dr} + r(1-v)\frac{d\sigma_\theta}{dr} = 0$$

$$\frac{1}{2}\frac{d\sigma_r}{dr} + \frac{d\sigma_\theta}{dr} = 0$$

$$\therefore \quad \frac{\sigma_r}{2} + \sigma_\theta = A \qquad\qquad (4.58b)$$

where A is a constant. Eliminating σ_θ between Eqs 4.56d and 4.58b:

$$2A - 3\sigma_r = r\frac{d\sigma_r}{dr}$$

and integrating:

$$\int \frac{dr}{r} = \int \frac{d\sigma_r}{2A - 3\sigma_r}$$

$$-3\ln r + \ln B = \ln(2A - 3\sigma_r)$$

where B is a further integration constant:

$$\frac{B}{r^3} = (2A - 3\sigma_r)$$

$$\therefore \quad \sigma_r = \frac{1}{3}\left(2A - \frac{B}{r^3}\right) \qquad\qquad (4.58c)$$

Writing $a = 2A/3$ and $b = B/3$ in Eqs 4.58b,c, the triaxial stress state becomes:

$$\sigma_r = a - \frac{b}{r^3} \qquad\qquad (4.59a)$$

$$\sigma_\theta = \sigma_\varphi = a + \frac{b}{2r^3} \qquad\qquad (4.59b)$$

The constants a and b in Eqs 4.59a,b follow from the boundary conditions. For example, with only internal pressure applied then: $\sigma_r = -p$ for $r = r_i$ and $\sigma_r = 0$ for $r = r_o$. Solving for the constants, a and b, Eqs 4.59a,b can be written in terms of the sphere's inner and outer radii as:

$$\sigma_r = \frac{pr_i^3}{r_o^3 - r_i^3}\left(1 - \frac{r_o^3}{r^3}\right) \qquad\qquad (4.59c)$$

$$\sigma_\theta = \sigma_\varphi = \frac{pr_i^3}{r_o^3 - r_i^3}\left(1 + \frac{r_o^3}{2r^3}\right) \qquad\qquad (4.59d)$$

Equation 4.59d shows that the hoop stress is greatest in tension at the inner radius $r = r_i$ and falls to its minimum at the outer radius $r = r_o$. Equation 4.59c confirms the two boundary conditions that were applied to these radii.

4.7.3 Change in Internal Volume

From Eq. 4.40, the volumetric strain is:

$$\frac{\Delta V}{V} = 3\left(\frac{\Delta r_i}{r_i}\right) + 3\left(\frac{\Delta r_i}{r_i}\right)^2 + \left(\frac{\Delta r_i}{r_i}\right)^3 \approx 3\left(\frac{\Delta r_i}{r_i}\right) \qquad (4.60a)$$

Taken with Eq. 4.57b and $\sigma_\theta = \sigma_\varphi$, the approximation in Eq. 4.60a provides:

$$\frac{\Delta V_i}{V_i} = 3\,\varepsilon_{\theta_i} = \frac{3}{E}\left[\sigma_\theta - v(\sigma_\theta + \sigma_r)\right]_{r=r_i} = \frac{3}{E}\left[\sigma_\theta(1 - v) - v\sigma_r\right]_{r=r_i} \qquad (4.60b)$$

where $u/r = \Delta r_i/r_i$ at the inner radius. Thus, we must substitute σ_θ and σ_r from Eqs 4.59c,d for $r = r_i$. This gives the change to the internal volume of a sphere when containing pressure p:

$$\Delta V_i = \frac{3pV_i}{E(R^3 - 1)}\left[(1 - 2v) + \left(\frac{1 + v}{2}\right)R^3\right] \qquad (4.60c)$$

In Eq. 4.60c, $V_i = 4\pi r_i^3/3$ is the initial, internal volume of the sphere and $R = r_o/r_i$ is its radius ratio. If a compressible fluid with bulk modulus K transmits the pressure then the change in the fluid volume will be, from Eq. 4.16a; $\delta V = -pV/K$. This additional volume is required if a compressible fluid has to fill the vessel completely under pressure.

4.7.4 Changes to the Inner and Outer Radii

The hoop strain in Eq. 4.51b supplies diameter (or radius) changes directly given its definition $\varepsilon_\theta = u/r = \Delta r/r$ where u is the radial displacement. Applying this to the inner radius:

$$\varepsilon_{\theta_i} = \frac{\Delta r_i}{r_i} = \frac{1}{E}[\sigma_\theta - v(\sigma_{r.} + \sigma_\varphi)]_{r=r_i} \qquad (4.61a)$$

when from Eqs 4.59c,d:

$$\Delta r_i = \frac{pr_i}{E(R^3 - 1)}\left[(1 - 2v) + \left(\frac{1 + v}{2}\right)R^3\right] = \frac{r_i}{3} \times \frac{\Delta V_i}{V_i} \qquad (4.61b)$$

where ΔV_i appears in Eq. 4.60c For the outer radius:

$$\varepsilon_{\theta_o} = \frac{\Delta r_o}{r_o} = \frac{1}{E}[\sigma_\theta - v(\sigma_r + \sigma_\varphi)]_{r=r_o} \qquad (4.62a)$$

and substituting from Eqs 4.59c,d:

$$\Delta r_o = \frac{3pr_o(1-v)}{2E(R^3-1)} \qquad (4.62b)$$

4.7.5 Thickness Change

The change to the wall thickness produced by internal pressure within a closed spherical vessel follows as:

$$\Delta t = (r_o + \Delta r_o) - (r_i + \Delta r_i)] - (r_o - r_i) \qquad (4.63a)$$

which identifies the difference between Eqs 4.62b and 4.61b:

$$\Delta t = \frac{pr_i}{E(R^3-1)}\left[\frac{3(1-v)}{2}R - (1-2v) - \frac{(1+v)}{2}R^3\right] \qquad (4.63b)$$

Alternatively, Δt follows from integrating the radial strain between the inner and outer radii. From Eq. 4.51a:

$$\varepsilon_r = \frac{du}{dr} = \frac{1}{E}[\sigma_r - v(\sigma_\theta + \sigma_\varphi)] \qquad (4.64a)$$

Substituting stress expressions from Eqs 4.59c,d:

$$\frac{du}{dr} = \frac{p}{E(R^3-1)}\left[(1-2v) - (1+v)\frac{r_o^3}{r^3}\right] \qquad (4.64b)$$

Integrating Eq. 4.64b with the appropriate limits gives the thickness change required:

$$\Delta t = \int_{r_i}^{r_o} du = \frac{p}{E(R^3-1)}\int_{r_i}^{r_o}\left[(1-2v) - (1+v)\frac{r_o^3}{r^3}\right]dr$$

$$= \frac{p}{E(R^3-1)}\left|(1-2v)r + \frac{(1+v)}{2}\frac{r_o^3}{r^2}\right|_{r_i}^{r_o}$$

$$= \frac{p}{E(R^3-1)}\left[(1-2v)(r_o - r_i) + \frac{(1+v)}{2}r_o(1 - r_o^2/r_i^2)\right]$$

and substituting $R = r_o/r_i$ recovers Eq. 4.63b.

4.7.5 Diametrical Force

When the spherical vessel is subjected to an external, vertical compressive force F aligned with the axis of symmetry in addition to the internal pressure, the stress state within the sphere's wall is no longer equi-triaxial. The radial and hoop stresses remain unchanged, being given by Eqs 4.59c,d, but the meridional stress is no longer equal to the hoop stress ($\sigma_\varphi \neq \sigma_\theta$). The former must account for the compressive stress in the wall due to F. Referring to the half-section in Fig. 4.19a, the meridional stress σ_φ within the annular area exposed by a horizontal plane distance h above the symmetry axis is shown.

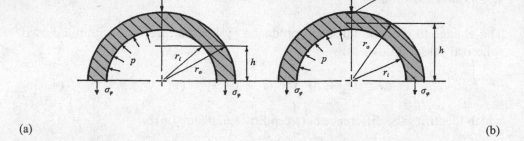

(a)   (b)

Figure 4.19 Sphere sections under diametral force

When the force is taken in isolation, the equilibrium condition for this plane becomes:

$$F = \sigma_\varphi \times \pi(r_o'^2 - r_i'^2) \tag{4.65a}$$

where:

$$r_o'^2 = r_o^2 - h^2 \text{ and } r_i'^2 = r_i^2 - h^2 \tag{4.65b,c}$$

Substituting Eqs 4.65b,c into Eq. 4.65a shows that σ_φ is constant within the annular section areas for $0 \le h \le r_i$:

$$\sigma_\varphi = \frac{F}{\pi(r_o^2 - r_i^2)} \tag{4.66a}$$

However, it can be seen from Fig. 4.19b that σ_φ becomes a variable as the top, solid section area changes with increasing h in the region $r_i \le h < r_o$

$$\sigma_\varphi = \frac{F}{\pi(r_o^2 - h^2)} \tag{4.66b}$$

and if we are avoid an infinite stress when $r_o = h$, the small area a over which F is applied (see Fig. 4.19b) will modify Eq. 4.66b to

$$\sigma_\varphi = \frac{F'a}{\pi(r_o^2 - h^2)} \tag{4.66c}$$

where $F' = F/a$ is the force intensity borne by a. In elasticity theory we may use the principle of superposition which states that the net stress under a combination of external loads is the sum of those stresses arising from the separate application of each load. Hence, the net stress in the meridional direction becomes the sum of the stresses from the internal pressure and from the diametrical force applied separately.

The latter contribution takes any one of the three forms given above. For example, in the hollow section (Fig. 4.19a) that contains most of the spherical vessel:

$$\sigma_\varphi = \frac{p r_i^3}{r_o^3 - r_i^3}\left(1 + \frac{r_o^3}{2 r^3}\right) - \frac{F}{\pi(r_o^2 - r_i^2)} \tag{4.67}$$

In fact, here the 'sum' becomes a difference because of the differing sense between each stress: tensile due to p and compressive due to F. It can be seen that volume, diameter and thickness changes are all influenced by the term in F within Eq. 4.67.

Exercises

Elastic Constants

4.1 A bar of metal, 50 mm diameter and 150 mm long, deforms elastically by the respective amounts -0.0065 mm and 0.080 mm under 75 kN in tension. Find E, G, K and v.
(Ans: $E = 71.62$ GPa, $G = 28.8$ GPa, $K = 46.63$ GPa, $v = 0.244$)

4.2 A 25 mm diameter steel bar 200 mm long increases by 0.01 mm under an axial tensile force of 50 kN and twists 0.82° under a torque of 220 Nm applied separately. Determine the steel's four elastic constants.
(Ans: $E = 203.7$ GPa, $G = 80.2$ GPa, $v = 0.27$, $K = 147.62$ GPa)

4.3 A solid circular bar 50 mm diameter and 150 mm long is subjected to a tensile force of 75 kN. The length increases by 0.08 mm and the diameter decreases by 0.0065 mm. Referring to § 4.3 and Example 4.4 (pp. 117–119) find Lamé's constants λ and μ.
(Ans 27.43 GPa, 28.8 GPa)

4.4 The elastic constants for a piece of steel are assumed to be: $E = 207$ GPa, $G = 80.6$ GPa. Find the apparent value of Poisson's ratio v and assess the % error in v if the assumed values of E and G are respectively 2% greater and 3% less than the true values. (Ans: 0.28, 29%)

4.5 A solid cylindrical bar of steel 38 mm diameter is subjected to a tensile force of 100 kN. The extension measured upon a 200 mm gauge length is 0.0864 mm and the corresponding reduction to its diameter is 0.0048 mm. Calculate E, v, G, K and λ.
(Ans: 205 GPa, 0.29, 79 GPa, 169.5 GPa, 116.8 GPa)

Extended Hooke's Law

4.6 A block of steel is subjected to triaxial principal stresses 77 and 61.75 MPa in tension and 92.5 MPa in compression. Find the micro-strains in the principal directions given that $E = 208$ GPa and $v = 0.286$. (Ans: 412μ, -635μ, 318μ)

4.7 A bar of metal is 50 mm × 75 mm in rectangular cross section and 250 mm long. Determine the changes in dimensions which take place when it is subjected to an axial tensile force of 300 kN, a compressive force of 3 MN on its 75 × 250 mm faces and a tensile force of 2 MN on its remaining 50 × 250 mm faces. Take $E = 210$ GPa and $v = 0.26$.
(Ans: -0.053 mm, 0.0646 mm, 0.0952 mm)

4.8 A rectangular section bar under an axial tensile stress σ_1 is to be prevented from contracting laterally. Show that the equal lateral stresses required are $\sigma_2 = \sigma_3 = v/(1 - v)$, and that the percentage reduction in axial strain is $200v^2/(1 - v)$. Calculate the corresponding values when $\sigma_1 = 15.5$ MPa, given $E = 207$ GPa and $v = 0.27$.
(Ans: 5.73 MPa, 20%)

4.9 An axial compressive force of 200 kN is applied to a 75 mm square section bar 200 mm long. Lateral stresses are applied to restrict lateral strains to 1/4 of what they would be if unrestricted. Given $E = 70$ GPa and $v = \frac{1}{3}$ find the lateral stresses and the restricted and unrestricted axial displacements.
(Ans: - 13.31 MPa, - 0.075 mm, and - 0.1016 mm)

4.10 If, for the bar with dimensions in Exercise 4.9, restricted lateral strains are to be one third of their unrestricted values under an axial compressive force of 225 kN, determine the necessary lateral stresses and the change in length of the bar. Take $E = 208$ GPa and $v = \frac{1}{4}$.
(Ans: 8.59 MPa, - 0.0335 mm)

4.11 A pressure vessel has hemi-spherical ends welded to a cylindrical body. Determine the thickness ratio required between the ends and the body when across the weld: (a) the axial strain is to be the same and (b) the hoop stress is to be the same. Take $v = \frac{1}{4}$.
(Ans: 3/2, 1/2)

4.12 A thin-walled cylinder 3 m long has an internal diameter of 75 mm and a thickness of 3 mm. Determine the changes in length, diameter and thickness under an applied internal pressure of 1.5 bar. Take $E = 210$ GPa and $v = 0.27$.
(Ans: $\Delta l = 0.615$ mm, $\Delta d = 0.0579$ mm, $\Delta t = -1.083 \times 10^{-3}$ mm)

Volumetric Strain

4.13 Find the change in volume of a 1 m long steel bar, with 75 mm square section, under an axial compressive force of 20 kN, given that $E = 207$ GPa and $G = 82$ GPa.
(Ans: - 45.95 mm^3)

4.14 A steel bar 25 mm diameter and 3 m long carries an axial tensile force of 60 kN. Find the extension, the reduction in diameter and the change in volume given $E = 201$ GPa and $v = 0.3$.

4.15 A bar 30 mm long and 20 mm diameter is subjected to an axial compressive load of 30 kN. The length and diameter change by -0.015 and 0.0027 mm, respectively. Determine the change in volume when the bar is immersed to a depth of 3 km in sea water of density of 815.5 kg/m^3.
(Ans: -163 mm^3)

4.16 A rectangular block is subjected to tensile forces of 120 kN on its 75 × 50 mm faces, 800 kN on its 50 × 250 mm faces and a compressive force of 900 kN on its 75 × 250 mm faces. Find the change in volume and the dimensional changes of the sides given that the elastic constants are $E = 105$ GPa and $v = 0.35$.

Thin-Walled Pressure Vessels

4.17 A closed copper tube 50 mm i.d., 1.25 m long and 1.25 mm thick is full of water under pressure. Determine the change in pressure when an additional 3275 mm³ of water is pumped in. Take $E = 104$ GPa and $v = 0.3$ for copper, $K = 2070$ MPa for water. (Ans: 1.56 MPa)

4.18 A 1 m long thin cylinder, with internal diameter 600 mm, is pressurised to 13.75 bar. Find the necessary wall thickness if the maximum tensile stress is not to exceed 82.5 MPa. What are the axial and circumferential strains and the increase in volume under this pressure? Take $E = 207$ GPa and $v = 0.3$. (Ans: 5 mm, 3.4×10^{-4}, 20.32×10^{4} mm³)

4.19 A thin-walled cylinder, 4.5 m long, 510 mm diameter and 5 mm thick, is prevented from changing it length when under an internal pressure of 82.5 bar. Determine the maximum tensile stress in the wall and the increase in the cylinder's internal volume under this pressure. Take $E = 165$ GPa and $v = 0.24$. (Ans: 413 MPa, 4.6×10^{6} mm³)

4.20 A thin-walled steel sphere with 600 mm mean radius and 25 mm thickness contains oil at atmospheric pressure. To what depth may the sealed sphere be immersed in sea water of density 1040 kg/m³ when the maximum compressive stress in the sphere wall is restricted to 92.5 MPa? Take $E = 207$ GPa, $v = 0.25$ for steel and $K = 3445$ MPa for oil. Hint: $\Delta V/V$ for oil and sphere are the same. (Ans: 1097 m)

4.21 A thin-walled steel sphere is subjected to an internal pressure of 5 bar. If the mean diameter is 2 m determine the necessary thickness using a factor of safety of 5 with a *UTS* of 450 MPa for the steel used.

4.22 A thin-walled cylindrical bottle with hemispherical ends is 150 mm diameter with a parallel length of 760 mm and constant thickness. It is to contain oil at a gauge pressure of 350 bar. Determine the required thickness for a safety factor of 2 if the vessel material can withstand a maximum tensile stress of 345 MPa. What additional volume of oil (for which $K = 3445$ MPa) must be pumped in to double the pressure? Take $E = 207$ GPa and $v = 0.28$ for the bottle material. (Ans: 15.22 mm, 182×10^{3} mm³)

4.23 A working pressure of 14 bar exists in a 2.4 m diameter cylinder. Find the necessary thickness for a safety factor of 6 when the ultimate strength of the steel is 465 MPa and the efficiency of the longitudinal joint is 75%. (Ans: 29 mm)

4.24 The plates of a cylindrical boiler are 12.5 mm thick and the internal pressure to which they are subjected is 14 bar. If the tensile stress in the material is limited to 137.5 MPa and the efficiency of the longitudinal and circumferential welded joints are 75% and 30%, respectively, determine the maximum permissible diameter of the boiler. (Ans: 1.5 m)

Thick-Walled Pressure Vessels

4.25 A thick-walled, steel cylinder, 203 mm inside diameter, is not to be stressed beyond 123.5 MPa under an internal pressure of 310 bar. Find a suitable external diameter and the magnitude of the % strains at the outer surface. Take $E = 207$ GPa and $v = 0.28$. (Ans: 262 mm, 0.039%, 0.0099%)

4.26 A thick-walled, cylindrical steel vessel 127 mm i.d. and 178 mm o.d., with closed ends, contains fluid at a pressure of 350 bar. Plot the distributions in circumferential, radial and axial stresses showing their values at the inner and outer diameters. What are the corresponding micro strains at each diameter? Take $E = 200$ GPa and $v = 0.28$.
(Ans: 107.6, –35, 36.29 MPa; 72.6, 0, 36.29 MPa; 536.2, –376.5, –283.1με; 312.1, –152.4, 79.8 με)

4.27 A cast-iron, closed-end, thick-cylinder, with i.d. of 114 mm and wall thickness 25 mm, withstands a safe internal pressure of 170 bar. What is the safe internal pressure a second cast iron cylinder can withstand with an i.d. of 152 mm and thickness 25 mm, when the greatest tensile stress is unaltered? What is the minimum tensile stress in each cylinder?
(Ans: 135 bar, 31.3 MPa, 34.7 MPa)

4.28 A pressure of 400 bar exists within the bore of a thick-walled pipe. If the maximum stress is not to exceed 120 MPa, determine the pipe's outer diameter. (Ans: 70.71 mm)

4.29 A closed, thick-cylinder, 150 mm i.d. with 75 mm wall thickness, is subjected to an internal gauge pressure of 700 bar. Calculate and plot, for each 25 mm of radius, the distribution of radial and hoop stress through the wall.
[Ans: σ_θ (MPa): 116.67, 75.83, 56.93, 46.66; σ_r (MPa) –70, –29.17, –10.27, 0; $\sigma_z = 23.33$ MPa]

4.30 Examine how the changes in volume, wall thickness, inner and outer radii are altered for a thick-walled cylinder whose ends are (i) prevented from expanding and (ii) open. Take the axial stress from Eqs 4.50a,b, respectively.

4.31 Show that when a thick-walled sphere is subjected to both internal and external pressures the radial, hoop and meridional stresses are given by:

$$\sigma_r = \frac{\left[-p_o + p_i (r_i/r_o)^3 + (p_o - p_i)(r_i/r)^3 \right]}{1 - (r_i/r_o)^3}, \quad \sigma_\theta = \sigma_\varphi = \frac{\left[-p_o + p_i (r_i/r_o)^3 - \tfrac{1}{2}(p_o - p_i)(r_i/r)^3 \right]}{1 - (r_i/r_o)^3}$$

4.32 A thick-walled steel bathasphere, 500 mm i.d., is to withstand an external fluid pressure of 1000 atmospheres whilst internally the pressure remains at 1 atmosphere. Find the thickness necessary if the maximum compressive stress in the vessel is limited to 230 MPa. Also find the internal volume change.

4.33 Find the changes in volume, wall thickness, inner and outer radii for a thick-walled sphere subjected to an internal pressure 500 bar. The original diameters of the sphere are 200 and 300 mm. Take $E = 210$ GPa and $v = 0.27$.

4.34 Examine how the changes to the internal volume, diameters and thickness of a thick-walled sphere under internal pressure are altered by the introduction of an additional diametrical compressive force.

4.35 A thick-walled sphere is subjected to diametrical compressive force F which influences the meridional stress according to Eq. 4.67. The radial and hoop 'pressure stresses' are given by Lamé's Eqs 4.59c,d. Derive the sphere's volume, diameter and thickness changes.

CHAPTER 5

BENDING AND SHEAR IN BEAMS

Summary: The two influences upon the design of a loaded beam lie in the distributions of shear force and bending moment along its length. These two distributions were shown within their corresponding diagrams constructed in Chapter 2 (see § 2.5, p. 47). The force and moment are the external 'loads' upon the beam each to be reacted by its cross section. The section does this by becoming stressed under each loading mode, distorting away from its original shape and position due to the strain within it. Here we shall examine the relationships that exist between stress and strain for elastic distortions arising from beam curvature and transverse shearing. The stress expressions provided by this analysis are crucial for designing load-bearing beams with the required degree of safety.

5.1 Direct Stress in Bending

The theory of bending enables the longitudinal bending stress σ, to be found anywhere in the uniform cross section of an initially straight beam when the bending moment M, at that section is known. The radius of curvature R in Fig. 5.1a defines the unstrained longitudinal fibre OO, which is referred to as the neutral axis (NA). Let the angle θ (radians) be subtended by a segment of the bent beam at its centre of curvature as shown.

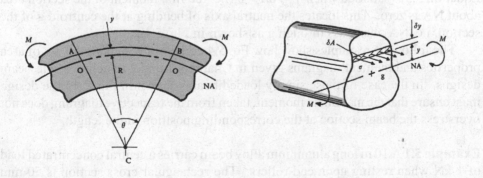

(a)
(b)

Figure 5.1 Beam in bending

The strain in the arc AB, at a positive distance y, from NA on the tensile side is:

$$\varepsilon = \frac{AB - OO}{OO} = \frac{(R + y)\theta - R\theta}{R\theta} = \frac{y}{R} \qquad (5.1a)$$

The tensile stress in arc AB is uniaxial and therefore Hooke's law applies directly: $\varepsilon = \sigma / E$. Combining this with Eq. 5.1a, the longitudinal bending strain becomes:

$$\varepsilon = \frac{\sigma}{E} = \frac{y}{R} \qquad (5.1b)$$

Now consider the equilibrium of moments for any cross section of a prismatic beam given in Fig. 5.1b. An elemental area δA of this cross section lies at the distance y from NA as shown. The applied moment M, is the resultant of all the elemental moments $\delta M = (\sigma \times \delta A)\, y$, produced by the action of σ on the area δA. It follows that for a section of area A:

$$M = \int_A \sigma y \, dA \qquad (5.2)$$

Substituting Eq. 5.1b into Eq. 5.2:

$$M = \frac{E}{R} \int_A y^2 \, dA = \frac{EI}{R} \qquad (5.3)$$

where the integral in Eq. 5.3 defines the second moment of area, I about NA. Combining Eqs 5.1b and 5.3 leads to the full bending theory

$$\frac{M}{I} = \frac{E}{R} = \frac{\sigma}{y} \qquad (5.4)$$

No axial force is applied to the beam and therefore the sum of all elemental forces $\delta F = \sigma \times \delta A$, normal to the cross section, must be zero. From Eq. 5.1b:

$$F = \int_A \sigma \, dA = \frac{E}{R} \int_A y \, dA = 0 \qquad (5.5)$$

Equation 5.5 is satisfied when $\int_A y\, dA = 0$, i.e. the first moment of the section area about NA is zero. This locates the neutral axis of bending at the centroid g of the section, i.e. NA must pass through g as shown in Fig. 5.1b.

The following examples show how Eq. 5.4 combines with the analyses of area properties and moment diagrams given in Chapters 1 and 2 to achieve safe beam designs. In the case of those simply loaded beams shown in Fig. 2.15 the design must ensure that the maximum moment, taken from the respective diagram, does not overstress the beam section at the corresponding position in the length.

Example 5.1 A 10 m long aluminium alloy beam carries a central concentrated load of 1 kN when resting upon end-rollers. The rectangular cross section is 50 mm wide and 100 mm deep. Determine, for the position in the length where the bending moment is a maximum, the beam's radius of curvature and the maximum tensile and compressive surface stresses. Estimate the value of the safety factor used in the design given a 0.1% proof stress value of 180 MPa (see Fig. 4.1) Take $E = 70$ GPa.

Figure 2.15a shows that a maximum sagging moment occurs at the beam centre:

$$M = \frac{WL}{4} = \frac{1 \times 10}{4} = 2.5 \text{ kNm}$$

The neutral axis of bending lies at the centre of the rectangle's depth for which the second moment of area is:

$$I = \frac{bd^3}{12} = \frac{50 \times 100^3}{12} = 4.167 \times 10^6 \text{ mm}^4$$

The beam's surface stresses, in tension at the bottom and compression at the top, attain their maxima with equal magnitudes under the peak moment, these being given by Eq. 5.4:

$$\sigma = \frac{My}{I} = \frac{(2.5 \times 10^6) \times (\pm 50)}{(4.176 \times 10^6)} = \pm 17.96 \text{ MPa}$$

for which the safety factor of 10 ($\approx 180/17.96$) applies to both surface stresses. The radius of curvature of the NA at the beam centre follows from Eq. 5.4 (in units of N and mm):

$$R = \frac{EI}{M} = \frac{(70 \times 10^3) \times (4.176 \times 10^6)}{(2.5 \times 10^6)} = 116.93 \times 10^3 \text{ mm}$$

or, should units of N and m be preferred to calculate R from Eq. 5.4:

$$R = \frac{EI}{M} = \frac{(70 \times 10^9) \times (4.176 \times 10^{-6})}{(2.5 \times 10^3)} = 116.93 \text{ m}$$

Example 5.2 Figure 5.2a shows the T-section of a 6 m long steel cantilever beam. A 5 kN concentrated load acts vertically upon the top flange at the free end. Find the maximum stresses and the curvature for the section at the position where the maximum hogging moment produces tension in the flange and compression in the web. Plot the stress distribution across the depth and establish the minimum safety factor for the beam. Take for the steel $E = 208$ GPa and a maximum allowable stress of 300 MPa.

The centroid position $\bar{y}$ of the T-section in Fig. 5.2a locates the neutral axis. Take first moments of area about the datum position X-X shown:

$$\sum_{i=1}^{2} A_i y_i = A\bar{y}$$

$$A_1 y_1 + A_2 y_2 = (A_1 + A_2)\bar{y}$$

$$(125 \times 25 \times 62.5) + (175 \times 25 \times 137.5) = [(125 \times 25) + (175 \times 25)]\bar{y}$$

$$\bar{y} = 106.25 \text{ mm}$$

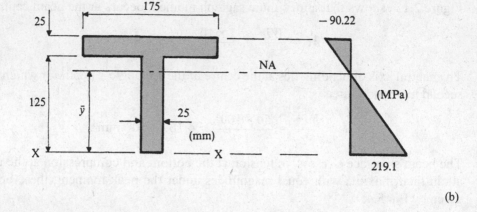

(a)   (b)

Figure 5.2 T section showing stress distribution

When three component rectangles divide this section's area, with each rectangle 'based' on the NA, $I = \sum bd^3/3$ becomes:

$$I = \frac{25 \times 106.25^3}{3} + \frac{175 \times 43.75^3}{3} - \frac{150 \times 18.75^3}{3} = 14.55 \times 10^6 \text{ mm}^4$$

The maximum bending moment occurs at the cantilever's fixed end, this being identified with Fig. 2.15c as:

$$M_{max} = WL = 6 \times 5 = 30 \text{ kNm}$$

The *maximum* compressive stress is reached in the web's bottom surface at the beam's fixed end ($y_c = 106.25$ mm):

$$\sigma_c = \frac{M_{max} y_c}{I} = \frac{(30 \times 10^6)(-106.25)}{(14.55 \times 10^6)} = -219.1 \text{ MPa}$$

The *maximum* tensile stress is reached at the flange's top surface at the fixed end ($y_t = 150 - 106.25 = +43.75$ mm):

$$\sigma_c = \frac{M_{max} y_t}{I} = \frac{(30 \times 10^6)(+43.75)}{(14.55 \times 10^6)} = 90.21 \text{ MPa}$$

Equation 5.4 provides the *minimum* curvature of the beam's neutral axis at the fixed-end. In units of $\underline{N}$ and $\underline{m}$:

$$R_{min} = \frac{EI}{M_{max}} = \frac{(208 \times 10^9) \times (14.55 \times 10^{-6})}{(30 \times 10^3)} = 100.8 \text{ m}$$

With σ in proportion to y for a given M and I, it follows that the stress will vary linearly with the depth between these maxima passing through the zero stress value at the NA (see Fig. 5.2b). The distribution shows that the *maximum* stress for the whole beam is 219.1 MPa in compression, for which the *minimum* safety factor is $S = 300/219.1 = 1.37$.

Example 5.3 A cast-iron girder, of unsymmetrical I-section (see Fig. 5.3), is simply supported at the ends of its 4.5 m length. Calculate the maximum uniformly distributed load that can be applied to this beam without the tensile stress exceeding 15 MPa. What is the maximum compressive stress and the minimum radius of curvature for this beam? What percentage of the bending moment supported by this beam is carried by the flanges? Take $E = 97$ GPa.

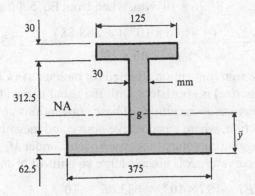

Figure 5.3 Unsymmetrical I-section

The techniques required to locate the centroid and calculate the second moment of this area were given in Chapter 1. For the centroid position, lying at height $\bar{y}$ from the base, we equate the first moments of the section's constituent rectangular areas to that for the whole area:

$$\sum_{i=1}^{3} A_i y_i = A\bar{y}$$

$$(375 \times 62.5 \times 31.25) + (312.5 \times 30 \times 218.75) + (125 \times 30 \times 390) = 36562.5\,\bar{y}$$

This gives $\bar{y} = 116.12$ mm. With a further series of rectangular, constituent areas all based upon the section's neutral axis (NA), we may take their sums and differences to build the second moment of the given area using $I = \sum bd^3/3$. Thus, taking the outer less two inner rectangular areas on each side of the NA:

$$I = 375(116.12)^3/3 - 345(53.62)^3/3 + 125(288.88)^3/3 - 95(258.88)^3/3$$

$$= (177 + 455.07)10^6 = 633.06 \times 10^6 \text{ mm}^4$$

Let w be the distributed loading in N per m length. The maximum sagging bending moment for the given loading is identified as the second of the standard beam loadings given in Chapter 2. Figure 2.15b gives this as:

$$M_{max} = -wL^2/8 = -w(4.5)^2/8 = -2.531w \text{ Nm} \qquad \text{(i)}$$

From Eq. 5.4, the maximum tensile stress occurs at the bottom edge of the lower flange where $y_t = -116.12$ mm. This gives σ in units of N and mm:

$$\sigma = \frac{My_t}{I} = \frac{(-2.531 \times 10^3 w)(-116.12)}{(633.06 \times 10^6)} = 15 \text{ MPa} \qquad \text{(ii)}$$

Note: 1 MPa = 1 N/mm^2. Hence, when solved, Eq. ii gives the uniform load distribution: $w = 32.3 \times 10^3$ N/m = 32.3 kN/m. The maximum compressive stress occurs under M_{max} at the flange's top surface where $y_c = 405 - 116.12 = 288.88$ mm. Equation i gives $M_{max} = -81.76 \times 10^3$ Nm, when from Eq. 5.4 (in units $\underline{N}$ and $\underline{mm}$):

$$\sigma = \frac{M_{max}\, y_c}{I} = \frac{(-81.76 \times 10^6)(+288.88)}{(633.06 \times 10^6)} = -37.31 \text{ MPa}$$

It is seen here how the sign convention adopted for moments in Chapter 2 (*hogging-positive, sagging-negative*) is consistent with the usual convention for stress (i.e. *tension-positive, compression-negative*). These conventions apply to a horizontal plane in the cross section, where y is positive above and negative below the NA.

The minimum sagging curvature (negative) occurs under M_{max}, this being found from Eq. 5.4. Beam curvature R is always large so units of $\underline{N}$ and $\underline{m}$ are preferred:

$$R = \frac{EI}{M} = \frac{97 \times 10^9 \times 633.06 \times 10^{-6}}{(-81.76 \times 10^3)} = -751 \text{ m}$$

Equation 5.2 provides the bending moment M_f carried by the two flanges as

$$M_f = \int_A \sigma y \, dA = \int_y \left(\frac{My}{I} \right) y \,(b \, dy) \qquad \text{(iii)}$$

where M is the applied moment referred to *any* position in the length. Hence the moment ratio M_f/M for this cross section follows as

$$\frac{M_f}{M} = \frac{1}{I} \left[375 \int_{53.62}^{116.12} y^2 \, dy + 125 \int_{258.88}^{288.88} y^2 \, dy \right]$$

$$= \frac{375\,(116.12^3 - 53.62^3) + 125\,(288.88^3 - 258.88^3)}{3 \times 633.06 \times 10^6} = 0.724$$

Here the signs allow for the change in sense of both σ and y above and below NA. The percentage ratio shows that 72.4% of the moment upon *every* cross section is supported by the flanges. It will be seen that the web plays an equally important role in supporting a large proportion of transverse shear forces (see Ex. 5.12, p.175).

5.2 Combined Bending and Direct Stress

Direct (or normal) stresses arise from the actions of bending, tension and compression. Where a combination of these actions arise the stresses that they produce may be superimposed since they are all of a similar nature. The following analyses apply to two common examples found in practice.

5.2.1 Axial Force and Bending Moment

Consider a beam carrying an axial force F, and a bending moment M. The direct stress produced by each action are longitudinal so they may be added to give the net stress:

$$\sigma = \pm \frac{F}{A} \pm \frac{My}{I} \qquad (5.6a)$$

The signs of the two terms in Eq. 5.6a must account for the sense of the stress at a point in the cross section. In the first term, the direct stress is either all tensile or all compressive. In the second term the sign of the bending stress alternates above and below the NA. Figures 5.4a–d illustrate how the stresses are superimposed under a hogging moment and a tensile force, i.e. positive M and F, respectively.

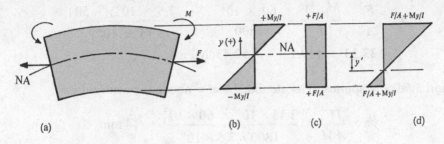

(a)   (b)   (c)   (d)

Figure 5.4 Beam under combined bending and direct stress

For the top surface Eq. 5.6a becomes:

$$\sigma = + F/A + My/I \qquad (5.6b)$$

and for the bottom surface, where y is negative:

$$\sigma = + F/A - My/I \qquad (5.6c)$$

The resultant stress in Fig. 5.4d, is the sum of the distributions due to bending in Fig. 5.4b and tension in Fig. 5.4c. Clearly, F has lowered the NA to a new position y' in Fig. 5.4d which can be found from setting Eq. 5.6c to zero:

$$y' = \frac{IF}{AM} \qquad (5.6d)$$

Example 5.4 A simply supported I-section beam, 2 m long, carries a central concentrated force of $W = 15$ kN in addition to a longitudinal tensile force $F = 60$ kN. Given the depth of section $= 100$ mm, area $= 1800$ mm^2 and $I = 3.15 \times 10^6$ mm^4, determine the maximum net tensile and compressive stresses and the position of the neutral axis.

Figure 2.15a provides that the maximum sagging bending moment:

$$M_{max} = -WL/4 = -(15 \times 10^3 \times 2)/4 = -7.5 \times 10^3 \text{ Nm}$$

Here the bending stress mirrors that shown in Fig. 5.4a. Thus, from Eq. 5.6a, with $y = +50$ mm for the top surface, the net stress attains a maximum in compression:

$$\sigma_c = \frac{F}{A} + \frac{M_{max}\, y}{I} = \frac{60 \times 10^3}{1800} + \frac{(-7.5 \times 10^6)(+50)}{3.15 \times 10^6}$$

$$= 33.33 - 119.1 = -85.77 \text{ MPa}$$

For the bottom surface, where $y = -50$ mm, Eq. 5.6a provides the maximum tensile stress, when written as:

$$\sigma_t = \frac{F}{A} + \frac{M_{max}\, y}{I} = \frac{60 \times 10^3}{1800} + \frac{(-7.5 \times 10^6)(-50)}{3.15 \times 10^6}$$

$$= 33.33 + 119.1 = -152.43 \text{ MPa}$$

Equation 5.6d now applies to locate the NA at y' above the centroid:

$$y' = \frac{IF}{AM} = \frac{3.15 \times 10^6 \times 60 \times 10^3}{1800 \times 7.5 \times 10^6} = 14 \text{ mm}$$

5.2.2 Eccentric Loading of a Short Column

Figure 5.5 shows a slice of a short column carrying an eccentric axial force F. Let a tensile force F act within the positive quadrant of the centroidal axes x and y. Let F act at a position with eccentricities k and h to axes x and y, respectively, as shown. Bending moments, $M_x = Fk$ and $M_y = Fh$, produced by the eccentricity, are both hogging about the positive x and y directions.

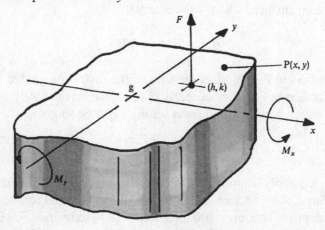

Figure 5.5 Eccentrically loaded column

The resultant tensile stress at any point $P(x, y)$ is the sum of a direct stress due to F and the bending stresses due to M_x and M_y. That is:

$$\sigma = F/A + M_x y/I_x + M_y x/I_y \qquad (5.7a)$$

The net stresses in other quadrants are also determined from Eq. 5.7a by substituting the appropriate signs for x and y. Equation 5.7a will also supply the net stress when F lies in another quadrant provided the correct signs are applied. However, it is much simpler to identify the load quadrant with the first (positive) quadrant of x, y and then substitute for the co-ordinates of $P(x, y)$ with their appropriate signs. Where F is applied along an axis of symmetry (a principal axis) only one bending stress term will appear. For example, when F lies upon the x-axis but is eccentric to y, then $M_x = 0$, $M_y = Fh$ and Eq. 5.7a becomes:

$$\sigma = \frac{F}{A} + \frac{(Fh)x}{I_y} \qquad (5.7b)$$

If the stress is not to change sign across the section then the extreme load position $F(h, 0)$ follows from putting $\sigma = 0$ in Eq. 5.7b.

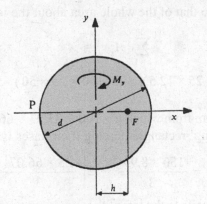

Figure 5.6 Circular section

For example, in the plan view of a solid circular column shown in Fig. 5.6, $\sigma_P = 0$ is ensured under a tensile force F when:

$$\sigma_P = \frac{F}{A} + \frac{M_y x}{I_y} = 0 \qquad (5.8a)$$

Hence, with $A = \pi d^2/4$, $I = \pi d^4/64$, $M_y = Fh$ and $x = -d/2$, Eq. 5.8a becomes:

$$\frac{F}{\pi d^2/4} + \frac{Fh(-d/2)}{\pi d^4/64} = 0 \qquad (5.8b)$$

Equation 5.8b shows that, provided h lies within a circle of radius $h = d/8$, σ will be wholly tensile. A similar radius applies to the extreme position for a compressive force F acting upon a circular column if σ is to be wholly compressive.

Example 5.5 The channel section of a short vertical column is given in Fig. 5.7. Determine the maximum compressive force F that may be applied at the position shown if the stress is nowhere to exceed 80 MPa. Where must F be placed along axis x if point Q is unstressed?

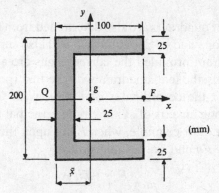

Figure 5.7 Channel section

The centroid position $\bar{x}$ is found from equating the sum of the first moments of three constituent rectangles to that of the whole area about the left vertical side:

$$\sum_{i=1}^{3} A_i y_i = A\bar{x}$$

$$(150 \times 25 \times 12.5) + 2(100 \times 25 \times 50) = 8750\bar{x}$$

This gives $\bar{x} = 33.93$ mm from which $h = 100 - 33.93 = 66.07$ mm. Now using $I = \sum bd^3/3$ for four 'building' rectangles having their bases lying on the y-axis:

$$I_y = \frac{200 \times 33.93^3}{3} - \frac{150 \times 8.93^3}{3} + 2\left(\frac{25 \times 66.07^3}{3}\right) = 7.3754 \times 10^6 \text{ mm}^4$$

The bending stress is tensile to the left of the y-axis, compressive to the right of the y-axis. Hence the greatest compressive stress ($\sigma = -80$ MPa) will occur all along the flange's vertical ends where $x = h = 66.07$ mm. Rearranging Eq. 5.7b:

$$F = \frac{\sigma}{1/A + hx/I_y} = \frac{-80}{1/8750 + (66.07 \times 66.07)/(7.3754 \times 10^6)}$$

$$= -113.29 \times 10^3 \text{ N} = -113.29 \text{ kN}$$

If the net stress at point Q ($x_Q = -33.93$ mm) is to be zero, Eq. 5.7b gives the new position h at which the force is to be applied:

$$\sigma_Q = 0 = 1/A + hx_Q/I_y$$

$$\therefore h = -\frac{I_y}{Ax_Q} = -\frac{7.3754 \times 10^6}{8750 \times (-33.93)} = 24.84 \text{ mm}$$

Example 5.6 A short column, with the rectangular section shown in Fig. 5.8, has a vertical compressive force $F = 200$ kN applied at a position with co-ordinates: $h = 25$ mm, $k = 15$ mm as shown. Calculate the resultant stress at each corner A, B, C and D.

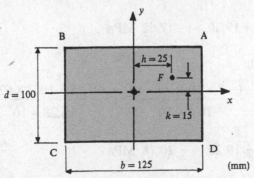

Figure 5.8 Rectangular section of a short column

The additional constant terms required for the application of Eq. 5.7a are:

$$A = bd = 125 \times 100 = 12.5 \times 10^3 \text{ mm}^2$$

$$I_x = bd^3/12 = 125(100)^3/12 = 10.4 \times 10^6 \text{ mm}^4$$

$$I_y = db^3/12 = 100(125)^3/12 = 16.28 \times 10^6 \text{ mm}^4$$

$$M_x = Fk = -200 \times 10^3 \times 15 = -3 \times 10^6 \text{ Nmm (sagging)}$$

$$M_y = Fh = -200 \times 10^3 \times 25 = -5 \times 10^6 \text{ Nmm (sagging)}$$

Here the moments are negative because Eq. 5.7a was based upon tensile bending stresses in quadrant 1 arising from hogging moments. Next, the corner co-ordinates, in mm, with their appropriate signs, must be substituted in turn. Thus, for corner A (62.5, 50), Eq. 5.7a gives:

$$\sigma_A = -\frac{200 \times 10^3}{12.5 \times 10^3} + \frac{(-3 \times 10^6) \times 50}{10.4 \times 10^6} + \frac{(-5 \times 10^6) \times 62.5}{16.28 \times 10^6}$$

$$= -16 - 14.42 - 19.20 = -49.62 \text{ MPa}$$

Corner B (−62.5, 50):

$$\sigma_B = -\frac{200 \times 10^3}{12.5 \times 10^3} + \frac{(-3 \times 10^6) \times 50}{10.4 \times 10^6} + \frac{(-5 \times 10^6) \times (-62.5)}{16.28 \times 10^6}$$

$$= -16 - 14.42 + 19.20 = -11.92 \text{ MPa}$$

Corner C (−62.5,−50):

$$\sigma_C = -\frac{200 \times 10^3}{12.5 \times 10^3} + \frac{(-3 \times 10^6) \times (-50)}{10.4 \times 10^6} + \frac{(-5 \times 10^6) \times (-62.5)}{16.28 \times 10^6}$$

$$= -16 + 14.42 + 19.20 = -17.62 \text{ MPa}$$

Corner D (62.5,−50):

$$\sigma_D = -\frac{200 \times 10^3}{12.5 \times 10^3} + \frac{(-3 \times 10^6) \times (-50)}{10.4 \times 10^6} + \frac{(-5 \times 10^6) \times 62.5}{16.28 \times 10^6}$$

$$= -16 + 14.42 - 19.20 = -20.78 \text{ MPa}$$

A material could then be chosen to provide a suitable degree of safety based upon its maximum allowable strengths. For example, the 0.1% proof stress for aluminium-copper alloys having trade names 'Duralumin' and 'Hiduminium' is ≈ 290 MPa. Allowing this value for both its compressive and tensile strengths gives safety factors of 290/49.62 = 5.8 in compression and 290/17.62 = 16.5 in tension.

5.3 Bending of Composite Beams

Where a beam is fabricated from more than one material the bending theory must be adapted to the particular cross section by ensuring that the strains between materials are compatible and that the force and moment equilibrium conditions are obeyed. These principles will now be applied to composite beam sections made from vertical and horizontal layers.

5.3.1 Balanced Vertical Layers

Vertical strips in materials A and B define the beam's cross section in Fig. 5.9. The strips are secured firmly at their common interfaces in a balanced arrangement, the simplest being where B is sandwiched between two identical strips A.

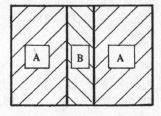

Figure 5.9 Composite beam section

Each material will attain the same radius of curvature under an applied moment M, when it follows from Eq. 5.4:

$$R = (EI/M)_A = (EI/M)_B \qquad\qquad (5.9a,b)$$

Equation 5.9a,b gives a relationship between the moments M_A and M_B supported by each material and the bending stress within them

$$M_A = M_B(EI)_A/(EI)_B \qquad\qquad (5.10a)$$

$$\sigma_A = (My/I)_A, \quad \sigma_B = (My/I)_B \qquad\qquad (5.10b,c)$$

The total moment M carried by the section, is the sum of the two moments:

$$M = M_A + M_B \qquad\qquad (5.11a)$$

Substituting Eq. 5.10a into Eq. 5.11a allows M to be expressed in terms of either M_A or M_B:

$$M = M_A[1 + (EI)_B/(EI)_A] \qquad\qquad (5.11b)$$

or

$$M = M_B[1 + (EI)_A/(EI)_B] \qquad\qquad (5.11c)$$

Example 5.7 Determine the maximum moment for the composite section shown in Fig. 5.10, given that the allowable stresses for steel (A) and wood (B) are 8.5 and 140 MPa, respectively. Take $E_A = 200$ GPa and $E_B = 8.3$ GPa.

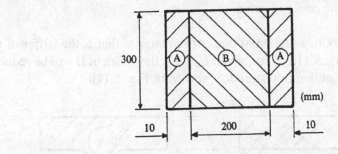

Figure 5.10 Composite section

The second moments of area are

$$I_B = (bd^3/12)_B = 200 \times 300^3/12 = 450 \times 10^6 \text{ mm}^4$$

$$I_A = 2(bd^3/12)_A = 2 \times 10 \times 300^3/12 = 45 \times 10^6 \text{ mm}^4$$

When the steel plates (A) and the wood core (B) attain their limiting stresses at the top and bottom edges, Eqs 5.10a and b give the corresponding bending moments:

$$M_A = (\sigma I/y)_A = 140 \times 45 \times 10^6/150 = 42 \text{ kNm},$$

$$M_B = (\sigma I/y)_B = 8.5 \times 450 \times 10^6/150 = 25.5 \text{ kNm}$$

When the wood is fully stressed, the steel moment is, from Eq. 5.10a, :

$$M_A = \frac{M_B(EI)_A}{(EI)_B} = \frac{25.5(200 \times 10^3 \times 45 \times 10^6)}{(8.3 \times 10^3 \times 450 \times 10^6)} = 61.45 \text{ kNm}$$

which indicates that the steel would become over-stressed. Alternatively, when the steel is fully stressed, Eq. 5.10a supplies the moment carried by the wood:

$$M_B = \frac{M_A(EI)_B}{(EI)_A} = = \frac{42(8.3 \times 10^3 \times 450 \times 10^6)}{(200 \times 10^3 \times 45 \times 10^6)} = 17.43 \text{ kNm}$$

Consequently, the wood is under-stressed to an acceptable value:

$$\sigma_B = (My/I)_B = (17.43 \times 10^6) \times 150/(450 \times 10^6) = 5.81 \text{ MPa}$$

The total moment becomes:

$$M = M_A + M_B = 42 + 17.43 = 59.43 \text{ kNm}$$

5.3.2 Horizontal Reinforcement Layers

It is convenient to work in a section of one entire material that is the stiffer of the two layers shown in Fig. 5.11a. Thus, given $E_A > E_B$, the width of B will be reduced within the equivalent section in material A, shown in Fig. 5.11b.

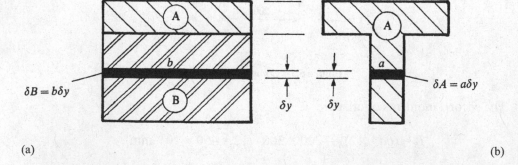

(a)   (b)

Figure 5.11 (a) Composite beam in A and B with (b) the equivalent section in A

Bending stresses σ_B and σ_A apply to elemental areas: $\delta A = b \times \delta y$ and $\delta B = a \times \delta y$ (shaded) for the original and equivalent parts of the section. One requirement is that when B is replaced by A material at the same depth suffers the same strain:

$$\varepsilon = (\sigma/E)_A = (\sigma/E)_B \therefore \sigma_B = \sigma_A E_B/E_A \qquad (5.12a)$$

Furthermore, the moments produced about the NA by the forces $\sigma_A \delta A$ and $\sigma_B \delta B$ must be the same between the original and equivalent sections:

$$\sigma_A(a\delta y)\, y = \sigma_B(b\delta y)\, y = \sigma_A(E_B/E_A)(b\delta y)\, y \qquad (5.12b)$$

Since the depth limits for y are identical, Eqs 5.12a,b are combined to give:

$$\sigma_A/\sigma_B = b/a = E_A/E_B \qquad (5.13a,b)$$

from which the equivalent width becomes

$$a = b(E_B/E_A) \qquad (5.13c)$$

Bending theory provides the stress anywhere in the equivalent section A:

$$\sigma_A = (My/I)_A \qquad (5.14)$$

The actual stress in material B follows from Eqs 5.13b and 5.14a as:

$$\sigma_B = \frac{E_B}{E_A}\sigma_A = \frac{E_B}{E_A}\left(\frac{M}{I}\right)_A y_{AB} \qquad (5.15a,b)$$

where y_{AB} refers to the position in A at which the stress in B is required. Dividing Eqs 5.15b by 5.14, the maximum σ_B is found for a given M_A as:

$$\frac{\sigma_B}{\sigma_A} = \frac{E_B\, y_{AB}}{E_A\, y_A} \qquad (5.16)$$

where y_{AB} becomes the position in A at which the stress in B attains its maximum value. A similar theory applies to beams reinforced with plates along their top and bottom surfaces. Equations 5.15a,b and 5.16 allow for a safe design based upon allowable stress levels for each material as shown in the following example.

Example 5.8 A timber beam is reinforced with steel plates along its top and bottom surfaces. Find the maximum allowable bending moment for the reinforced section in Fig. 5.12a given maximum permissible stresses of 150 MPa for steel (A) and 8.2 MPa for timber (B). Their respective elastic moduli are 210 GPa and 8.5 GPa.

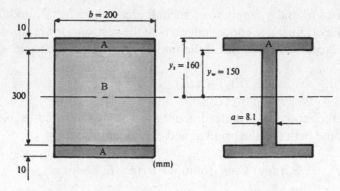

(a)   (b)

Figure 5.12 Equivalent section in steel

Using Eq. 5.13c, first replace the timber in Fig. 5.12a by the equivalent width in steel: $a = 200 \times 8.5/210 = 8.1$ mm. Using $I = \sum bd^3/12$ for the equivalent section in steel (see Fig. 5.12b):

$$I_A = 200 \times 320^3/12 - 191.91 \times 300^3/12 = 114.35 \times 10^6 \text{ mm}^4$$

When at the outer fibres the steel (A) has been fully stressed, Eq. 5.14 gives the moment as:

$$M_A = (\sigma I/y)_A = 150 \times (114.35 \times 10^6)/160 = 107.2 \text{ kNm}$$

If the timber (B) is to be fully stressed at the interface position then Eq. 5.15b gives the corresponding moment that would need to be applied to the equivalent section in Fig. 5.12b. This is:

$$M_A = \sigma_B \left(\frac{E_A}{E_B} \right) \frac{I_A}{y_{AB}} = 8.2 \left(\frac{210}{8.5} \right) \frac{114.35 \times 10^6}{150} = 154.44 \text{ kNm}$$

Selecting the lesser M_A value will ensure that the steel is fully stressed while the timber is under-stressed. Equation 5.16 provides the maximum stress value reached in the timber:

$$\sigma_B = \sigma_A \left(\frac{E_B}{E_A} \right) \frac{y_{AB}}{y_A} = 150 \left(\frac{8.5}{210} \right) \frac{150}{160} = 5.69 \text{ MPa}$$

Alternatively, the stress in the steel at the interface position is

$$\sigma_A = (My/I)_A = (107.2 \times 10^6 \times 150)/(114.35 \times 10^6) = 140.62 \text{ MPa}$$

when, from Eq. 5.12a, the stress in the timber at this position becomes

$$\sigma_B = \sigma_A E_B/E_A = 140.62 \times 8.5/210 = 5.69 \text{ MPa}$$

5.3.3 Layered Horizontal Strips

In the case of a composite beam section, made up of two or more thin horizontal layers, their individual neutral axes approximate to a common radius of curvature. The approximation applies irrespective of whether or not the strips are bonded along their interfaces as the following example shows.

Example 5.9 Rectangular strips (40 × 10 mm) of steel and brass form the composite rectangular section (40 × 20 mm) of a beam (see Fig. 5.13a). The steel lies above the brass and the beam is simply supported over a length of 1 m. Given that the allowable stresses for each material are 105 and 70 MPa, respectively, find the maximum central concentrated load W when: (a) each beam can bend independently and (b) they are bonded along the interface. Take, for steel $E = 207$ GPa and for brass $E = 84$ GPa.

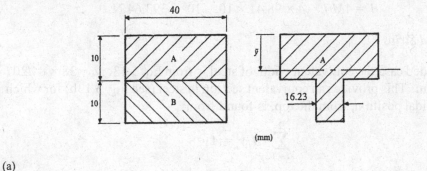

Figure 5.13 Rectangular composite section

(a) Separated, Unbonded Strips

Let A refer to the steel and B to the brass in Fig. 5.13a. When the strips rest upon one another they may be treated as separate beams each with a second moment of area:

$$I_A = I_B = \text{bd}^3/12 = 40 \times 10^3/12 = 3\tfrac{1}{3} \times 10^3 \text{ mm}^4$$

The maximum moments corresponding to the given allowable stresses are then given by Eq. 5.4:

$$M_A = (\sigma\, I/y)_A = 105 \times /5 = 70 \text{ Nm}$$

$$M_B = (\sigma\, I/y)_B = 70 \times 3\tfrac{1}{3} \times 10^3/5 = 46.67 \text{ Nm}$$

Assuming a common interface radius of curvature, with $I_A = I_B$, Eqs 5.9a,b give:

$$M_A = M_B(E_A/E_B) = 46.67(207/84) = 115 \text{ Nm}$$

This moment value shows that when brass B is fully stressed the steel A would be over-stressed. Therefore, the requirement is that the steel be fully stressed under $M_A = 70$ Nm and the brass be under-stressed. The moment carried by the brass becomes:

$$M_B = M_A(E_B/E_A) = 70\,(84/207) = 28.41 \text{ Nm}$$

The maximum stress reached in the brass is:

$$\sigma_B = (My/I)_B = (28.41 \times 10^3) \times 5/(3\tfrac{1}{3} \times 10^3) = 42.62 \text{ MPa}$$

The total moment carried by the section is

$$M = M_A + M_B = 70 + 28.41 = 98.41 \text{ Nm}$$

from which the central load is found:

$$W = 4M/L = 4 \times 98.41 \times 10^3/10^3 = 393.64 \text{ N}$$

(b) Bonded Strips

In the bonded case the equivalent width of steel is from Eq. 5.13c: $a = 38 \times 84/207 = 16.23$ mm. This provides an equivalent section in steel (see Fig. 5.13b) for which the centroidal position, from the top, is found from:

$$\sum_{i=1}^{2} A_i y_i = A\bar{y}$$

$$(40 \times 10 \times 5) + (16.23 \times 10 \times 15) = (400 + 162.3)\,\bar{y}$$

This gives $\bar{y} = 7.89$ mm. Using $I = \sum bd^3/3$ for the second moment of area:

$$I = (40 \times 7.89^3/3) + (23.77 \times 2.11^3/3) + (16.23 \times 12.11^3/3) = 16.23 \times 10^3 \text{ mm}^4$$

Re-arranging Eq. 5.14 gives the bending moment under which the steel becomes fully stressed along its upper fibres:

$$M_A = (\sigma I/y)_A = 105 \times 16.23 \times 10^3/7.89 = 216 \text{ Nm}$$

Correspondingly, Eq. 5.15b provides the maximum stress in the brass:

$$\sigma_B = \left(\frac{E_B}{E_A}\right)\left(\frac{My}{I}\right)_A = \left(\frac{84}{207}\right)\left(\frac{216 \times 10^3 \times 12.11}{16.23 \times 10^3}\right) = 65.4 \text{ MPa}$$

This shows that the brass is under-stressed, a condition which governs the maximum central load allowed:

$$W = 4M/L = 4 \times 216 \times 10^3/10^3 = 864 \text{ N}$$

in which the influence that the bonding ha upon load capacity is revealed.

5.3.4 Bi-Metallic Strip

Consider two thin, initially straight strips of dissimilar metals, 1 and 2, joined together along their longer faces to form a beam. An increase in the temperature (ΔT) will result in bending and deflection of the beam. This principle is employed in a thermostat. If we take elastic moduli $E_1 > E_2$ then the linear expansion coefficients must obey $\alpha_1 < \alpha_2$. The beam will then bend, inducing strip forces and moments, each having the sense shown in Fig. 5.14a.

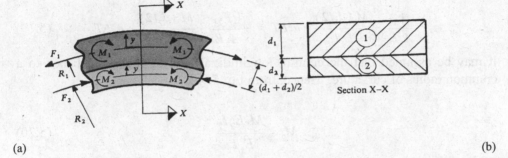

(a)   (b)

Figure 5.14 Bi-metallic strip

The longitudinal forces F_1 and F_2 arise due to restrained thermal expansion. The forces are tensile and compressive respectively because the fibres within strips 1 and 2 are stretched by more and less amounts than would be their free expansions. The problem is *statically indeterminate* and therefore its solution require a match to both its equilibrium and compatibility conditions as follows.

(a) Equilibrium

In the absence of any external axial force, horizontal force equilibrium upon section X–X in Fig. 5.14b requires that the net axial force is zero:

$$F_1 - F_2 = 0 \qquad\qquad (5.17a)$$

and therefore

$$F_1 = F_2 = F \qquad\qquad (5.17b)$$

The equal forces F produce a couple $F(d_1 + d_2)/2$ to each side of X–X as shown. As there are no moments applied externally, the couple must be resisted by moments M_1 and M_2 acting in the opposite direction to the left and right of X–X:

$$M_1 + M_2 = F(d_1 + d_2)/2 \qquad\qquad (5.18)$$

(b) Compatibility

The net axial strain within each strip is composed of three components:

(i) a *direct strain* due to F:

$$\varepsilon_i = \frac{\sigma}{E} = \frac{F}{AE}$$

(ii) a *bending strain* due to M:

$$\varepsilon_{ii} = \frac{\sigma}{E} = \frac{My}{EI}$$

(iii) a *temperature strain* due to ΔT:

$$\varepsilon_{iii} = \frac{\Delta l}{l} = \frac{\alpha l \Delta T}{l} = \alpha \Delta T$$

The F and M directions shown in Fig. 5.14a will ensure that the net strain in 1 and 2 is compatible at the interface:

$$\frac{F_1}{A_1 E_1} + \frac{M_1(d_1/2)}{E_1 I_1} + \alpha_1 \Delta T = -\frac{F_2}{A_2 E_2} - \frac{M_2(d_2/2)}{E_2 I_2} + \alpha_2 \Delta T \qquad (5.19)$$

It may be assumed that the central, longitudinal axis of each strip bends to a common radius of curvature. Hence, from Eq. 5.9a:

$$M_2 = \frac{M_1 E_2 I_2}{E_1 I_1} \qquad (5.20)$$

Combining Eqs 5.18 and 5.20:

$$M_1 = \frac{F E_1 I_1 (d_1 + d_2)}{2(E_1 I_1 + E_2 I_2)} \qquad (5.21)$$

Equations 5.17b–5.21 allow the solution to F, M_1 and M_2 simultaneously, thereby satisfying both equilibrium and compatibility conditions. While general expressions for each may be derived in this way, it is simpler to obtain specific solutions after having substituted the numbers as the following example will show.

The axial stress within each strip arises from mechanical effects only, i.e. due to the combination of F and M. In general, the axial stress, which applies to a distance y taken above or below the two central axes, is found from

$$\sigma = \pm F/A \pm My/I \qquad (5.22a)$$

The signs attach to tensile and compressive forces and hogging and sagging moments appropriately. For example, in strip 1, Eq. 5.22a gives:

$$\sigma = F_1/A_1 \pm M_1 y/I_1 \qquad (5.22b)$$

and in strip 2, Eq. 5.22a becomes:

$$\sigma = -F_2/A_2 \pm M_2 y/I_2 \qquad (5.22c)$$

The maximum deflection of a bi-metallic strip will depend upon the manner of its support. Figure 5.15 shows the central deflection δ_c, of the strip interface when its length l is simply supported at its ends:

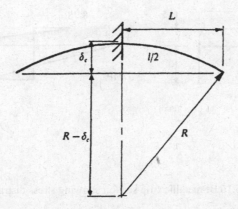

Figure 5.15 Strip deflection for bi-metallic beam

The beam's circular curvature R obeys Pythagoras's theorem:

$$R^2 = (R - \delta_c)^2 + (l/2)^2$$

Neglecting the relatively small quantity $(\delta_c)^2$ gives an acceptable approximation for the deflection:

$$\delta_c \simeq \frac{l^2}{8R} \tag{5.23a}$$

where R follows from bending theory as:

$$R = R_1 = R_2 = (EI/M)_1 = (EI/M)_2 \tag{5.23b,c}$$

The maximum end-deflection δ_e, at the free end of a bi-metallic cantilever of length L is found from putting $l = 2L$ in Eq. 5.23a (see Fig. 5.15). This gives:

$$\delta_e \simeq \frac{L^2}{2R} \tag{5.23d}$$

That is, for a similar curvature, the deflection at the end of a cantilever is the same as that at the centre of a beam simply supported over twice the length.

Example 5.10 Calculate the stresses induced at the free and interface surfaces when the copper-steel, bi-metallic strip in Fig. 5.16a is subjected to a temperature rise of 60°C. What is the maximum deflection when the strip is mounted as a cantilever 350 mm long? For steel: $E = 207$ GPa and $\alpha = 11 \times 10^{-6}/°C$ and for copper: $E = 103$ GPa and $\alpha = 17.5 \times 10^{-6}/°C$.

For consistency with the theory given above, identify 1 with steel and 2 with copper such that $E_1 > E_2$ and $\alpha_1 < \alpha_2$. The properties of their areas are:

$$A_1 = 30 \text{ mm}^2, \quad I_1 = b_1 d_1^3/12 = 20 \times 1.5^3/12 = 5.625 \text{ mm}^4$$

$$A_2 = 60 \text{ mm}^2, \quad I_2 = b_2 d_2^3/12 = 20 \times 3^3/12 = 45 \text{ mm}^4$$

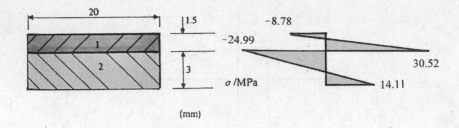

(mm)

(a)   (b)

Figure 5.16 Bi-metallic strip section showing stress distribution

Equation 5.21 gives M_1 in terms of F:

$$M_1 = \frac{F \times 207 \times 5.625\,(1.5 + 3)}{2[(207 \times 5.625) + (103 \times 45)]} = 0.4518F$$

when, from Eq. 5.20, we find M_1 in terms of F:

$$M_2 = \frac{0.4518F \times 103 \times 45}{207 \times 5.625} = 1.7983F$$

Because the directions of F and M correspond with those given in Fig. 5.14a, we may substitute directly into Eq. 5.19:

$$F\left(\frac{1}{A_1E_1} + \frac{1}{A_2E_2}\right) + F\left(\frac{0.4518d_1}{2E_1I_1} + \frac{1.7983d_2}{2E_2I_2}\right) = (\alpha_1 - \alpha_2)\Delta T$$

from which:

$$F = \frac{(17.5 - 11)\,60 \times 10^{-6}}{(0.3228 \times 10^{-6}) + (0.873 \times 10^{-6})} = 326.13\ \text{N}$$

giving

$$M_1 = 147.35\ \text{Nmm}, \quad M_2 = 586.48\ \text{Nmm}$$

The stresses in each strip follow from Eq. 5.22a. At the interface, the position of the sagging moment M_1 produces tension in the steel. Equation 5.22b gives:

$$\sigma_1 = \frac{F}{A_1} + \frac{M_1 d_1}{2I_1} = \frac{326.13}{30} + \frac{147.35 \times 1.5}{2 \times 5.625}$$

$$= 10.87 + 19.65 = 30.52\ \text{MPa}$$

At the steel's outer surface, where M_1 produces compression:

$$\sigma_1 = 10.87 - 19.65 = -8.78\ \text{MPa}$$

At the copper interface, where M_2 produces compression, Eq. 5.22c gives:

$$\sigma_2 = -\frac{F}{A_2} + \frac{M_2 d_2}{2I_2} = -\frac{326.13}{60} - \frac{586.48 \times 3}{2 \times 45}$$

$$= -5.44 - 19.55 = -24.99 \text{ MPa}$$

At the free copper surface, where M_2 produces tension:

$$\sigma_2 = -5.44 + 19.55 = 14.11 \text{ MPa}$$

The resulting distribution in stress through the section is shown in Fig. 5.16b. Within the stress distribution the net leftward and rightward forces within each strip must balance. Strain compatibility at the interface is accommodated by the *stress discontinuity* which arises at this position.

To find the maximum deflection the curvature is required from either Eqs 5.23b,c:

$$R = (EI/M)_1 = 207 \times 10^3 \times 5.625/147.35 = 7.902 \text{ m}$$

$$R = (EI/M)_2 = 103 \times 10^3 \times 45/586.48 = 7.903 \text{ m} \checkmark$$

Substituting into Eq. 5.23d reveals the deflection for activating a switch (say):

$$\delta_e = \frac{350^2}{2 \times 7.902 \times 10^3} = 7.75 \text{ mm}$$

5.4 Shear Stress Due to Shear Force

So far we have dealt with the stresses that arise from bending moments in various types of loaded beams. When the moments vary with the length there are accompanying transverse shear forces which in turn induce shear stress in the direction of the length that vary throughout the depth between adjacent layers. To demonstrate this, when a ream of A4 paper is bent by hand along its length the sheets of paper slide over one another by a miniscule amount. The prevention of this sliding in a solid beam section sets up a shear stress between its layers. Figure 5.17a shows an element δz, in the length of a beam over which the shear force and bending moment vary by δF and δM respectively. These produce corresponding variations in the shear and bending stresses of $\delta\tau$ and $\delta\sigma$ as shown upon the layer $\delta z \times \delta y \times x$ lying at a distance y from the neutral axis (NA in Fig. 5.17b).

Since there is no adjacent layer to one side of each free surface the shear stress is known to be zero. Hence the shear stress increases as the distance y from the NA decreases. Thus, for the variations shown upon the beam element in Fig. 5.17b, $\delta\tau$ is positive in the negative y-direction. In contrast the bending stress is zero at the NA so that $\delta\sigma$ is positive in the positive y-direction.

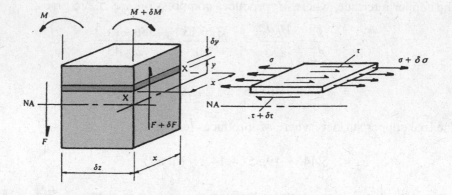

(a)   (b)

Figure 5.17 Flexural shear in a beam element

In order to find an expression for the shear stress τ, acting upon layer X–X, it is only necessary to express horizontal force equilibrium over δz between the opposing complementary shear stress and the tensile bending stress σ, arising from the hogging moments M and $M + \delta M$ as shown. This gives:

$$(\sigma + \delta\sigma)x\delta y + \tau x\delta z = \sigma x\delta y + (\tau + \delta\tau)x\delta z$$

$$x\,\delta y\,\delta\sigma = x\,\delta z\,\delta\tau$$

$$\therefore \delta\tau = \frac{x\,\delta y\,\delta\sigma}{x\,\delta z} \tag{5.24}$$

Now Eq. 5.4 gives: $\delta\sigma = \delta My/I$, within which $\delta M = F\delta z$ (see Eq. 2.13b). Substituting these relationships into Eq. 5.24 gives:

$$\delta\tau = \frac{F}{Ix}(yx\delta y) \tag{5.25}$$

Integrating Eq. 5.25 gives the longitudinal shear stress τ acting across the beam width X–X at a distance y from the NA where $x = x(y)$ (see Fig. 5.18a):

$$\tau = \frac{F}{Ix}\int_{y}^{y_1} y\,x(y)\,\mathrm{d}y \tag{5.26a}$$

where y_1 defines the top surface where the shear stress is zero. The width x, outside the integral applies to the chosen position at height y above the NA. The width function $x(y)$, accounts for variable breadth within the shaded area shown. Otherwise, for a section with constant breadth, the x's in Eq. 5.26a will cancel.

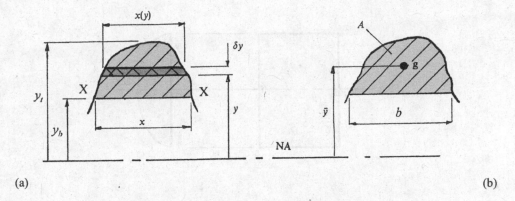

Figure 5.18 Effective section area in shear

Figure 5.18b interprets Eq. 5.26a geometrically. The integral defines the first moment of the section's shaded area A, about the centroidal neutral axis (NA) in Fig. 5.18b. Hence, an alternative expression for τ is:

$$\tau = \frac{F(A\bar{y})}{Ib} \tag{5.26b}$$

where b replaces x as the section's width at X–X and $\bar{y}$ is the distance between the centroid g of area A, and the NA. Note that I in Eq. 5.26b is the second moment of area for the whole cross-section about NA and, conventionally, F is positive when acting vertically downward. It is in the nature of *complementary shear*, considered in § 4.2.1, (p. 114) that τ found from either Eqs 5.26a,b also defines the distribution of vertical shear stress. Thus, as expected, the complementary shear stress τ is aligned with the transverse shear force F that is the resultant of the τ distribution:

$$F = \int_{y_b}^{y_t} \tau \, (b \times \mathrm{d}y) \tag{5.27}$$

where y_t and y_b are the distances of the top and bottom surfaces from the NA. Substituting either Eqs 5.26a,b into Eq 5.27 confirms expressions for I in their respective forms:

$$I = \int_A y^2 \mathrm{d}A = \int_{y_b}^{y_t} y^2 (x \mathrm{d}y) = \int_{y_b}^{y_t} (A\bar{y}) \mathrm{d}y$$

When applied to the web of an I-beam, Eq. 5.27 does not account for the contribution from the two flanges (see Example 5.12).

Example 5.11 Establish the variation in τ through the depth of a beam with rectangular section $b \times d$ (Fig. 5.19a) and show that the maximum shear stress occurs at the neutral axis, being given by $\tau_{max} = 3F/2bd$.

A simply supported timber beam of length $l = 3$ m is to be designed to carry a uniformly distributed load, $w = 20$ kN/m. Safe working stresses parallel to the timber's grain are: $\sigma_{max} = 10$ MPa in bending and $\tau_{max} = 0.75$ MPa in shear. If the depth of the rectangular beam section is $d = 300$ mm, what should be its width?

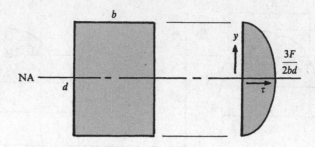

(a)   (b)

Figure 5.19 Shear stress distribution for a rectangular section

An expression showing the manner in which the shear stress varies with this section's depth follows from substituting $I = bd^3/12$, $x = b$, $y_1 = d/2$ into Eq. 5.26a but leaving the lower limit as the variable y:

$$\tau = \frac{12F}{bd^3} \int_{y}^{d/2} y\,dy$$

$$= \frac{6F}{bd^3} \left|y^2\right|_{y}^{d/2} = \frac{6F}{bd^3}\left(\frac{d^2}{4} - y^2\right) \tag{i}$$

Equation (i) reveals a parabolic variation of τ with y, as shown in Fig. 5.19b. Clearly, the maximum value, $\tau_{max} = 3F/2bd$, occurs at the NA, where $y = 0$. This result is confirmed when Eq. 5.26b is applied to the NA, for the upper-half section:

$$A\bar{y} = (b \times d/2) \times d/4 = bd^2/8$$

giving
$$\tau_{max} = \frac{F \times bd^2/8}{b \times bd^3/12} = \frac{3F}{2bd} \tag{ii}$$

Now, Fig. 2.15b shows that the magnitudes of the maximum shear force and bending moment for this beam are $wL/2$ and $wL^2/8$, respectively. Substituting the first into Eq. (ii) for a design based upon the allowable shear stress gives:

$$\tau_{max} = \frac{3F_{max}}{2bd} = \frac{3(wL/2)}{2bd}$$

from which the breadth is found

$$b = \frac{3wL}{4\tau_{max}d} = \frac{3 \times 20 \times 3 \times 10^3}{4 \times 0.75 \times 300} = 200 \text{ mm}$$

For a design based upon bending Eq. 5.4 gives:

$$\sigma_{max} = \frac{M_{max}y}{I} = \frac{(wL^2/8)(d/2)}{bd^3/12}$$

for which the breadth required is:

$$b = \frac{3wL^2}{4\sigma_{max}d^2} = \ = \frac{3 \times 20 \times (3 \times 10^3)^2}{4 \times 10 \times 300^2} = 150 \text{ mm}$$

We see that shear dictates the design of this timber beam from the selection of the greater breadth.

Example 5.12 The unsymmetrical I-section shown in Fig. 5.20a is subjected to a downward vertical shear force of 30 kN. Establish the variation in shear stress through the depth, showing its value at the neutral axis (NA) and at the web-flange intersection. What percentage of the shear force is carried by the web?

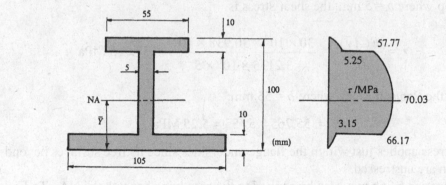

(a)   (b)

Figure 5.20 I-section under shear

The NA position (from the bottom) is found from three constituent rectangles:

$$\sum_{i=1}^{3} A_i y_i = A\bar{y}$$

$$(105 \times 10 \times 5) + (80 \times 5 \times 50) + (55 \times 10 \times 95) = (1050 + 400 + 550)\bar{y}$$

This gives $\bar{y}$ = 38.75 mm. Using $I = \sum bd^3/3$ for four 'building' rectangles with their bases lying on NA, the second moment of area becomes:

$$I = (105 \times 38.75^3/3) - (2 \times 50 \times 28.75^3/3) + (55 \times 61.25^3/3) - (2 \times 25 \times 51.25^3/3)$$
$$= 3.2135 \times 10^6 \text{ mm}^4$$

When applying Eq. 5.26b to the NA, the first moment of the whole area *above* the NA is required. Thus, for the top flange and the upper part of the web:

$$A\bar{y} = (55 \times 10 \times 56.25) + (5 \times 51.25 \times 25.625) = 37.504 \times 10^3 \text{ mm}^3$$

The NA is only 'neutral' to bending stress. For shear stress the NA marks the position of its maximum value:

$$\tau_{max} = \frac{F(A\bar{y})}{Ib} = \frac{30 \times 10^3 \times 37.504 \times 10^3}{3.2135 \times 10^6 \times 5} = 70.025 \text{ MPa}$$

It is a property of the centroid that $A\bar{y}$ is the same for areas both above and below the NA irrespective of whether the section is axi-symmetric or not. To find the shear stress at the *upper web-flange intersection*, the first moment of the flange area lying *above* the intersection is required:

$$A\bar{y} = 55 \times 10 \times 56.25 = 30.938 \times 10^3 \text{ mm}^3$$

The fact that the breadth b at the intersection is either 5 mm or 55 mm reveals how a stress discontinuity in the shear stress distribution arises at this position. Along the web top where $b = 5$ mm, the shear stress is

$$\tau = \frac{F(A\bar{y})}{Ib} = \frac{30 \times 10^3 \times 30.938 \times 10^3}{3.2135 \times 10^6 \times 5} = 57.765 \text{ MPa}$$

Along the flange bottom, where $b = 55$ mm:

$$\tau = 55.765 \times 5/55 = 5.25 \text{ MPa}$$

This stress applies just within the flange undersides since its free surfaces beyond the join are unstressed.

Equation 5.26b may also be applied to the area lying beneath the NA. To find the shear stresses at the *lower web-flange intersection*, the first moment of the area beneath the intersection is:

$$A\bar{y} = 105 \times 10 \times (38.75 - 5) = 35.438 \times 10^3 \text{ mm}^3$$

so that at the web bottom, Eq. 5.26b provides the shear stress:

$$\tau = \frac{F(A\bar{y})}{Ib} = \frac{30 \times 10^3 \times 35.438 \times 10^3}{3.2135 \times 10^6 \times 5} = 66.17 \text{ MPa}$$

and in the flange at this position, where $b = 105$ mm:

$$\tau = 66.16 \times 5/105 = 3.15 \text{ MPa}$$

The shear stress values are plotted to reveal the distribution shown in Fig. 5.20b. The area 'beneath' this distribution will equal the transverse shear force applied to the cross section. The area beneath the web distribution gives the greater part of the force carried by the web. This area is given by Eq. 5.27:

$$F_w = b\int_{-28.75}^{51.25} \tau \, dy \tag{i}$$

where $b = 5$ mm and τ is the shear stress expression at a position y in the web. The $\tau = \tau(y)$ expression is found from applying Eq. 5.26a with limits of integration for the 'T'-area lying above the web's y-position:

$$\tau = \frac{F}{5I} \left[55 \int\limits_{51.25}^{61.25} y\,dy + 5 \int\limits_{y}^{51.25} y\,dy \right] = \frac{F}{I} \left(7500.8 - \frac{y^2}{2} \right) \qquad \text{(ii)}$$

Substituting Eq. ii into Eq. i:

$$\frac{F_w}{F} = \frac{b}{I} \int\limits_{-28.75}^{51.25} \left(7500.8 - \frac{y^2}{2} \right) dy \qquad \text{(iii)}$$

Evaluating the integral in Eq. iii shows $F_w/F = 0.93$, i.e. the web carries 93% of the shear force. Also, when Eq. 5.2 is used to convert to moments the area beneath the linear bending stress distribution it was found in Ex. 5.3 that the flange carried more than 70% of the bending moment (in Exercises 5.15 and 5.16: $M_f/M > 90\%$). This explains why the I-section is so efficient in providing the desired strengths in shear and bending whilst reducing material to a minimum.

5.5 Shear Flow in Thin-Walled Sections

An interesting feature of shear stress arising from transverse shear forces applied to beams is the concept of shear flow within cross-sections having thin walls. Shear flow is the product of the shear stress and the wall thickness $q = \tau t$ and hence Eq. 5.26 may be written in the alternative form,

$$q = \frac{F_y D_x}{I_x} \qquad \text{(5.28a)}$$

where D_x and I_x are, respectively, the first and second moment of the section area about the x-axis. Both moments derive from integrating an element of the wall area $t \times \delta s$ (t is thickness, s is the mean perimeter length) about the x-axis:

$$D_x = \int_s yt\,ds \quad \text{and} \quad I_x = \int_s y^2 t\,ds \qquad \text{(5.28b,c)}$$

Within Eqs 5.28b,c, the integration for I_x must extend over the whole cross section but that for D_x applies only to the area that contains s. If the beam is not to twist we must identify F_y with a vertical force applied at the *shear centre* of thin-walled sections whether they be open or closed. The shear centre will lie along or at the intersection between the section's axes of symmetry. If its position is not obvious it will need to be calculated as the following examples will show. The shear flow q, as found from Eq. 5.28a, will be seen to vary with the dimension s directed along the section's mid-wall perimeter. The origin for s is taken to lie at a free surface in an open section and, where it exists, at the intersection between the wall and an axis of symmetry for a closed section.

Where a second, horizontal shear force F_x acts at the shear centre the respective shear flows may be added from the separate applications of forces F_x and F_y. Adopting this *principle of superposition* the net shear flow is given by:

$$q = \frac{F_y D_x}{I_x} + \frac{F_x D_y}{I_y} \tag{5.29a}$$

where, in addition to Eqs 5.28b,c:

$$D_y = \int_s xt\,ds \quad \text{and} \quad I_y = \int_s x^2 t\,ds \tag{5.29b,c}$$

The derivation of Eqs 5.28a and 5.29a requires that x and y are the section's principal axes and that, conventionally, positive shear forces F_x and F_y are applied in opposite directions to positive x and y.

5.5.1 Open Sections

(a) Channel

Consider the thin-walled channel section shown in Fig. 5.21a. The shear force is placed at the shear centre E, whose position e is to be found along the x-axis of symmetry. Also to be established are expressions for the shear flow q-distributions around the section's web and flanges, shown in Fig. 5.21b.

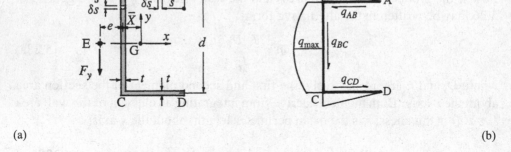

(a)   (b)

Figure 5.21 Shear flow in a uniformly thin channel with vertical force at E

Because x is a symmetry axis, the centroidal axes x and y are the section's *principal axes* and Eq. 5.28a provides the shear flow directly. The position $\bar{X}$ of its centroid G is found from taking first moments of the three rectangular areas about the vertical edge BC:

$$(dt)t/2 + 2at(a/2) = (2a + d)t\bar{X} \tag{5.30a}$$

from which

$$\frac{\bar{X}}{a} = \frac{(a/d)}{1 + 2(a/d)}.$$

which is a good approximation when $t \ll a$. With the origin of the axes x and y at the centroid, as shown, I_x is the only second moment of area required:

$$I_x = 2\left[at^3/12 + at(d/2)^2\right] + td^3/12 \approx (td^2/2)(a + d/6) \quad (5.30b)$$

Taking the origin for s at A, the shear flow in flange AB becomes:

$$q_{AB} = \frac{F_y}{I_x}\int_0^s yt\,ds = \frac{F_y}{I_x}\int_0^s (d/2)t\,ds = \frac{F_y s}{d(a + d/6)} \quad (5.31)$$

Equation 5.31 gives a linear distribution having its maximum at B, i.e. q_B for $s = a$ (as shown in Fig. 5.21b). This q_B-value reappears in the shear flow integral for the web BC as follows:

$$q_{BC} = \frac{F_y}{I_x}\int_0^s yt\,ds + q_B = \frac{F_y}{I_x}\int_0^s (d/2 - s)t\,ds + \frac{F_y a}{d(a + d/6)} \quad (5.32a)$$

where the origin for s is at B. Integrating Eq. 5.32a and substitution for I_x gives:

$$q_{BC} = \frac{F_y t}{I_x}\left|\frac{ds}{2} - \frac{s^2}{2}\right|_0^s + \frac{F_y a}{d(a + d/6)} = \frac{F_y[s(1 - s/d) + a]}{d(a + d/6)} \quad (5.32b)$$

Equation 5.32b describes the parabolic distribution shown in Fig. 5.21b, this revealing that the maximum shear flow occurs at the neutral axis ($s = d/2$):

$$q_{max} = \frac{F(a + d/4)}{d(a + d/6)} = \frac{3F[1 + 4(a/d)]}{2d[1 + 6(a/d)]} \quad (5.32c)$$

Equation 5.32c provides a maximum shear stress $\tau_{max} = q_{max}/t$, which is identical to that which would be found at the web-centre for a uniform I-beam with flange lengths a and web depth d. However, the corresponding shear flow analyses must refer to the condition that F_y is applied at the shear centre E (see Fig. 5.21a). While E coincides with the centroid for an I-section, it lies outside the channel section, distance e from the web as shown. The requirement that F_y does not twist the section is met when the torque applied to the section $T = F_y e$ is equilibrated by the net torque due to the q distributions in each limb. Such a *static equivalence* applies to *any* point that we care to choose, so by taking the corner point C, for example, only the shear flow in AB is influential. That is, taking moments at C:

$$\sum T_C = F_y e - d\int_0^a q_{AB}\,ds = 0 \quad (5.33a)$$

and substituting q_{AB} from Eq. 5.31, e is found from Eq. 5.33a as:

$$F_y e - \frac{F_y d}{d(a + d/6)}\int_0^a s\,ds = 0, \quad \therefore \frac{e}{a} = \frac{3(a/d)}{1 + 6(a/d)} \quad (5.33b)$$

If F_y is not applied at E then shear stress arises from both shear and torsion. Such a combination arises when F_y acts at the centroid G or along the web. The reader is referred to the specialist chapters within [1, 2] for these analyses.

(b) Semi-Circular Channel

Let perpendicular shear forces F_x and F_y be applied at the shear centre E of a semi-circular channel, with mean radius r and uniform thickness t, as shown in Fig. 5.22.

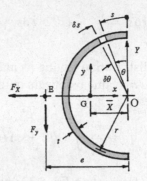

Figure 5.22 Semi-circular channel with shear forces applied at E

The position $\overline{X}$ of the centroid G is found from summing the first moments of an elemental area $t\delta s$ about the Y-axis. With $\delta s = r\delta\theta$, this gives:

$$\left[2rt \int_0^{\pi/2} d\theta \right] \overline{X} = 2r^2 t \int_0^{\pi/2} \sin\theta \, d\theta, \qquad \therefore \ \overline{X} = \frac{2r}{\pi} \tag{5.34a}$$

The second moments of the elemental area $t\delta s$ about the x-and y-axes are:

$$I_x = 2r^3 t \int_0^{\pi/2} \cos^2\theta \, d\theta, \qquad \therefore \ I_x = \frac{\pi r^3 t}{2} \tag{5.34b}$$

$$I_Y = 2r^3 t \int_0^{\pi/2} \sin^2\theta \, d\theta, \qquad \therefore \ I_Y = \frac{\pi r^3 t}{2} \tag{5.35a}$$

Transferring I_Y in Eq. 5.35a, to the parallel, centroidal axis y:

$$I_y = I_Y - A\overline{X}^2 = \frac{\pi r^3 t}{2} - (\pi r t)(2r/\pi)^2 = r^3 t (\pi/2 - 4/\pi) \tag{5.35b}$$

The first moment of the elemental area $t \times \delta s$ at distance s from the top free edge, about the centroid's y-axis is:

$$D_x = \int_0^s yt \, ds = r^2 t \int_0^\theta \cos\theta \, d\theta = r^2 t \sin\theta \tag{5.36a}$$

and D_y for axis x, with $x = \overline{X} - r\sin\theta$ becomes:

$$D_y = \int_0^s xt \, ds = r^2 t \int_0^\theta (2/\pi - \sin\theta) \, d\theta = r^2 t (2\theta/\pi + \cos\theta - 1) \tag{5.36b}$$

Substituting the first and second moment of Eqs 5.34b, 5.35b and 5.36a,b into Eq. 5.29a gives the net shear flow expression:

$$q_\theta = \frac{2}{r\pi} \left[F_y \sin\theta + F_x \frac{(2\theta/\pi + \cos\theta - 1)}{(1 - 8/\pi^2)} \right] \tag{5.37a}$$

Case I Equal Forces

When $F_x = F_y = F$ in Eq. 5.37a, the q_θ distribution in Fig. 5.23a applies.

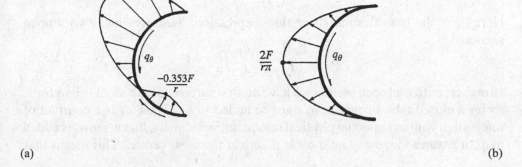

(a)   (b)

Figure 5.23 Shear flows in a semi-circular tube for cases I and II

The maximum and minimum q_θ shown are found from applying $dq_\theta/d\theta = 0$ to Eq. 5.37a. They are, respectively:

$$\theta = 49.45°, \; q_\theta = 1.154F/r$$

$$\theta = 152.01°, \; q_\theta = -0.353F/r$$

$$\theta = 119.3°, \; q_\theta = 0.$$

Case II Single Vertical Force

Here $F_x = 0$, when Eq. 5.37a reduces to:

$$q_\theta = \frac{2F_y}{r\pi} \sin\theta \qquad (5.37b)$$

which has its maximum $q_\theta = 2F/r\pi$ for $\theta = 90°$, as shown in Fig. 5.23b. The position e of the shear centre E in Fig. 5.22 follows from taking moments about the centre O. The shear flow must react to the applied torque $F_y e$. Firstly, q_θ from Eq. 5.37b is multiplied by $\delta s = r \times \delta\theta$ and then by r to give:

$$F_y e = \int_0^\pi q_\theta r^2 d\theta = \frac{2F_y r^2}{r\pi} \int_0^\pi \sin\theta \, d\theta \qquad (5.38a)$$

from which the position e from O of the shear centre E is found as

$$e = 4r/\pi \qquad (5.38b)$$

In taking moments about O it is seen that Eqs 5.38a,b apply to both distributions in Figs 5.23a,b. It follows that the shear centre is a property of the cross section being independent of the forces applied to it.

5.5.2 Thin-Walled, Closed Tubes

The net shear flow in a closed, thin-walled tube under a single vertical force F_y applied through the shear centre E, is written as

$$q = q_b + q_E \qquad (5.39a)$$

Here q_b is the base flexural shear flow expression, used previously with open sections:

$$q_b = \frac{F_y D_{xb}}{I_x} = \frac{F_y}{I_x} \int_A y \, dA \qquad (5.39b)$$

However, unlike an open section we have no free surface to take as an origin for s, so for a closed tube a constant q_E must be added to q_b. Thus, q_E is a constant of integration with an important physical interpretation; namely, that when q_E is added to q_b, it ensures the *rate of twist* $\delta\theta/\delta z$ is zero at the shear centre. This means that

$$\oint q_b \frac{ds}{t} + q_E \oint \frac{ds}{t} = 0 \qquad (5.40a)$$

which allows q_E to be found (see Eq. 7.34a, p. 266). If the position of the shear centre E is not obvious it may be found from combining Eq. 5.40a with the counterbalance required between the torque from the net shear flow $q = q_b + q_E$ and the torque with F_y applied at E (see Fig. 5.24).

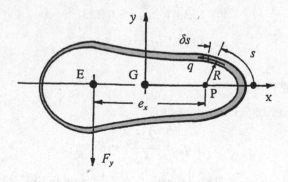

Figure 5.24 Shear centre for a thin-walled closed tube

Each torque is referred to a convenient reference point P. This torque balance is written as:

$$F_y e_x = \oint q R \, ds = \oint q_b R \, ds + q_E \oint R \, ds \qquad (5.40b,c)$$

where e_x is the horizontal distance required between E and P. Within the first path integral in Eq. 5.40b, R is a perpendicular distance from P to the net incremental force $q \, \delta s$. Alternatively, splitting this integral in Eq. 5.40c recognises that q_E is a constant when $\oint R ds$ becomes twice the area enclosed by the wall's mean centre line. The following examples, employing relatively simple sections, will demonstrate the use of Eqs 5.39a,b and 5.40a,c.

(a) Thin-Walled Circular Tube

Let a vertical force F_y be applied at the centre of a thin, circular tube of mean wall radius r and uniform wall thickness t, as shown in Fig. 5.25a. As F_y lies on the vertical centre line of the tube cross section, it passes through the shear centre E. The latter obviously coincides with the circle's centre (centroid), which is also the centre of twist, as it has no rotation under torsion.

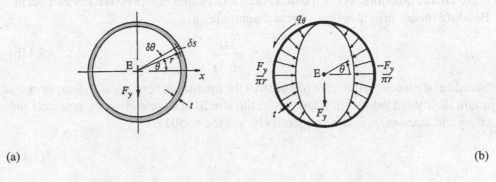

(a)   (b)

Figure 5.25 Circular tube under a central, vertical shear force

The first moment of the elemental area $t\delta s$ of the tube wall about the x-axis is:

$$D_{xb} = \int y(t\,ds) = 2tr^2 \int_0^{\pi} \sin\theta\,d\theta = r^2 t(1 - \cos\theta) \qquad (5.41a)$$

where $\delta s = r \times \delta\theta$. The second moment of this area about the x-axis is:

$$I_x = \int y^2(t\,ds) = 2tr^3 \int_0^{\pi} \sin^2\theta\,d\theta = \pi r^3 t \qquad (5.41b)$$

Substituting Eqs 5.41a,b into Eq. 5.39b gives q_b as:

$$q_b = \frac{F_y r^2 t(1 - \cos\theta)}{\pi r^3 t} = \frac{F_y}{\pi r}(1 - \cos\theta) \qquad (5.42a)$$

Next, we find q_E from applying Eq. 5.40a to the q_b distribution in Eq. 5.42a. This gives:

$$\frac{F_y}{\pi t r} \int_0^{2\pi} (1 - \cos\theta)\,r\,d\theta + \frac{q_E}{t} \int_0^{2\pi} r\,d\theta = 0$$

from which the q_E distribution is found:

$$\frac{F_y}{\pi t}\left|\theta - \sin\theta\right|_0^{2\pi} + \frac{2\pi r q_E}{t} = 0, \qquad \therefore \quad q_E = -\frac{F_y}{\pi r} \qquad (5.42b)$$

Adding Eqs 5.42a,b gives the net shear flow distribution $q_\theta = q_b + q_E$, as

$$q_\theta = \frac{F_y}{\pi r}(1 - \cos\theta) - \frac{F_y}{\pi r} = -\frac{F_y}{\pi r}\cos\theta \qquad (5.43)$$

The net shear flow distribution from Eq. 5.43 is shown in Fig. 5.25b. There appear two maxima, $q_{max} = \mp (F_y)/(\pi r)$, at $\theta = 0°$ and $180°$ respectively, as shown. At each orientation the magnitude of the maximum shear stress in the tube becomes:

$$\tau_{max} = \frac{F_y}{\pi r t} \qquad (5.44a)$$

When $\theta = 90°$ and $270°$ it is seen from Eq. 5.43 that $\tau = 0$ but for a tube in bending these are the positions where the maximum tensile and compressive stresses occur. Bending theory provides these stress magnitudes as:

$$\sigma_{max} = \frac{Mr}{\pi r^3 t} \qquad (5.44b)$$

Should a beam design criterion require that the maximum bending and shear stresses attain their yield values simultaneously, the von Mises yield criterion connects the two yield stresses, σ_y and τ_y, respectively, as (see p. 501):

$$\tau_y = \frac{\sigma_y}{\sqrt{3}} \qquad (5.45)$$

If the tube is mounted as a cantilever with an end load W, then $|F_y| = W$ and $|M| = WL$ in Eqs 5.44a,b when from Eq. 5.45:

$$\frac{W}{\pi r t} = \frac{WL}{\sqrt{3}\,\pi r^2 t}, \qquad \therefore\ L = \sqrt{3}r \qquad (5.46a)$$

It follows that for a short cantilever, where $L < \sqrt{3}r$, shear dictates the design. When $L > \sqrt{3}r$ a bending design applies to the cantilever. If a similar design criterion is applied when the loading is distributed uniformly, we have $|F_y| = wL$ and $|M| = wL^2/2$. Equation 5.45 gives the critical length as:

$$\frac{wL}{\pi r t} = \frac{(wL^2)r}{2\sqrt{3}\,\pi r^3 t}, \qquad \therefore\ L = 2\sqrt{3}r \qquad (5.46b)$$

For the two cantilevers, both Eqs 5.46a and b apply to the tubular cross section at the encastré fixing. The reader should now show for simply supported, thin-walled tubes of length L carrying (i) a central concentrated load W and (ii) uniformly distributed loading w/unit length, that the critical lengths are each $L = 4\sqrt{3}r$. Though the result is similar beam (i) refers to the central section, where both F and M are maximum but for beam (ii) the analysis refers to the centre for bending and the ends for shear, i.e. the respective yield stresses are reached within cross sections at these positions in the length.

(b) Rectangular Tube

Let a vertical shear force F_y be applied along the vertical centre line of a rectangular tube having breadth b and depth d with uniform wall thicknesses t, in Fig. 5.26a.

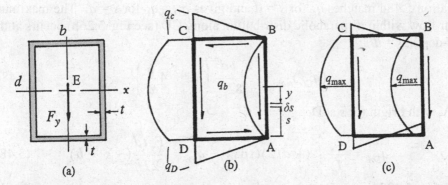

Figure 5.26 Rectangular tube showing shear flows q_b, and q_{net} with F_y applied at the shear centre

The second moment of this area about the x-axis is:

$$I_x = 2\left[\frac{bt^3}{12} + bt\left(\frac{d}{2}\right)^2\right] + 2\left(\frac{td^3}{12}\right) \approx \frac{td^3}{6}\left[1 + 3\left(\frac{b}{d}\right)\right] \qquad (5.47)$$

The q_b shear flow distributions within each side follow from Eq. 5.39b:

A–B, with origin for s at A:

$$q_{AB} = \frac{F_y t}{I_x}\int_0^s (s - d/2)\mathrm{d}s = \frac{F_y t}{2I_x}(s^2 - ds) \qquad (5.48a)$$

Equation 5.48a gives $q_A = q_B = 0$ for $s = 0$ and d, respectively. The maximum in the parabolic distribution occurs for $s = d/2$:

$$q = -\frac{F_y t d^2}{8I_x}$$

in which the minus sign implies q opposes the s-direction (see Fig. 5.26b).

B–C, with origin for s at B:

$$q_{BC} = \frac{F_y}{I_x}\int_0^s \frac{d}{2}(t\,\mathrm{d}s) + q_B = \frac{F_y dts}{2I_x} \qquad (5.48b)$$

Equation 5.48b matches $q_B = 0$ for $s = 0$. Figure 5.26a shows a maximum at C in the linear distribution where $s = b$:

$$q_C = \frac{F_y tdb}{2I_x} \qquad (5.48c)$$

C–D, with origin for s at C:

$$q_{CD} = \frac{F_y}{I_x}\int_0^s (d/2 - s)t\,\mathrm{d}s + q_C = \frac{F_y t}{2I_x}(ds - s^2) + \frac{F_y tbd}{2I_x} \qquad (5.48d)$$

Equation 5.48d matches q_C for $s = 0$ and gives $q_D = q_C$ for $s = d$. The maximum shear flow within its parabolic distribution along CD (see Fig. 5.26b) occurs at the half-depth, $s = d/2$:

$$(q_{CD})_{max} = \frac{F_y t d^2}{8 I_x} \left(1 + 4 \frac{b}{d} \right) \tag{5.48e}$$

D–A, with origin for s at D:

$$q_{DA} = \frac{F_y}{I_x} \int_0^s (-d/2)(t\,ds) + q_D = \frac{F_y t d}{2 I_x}(-s + b) \tag{5.48f}$$

Equation 5.48f gives $q_A = 0$ for $s = b$ and matches q_D for $s = 0$, with q_D being the maximum in the linear distribution, as shown in Fig. 5.26b.

Since F_y acts at the shear centre (the centroid) this requires that q_E be found from Eq. 5.40a:

$$\left[\int_s q_{AB}\,ds/t + \int_s q_{BC}\,ds/t + \int_s q_{CD}\,ds/t + \int_s q_{DA}\,ds/t \right] + q_E \oint ds/t = 0 \tag{5.49a}$$

Substituting the q_b distributions (i.e. q_{AB}, q_{BC} etc.) from Eqs 5.48a–f into Eq. 5.49a:

$$\frac{F_y}{2 I_x} \int_0^d (s^2 - ds)\,ds + \frac{F_y d}{2 I_x} \int_0^b s\,ds + \frac{F_y}{2 I_x} \int_0^d (ds - s^2)\,ds + \frac{F_y bd}{2 I_x} \int_0^d ds$$

$$- \frac{F_y d}{2 I_x} \int_0^b s\,ds + \frac{F_y bd}{2 I_x} \int_0^b ds + 2 q_E \left(\frac{b}{t} + \frac{d}{t} \right) = 0$$

from which

$$q_E = - \frac{F_y bdt}{4 I_x} \tag{5.49b}$$

Equation 5.49b describes the constant shear flow in the walls of the tube that is to be added to the q_B distribution in Fig. 5.26b. This results in the net shear flow shown in Fig. 5.26c. In particular, adding q_E from Eq. 5.49b to Eq. 5.48e leads to the equal maximum, net shear flows at the mid-sides of AB and CD:

$$(q_{CD})_{max} = \frac{F_y t d^2}{8 I_x} \left(1 + 2 \frac{b}{d} \right) = - (q_{AB})_{max} \tag{5.50a}$$

Similarly, adding Eq. 5.49b to Eq. 5.48c for the maximum shear flow at C in the side BC (and at D in side DA):

$$(q_{BC})_{max} = \frac{F_y bdt}{4 I_x} = (q_{DA})_{max} \tag{5.50b}$$

Net shear flows of similar magnitude to Eq. 5.50b apply to points A and B (with a sign change). Substituting for I_x from Eq. 5.47 into Eq. 5.50a, the greatest shear stress ($\tau = q/t$), which lies at mid-side CD, is

$$(\tau_{CD})_{max} = \frac{3\,F_y\,[1\,+\,2\,(b/d)]}{4\,t\,d\,[1\,+\,3\,(b/d)]} \qquad (5.50c)$$

Equation 5.50b gives the lesser maximum stress at C in the side BC:

$$(\tau_{BC})_{max} = \frac{3\,F_y\,b}{2\,t\,d^2\,[1\,+\,3\,(b/d)]} \qquad (5.50d)$$

The net distribution in Fig. 5.26b is symmetrical and therefore Eq. 5.50c applies to the centre of both vertical walls and Eq. 5.50d applies to each corner. A shear design requires that Eq. 5.50c should not exceed an allowable shear stress for the tube material. In the case of a *thin-walled square tube* of side length a and wall thickness t, Eqs 5.50c,d give the maximum shear stress at the corresponding positions in the vertical and horizontal walls:

$$(\tau_{CD})_{max} = \frac{9\,F_y}{16\,at}, \qquad (\tau_{BC})_{max} = \frac{3\,F_y}{8\,at} \qquad (5.51a,b)$$

Hence, the basis of a safe elastic design lies with Eq. 5.51a with the maximum shear stress being 50% greater than that from Eq. 5.51b. For equal maxima: $t_{BC}/t_{CD} = 2/3$.

References

[1] Rees, D. W. A. *Mechanics of Solids and Structures*, 2nd Ed., I. C. Press, 2015.
[2] Rees, D. W. A. *Mechanics of Optimal Structural Design, Minimum Weight Structures*, Wiley, 2009.

Exercises

Bending of Beams

5.1 An 8 kN concentrated load acts vertically downwards along the vertical centre line of the T-section in Fig. 5.2a at the centre of a 15 m long, simply supported steel beam. Find the maximum stresses and the curvature for the section at the position where the maximum sagging moment produces compression in the flange and tension in the web. Plot the stress distribution across the depth and establish the minimum safety factor for the beam. Take, for the steel: $E = 208$ GPa and a maximum allowable stress of 300 MPa.
(Ans: 219.1, –90.2, 100.8 m, 1.37)

5.2 A T-section beam 6 m long is simply supported at its ends with the flange uppermost. The width of the flange is 150 mm and its depth is 25 mm. The web thickness is 25 mm and its depth is 125 mm. Calculate the greatest distributed loading the beam can support when the maximum tensile and compressive stresses are limited to 55 and 31 MPa respectively.
(Ans: 1.16 kN/m)

5.3 A light wooden bridge is supported by six parallel timber beams each being 300 mm deep and 200 mm wide. Each beam may be considered as simply supported over a 4.5 m span. If the allowable bending stress in the timber is 5.5 MPa calculate the greatest distributed loading the bridge can support.
(Ans: 4057 kg/m)

5.4 Estimate the greatest floor loading that can be carried over a 4 m span when it is simply supported on parallel timber joists 170 mm deep × 45 mm wide spaced at 400 mm intervals.
(Ans: 1.9 kN/m^2)

5.5 A 1.2 m length circular shaft is supported in bearings at its ends and carries concentrated loads of 20 and 30 kN at respective distances of 0.45 and 0.75 m from the left-hand end. If the maximum stress in the material is to be 60 MPa at all points in the length determine the shaft profile.

5.6 The section in Fig. 5.27 is simply supported as a beam over 900 mm and carries a central concentrated load of 2 t. Find the maximum bending moment and the greatest stress in the material.
(Ans: 0.45 tm, 87.59 MPa)

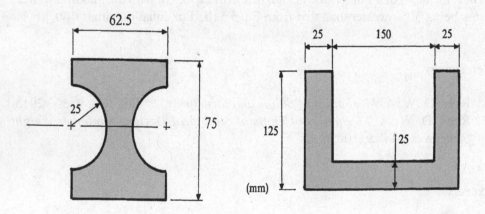

Figure 5.27   **Figure 5.28**

5.7 The channel section in Fig. 5.28 is 3 m long and simply supported at its ends with the legs uppermost. Calculate the greatest distributed loading the beam can carry if the tensile and compressive stresses are limited to 31 and 55 MPa respectively. Sketch the distribution of bending stress across the section.
(Ans: 872 kg/m)

5.8 The I-section in Fig. 5.29 is to carry a total load of 20 tonne uniformly distributed over its 7 m simply supported length. If the allowable bending stress is 150 MPa find the dimensions b and d, where $b = d/2$.

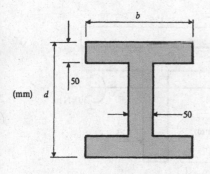

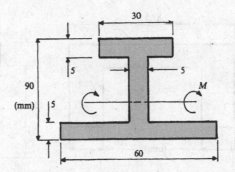

Figure 5.29   **Figure 5.30**

5.9 The beam section in Fig. 5.30 is subjected to a bending moment M, acting in the sense shown. If the maximum tensile and compressive stresses are limited to 70 and 95 MPa, respectively, calculate the greatest value of M.
(Ans: 1772 Nm)

5.10 The channel section in Fig. 5.31, when mounted as a cantilever beam of length 4 m, is to carry a uniformly distributed loading of 200 N/m. Calculate the maximum tensile and compressive stresses in the beam.
(Ans: 94.7 MPa, 150.8 MPa)

5.11 An I-beam (see Fig. 5.32) is simply supported over a 3.6 m length. If the tensile and compressive stresses are limited to 31 and 23 MPa respectively, what is the greatest distributed loading the beam can support?
(Ans: 1.1 kN/m)

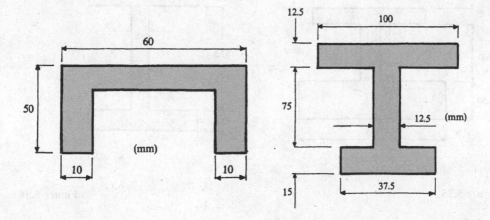

Figure 5.31   **Figure 5.32**

5.12 A cantilever beam, 1.8 m long, with the inverted channel section in Fig. 5.33 carries a uniformly distributed loading of 290 N/m together with a concentrated load of 220 N at a point 0.6 m from the free end. Determine the maximum tensile and compressive stresses due to bending.

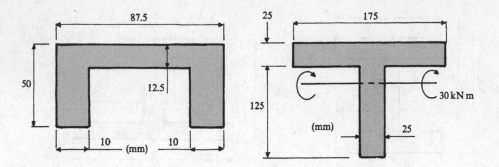

Figure 5.33   **Figure 5.34**

5.13 Calculate, for the T-section shown in Fig. 5.34, the maximum bending stress and the radius of curvature when an applied moment of 30 kNm causes compression in the flange. Take $E = 208$ GPa.
(Ans: 219.1 MPa, 100.8 m)

5.14 A cast-iron beam 9 m long with the section given in Fig. 5.35 is simply supported at its ends. Given that the density of cast iron is 6900 kg/m³, calculate the maximum tensile and compressive stresses in the beam due to its own weight.
(Ans: −29.9, 41.4 MPa)

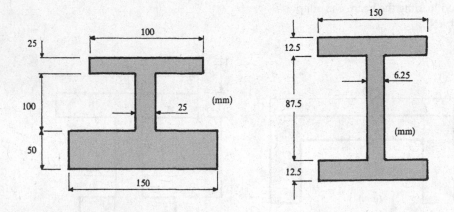

Figure 5.35   **Figure 5.36**

5.15. Find the maximum moment for the I-section in Fig. 5.36 when the working stress is limited to 77 MPa. Find the radius of curvature under this moment. What percentage of the moment is carried by the two flanges? Take $E = 207$ GPa.
(Ans: 13.38 kNm, 151.2 m, 96.4%)

5.16 If in Example 5.12 (p. 175) the vertical force of 30 kN is applied to the centre of a 1 m long, simply supported beam, show that 91.8% of the maximum bending moment is carried by the flanges. Is the percentage altered for an off-centre position of this force?

Combined Bending and Direct Stress

5.17 A cantilever 125 mm long supports a force of 10 kN inclined at 15° to its horizontal axis at the free end. If the section is rectangular 50 mm wide and 25 mm deep, determine the net stress distribution across the vertical section at the fixed end.

5.18 A simply supported beam, of rectangular section, 75 mm deep with cross-sectional area 1850 mm^2 and second moment of area 300×10^4 mm^4, carries a uniformly distributed load of 45 kN/m over a 1.5 m length. Determine the distribution of stress at mid-span when the beam supports an additional axial compressive force of 50 kN.

5.19 A cast-iron column, 150 mm o.d. and 112.5 mm i.d., carries a vertical load of 20 tonne. Find the maximum allowable eccentricity of the load if the maximum tensile stress is not to exceed 30 MPa. What is the maximum compressive stress?
(Ans: 65 mm, 80.7 MPa)

5.20 Figure 5.37 shows the section of a hydraulic riveter. Determine the maximum tensile and compressive stresses in the section when a compressive load of 30 tonne is applied at the anvil centre (C).
(Ans: 829.4 MPa, −929.5 MPa)

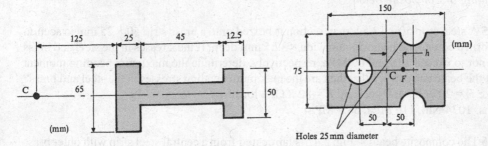

Figure 5.37	**Figure 5.38**

5.21 The cross section of a short column is shown in Fig. 5.38. For what value of h, defining the position of a compressive point force F, will the tensile stress at the section's left edge be zero? Also find the maximum compressive stress when $F = 24$ tonne.
(Ans: 24.2 mm, 44.97 MPa)

5.22 A cast-iron column 200 mm in external diameter and 165 mm in internal diameter carries an eccentric compressive load P, at 330 mm from the axis of the column together with an axial compressive load of 800 kN. Calculate the maximum value of P if the limiting tensile and compressive stresses in the material are 31 and 123.5 MPa, respectively.
(Ans: 62.78 kN)

5.23 A compressive force F acts normally to the section given in Fig. 5.39. Determine the maximum eccentricity h if there is to be no tensile stress across the section. What is F if the maximum compressive stress is 92.5 MPa at this h value? Establish the area within which F must lie to avoid tensile stress in the section.
(Ans: 25 mm, 976 kN)

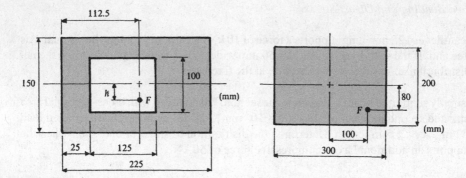

Figure 5.39   **Figure 5.40**

5.24 The short column of rectangular cross section in Fig. 5.40 has a normal vertical compressive force $F = 50$ kN, applied eccentrically to both principal axes. Calculate the resultant stress at each corner of the section.

Bending of Composite Beams

5.25 A steel strip 50×12.5 mm in section is brazed onto a brass strip 50×25 mm in section to form a compound beam 50 mm wide $\times$ 37.5 mm deep. If the stresses in the steel and brass are not to exceed 138 and 69 MPa, respectively, determine the maximum bending moment that the beam can carry. What then are the maximum bending stresses in the steel and brass? Take $E = 207$ GPa for steel and $E = 82.8$ GPa for brass.
(Ans: 1034 Nm, 110 MPa, 68.9 MPa)

5.26 The composite beam in Fig. 5.41 is fabricated from a central steel strip with outer brass strips. If the allowable stresses for each material are 110 and 80 MPa, respectively, determine the maximum load that could be applied to this beam at the centre of its 2 m, simply supported length. What are the maximum stresses actually achieved in each material? For steel $E = 210$ GPa and for brass $E = 85$ GPa.

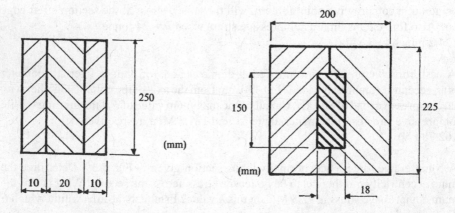

Figure 5.41   **Figure 5.42**

5.27 The composite beam in Fig. 5.42 is made from timber joists reinforced with a central steel strip. Calculate the moment of resistance for the section when the maximum bending stress in the timber of 8.25 MPa. What then is the maximum bending stress in the steel? Take $E = 207$ GPa for steel and $E = 12.4$ for timber.
(Ans: 21 kNm, 91.3 MPa)

5.28 A timber beam, 200 mm wide and 300 mm deep, is reinforced along its bottom edge by a steel plate 200 mm wide and 12.5 mm thick to give a composite beam 312.5 mm deep × 200 mm wide. Calculate the maximum stresses in the steel and timber when the beam carries a uniformly distributed load of 14.6 kN/m over a simply supported span of 6 m. Take E(steel) $= 20E$(timber).
(Ans: 106 MPa, 12.8 MPa)

5.29 A timber beam, 75 mm wide × 150 mm deep, is reinforced by fixing aluminium strips 75 mm wide on to its top and bottom surfaces for the whole of its length. The moment of resistance of the composite beam is to be four times that of the timber alone whilst maintaining the same value of maximum bending stress in the timber. Determine the thickness of the aluminium strips and the ratio of the maximum bending stresses in each material. Take E (aluminium) $= 7.15$ E (timber).
(Ans: 9.3 mm, 8.03)

5.30 A lightweight glass-fibre beam, 400 mm long and 10 mm wide is reinforced with carbon-fibre of the same width along its outer edges to occupy 20% of the total beam volume. If the simply supported length of the beam is 20 times its total depth, determine the maximum central concentrated load that the beam can carry when the allowable bending stresses for each material are 208 MPa for glass fibre and 1040 MPa for carbon fibre. The respective moduli are 16 GPa and 124 GPa.
(Ans: 3.84 kN)

5.31 A bi-metallic strip in a temperature controller consists of brass bonded to steel, each material having the same rectangular cross section and length. Calculate the stresses set up at the outer and interface surfaces in the longitudinal direction for an increase in temperature of 50°C in the strip. Neglect distortion and any transverse stress. For steel: $\alpha = 11 \times 10^{-6}$/°C, $E = 207$ GPa and for brass: $\alpha = 20 \times 10^{-6}$/°C, $E = 103$ GPa.
(Ans: steel: -25.4, 42.3 MPa; brass: -25.29, 8.39 MPa)

5.32 A thermostatic control-unit employs a bi-metallic cantilevered strip to operate a switch at its free end. If the metal strips are each 100 mm long and 15 mm × 1 mm in cross section, calculate the common radius of curvature and deflection at the free end for a temperature rise of 50°C. Take $E = 200$ GPa, $\alpha = 10 \times 10^{-6}$/°C for one material and $E = 150$ GPa, $\alpha = 20 \times 10^{-6}$/°C for the other material.
(Ans: 2.68 m, 1.87 mm)

Shear Stress Due to Shear Force

5.33 Derive expressions for the maximum shear stress due to shear force F in each of the sections given in Fig. 5.43a,b,c. Find the ratio between the maximum shear stress and the mean shear stress in each case.
(Ans: $3F/2bd$, $4F/3\pi R^2$, F/a^2, 3/2, 4/3, 1)

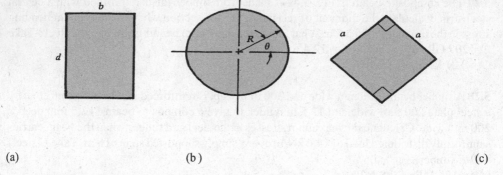

(a)   (b)   (c)

Figure 5.43

5.34 Draw the distribution of shear stress when each section in Figs 5.44 a and b, is subjected to a transverse shear force of 30 kN. Indicate τ at the flange-web interface and at the NA. [Ans:(a) 4.12, 28.87, 36.2 MPa; (b) 4.35, 15.24, 21.05 MPa]

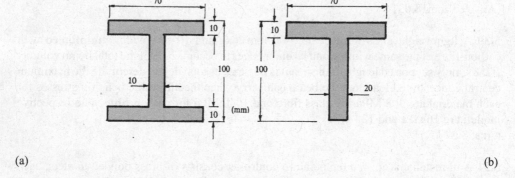

(a)   (b)

Figure 5.44

5.35 An I-beam, with the section in Fig. 5.44a, is simply supported over a span of 3.65 m and carries a total uniformly distributed load of 2 t over its length. Find the maximum shear stress due to shear force and state where in the length and cross section this occurs. (Ans: 118.4 MPa)

5.36 A cantilever 1.5 m long is formed from bolting two 150 × 100 mm timbers together to give a cross section 150 mm wide × 200 mm deep. Clamping bolts of 12.5 mm diameter are spaced 150 mm apart along he beam. Calculate the shear stress in each bolt caused by a concentrated vertical load of 6.5 kN applied at the free end. (Ans: 65 MPa)

5.37 Find, based upon both bending and shear designs, a suitable mean radius for a short cantilever tube 50 mm long with a 1 mm wall thickness in 5083 wrought, aluminium alloy, with a 0.2% proof stress of 235 MPa. The tube is to carry a 5 kN end load with a safety factor of 2. What radius ensures an equal influence of bending and shear upon the design? (Ans: 26.02 mm, 23.46 mm, 28.87mm)

CHAPTER 6

SLOPE AND DEFLECTION OF BEAMS

Summary: Various methods are available for calculating the deflected shape of a transversely loaded beam. The flexure equation uses the beam's curvature to express its slope and deflection in the length. The moment area method interprets an integration of the said equation within the area of its bending moment diagram. Integrating by Macaulay's step function method accounts for discontinuities in the bending moment diagram. When Hooke's law applies so too do the superposition principle and Maxwell's reciprocal law. Each is used to connect many applied loads to the elastic displacements produced as a sum of their separated effects.

6.1 Differential Equation of Flexure

The basic flexure equation enables integrations for the slope and deflection due to bending of an initially straight beam. Consider the elemental length AB along a deflected beam's neutral in Fig. 6.1a.

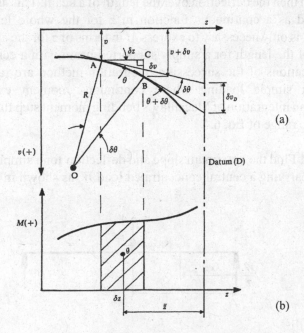

Figure 6.1 Beam element deflection

Let the deformed beam lie in axes having z aligned with the undeflected neutral axis such that the *positive dowward* deflections of A and B are v and $v + \delta v$, respectively. The hogging curvature shown is that adopted earlier for constructing positive M-diagrams (see Fig. 2.14, p. 47). Arc AB subtends an angle $\delta\theta$ at its centre with a *radius of curvature R*. The approximation is made that the slope θ at point A is given within the triangle ABC:

$$\tan\theta \approx \theta \ (\text{rad}) \approx \frac{\delta v}{\delta z} \tag{6.1}$$

Also, from the segment AOB, the change in slope $\delta\theta$ between A and B may be approximated as:

$$\delta\theta = \frac{AB}{R} \approx \frac{\delta z}{R} \tag{6.2}$$

Provided the displacement and slope are small, Eqs 6.1 and 6.2 may be combined as $\delta z \to 0$:

$$\frac{1}{R} = \frac{d\theta}{dz} = \frac{d}{dz}\left(\frac{dv}{dz}\right) = \frac{d^2 v}{dz^2} \tag{6.3}$$

Taking, from Eq. 5.3, $R = EI/M$ within Eq. 6.3 leads to the *flexure equation*:

$$EI\frac{d^2 v}{dz^2} = M \tag{6.4}$$

The convention for moments adopted here avoids the minus sign that may be seen accompanying M as given elsewhere. Successive integration of Eq. 6.4 will yield the slope and then the deflection over the length of a beam. This assumes that M can be expressed as a continuous function in z for the whole length. In cases of symmetry, it is only necessary to express M in terms of z for the appropriate region, e.g. over half the length for a simply supported beam with a central concentrated load. Applications of the successive integration method are rather restricted to beams with simple loading where continuous moment expressions apply. Fortunately an integration of Macaulay's bending moment step function is available to extend the range of Eq. 6.4.

Example 6.1 Find the maximum slope and deflection for a simply supported beam of length L carrying a central concentrated load W, as shown in Fig. 6.2.

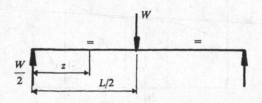

Figure 6.2 Simply supported beam with central load

Because of symmetry Eq. 6.4 need only be applied to the half-length: $0 \le z \le L/2$. Taking moments to the left of the section at z within this region gives $M = -\ Wz/2$, (sagging) so that:

$$EI\ d^2v/dz^2 = M = -\ Wz/2 \text{ (sagging)}$$

$$EI\ dv/dz = -\ Wz^2/4 + C_1$$

$$EI\ v = -\ Wz^3/12 + C_1z + C_2$$

The integration constants are determined from the following *boundary conditions*, which express the known zero slope at the centre ($z = L/2$) and zero deflection at the left end ($z = 0$):

$$dv/dz = 0 \text{ when } z = L/2, \therefore\ C_1 = WL^2/16$$

$$v = 0 \text{ when } z = 0, \quad \therefore\ C_2 = 0$$

Conversely, the maximum slope and deflection occur at the ends and centre, respectively:

$$dv/dz = C_1/EI = WL^2/16EI$$

$$v = (-\ WL^3/96 + WL^3/32)/EI = WL^3/48EI$$

Example 6.2 Derive expressions for the maximum slope and deflection of a cantilever beam of length L, when it carries a concentrated load W at its free end (see Fig. 6.3).

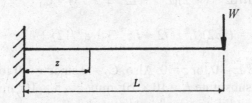

Figure 6.3 Cantilever beam with a concentrated, free-end load

Selecting the origin at the fixed end ensures that the integration constants are zero. Thus, applying Eq. 6.4 to the right of the section at z:

$$EI\ d^2v/dz^2 = M = W\ (L - z) \text{ (hogging positive, for } 0 \le z \le L)$$

$$EI\ dv/dz = W\ (Lz - z^2/2) + C_1$$

$$EI\ v = W\ (Lz^2/2 - z^3/6) + C_1z + C_2$$

Clearly $C_1 = 0$ and $C_2 = 0$ given $dv/dz = v = 0$ for $z = 0$. The maximum slope and deflection both occur at the free end $(z = L)$:

$$dv/dz = WL^2/2EI$$

$$v = W(L^3/2 - L^3/6)/EI = WL^3/3EI$$

Example 6.3 The cantilever beam in Fig. 6.4 carries a total weight of 20 kN evenly distributed over its 10 m length. Working from general expressions determine the maximum slope and deflection and the slope and deflection for a point 5 m from the fixed end. Take $E = 207$ GPa, $I = 10^9$ mm^4.

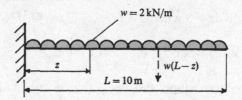

Figure 6.4 Cantilever beam with uniformly distributed loading

Integrating the flexure Eq. 6.4, with origin at the fixing, gives:

$$EI\, d^2v/dz^2 = (w/2)(L - z)^2 \text{ (hogging for } 0 \leq z \leq L)$$

$$EI\, dv/dz = (w/2)(L^2z - Lz^2 + z^3/3) + C_1 \tag{i}$$

$$EI\, v = (w/2)(L^2z^2/2 - Lz^3/3 + z^4/12) + C_1z + C_2 \tag{ii}$$

Now $C_1 = 0$ since $dv/dz = 0$ for $z = 0$. Also, $C_2 = 0$ since $v = 0$ for $z = 0$. Working in units of <u>N and mm</u>, substitute $L = 10 \times 10^3$ mm, $z = 5 \times 10^3$ mm and $w = 2$ N/mm $(= 2$ kN/m$)$ in Eqs i and ii, to give the slope and deflection:

$$dv/dz = w(L^2z - Lz^2 + z^3/3)/(2EI)$$

$$= \frac{2[(10^2 \times 5) - (10 \times 5^2) + (5^3/3)] \times 10^9}{(2 \times 207 \times 10^3 \times 10^9)} = 0.00141^c \text{ (rad)}$$

$$y = w(L^2z^2/2 - Lz^3/3 + z^4/12)/(2EI)$$

$$= \frac{2[(10^2 \times 5^2/2) - (10 \times 5^3/3) + (5^4/12)] \times 10^{12}}{(2 \times 207 \times 10^3 \times 10^9)} = 4.28 \text{ mm}$$

The maximum slope and deflection occur at the fixed end, where $z = 10$ m. Again, in units of N and mm:

$$\frac{dv}{dz} = \frac{wL^3}{6EI} = \frac{2 \times (10 \times 10^3)^3}{6 \times (207 \times 10^3) \times 10^9} = 0.00161^c$$

$$v = \frac{wL^4}{8EI} = \frac{2 \times (10 \times 10^3)^4}{8 \times (207 \times 10^3) \times 10^9} = 12.08 \text{ mm}$$

6.2 Mohr's Theorems

Referring again to the beam in Fig. 6.1a, it follows further from Eqs 6.1 and 6.2 that the change in slope between A and B, over the incremental length δz, is

$$\delta\theta = \frac{\delta z}{R} = \frac{1}{EI}(M\delta z) \qquad (6.5a)$$

Equation 6.5a allows the change in deflection δv_D between the tangents at A and B to be referred to any arbitrary datum D as shown, so that:

$$\delta v_D = \delta\theta \times \bar{z} = \frac{\delta z}{R} \times \bar{z} = \frac{1}{EI}(M\delta z) \times \bar{z} \qquad (6.5b)$$

Now in Eqs 6.5a,b, $(M\delta z)$ is an elemental area of the M-diagram, shown shaded in Fig. 6.1b. Also, $\bar{z}$ appears in Fig. 6.1b as the distance between the datum and the centroid g of the shaded area. Consequently, for a beam of uniform cross section, where I is constant, Eqs 6.5a,b may be interpreted as follows:

$$\delta\theta = (1/EI)[Area \ of \ the \ M\text{-}diagram \ over \ \delta z] \qquad (6.6a)$$

$$\delta v_D = (1/EI)[Moment \ of \ area \ of \ the \ M\text{-}diagram \ over \ \delta z] \qquad (6.6b)$$

For a beam of non-uniform cross section, where I is variable, Eqs 6.5a,b may be interpreted as

$$\delta\theta = (1/E) \ [Area \ of \ the \ M/I\text{-}diagram \ over \ \delta z] \qquad (6.6c)$$

$$\delta v_D = (1/E)[Moment \ of \ area \ of \ the \ M/I\text{-}diagram \ over \ \delta z] \qquad (6.6d)$$

Equations 6.6a–d express Mohr's first and second theorems; more commonly referred to as the *moment-area equations*. They can be applied to find the slope and deflection for a beam at a given position in its length, accounting for the manner in which the beam is supported. The simply supported beams in Examples 6.1–6.3, are *statically determinate*, which means that their reactions can be found from two equations of equilibrium, which express the beam's force and moment balance. The propped cantilever and a beam with encastré ends, given in the examples that follow, are *statically indeterminate*. They require additional slope and deflection equations to solve for the number of excess or *redundant* reactions. The latter arise from propping and end-fixing a beam that could otherwise support the applied loading from end-fixing alone but with greater deflection. It will now be shown, firstly, how the moment-area method is well suited to cantilever beams, including those with props, when the origin for z coincides with the fixed end. Because the slope and deflection are zero at this end, it follows that the change in slope and defection between the origin and the chosen datum will equal the true values for the cantilever. When applied to simply supported beams, the theorems require both the slope and deflection for the origin. Encastré beams require that Mohr's theorems be applied to net moment diagrams that express the difference between their free and fixing moments.

6.2.1 Cantilever

Both the slope and deflection at the fixed end of a cantilever are zero. By taking this end as the origin O, Eqs 6.5a,b may be integrated directly to give absolute values of slope and deflection at any datum position z in the length. Hogging moments produce positive change in slope with v measured downwards. Therefore, the areas of the M-diagram are positive in the following equations for the slope and deflection at length position z:

$$\theta = (1/EI)[Area\ of\ M\text{-}diagram\ from\ O\ to\ z] \qquad (6.7)$$

$$v = (1/EI)[Moment\ of\ area\ of\ M\text{-}diagram\ from\ O\ to\ z\ about\ z] \qquad (6.8)$$

With the origin fixed at O, we may shift the datum to wherever the deflection is required. In particular, taking the datum to coincide with the free end gives the maximum slope and deflection, as the following examples will show.

Example 6.4 Using the moment-area method, determine expressions for the maximum slope and deflection of a cantilever beam carrying uniformly distributed loading w/unit length (see Fig. 6.5a) .

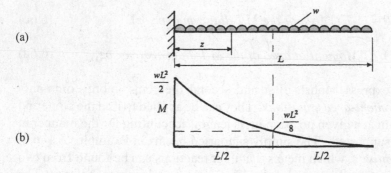

Figure 6.5 Moment area for a cantilever

Figure 6.5b shows the bending moment diagram in which $M_{max} = wL^2/2$. With origin at the fixed end Eqs 6.7 and 6.8 give θ and v directly. Since maximum values of θ and v are required, the datum is placed at the free end. This gives:

$$\theta = (1/EI)[\text{Area of M-diagram between } z = 0 \text{ and } z = L]$$

$$= \frac{1}{EI}\left[M_{max} \times \frac{L}{3}\right] = \frac{1}{EI}\left[\frac{\dot{w}L^2}{2} \times \frac{L}{3}\right] = \frac{\dot{w}L^3}{6EI}$$

$$v = (1/EI)[\text{Moment of area of M-diagram between 0 and L about the datum}]$$

$$= \frac{1}{EI}\left[M_{max} \times \frac{L}{3} \times \frac{3L}{4}\right] = \frac{1}{EI}\left(\frac{wL^2}{2}\right)\left(\frac{L}{3}\right)\left(\frac{3L}{4}\right) = \frac{wL^4}{8EI}$$

The central moment allows θ and v to be found at mid-position (see Exercise 6.10).

Example 6.5 The cantilever shown in Fig. 6.6a is propped at point P to prevent deflection from occurring at that point. Find the slope and deflection at the free-end given $I = 50 \times 10^3$ mm^4 and $E = 200$ GPa.

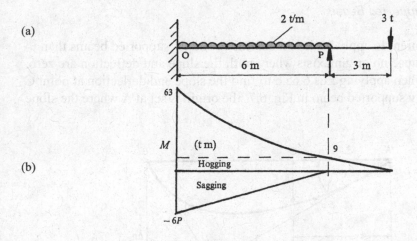

Figure 6.6 Propped cantilever

The bending moment diagrams for the applied loading and for the prop are separated in Fig. 6.6b. These appear as their respective hogging and sagging moment diagrams shown. With origin at O and datum at P, it follows from Eq. 6.8 that $v = 0$. Retaining the units of the metric tonne (10^3 kg) and metres (t and m):

$$v = (1/EI)[\text{Moment of area of M-diagrams from 0 to P about P}] = 0$$

$$= \frac{1}{EI}\left[\left(6 \times \frac{54}{3}\right)\left(\frac{3}{4} \times 6\right) + (9 \times 6 \times 3) - \left(6P \times \frac{6}{2}\right)\left(\frac{2}{3} \times 6\right)\right] = 0$$

from which $P = 9$ t (9000 kg). With origin for z at O, Eqs 6.7 and 6.8 provide the slope and deflection for a datum at the free end:

$$\theta = (1/EI)[Area\ of\ Net\ M\text{-}diagrams\ between\ z = 0\ and\ z = 9\ m]$$

$$= \frac{1}{EI}\left[\left(54 \times \frac{6}{3}\right) + (9 \times 6) + \left(9 \times \frac{3}{2}\right) - \left(6 \times 9 \times \frac{6}{2}\right)\right]$$

$$= \frac{13.5\ (tm)}{EI} = \frac{13.5 \times 10^3 \times 9.81 \times 10^3}{(200 \times 10^3) \times (50 \times 10^3)} = 0.0132\ rad$$

$$v = (1/EI)[Moment\ of\ area\ of\ Net\ M\text{-}diagrams\ between\ z = 0\ and\ z = 9\ m]$$

$$v = \frac{1}{EI}\left\{\frac{9 \times 3 \times 2 \times 3}{2 \times 3} + (9 \times 6 \times 6) + \left(54 \times \frac{6}{3}\right)\left[3 + \left(\frac{3}{4} \times 6\right)\right] - \left(6 \times 9 \times \frac{6}{2}\right)\left[3 + \left(\frac{2}{3} \times 6\right)\right]\right\}$$

$$= \frac{27(tm^2)}{EI} = \frac{27 \times 10^3 \times 9.81 \times 10^6}{(200 \times 10^3) \times (50 \times 10^3)} = 26.49\ mm$$

6.2.2 Simply Supported Beams

It is less convenient to apply Mohr's theorems to simply supported beams than to cantilevers because no origin exists where both the slope and deflection are zero. For example, when applying Eqs 6.6a,b to find the slope and deflection at point C along the simply supported beam in Fig. 6.7, the origin is set at A where the slope is not zero.

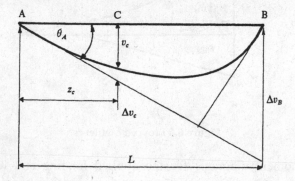

Figure 6.7 Deflection of a simply supported beam

Firstly, with the origin at A the datum is set at C. Under sagging moments, Eq. 6.6a gives the negative change in slope from A to C as:

$$\theta_C - \theta_A = - (1/EI)(Area\ of\ M\text{-}diagram\ between\ A\ and\ C) \qquad (6.9)$$

Figure 6.7 shows that when the deflections are small, the slope at A is,

$$\theta_A \simeq \frac{\Delta v_B}{L} \tag{6.10}$$

Mohr's second theorem (Eq. 6.6b) gives the intercept between tangents at A and C:

$$\Delta v_C = (1/EI)(Moment\ of\ area\ of\ M\text{-}diagram\ from\ A\ to\ C\ about\ C) \tag{6.11}$$

Now, by taking the datum to lie at B, Eq. 6.6b provides the intercept Δv_B, made by the tangents at A and B as shown:

$$\Delta v_B = (1/EI)(Moment\ of\ area\ of\ M\text{-}diagram\ from\ A\ to\ B\ about\ B) \tag{6.12}$$

Combining Eqs 6.9–6.12 embodies Mohr's two theorems for the slope at C:

$$\theta_C = (1/L)(1/EI)[Moment\ of\ area\ of\ M\text{-}diagram\ from\ A\ to\ B\ about\ B]$$
$$- (1/EI)(Area\ of\ M\text{-}diagram\ between\ A\ and\ C) \tag{6.13}$$

The true deflection at C follows from Eqs 6.10 and 6.11 as:

$$v_C = \theta_A z_C - \Delta v_C = \frac{z_C}{L} \Delta v_B - \Delta v_C \tag{6.14}$$

Substituting Eqs 6.11 and 6.12 into Eq. 6.14 leads to the deflection at C:

$$v_C = (z_C/L)(1/EI)(Moment\ of\ area\ of\ M\text{-}diagram\ from\ A\ to\ B\ about\ B)$$
$$- (1/EI)(Moment\ of\ area\ of\ M\text{-}diagram\ from\ A\ to\ C\ about\ C) \tag{6.15}$$

Equations 6.13 and 6.15 are applied to simply-supported beams with concentrated loading in the two examples that follow.

Example 6.6 Find expressions for the slope and deflection beneath the offset load W, applied to the simply supported beam AB in Fig. 6.8a.

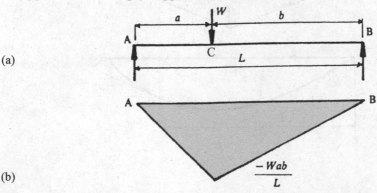

(a)

(b)

Figure 6.8 Beam with concentrated load offset from centre

Here C lies at the position of the concentrated load W. Applying Eqs 6.13 and 6.15 to the bending moment diagram in Fig. 6.8b gives:

$$EI\theta = (1/L)[Moment\ of\ area\ of\ M\text{-}diagram\ from\ A\ to\ B\ about\ B]$$
$$- [Area\ of\ M\text{-}diagram\ from\ A\ to\ W]$$

$$= (1/L)[(Wa^2b/2L)(b + a/3) + (Wab^2/2L)(2b/3)] - Wa^2b/2L$$

$$= (Wab/2L^2)[ab + a^2/3 + 2b^2/3 - aL]$$

$$= (Wab/3L^2)(b^2 - a^2) = Wab(b - a)/3L$$

$$EIv = (a/L)[Moment\ of\ area\ of\ M\text{-}diagram\ from\ A\ to\ B\ about\ B]$$
$$-[Moment\ of\ area\ of\ M\text{-}diagram\ from\ A\ to\ W\ about\ W]$$

$$= (a/L)[Wa^2b/2L)(b + a/3) + (Wab^2/2L)(2b/3)] - (Wa^2b/2L)(a/3)$$

$$= Wa^2b^2/3L$$

Had a negative, maximum sagging moment $-Wab/L$ been substituted in the θ and v expressions above it have would served to change their signs.

Example 6.7 Calculate the slope and deflection at point C, 40 mm from B, for the centrally loaded, simply-supported beam shown in Fig. 6.9a. Take the product $EI = 9 \times 10^6$ Nmm^2 to be constant throughout the length.

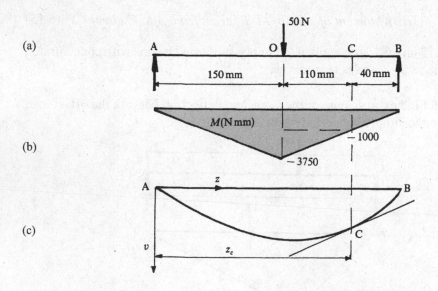

Figure 6.9 Centrally loaded, simply-supported beam

Figures 6.9b and c are the bending moment and deflection diagrams, respectively. Applying Eq. 6.13 to Fig. 6.9b gives the slope at C:

$$\theta_C = (1/300)(1/EI)(3750 \times 150 \times 150)$$

$$- (1/EI)[(150 \times 3750/2) + (110 \times 1000) + (2750 \times 110/2)]$$

$$= (0.28125 - 0.5425) \times 10^6/(9 \times 10^6) = -0.02903^c \text{ (rad)}$$

This slope ($\theta = dv/dz$) is negative for the positive z, v co-ordinates indicated. Note that since the slope at the centre, O, is zero, Eq. 6.9 could also be applied with O as the origin. This gives the negative change in slope from O to C as:

$$\theta_C - \theta_O = -(1/EI)(\textit{Area of M-diagram between O and C})$$

$$\theta_C = -(1/EI)[(110 \times 1000) + (2750 \times 110/2)$$

$$= -261.25 \times 10^3 \text{ (Nmm}^2)/EI$$

$$= -(0.26125 \times 10^6)/(9 \times 10^6) = -0.02903^c$$

Equation 6.15 gives the deflection at C:

$$v_C = (260/300)(1/EI)(3750 \times 150 \times 150)$$

$$- (1/EI)[(3750 \times 150/2)(110 + 50) + (1000 \times 110 \times 55)$$

$$+ (2750 \times 110/2)(2 \times 110/3)] = (73.125 - 62.142) \times 10^6 \text{ (N/mm}^3)/EI$$

$$= 10.983 \times 10^6/(9 \times 10^6) = 1.22 \text{ mm}$$

Example 6.8 Find expressions for the maximum slope and deflection for the beam with uniformly distributed loading shown in Fig. 6.10a.

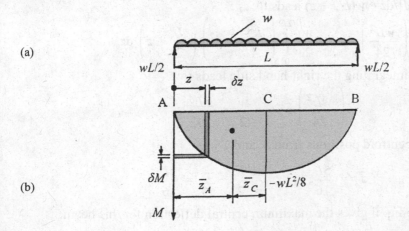

Figure 6.10 Simply supported beam with uniformly distributed loading

Combining Eqs 6.10 and 6.12 gives the maximum slope at end support A:

$$\theta_A = (1/L)(1/EI)[\text{Moment of area of M-diagram from A to B about B}] \quad (i)$$

in which the full area enclosed within the M-diagram is found from:

$$A = \int_0^L |M(z)| \, dz = \frac{w}{2} \int_0^L z(L - z) \, dz$$

$$= \frac{w}{2} \left| \frac{Lz^2}{2} - \frac{z^3}{3} \right|_0^L = \frac{wL^3}{12}$$

where $M(z) = -(wz/2)(L-z)$ is the sagging moment at position z in Fig. 6.10b. Substituting into Eq. i:

$$\theta_A = \frac{1}{L} \times \frac{1}{EI} \times \frac{wL^3}{12} \times \frac{L}{2} = \frac{wL^3}{24EI}$$

With the datum C at the centre, $z_C = L/2$, when Eq. 6.15 gives the maximum displacement:

$$v_C = (1/2EI)(\text{Moment of area of M-diagram from A to B about B})$$

$$- (1/EI)(\text{Moment of area of M-diagram from A to C about C}) \quad (ii)$$

Equation ii requires the position of the centroid of the half area from the centre. It is worth remembering that the parabola of height h ($= M_{max}$) and length b ($= L/2$) encloses an area of $\frac{2}{3}bh$ ($= wL^3/24$) with its centroidal position at $\frac{3}{8}b$ ($= 3L/16$). Alternatively, working from first principles, the position of the centroid $\bar{z}_A$ in Fig. 6.10b for the half area follows from equating its first moments with the origin at A:

$$\left(\frac{wL^3}{24} \right) \bar{z}_A = \frac{1}{2} \int_0^{M_{max}} \left(\frac{L}{2} - z \right) dM \times \left(\frac{L}{2} + z \right)$$

Substituting $d|M|/dz = w(L/2 - z)$ leads to:

$$\left(\frac{wL^2}{24} \right) \bar{z}_A = \frac{w}{2} \int_0^{L/2} \left(\frac{L^2}{4} - z^2 \right) \left(\frac{L}{2} - z \right) dz$$

Expanding then integrating the right-hand side leads to:

$$\left(\frac{wL^2}{24} \right) \bar{z}_A = \frac{w}{2} \times \frac{5L^4}{192}$$

from which the centroid positions from A and C are

$$\bar{z}_A = \frac{5L}{16}, \quad \bar{z}_C = \frac{L}{2} - \bar{z}_A = \frac{3L}{16}$$

Substituting into Eq. ii gives the maximum central deflection for this beam:

$$v_C = \frac{1}{2EI} \left(\frac{wL^3}{12} \times \frac{L}{2} \right) - \frac{1}{EI} \left(\frac{wL^3}{24} \times \frac{3L}{16} \right) = \frac{5wL^4}{384EI}$$

Finally, note that it is possible to obtain the second term in Eq. ii directly from a first moment integral for a datum that does not require the centroid position:

$$\int_0^{M_{max}} (L/2 - z)\, \mathrm{d}M \times (L/2 - z)/2 = (w/2) \int_0^{L/2} (L/2 - z)^3 \mathrm{d}z = wL^4/128$$

6.2.3 Encastré Beams

Both ends of an encastré beam are built-in (see Fig. 6.11a). The slope and deflection at each end is zero. Figures 6.11b,c show how the net bending moment diagram for a central load is derived from its component diagrams. These include the *free moment diagram* (sagging) for when the applied loading is simply supported, and the *fixing moment diagram* (hogging), determined by the fixed-end moments.

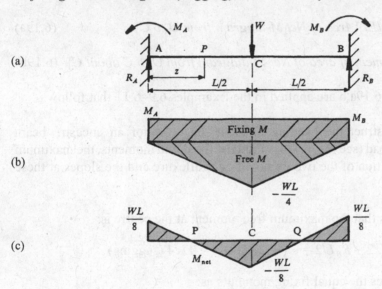

Figure 6.11 Free and fixing moments for an encastré beam with central load

Applying Eq. 6.15 with the origin at A and the datum at B in Fig. 6.11a:

$v_B = (1/EI)[$*Moment of area of Fixing M-diagram from A to B about B*
  $-$ *Moment of area of Free M-diagram from A to B about B*$] = 0$

With $v_B = 0$, it follows immediately that for the full span of an encastré beam:

Moment of area of Free M-Diagram = Moment of area of Fixing M-Diagram   (6.17a)

with each moment taken from A-B about B. Now apply Eq. 6.13 to Fig. 6.11a, with the origin at A and the datum at B. Given $\theta_B = 0$ and equality must apply also to the areas of an encastré beam's moment diagrams:

Area of Free M-Diagram = Area of Fixing M-Diagram   (6.17b)

Equations 6.17a,b are used to find the net support reactions, R_A and R_B. These reactions are composed of the free reactions R_1 and R_2 plus fixed-end reactions $R = (M_A - M_B)/L$. The latter arise from a couple $(M_A - M_B)$ that appears when the fixing moments differ $(M_A \neq M_B)$. Assuming $M_A > M_B$ (as in Fig. 6.13) the support reactions become:

$$R_A = (R_1 + R) = R_1 + (M_A - M_B)/L \qquad (6.18a)$$

$$R_B = (R_2 - R) = R_2 - (M_A - M_B)/L \qquad (6.18b)$$

Taking the origin O at the left-hand fixed end, the absolute values of the slope and deflection at any datum C, position z in the length, of an encastré beam follow from integrating Mohr's theorems. These give similar expressions to those for a cantilever except that *net* moments are required:

$$\theta_C = (1/EI)[\text{Area of Net M-diagram from O to C}] \qquad (6.19a)$$

$$v_C = (1/EI)[\text{Moment of area of Net M-diagram from O to C about C}] \qquad (6.19b)$$

Equations 6.17a,b–6.19a,b are applied in the Examples 6.9–6.11 that follow.

Example 6.9 Construct the bending moment diagrams for an encastré beam carrying a central load (see Fig. 6.11a). Find the fixed-end moments, the maximum deflection, the position of the two points of contraflexure and the slopes at these two points.

Figure 6.11b shows that the maximum free moment at the centre is:

$$R_A L/2 = R_B L/2 = -(W/2)(L/2) = -WL/4 \text{ (sagging)}$$

Equation 6.17b gives the equal fixing moments as:

$$(WL/4)(L/2) = M_A L, \quad \therefore M_A = WL/8 = M_B \text{ (hogging)}$$

At the centre, the net moment (= fixing – free) in Fig. 6.11c becomes:

$$WL/8 - WL/4 = -WL/8$$

The maximum deflection at centre, C, is given by Eq. 6.19b as:

$$v_{max} = (1/EI)[\text{Moment of area of Net M-diagram from A to C about C}]$$

$$= (1/EI)[\text{Moment of area of Fixing M-diagram from A to C about C}$$

$$- \text{Moment of area of Free M-diagram from A to C about C}]$$

$$= \frac{1}{EI}\left[\left(\frac{L}{2} \times \frac{WL}{8} \times \frac{L}{4}\right) - \left(\frac{WL}{4} \times \frac{L}{4} \times \frac{1}{3} \times \frac{L}{2}\right)\right] = \frac{WL^3}{192EI}$$

The position z for the *point of contraflexure* P is found from setting $M_{net} = 0$:

$$M_{net} = 0 = M_A - R_A z, \quad \therefore z = \frac{M_A}{R_A} = \frac{WL}{8} \times \frac{2}{W} = \frac{L}{4}$$

The slope at $z = L/4$ follows from applying Eq. 6.19a, either to Fig. 6.11b:

$$\theta = (1/EI)[Area\ of\ Fixing\ M\text{-}diag\ from\ A\text{-}P - Area\ of\ Free\ M\text{-}diag\ from\ A\text{-}P]$$

$$= \frac{1}{EI}\left[\frac{L}{4} \times \frac{WL}{8} - \frac{L}{8} \times \frac{WL}{8}\right] = \frac{WL^2}{64EI}$$

or, to Fig. 6.11c:

$$\theta = (1/EI)[Area\ of\ Net\ M\text{-}diagram\ from\ A\ to\ P]$$

$$= \frac{1}{EI}\left[\frac{1}{2} \times \frac{WL}{8} \times \frac{L}{4}\right] = \frac{WL^2}{64EI} \checkmark$$

The position of the second point of contraflexure Q is again found from setting $M_{net} = 0$ when $z > L/2$:

$$M_{net} = 0 = M_A + \left(z - \frac{L}{2}\right)W - R_A z, \quad \therefore z = \frac{WL/2 - M_A}{W - R_A} = \frac{WL/2 - WL/8}{W - W/2} = \frac{3L}{4}$$

The slope at $z = 3L/4$ follows from applying Eq. 6.19a to either Fig. 6.11b or c:

$$\theta = (1/EI)[Area\ of\ Fixing\ M\text{-}diag\ from\ A\text{-}Q - Area\ of\ Free\ M\text{-}diag\ from\ A\text{-}Q]$$

$$= \frac{1}{EI}\left\{\left(\frac{3L}{4} \times \frac{WL}{8}\right) - \left[\left(\frac{1}{2} \times \frac{WL}{4} \times \frac{L}{2}\right) + \frac{1}{2}\left(\frac{WL}{4} + \frac{WL}{8}\right)\frac{L}{4}\right]\right\} = -\frac{WL^2}{64EI}$$

or,

$$\theta = (1/EI)[Area\ of\ net\ M\text{-}diagram\ from\ A\ to\ Q]$$

$$= \frac{1}{EI}\left[\frac{1}{2} \times \frac{WL}{8} \times \frac{L}{4} - \frac{1}{2} \times \frac{WL}{8} \times \frac{L}{2}\right] = -\frac{WL^2}{64EI}$$

Example 6.10 Construct the bending moment diagrams for an encastré beam carrying uniformly distributed loading (see Fig. 6.12a). Find the fixed-end moments, the maximum deflection, the position of the two points of contraflexure and the slopes at these two points. What load would a prop support when placed at the centre to eliminate the deflection there? What then is the position and magnitude of the maximum deflection?

Firstly apply Eq. 6.17b by equating the areas in Fig. 6.12b. Using the free moment area expression, derived in Example 6.8, gives the equal fixing moments as:

$$\frac{wL^3}{12} = M_A L, \quad \therefore M_A = \frac{wL^2}{12} = M_B \text{ (hogging)} \tag{i}$$

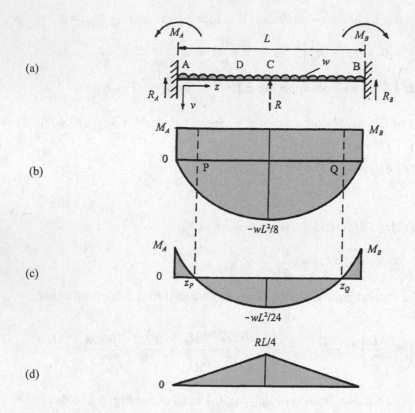

Figure 6.12 Free and fixing moments for an encastré beam with uniformly distributed loading

Hence, at the centre, C, the net moment (= fixing – free) in Fig. 6.11c becomes,

$$M_C = \frac{wL^2}{12} - \frac{wL^2}{8} = -\frac{wL^2}{24} \tag{ii}$$

The maximum deflection at the centre, C, is found from Eq. 6.19b as:

v_{max} = (1/EI)[*Moment of area of Fixing M-diag from A-C about C*
     – *Moment of area of Free M-diagram from A-C about C*]

$$= \frac{1}{EI}\left[\left(\frac{L}{2} \times \frac{wL^2}{12} \times \frac{L}{4}\right) - \left(\frac{wL^3}{24} \times \frac{3L}{16}\right)\right] = \frac{wL^4}{384EI}$$

The positions z_P and z_Q of *both* points of contraflexure P and Q are found from setting the net moment to zero:

$$M_{P,Q} = 0 = M_A - R_A z_{P,Q} + \frac{w z_{P,Q}^2}{2}$$

Substituting for M_A from Eq. i and $R_A = wL/2$, leads to a quadratic equation:

$$z_{P,Q}^2 - Lz_{P,Q} + L^2/6 = 0$$

for which the roots identify with positions z_P and z_Q as

$$z_P = \frac{L}{2\sqrt{3}}(\sqrt{3} - 1), \quad z_Q = \frac{L}{2\sqrt{3}}(\sqrt{3} + 1)$$

The free moment magnitudes at each position must follow:

$$|M_P| = |M_Q| = M_A = wL^2/12$$

The slope at P follows from applying Eq. 6.19a:

$$\theta_P = (1/EI)[Area\ of\ Fixing\ M\text{-}diag\ from\ A\text{-}P - Area\ of\ Free\ M\text{-}diag\ from\ A\text{-}P]$$

$$= \frac{1}{EI}\left[M_A z_P - \frac{2}{3}M_A z_P\right] = \frac{M_A z_P}{3EI}$$

and substituting for M_A and z_P:

$$\theta_P = \frac{1}{3EI} \times \frac{wL^2}{12} \times \frac{L}{2\sqrt{3}}(\sqrt{3} - 1) = \frac{wL^3}{72EI}\left(1 - \frac{1}{\sqrt{3}}\right)$$

The slope at Q also follows from applying Eq. 6.19a:

$$\theta_Q = (1/EI)[Area\ of\ fixing\ M\text{-}diag\ from\ A\text{-}Q - Area\ of\ free\ M\text{-}diag\ from\ A\text{-}Q]$$

$$= \frac{1}{EI}\left[M_A z_Q - \left(\frac{wL^3}{12} - \frac{2}{3}M_A z_P\right)\right]$$

and substituting for M_A, z_Q and z_P:

$$\theta_Q = \frac{1}{EI} \times \frac{wL^3}{12}\left[\frac{1}{2}\left(1 + \frac{1}{\sqrt{3}}\right) - 1 + \frac{1}{3}\left(1 - \frac{1}{\sqrt{3}}\right)\right] = -\frac{wL^3}{72EI}\left(1 - \frac{1}{\sqrt{3}}\right)$$

in which the signs of positive θ_P and negative θ_Q conform to the positive z, v co-ordinates shown in Fig. 6.12a.

With the introduction of a prop at the centre both the fixing and net moments are altered by the additional free moment diagram of Fig. 6.12d. The prop reaction R and fixing moments $M_A = M_B$ follow from applying Eqs 6.19a,b from A to the centre C where both the slope and deflection become zero:

Area of Fixing M-diagram - Area of Free M-diagrams (from A to C) = 0

$$\left(M_A \times \frac{L}{2}\right) - \left[\left(\frac{1}{2} \times \frac{wL^3}{12}\right) - \left(\frac{1}{2} \times \frac{RL}{4} \times \frac{L}{2}\right)\right] = 0$$

$$M_A = \frac{wL^2}{12} - \frac{RL}{8} \tag{i}$$

Moment of area of Fixing M-diag from A-C about C

− Net Moment of area of Free M-diagrams from A-C about C = 0

$$\left(M_A \times \frac{L}{2} \times \frac{L}{4} \right) - \left[\left(\frac{1}{2} \times \frac{wL^3}{12} \times \frac{3}{8} \times \frac{L}{2} \right) - \left(\frac{1}{2} \times \frac{RL}{4} \times \frac{L}{2} \times \frac{1}{3} \times \frac{L}{2} \right) \right] = 0$$

$$M_A = \frac{wL^2}{16} - \frac{RL}{12} \tag{ii}$$

Equating i and ii gives:

$$R = \frac{wL}{2}, \quad M_A = \frac{wL^2}{48} \tag{iii}$$

The position of the maximum deflection now lies at a position D, where $\theta = 0$, within the half length, between the end fixing and the prop. The fixing moment at the position z_D is M_A. The free moments at this position are $Rz_D/2$ (sagging) due to the prop and $w \times z_D(L - z_D)/2$ (hogging) due to the distributed loading. Applying Eq. 6.19a to the three respective areas enclosed within $0 \le z \le z_D$:

$$\theta_D = 0 = M_A z_D - \left[\frac{2z_D}{3} \times \frac{wz_D}{2}(L - z_D) - \left(\frac{Rz_D}{2} \times \frac{z_D}{2} \right) \right] \tag{iv}$$

Substituting for M_A and R from Eq. iii into Eq. iv leads to the quadratic:

$$16z_D^2 - 10Lz_D + L^2 = 0 \quad \therefore \quad (8z_D - L)(2z_D - L) = 0$$

for which the respective roots $z_D = L/8$ and $z_D = L/2$ locate both the position of maximum deflection and confirm zero slope at the prop. The maximum deflection at the lesser z_D follows from Eq. 6.19b as:

$$EIv_{max} = \left(M_A z_D \times \frac{z_D}{2} \right) - \left[\left(\frac{2z_D}{3} \times \frac{wz_D}{2}(L - z_D) \times \frac{3z_D}{8} \right) - \left(\frac{Rz_D}{2} \times \frac{z_D}{2} \times \frac{z_D}{3} \right) \right]$$

$$EIv_{max} = \frac{M_A z_D^2}{2} - \frac{wz_D^3}{8}(L - z_D) + \frac{Rz_D^3}{12} \tag{v}$$

Substituting $z_D = L/8$ into Eq. v with M_A and R from Eq. iii leads to a much reduced maximum deflection for the propped beam:

$$v_{max} = \frac{wL^4}{2^{15}EI}$$

The numerical factor in the denominator is 32768 in contrast to a denominator of 384 for the un-propped beam.

Example 6.11 Establish the net bending moment diagram, the position of the points of contraflexure and the maximum deflection for the encastré beam with an off-centre concentrated load shown in Fig. 6.13a.

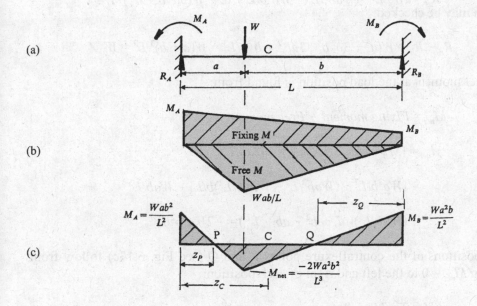

Figure 6.13 Encastré beam with concentrated load offset from centre

For the free M-diagram, the support reactions R_1 and R_2 give its maximum moment:

$$aW = LR_2, \quad \therefore R_2 = aW/L$$

$$R_1 + R_2 = W, \quad \therefore R_1 = bW/L$$

$$\therefore M_{max} = aR_1 = bR_2 = Wab/L$$

Applying Eqs 6.17a,b to Fig. 6.13b:

$$(L/2)(Wab/L) = (M_A + M_B)L/2$$

$$\therefore M_A + M_B = Wab/L \tag{i}$$

$$(Wab/L)(b/2)(2b/3) + (Wab/L)(a/2)(b + a/3) = M_B L(L/2) + (M_A - M_B)(L/2)(2L/3)$$

$$2M_A + M_B = (Wab/L^3)(a^2 + 3ab + 2b^2) \tag{ii}$$

Subtract Eq. i from Eq. ii:

$$M_A = (Wab/L^3)(a^2 + 3ab + 2b^2) - Wab/L = Wab^2/L^2$$

$$M_B = Wab/L - Wab^2/L^2 = Wa^2b/L^2$$

Substituting M_A and M_B into Eqs 6.18a,b:

$$R_A = bW/L + (Wab^2/L^2 - Wa^2b/L^2)/L = W(3ab^2 + b^3)/L^3$$

$$R_B = aW/L - (Wab^2/L^2 - Wa^2b/L^2)/L = W(3a^2b + a^3)/L^3$$

which may be checked:

$$R_A + R_B = W(a^3 + 3a^2b + 3ab^2 + b^3)/L^3 = W(a + b)^3/L^3 = W \checkmark$$

The net moment at the load position is found from:

$$M_{net} = Fixing\ moment - Free\ moment$$

$$M_{net} = [M_B + (M_A - M_B)b/L] - Wab/L$$

$$= [Wa^2b/L^2 + (Wab^2/L^2 - Wa^2b/L^2)b/L] - Wab/L$$

$$= (Wab/L^3)(aL + b^2 - ab - L^2) = -2Wa^2b^2/L^3$$

The positions of the contraflexure points P and Q (see Fig. 6.13c) follow from setting $M_{net} = 0$ to the left and right of each position:

$$M = 0 = M_A - R_A z_P$$

$$z_P = M_A/R_A = (Wab^2/L^2)L^3/[W(3ab^2 + b^3)] = aL/(b + 3a)$$

$$M = 0 = M_B - R_B z_Q$$

$$z_Q = M_B/R_B = (Wa^2b/L^2)L^3/[W(3a^2b + a^3)] = bL/(a + 3b)$$

The maximum deflection occurs not beneath the load, but at a position C in the length, z_C from A, where $\theta_C = 0$. Assuming $z_C > a$, Eq. 6.19a and Fig. 6.13c give:

$$\frac{Wab^2 z_P}{2L^2} - \frac{2Wa^2b^2(a - z_P)}{2L^3} - \frac{2Wa^2b^2(z_C - a)}{2L^3}\left[1 + \frac{(L - z_C - z_Q)}{(b - z_Q)}\right] = 0$$

Substituting z_P and z_Q, after some manipulation one root of the quadratic in z_C is:

$$z_C = \frac{(a + b)^2}{a + 3b}$$

The maximum deflection, which occurs at C, follows from:

$$v_{max} = (1/EI)[Moment\ of\ Net\ M\text{-}diagram\ from\ A\ to\ C\ about\ C]$$

Dividing the net moment diagram (Fig. 6.13b) into one rectangular and three triangular areas, then taking their moments about C, leads to:

$$v_{max} = \frac{Wa^2b^2}{12EI(a + 3b)^2(b + 3a)^2L^3} \times \left[2L^3(a + 3b)(8a^2 + 3b^2 + 9ab)\right]$$

$$- 8a^2(a + 3b)(2a^3 + 3b^3 + 6ab^2 - 3a^2b) - 6bL(b - a)^2(b + 3a)^2 \Big]$$

which we may check from setting $a = b = L/2$ to recover the maximum displacement expression found previously for Fig. 6.11a when the load is central. This confirms $v_{max} = WL^3/192EI$.

6.3 Telescopic Cantilever

There are many portable lifting machines such as cranes and raised platforms that achieve the desired reach with a telescoping arm. The arm may be idealised as, say, a two-section cantilever, fixed at one end where it is built into the vehicle chassis, with the other end remaining free and capable of extending to the desired length. In its lifting application the load may be assumed concentrated at the free end of the cantilever but in a design with added safety an allowance would be made for self-weight, wind loading and accidental knocks that all increase the end deflection. Similar loadings are imposed upon other telescopic structures including hydraulic jacks, masts, booms and fishing rods. The cantilever slope and deflection analyses that follow are common to these structures when constructed from two tubular cross sections of equal length L with a common overlapping length αL.

6.3.1 Geometry and Self-Weight

Figures 6.14a,b show the tubular cross sections of two telescoping arrangements: (a) one having a sliding fit between circular tubes, and (b) another with pressure pads to lessen the friction between square tubes.

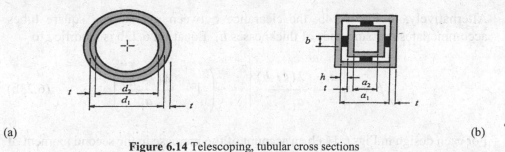

(a)   (b)

Figure 6.14 Telescoping, tubular cross sections

The constant $\gamma < 1$ in Fig. 6.14 defines the ratio between the uniformly distributed self-weights for the individual beam lengths:

$$\gamma = \frac{w_2}{w_1} = \frac{W_2/L_2}{W_1/L_1} = \frac{\rho_2 g A_2}{\rho_1 g A_1} \tag{6.20a}$$

where $W = \rho g A L$ is the total weight of each section. Hence, where the lengths L_1 and L_2 are the same γ becomes equal to their total weight ratio. Moreover, where the two sections are made from a common material with similar density, Eq. 6.20a simplifies to:

$$\gamma = \frac{A_2}{A_1} \tag{6.20b}$$

A further constant $\beta < 1$ is used to define the ratio between the second moments of area for the free-end and fixed-end sections. This ratio appears in terms of the section areas (A) and their radii of gyration (k) as:

$$\beta = \frac{I_2}{I_1} = \frac{A_2 \, k_2^2}{A_1 \, k_1^2} \tag{6.21a}$$

Substituting Eq. 6.20b into Eq. 6.21a it is seen how the radii of gyration connect the two constants β and γ under these conditions:

$$\beta = \gamma \left(\frac{k_2}{k_1} \right)^2 \tag{6.21b}$$

For example, with telescoping, thin-walled, circular or square tubes in Fig. 6.14a, Eq. 6.21b gives, respectively:

$$\beta = \gamma \left(\frac{d_2}{d_1} \right)^2, \qquad \beta = \gamma \left(\frac{a_2}{a_1} \right)^2 \tag{6.22a,b}$$

If the two circular tubes are to fit closely as in Fig. 6.14a then the outside diameter of the free-end section is less than that for the fixed-end section by two thicknesses. Equation 6.22a gives

$$\beta = \gamma \left(1 - \frac{2t}{d_1} \right)^2 \approx \gamma \left(1 - \frac{4t}{d_1} \right) \tag{6.23a}$$

Alternatively, in Fig. 6.14b, the clearance between overlapping square tubes accommodates pressure pads of thicknesses h. Equation 6.22b is modified to

$$\beta = \gamma \left[\frac{a_1 - 2(t+h)}{a_1} \right]^2 \approx \gamma \left(1 - \frac{4(t+h)}{a_1} \right) \tag{6.23b}$$

For each design in Figs 6.14a,b an account of the increase in the second moment of area within the overlap is required:

$$I_1 + I_2 = (1 + \beta)I \tag{6.24a}$$

together with the increase in the distributed weight as:

$$w_1 + w_2 = (1 + \gamma)w \tag{6.24b}$$

where $I = I_1$ and $w = w_1$ in Eqs 6.24a,b refer to the larger values for the fixed-end section.

6.3.2 Slope and Deflection Under Combined Loading

Figure 6.15 shows the combination of a concentrated end load P with distributed self weights: w, $(1 + \gamma)w$ and γw across the fixed-end, overlapping and free end sections, respectively, given in § 6.3.1 above.

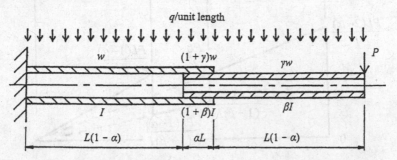

Figure 6.15 Two-section telescopic cantilever showing combined loadings

In addition, wind pressure Q/unit area loading increases the self-weights by constant amount q/unit length. An accidentally impacted end-load may be added to P as an equivalent static load (see § 10.4, p. 406). A separate analysis of the increased end-deflection is required should an impact occur elsewhere in the length whilst the cantilever bears the end load it was designed for. This procedure, whereby deflections arising from the various sources of loading are added, assumes that the principle of superposition applies. This principle, which pre-supposes a linear elastic (Hookean) structure, enables both the slope and deflections due to the combination of concentrated and distributed loading to be found as a sum of contributions from applying each load separately. Mohr's first and second theorems (see Eqs 6.6c and d) will provide the maximum slope and deflection for a variable section cantilever under combined loading, most expediently. The variable I that Mohr's theorems must provide for in Fig. 6.15 are I, $(1 + \beta)I$ and βI within the three sections of its length. Thus, the M/I diagrams that follow (see Figs 6.16–6.18) apply to where concentrated, distributed and wind loadings are applied separately. Discontinuities appear within the ordinates at the length positions where there are abrupt changes to I. The slope and deflection Eqs 6.6c and d employ, respectively, the area and moment of area of each M/I diagram as follows.

6.3.3 Concentrated End Load

(a) End-Slope due to P

The change in slope between the fixed and free ends for Fig. 6.16a is found from the full area of the M/I-diagram given in Fig. 6.16b. Since the fixed-end slope is zero, Eq. 6.6c is applied to provide the end slope from the three areas that constitute this diagram. The areas are found from taking the sums and differences of five triangles, each having its apex at the free end, within the following three terms:

(a)

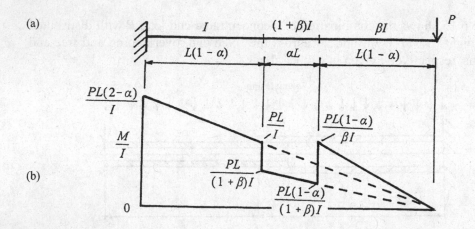

(b)

Figure 6.16 M/I diagram for a two-section, telescopic cantilever with concentrated end load P

$$\theta = \frac{1}{E}\left\{\left[\frac{PL(1-\alpha)}{\beta I} \times \frac{L(1-\alpha)}{2}\right]_1 + \left[\frac{PL}{(1+\beta)I} \times \frac{L}{2} - \frac{PL(1-\alpha)}{(1+\beta)I} \times \frac{L(1-\alpha)}{2}\right]_2 \right.$$

$$\left. + \left[\frac{PL(2-\alpha)}{I} \times \frac{(2-\alpha)L}{2} - \frac{PL}{I} \times \frac{L}{2}\right]_3\right\} \tag{6.25a}$$

Each of the three area terms in the end slope Eq. 6.25a have the common factor $PL^2/2EI$, allowing a simplified slope expression:

$$\theta = \frac{PL^2}{2EI}\left[\frac{(1-\alpha)^2}{\beta} + \frac{1}{1+\beta} - \frac{(1-\alpha)^2}{(1+\beta)} + (2-\alpha)^2 - 1\right] \tag{6.25b}$$

(b) End Deflection due to P

The end deflection is found from the sum of the first moments of the three areas that constitute the M/I-diagram in Fig. 6.16b. Equation 6.6d requires that each moment of each area be referred to the free end to provide its deflection. These areas in turn are made up from the moments of five right-triangular areas with a common apex at the free end:

$$v = \frac{1}{E}\left\{\left[\frac{PL(1-\alpha)}{\beta I} \times \frac{L(1-\alpha)}{2} \times \frac{2L(1-\alpha)}{3}\right]_1\right.$$

$$+ \left[\frac{PL}{(1+\beta)I} \times \frac{L}{2} \times \frac{2L}{3} - \frac{PL(1-\alpha)}{(1+\beta)I} \times \frac{L(1-\alpha)}{2} \times \frac{2L(1-\alpha)}{3}\right]_2$$

$$\left. + \left[\frac{PL(2-\alpha)}{I} \times \frac{L(2-\alpha)}{2} \times \frac{2L(2-\alpha)}{3} - \frac{PL}{2} \times \frac{L}{2} \times \frac{2L}{3}\right]_3\right\} \tag{6.26a}$$

Each of the three terms in the end-deflection Eq. 6.26a have the common factor $PL^3/3EI$ within its simplification:

$$v = \frac{PL^3}{3EI}\left[\frac{(1-\alpha)^3}{\beta} + \frac{1}{1+\beta} + \frac{(1-\alpha)^3}{1+\beta} + (2-\alpha)^3 - 1\right] \qquad (6.26b)$$

These closed solutions for end slope and deflection may be checked from setting $\alpha = 0$ and $\beta = 1$ in Eqs 6.25b and 6.26b. This gives a cantilever of uniform section I with length $2L$ for which the free end slope and deflection agree with the expression expected:

$$\theta = \frac{P(2L)^2}{2EI}, \qquad v = \frac{P(2L)^3}{3EI}$$

Also, when $\alpha = 1$ this gives a cantilever of length L with a uniform second moment of area $I(1 + \beta)$. Equations 6.25b and 6.26b show that the maximum slope and deflection at the free-end take the standard forms

$$\theta = \frac{PL^2}{2E(1+\beta)I}, \qquad v = \frac{PL^3}{3E(1+\beta)I}$$

Equations 6.26a,b also apply when an impacted end load provided P is substituted as an equivalent static load. The latter is found from Eq. 10.47a which should be applied in the manner of Example 10.13 (part c).

6.3.4 Self-Weight Distributions

(a) End Slope Due to w

The change in slope between the fixed and free ends in Fig. 6.17a is found from the full area of its M/I diagram, given in Fig. 6.17b. Equation 6.6c is applied to provide the end slope from the three parabolic areas 1, 2 and 3 that constitute the diagram given in Fig. 6.17b. These three areas are built from the spandrel of five parabolic areas, having their common apex at the free end, as follows:

$$\theta = \frac{1}{E}\left\{\left[\frac{\gamma w L^2(1-\alpha)^2}{2\beta I} \times \frac{L(1-\alpha)}{3}\right]_1\right.$$

$$+ \left[\frac{wL^2(\alpha^2+\gamma)}{2(1+\beta)I} \times \frac{L}{3} - \frac{\gamma w L^2(1-\alpha)^2}{2(1+\beta)I} \times \frac{L(1-\alpha)}{3}\right]_2$$

$$\left. + \left[\frac{wL^2[1+\gamma(3-2\alpha)]}{2I} \times \frac{L(2-\alpha)}{3} - \frac{wL^2(\alpha^2+\gamma)}{2I} \times \frac{L}{3}\right]_3\right\} \qquad (6.27a)$$

The three terms in the end slope Eq. 6.27a have the common factor $wL^3/6EI$ simplifying it to:

$$\theta = \frac{wL^3}{6EI}\left\{\frac{\gamma(1-\alpha)^3}{\beta} + \frac{(\alpha^2+\gamma)}{(1+\beta)} - \frac{\gamma(1-\alpha)^3}{(1+\beta)} + [1+\gamma(3-2\alpha)](2-\alpha) - (\alpha^2+\gamma)\right\} \qquad (6.27b)$$

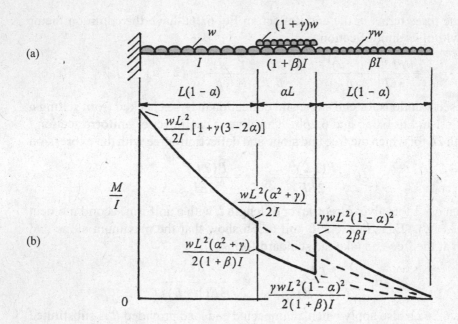

Figure 6.17 M/I-diagram for the self-weight w/unit length for a two-section, telescopic cantilever

(b) End-Deflection Due to w

The end deflection is found from the sum of the first moments of the three areas 1, 2 and 3, that form the M/I-diagram in Fig. 6.17b. Referring the moment of areas 2 and 3 to the free end requires that the moments of the areas beneath the dotted lines be subtracted in the following manner:

$$v = \frac{1}{E} \left\{ \left[\frac{\gamma w L^2 (1-\alpha)^2}{2\beta I} \times \frac{L(1-\alpha)}{3} \times \frac{3L(1-\alpha)}{4} \right]_1 \right.$$

$$+ \left[\frac{w L^2 (\alpha^2 + \gamma)}{2(1+\beta)I} \times \frac{L}{3} \times \frac{3L}{4} - \frac{\gamma w L^2 (1-\alpha)^2}{2(1+\beta)I} \times \frac{L(1-\alpha)}{3} \times \frac{3L(1-\alpha)}{4} \right]_2$$

$$+ \left[\frac{w L^2 [1+\gamma(3-2\alpha)]}{2I} \times \frac{L(2-\alpha)}{3} \times \frac{3L(2-\alpha)}{4} - \frac{w L^2 (\alpha^2 + \gamma)}{2I} \times \frac{L}{3} \times \frac{3L}{4} \right]_3 \right\} \quad (6.28a)$$

The three terms in this end deflection Eq. 6.28a are simplified with the common factor $wL^4/8EI$:

$$v = \frac{wL^4}{8EI} \left\{ \frac{\gamma(1-\alpha)^4}{\beta} + \frac{\alpha^2 + \gamma}{1+\beta} - \frac{\gamma(1-\alpha)^4}{1+\beta} + [1+\gamma(3-2\alpha)](2-\alpha)^2 - (\alpha^2 + \gamma) \right\}$$

$$(6.28b)$$

The closed solutions for self-weight slope and deflection may be checked from setting $\alpha = 0$ and $\beta = \gamma = 1$ in Eqs 6.27b and 6.28b. This gives a cantilever of uniform section I with length $2L$ for which the free-end slope and deflection are known:

$$\theta = \frac{w(2L)^3}{6EI}, \qquad v = \frac{w(2L)^4}{8EI}$$

Also, when $\alpha = 1$, this gives a cantilever of length L with a uniform second moment of area $(1 + \beta)I$. The maximum (free-end) slope and deflection, as found from Eqs 6.27b and 6.28b, again take the standard forms:

$$\theta = \frac{w(1 + \gamma)L^3}{6E(1 + \beta)I}, \qquad v = \frac{w(1 + \gamma)L^4}{8E(1 + \beta)I}$$

6.3.5 Wind Pressure Distributions

(a) End Slope Due to q

The most severe condition arising from wind loading is where a uniform pressure Q acts normally to the top surface (see Fig. 6.18a). The pressure adds to the self-weights by a constant force/unit length $q = Qa$ where a is the section breadth. Two breadths a_1 and a_2 are exposed to the wind so that a loading $q_1 = Qa_1$ is applied to the fixed end, which now includes the overlap and $q_2 = Qa_2$ is applied to the free end. If we write these as a ratio $\mu = q_2/q_1 = a_2/a_1$ then the M/I-diagram is that shown in Fig. 6.18b. The change in slope between the fixed and free ends is found from the full area of the M/I-diagram. Equation 6.6c is applied to provide the end slope from the three parabolic areas that constitute the diagram given in Fig. 6.18b:

$$\theta = \frac{1}{E}\left\{ \left[\frac{\mu q L^2 (1 - \alpha)^2}{2\beta I} \times \frac{L(1 - \alpha)}{3} \right]_1 \right.$$

$$+ \left[\frac{qL^2[\alpha^2 + \mu(1 - \alpha^2)]}{2(1 + \beta)I} \times \frac{L}{3} - \frac{\mu q L^2 (1 - \alpha)^2}{2(1 + \beta)I} \times \frac{L(1 - \alpha)}{3} \right]_2$$

$$\left. + \left[\frac{qL^2[1 + \mu(1 - \alpha)(3 - \alpha)]}{2I} \times \frac{L(2 - \alpha)}{3} - \frac{qL^2[\alpha^2 + \mu(1 - \alpha^2)]}{2I} \times \frac{L}{3} \right]_3 \right\} \qquad (6.29a)$$

The three lines in the end-slope Eq. 6.29a have the common factor $qL^3/6EI$, this allowing for a simplified expression:

$$\theta = \frac{qL^3}{6EI}\left\{ \frac{\mu(1 - \alpha)^3}{\beta} + \frac{\alpha^2 + \mu(1 - \alpha^2)}{(1 + \beta)} - \frac{\mu(1 - \alpha)^3}{(1 + \beta)} \right.$$

$$\left. + [1 + \mu(1 - \alpha)(3 - \alpha)](2 - \alpha) - [\alpha^2 + \mu(1 - \alpha^2)] \right\} \qquad (6.29b)$$

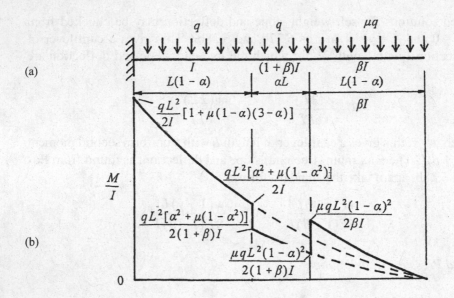

Figure 6.18 M/I-diagram for a two-section, telescopic cantilever under wind loadinq q/unit length

(b) End Deflection Due to q

The end deflection is found from the sum of the first moments of the three areas 1, 2 and 3, that constitute the M/I diagram in Fig. 6.18b. Referring the moment of areas 2 and 3 to the free end requires the moments of the areas beneath the dotted lines are subtracted as follows:

$$v = \frac{1}{E}\left\{\left[\frac{\mu q L^2(1-\alpha)^2}{2\beta I} \times \frac{L(1-\alpha)}{3} \times \frac{3L(1-\alpha)}{4}\right]_1\right.$$

$$+\left[\frac{qL^2[\alpha^2+\mu(1-\alpha^2)]}{2(1+\beta)I} \times \frac{L}{3} \times \frac{3L}{4} - \frac{\mu q L^2(1-\alpha)^2}{2(1+\beta)I} \times \frac{L(1-\alpha)}{3} \times \frac{3L(1-\alpha)}{4}\right]_2$$

$$+\left[\frac{qL^2[1+\mu(1-\alpha)(3-\alpha)]}{2I} \times \frac{L(2-\alpha)}{3} \times \frac{3L(2-\alpha)}{4}\right.$$

$$\left.\left.-\frac{qL^2[\alpha^2+\mu(1-\alpha^2)]}{2I} \times \frac{L}{3} \times \frac{3L}{4}\right]_3\right\} \tag{6.30a}$$

The four lines in this end-deflection Eq. 6.30a have the common factor $qL^4/8EI$ allowing for its simplification:

$$v = \frac{qL^4}{8EI}\left\{\frac{\mu(1-\alpha)^4}{\beta} + \frac{\alpha^2+\mu(1-\alpha^2)}{(1+\beta)} - \frac{\mu(1-\alpha)^4}{(1+\beta)}\right.$$

$$\left.+\left[1+\mu((1-\alpha)(3-\alpha)](2-\alpha)^2 - [\alpha^2+\mu(1-\alpha^2)]\right\} \tag{6.30b}$$

Setting $\alpha = 0$ and $\beta = \mu = 1$ in Eqs 6.29b and 6.30b gives a cantilever of uniform section I with length $2L$ for which the free-end slope and deflection are expected as:

$$\theta = \frac{q(2L)^3}{6EI}, \qquad v = \frac{q(2L)^4}{8EI}$$

Also when $\alpha = 1$ this gives a cantilever of length L with a uniform second moment of area $(1 + \beta)I$ within a full length overlap. Here Eqs 6.29b and 6.30b confirm the standard forms for the maximum slope and deflection under a constant wind loading q/unit length:

$$\theta = \frac{qL^3}{6E(1+\beta)I}, \qquad v = \frac{qL^4}{8E(1+\beta)I}$$

6.3.6 Net Slope and Deflection

The superposition principle allows the slope and end deflection found separately for the concentrated load, self-weight and wind-loading above to be added when these loads act together upon the beam. Hence the net slope becomes the sum of Eqs 6.25b, 6.27b and 6.29b and the net deflection the sum of Eqs 6.26b, 6.28b and 6.30b. This analysis demonstrates the convenience of adopting Mohr's theorems in combination with this principal to provide a solution to cantilever deflection that might prove intractable by alternative analyses. For example, the Macaulay method that follows becomes so unwieldy when applied to this problem that a numerical analysis may be preferred. However, Macaulay's method does have an advantage in its provision of the full slope and deflection profiles should they be required.

6.4 Macaulay's Method

With successive integration of the flexure Eq. 6.4 discontinuities will arise in the moment expression when z passes points of concentrated forces, moments and abrupt changes in distributed loading. If Eq. 6.4 is applied to regions in z where M (z) is continuous, then the number of differential equations and integration constants to be solved increases. Fortunately, the amount of work involved may be lessened when $M(z)$ in Eq. 6.4 is replaced with a step function $M[z - a]$

$$EI\frac{d^2v}{dz^2} = M[z-a] \tag{6.31}$$

in which a defines positions at which discontinuities arise. Equation 6.31 permits successive integration of a single-step function for simply supported and single-span encastré beams. The following rules must be applied:

(a) Take the origin for z at the left-hand end.
(b) Where necessary, extend and counterbalance uniformly distributed loading to the right-hand end.

(c)   Let concentrated moments lie at the position $[z - a]^0$.

(d)   Establish the function, $M[z - a]$, in the furthest right portion of the beam. Hogging moments are positive.

(e)   Integrate such terms as $[z - a]$ in the form $\frac{1}{2}[z - a]^2$. These terms are to be ignored when, in substituting values for z, the value of the bracket [ ] becomes negative.

(f)   Apply the known slope and deflection values to find the two constants of integration.

6.4.1 Simply Supported Beam

The constants that arise from integrating the step function are found from applying the condition that the displacements are zero at the support points. The position of maximum deflection, which corresponds to the position of zero slope, is assumed to lie within a selected range for z. The solution will confirm whether the assumption was correct.

Example 6.12 Find the position and magnitude of the maximum deflection and the deflection beneath each concentrated load applied for the beam in Fig. 6.19. Take $E = 207$ GPa and $I = 10^{10}$ mm^4. Note: 1 t $= 10^3$ kg.

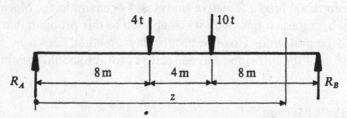

Figure 6.19 Simply supported beam with concentrated loading

The left-hand reaction R_A is found from applying moment equilibrium at B:

$$(8 \times 10) + (12 \times 4) = 20R_A, \quad \therefore \quad R_A = 6.4 \text{ t}$$

($R_B = 7.6$ t is not required). The datum for $M[z - a]$ lies at distance z from the left-hand end as shown. With successive integration, Eq. 6.31 becomes:

$$EI\, d^2v/dz^2 = -6.4z + 10[z - 12] + 4[z - 8]$$

$$EI\, dv/dz = -6.4z^2/2 + 10[z - 12]^2/2 + 4[z - 8]^2/2 + C_1 \tag{i}$$

$$EI\, v = -3.2z^3/3 + 5[z - 12]^3/3 + 2[z - 8]^3/3 + C_1z + C_2. \tag{ii}$$

Substituting $v = 0$ when $z = 0$ in Eq. ii gives $C_2 = 0$ (ignore [-]). Substituting $v = 0$ when $z = 20$ m gives $C_1 = 326.25$ (t m^2). Eq. ii becomes:

$$EI\,v = -\ 1.066\,z^3 + 5[\ z - 12]^3/3 + 2[\ z - 8]^3/3 + 326.25\,z \qquad \text{(iii)}$$

Substituting $z = 8$ m in Eq. iii:

$$EI\,v = -\ 1.066(8)^3 + 326.25(8) = 2064.21\ \text{tm}^3$$

Converting units of tonne $\underline{t}$ to $\underline{N}$ and $\underline{m}$ to $\underline{mm}$ gives

$$v = (2064.21 \times 10^3 \times 9.81 \times 10^9)/(207 \times 10^3 \times 10^{10}) = 9.783\ \text{mm}$$

Substituting $z = 12$ m in Eq. iii, then using proportion:

$$EI\,v = -\ 1.066(12)^3 + 2[4]^3/3 + 326.25(12) = 2115.62\ \text{tm}^3$$

$$v = (2115.62/2064.21)9.783 = 10.03\ \text{mm}$$

Assume v_{max} lies between the loads, $8 \le z \le 12$, and set $dv/dz = 0$ in Eq. i:

$$-\ 3.2z^2 + 2[z - 8]^2 + 326.25 = 0 \qquad \text{(iv)}$$

The solution to Eq. iv is $z = 10.3$ m, confirming that the assumed position is correct. Substituting into Eq. ii, the maximum displacement is found proportionally:

$$EI\,v_{max} = -\ 3.2(10.3)^3/3 + 2[2.3]^3/3 + 326.25(10.3) = 2203.64\ \text{tm}^3$$

$$\therefore v_{max} = (2203.63/2064.21)9.783 = 10.44\ \text{mm}$$

Example 6.13 Find the slope at the left support and the position and magnitude of the maximum deflection for the beam shown in Fig. 6.20. Take $EI = 150$ MNm^2.

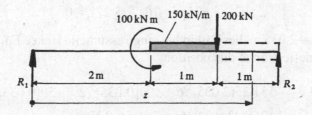

Figure 6.20 Simply supported beam with mixed loading

Firstly, the left-hand reaction R_1 is calculated from moment equilibrium:

$$(200 \times 1) + (150 \times 1 \times 1.5) + 100 = 4R_1, \ \therefore R_1 = 131.25\ \text{kN}$$

($R_2 = 218.75$ kN). In applying the rules (b)–(d) listed above to the distributed loading and concentrated moments shown, Eq. 6.31 becomes:

$$EI \, d^2v/dz^2 = -13.125z + 100[z - 2]^0 + 200[z - 3] + 150[z - 2]^2/2$$
$$- 150[z - 3]^2/2$$

$$EI \, dv/dz = -131.25z^2/2 + 100[\, z - 2] + 200[\, z - 3]^2/2 + 150[\, z - 2]^3/6$$
$$, - 150[\, z - 3]^3/6 + C_1$$

(i)

$$EI \, v = -131.25 \, z^3/6 + 100[\, z - 2]^2/2 + 200[\, z - 3]^3/6 + 150[\, z - 2]^4/24$$
$$- 150[\, z - 3]^4/24 + C_1z + C_2$$

(ii)

In Eq. ii, the boundary conditions are: $v = 0$ when $z = 0$, which gives $C_2 = 0$ and $v = 0$ when $z = 4$ m, which leads to:

$$0 = -131.25(4)^3/6 + 100[2]^2/2 + 200[1]^3/6 + 150[2]^4/24 - 150[1]^4/24 + 4C_1 \quad \text{(iii)}$$

Equation iii yields $C_1 = 268.23$ kNm². Substituting $z = 0$ in Eq. i, gives the slope at the left-hand support:

$$dv/dz = C_1/EI = 268.23/(150 \times 10^3) = 1.788 \times 10^{-3} \text{ c}$$

Assume that the maximum deflection lies in the region $0 \leq z \leq 2$ and set $dv/dz = 0$, in Eq. i:
$$EI \, dv/dz = 0 = 131.25 \, z^2/2 + 268.23$$

from which $z = 2.022$ m. This solution disagrees with the assumed position and so we must assume next that v_{max} lies in the region $2 \leq z < 3$. Correspondingly, Eq. i gives zero slope for this condition:

$$EI \, dv/dz = 0 = -131.25 \, z^2/2 + 100[z - 2] + 150[z - 2]^3/6 + 268.23$$

This gives :
$$z^3 - 8.625z^2 + 16z - 5.271 = 0$$

The solution, $z = 2.035$ m, lies within the range assumed. Hence Eq. ii will provide the maximum deflection at this position:

$$EI \, v = -131.25(2.035)^3/6 + 100[0.035]^2/2 + 150[0.035]^4/24$$
$$+ (268.23 \times 2.035) = 361.56 \text{ (kNm}^3)$$

$$\therefore v_{max} = 361.56/(150 \times 10^3) = 2.41 \times 10^{-3} \text{ m} = 2.41 \text{ mm}$$

6.4.2 Encastré Beam

The Macaulay method may be extended to find the slopes and deflections for encastré beams. A complete solution to the support reactions and fixing moments involves both Eqs 6.18a,b as the following example shows.

Example 6.14 Determine the support reactions, the fixing moments and the maximum deflection for the encastré beam in Fig. 6.21. Take $EI = 150$ MNm^2.

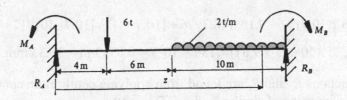

Figure 6.21 Encastré beam with mixed loading

The left-hand reaction R_A, and fixing moment, M_A, must now appear in the flexure Eq. 6.31 as constants additional to those arising from the integration. All constants are found from knowing that zero slope and deflection apply to both the left-hand and right-hand end fixings.

$$EI \, d^2v/dz^2 = M_A - R_A z + 6[z - 4] + 2[z - 10]^2/2$$

$$EI \, dv/dz = M_A z - R_A z^2/2 + 6[z - 4]^2/2 + 2[z - 10]^3/6 + C_1 \tag{i}$$

$$EIv = M_A z^2/2 - R_A z^3/6 + 6[z - 4]^3/6 + 2[z - 10]^4/24 + C_1 z + C_2 \tag{ii}$$

The boundary conditions are $dv/dz = 0$ and $v = 0$ for $z = 0$. It follows from Eqs i and ii that $C_1 = C_2 = 0$, which will always apply to an encastré beam. Also, $dv/dz = 0$ and $v = 0$ when $z = 20$ m. From Eqs i and ii:

$$0 = 20M_A - R_A (20)^2/2 + 3[16]^2 + [10]^3/3$$

$$\therefore 10R_A - M_A = 55.065 \tag{iii}$$

$$0 = (20)^2 M_A/2 - (20)^3 R_A/6 + [16]^3 + [10]^4/12$$

$$\therefore 6.667R_A - M_A = 24.617 \tag{iv}$$

Subtracting Eq. iv from Eq. iii gives $R_A = 9.135$ t and $M_A = 36.2$ tm. Assume that v_{max} lies in the region: $4 \le z < 10$ (m), with $dv/dz = 0$ in Eq. i:

$$36.2z - 9.135 \, z^2/2 + 3[z - 4]^2 = 0,$$

$$z^2 - 7.796 \, z - 30.67 = 0 \tag{v}$$

The solution to Eq. v is: $z = 10.67$ m, which indicates that the assumed position was incorrect. Next, assume that v_{max} lies in the region: $10 \le z < 20$, Eq. i gives:

$$z^3 - 34.703z^2 + 336.6 \, z - 856 = 0 \tag{vi}$$

The solution to Eq. vi is $z = 10.7$ m, which lies within range. Despite being only slightly different from that found previously, Eq. ii will now provide the maximum deflection more accurately:

$$EI\,v_{max} = 36.2(10.7)^2/2 - 9.135(10.7)^3/6 + [10.7-4]^3 + [10.7-10]^4/12 = 508.95 \text{ tm}^3$$

$$\therefore v_{max} = (508.95 \times 9.81)/(150 \times 10^3) = 0.0333 \text{ m} = 33.33 \text{ mm}$$

The free reactions R_1 and R_2 are found from applying equilibrium conditions to a simple end-supporting of similar loading. This gives:

$$(4 \times 6) + (10 \times 2 \times 15) = 20R_2$$

$$\therefore R_2 = 16.2 \text{ t and } R_1 = (6 + 20) - 16.2 = 9.8 \text{ t}$$

The right-hand support reaction and fixing moment are found from Eqs 6.18a,b:

$$M_B = M_A - L(R_A - R_1) = 36.2 - 20(9.135 - 9.8) = 49.5 \text{ tm}$$

$$R_B = R_2 - (M_A - M_B)/L = 16.2 - (36.2 - 49.5)/20 = 16.87 \text{ t}$$

Check: $M_B = 36.2 - (20 \times 9.135) + (16 \times 6) + (10 \times 2 \times 5) = 49.5 \text{ tm}$ ✓

6.5 Superposition Principle

The essential condition for each of the following forms of superposition principle is that a structure must obey Hooke's law within its elastic range. The most common form of this principle applies to the deflection arising from external loading which states that:

The total elastic displacement at a point in a structure under any load combination is the sum of the displacements which occur at that point as each load is applied to its position independently and in any sequence.

It was mentioned earlier in § 6.3.6 how this form of superposition could be used to simplify the solutions to the free-end slope and deflection of a telescopic cantilever under various combined loadings.

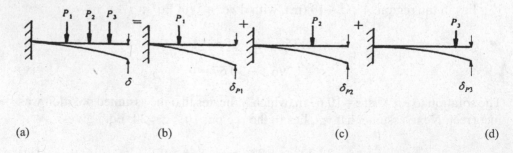

(a)   (b)   (c)   (d)

Figure 6.22 Superposition principle

As an illustrative example, let us apply this principle to find the end deflection δ for the cantilever beam loaded as shown in Fig. 6.22a. We may sum the end deflections δ_{P1}, δ_{P2} and δ_{P3} from each load applied independently to its respective position in Figs 6.22b–d). This gives:

$$\delta = \delta_{P1} + \delta_{P2} + \delta_{P3} \tag{6.32}$$

In Eq. 6.32 the principle has been applied to load–displacement but it can also be applied to connect the following pairs: load–stress, load–strain, displacement–stress and displacement–strain, provided the relationship between each pair is linear. Within each coupling the external action (cause) precedes the internal reaction (effect) so that the first two would modify the principle to read as:

The total stress (strain) at a point in a structure under any load combination is the sum of the stress (strain) which occur at that point as each load is applied independently.

Similarly, the last two are interpreted within a similar superposition principle with displacement as the cause of stress and strain. The uncoupling of load allows the displacement of a structure to arise from other sources the most common being its temperature changes with an exposure to the environment. Then:

The total stress (strain) at a point in a structure under any displacement combination is the sum the stress (strain) which occur at that point as each displacement is imposed independently.

Each form rests with the linear elastic response that metallic structures display when loaded elastically. This also means that the displacement, stress and strain, which the loading produces, will all disappear when the load is removed. Non-metallic materials and particularly elastomers, while having a similar capacity to recover, are non-linear. Metals when loaded beyond the elastic limit become non-linear and will remain permanently set when unloaded. The stress versus strain plot is the usual indicator of the material response and when in each case given it is non-linear, then none of the forms of superposition principle stated above will apply.

One further notable exception to the principal of superposition applies to the lateral displacement v of a long thin strut carrying an axial compressive force, P (see Fig. 8.1, p. 288). Here Eq. 8.1 governs the relationship between P and v:

$$EI \, d^2v/dz^2 + Pv = 0 \tag{6.33a}$$

The solution to Eq. 6.33a is:

$$v = A \sin\sqrt{(P/EI)}\, z + B \cos\sqrt{(P/EI)}\, z \tag{6.33b}$$

Equation 6.33b shows that for any position z in the length, the elastic displacement v is dependent upon P non-linearly and hence superposition will not apply. To show this let v_1 and v_2 be the lateral displacements of the strut, at a common length position z, corresponding to respective axial loads P_1 and P_2. We *cannot* say that when an axial load P_2 is superimposed upon P_1, the displacement will be the following sum $v_1 + v_2$, from Eq. 6.23:

$$v_1 + v_2 = A \sin\sqrt{(P_1/EI)}\, z + B \cos \sqrt{(P_1/EI)}\, z$$
$$+ A \sin\sqrt{(P_2/EI)}\, z + B \cos\sqrt{(P_2/EI)}\, z \qquad (6.34a)$$

However, we *can* say that when P_2 is superimposed upon P_1 the strut displacement will be:

$$v = A\sin \sqrt{(P_1 + P_2)/(EI)}\, z + B\cos \sqrt{(P_1 + P2)/(EI)}\, z \qquad (6.34b)$$

Clearly, Eqs 6.34a and 6.34b are not the same. Mathematically, the strut problem is non-linear because the bending moment at position z depends upon both the axial loading P and the lateral displacement v, i.e. $M = Pv$. Linear superposition applies to a beam under lateral loading because the bending moment does not depend upon the displacement directly but upon the loading and its position z, i.e. $M = M(z)$.

Example 6.15 Use superposition to establish the total strain ε and displacement δ for a bar of area A and length L when axial tensile forces F_1 and F_2 are superimposed upon one another.

The independent strain–force relationships are linear:

$$\varepsilon_1 = F_1/AE, \qquad \varepsilon_2 = F_2/AE$$

So too are the independent displacement–force relationships:

$$\delta_1 = F_1 L/AE, \qquad \delta_2 = F_2 L/AE$$

From superposition:

$$\varepsilon = \varepsilon_1 + \varepsilon_2 = \varepsilon_2 + \varepsilon_1 = (F_1 + F_2)/AE$$

$$\delta = \delta_1 + \delta_2 = \delta_2 + \delta_1 = (F_1 + F_2)L/AE$$

Example 6.16 Show that the deflection by Macaulay at a point in the length of a beam under the loading shown in Fig. 6.20 (see Example 6.13) conforms to the principle of superposition.

'Load' symbols replace their numerical magnitudes: $R_1 = 131.25$ kN, $M = 100$ kNm, $F = 200$ kN and $w = 150$ kN/m. The Macaulay deflection Eq. ii in Example 6.13 may then be written as:

$$EI\,v = -\,R_1 z^3/6 + (M/2)[z - 2]^2 + (F/6)[z - 3]^3$$

$$+ (w/24)[z - 2]^4 - (w/24)[z - 3]^4 + C_1 z \tag{i}$$

where the reaction and integration constant are:

$$R_1 = (F + M + 3w/2)/4 \tag{ii}$$

$$C_1 = 8R_1/3 - M/2 - F/24 - 15w/96 \tag{iii}$$

Equations i–iii show that v depends upon linear terms in the loads R_1, M, F and w. This is the condition upon which superposition depends. Thus, v may be found from summing the v's at a given position as each load is applied independently. The contribution that each load makes to the displacement appears within the total displacement Eq. i. This shows that when the force F acts alone $R_1 = F/4$ and $C_1 = 5F/8$ giving its displacement contribution as:

$$EI\,v_F = -\,(F/4)\,z^3/6 + (F/6)\,[z - 3]^3 - (5F/8)\,z$$

When the distributed load w acts alone $R_1 = 3w/8$ and $C_1 = 27w/32$, giving its displacement contribution as:

$$EI\,v_w = -\,w\,z^3/16 + (w/24)[z - 2]^4 - (w/24)[z - 3]^4 + (27w/32)z$$

Finally, when the concentrated moment M acts alone $R_1 = M/4$ and $C_1 = M/6$, giving its displacement contribution as:

$$EI\,v_M = -\,M\,z^3/24 + (M/2)[z - 2]^2 + (M/6)z$$

Superposition states that when the three loads are applied together in any sequence the displacement beneath them follows as a linear summation $v = v_F + v_w + v_M$.

6.5.1 Load Equivalents

Examples 6.15 and 6.16 show how strain and displacement may be found from the superposition principle. This principle also applies between the equivalent of each asymmetrical loading, shown in Figs 6.23a–c. When stress, strain and displacement are required under each of these loadings they may be found from summing under their equivalent loading. The latter is composed of symmetric and skew-symmetric components. For example, in Fig. 6.23a the beam's distributed loading shown is equivalent to the sum of uniform and linearly varying load distributions:

$$w(z) = w_o + cz \quad (c \text{ is a constant})$$

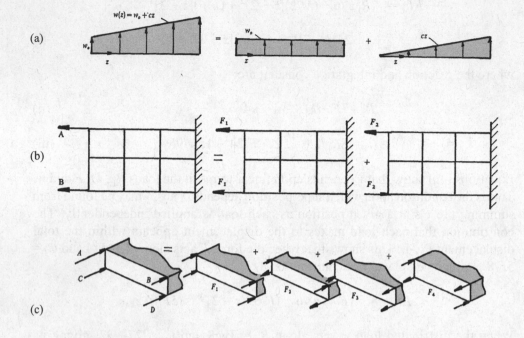

Figure 6.23 Superposition of symmetric and skew-symmetric loadings

For the pin-jointed, cantilever frame in Fig. 6.23b, equivalent equal loads F_1 and F_2 replace the differing applied loads A and B, such that

$$A = F_1 + F_2, \quad B = F_1 - F_2$$

Solving simultaneously, the equivalent loads become

$$F_1 = \tfrac{1}{2}(A + B), \quad F_2 = \tfrac{1}{2}(A - B)$$

Similarly, four different corner loads A, B, C and D in Fig. 6.23c are equivalent to the sum of equal corner loadings:

$$
\begin{aligned}
A &= F_1 + F_2 + F_3 + F_4 \\
B &= F_1 + F_2 - F_3 - F_4 \\
C &= F_1 - F_2 + F_3 - F_4 \\
D &= F_1 - F_2 - F_3 + F_4
\end{aligned}
\quad \rightarrow \quad
\begin{bmatrix} F_1 \\ F_2 \\ F_3 \\ F_4 \end{bmatrix}
=
\begin{bmatrix}
1 & 1 & 1 & 1 \\
1 & 1 & -1 & -1 \\
1 & -1 & 1 & -1 \\
1 & -1 & -1 & 1
\end{bmatrix}^{-1}
\begin{bmatrix} A \\ B \\ C \\ D \end{bmatrix}
\tag{6.35}
$$

in which equivalent loading F_i ($i = 1, 2, 3, 4$) follow from the simultaneous solution to the four equations, indicated by the inverse matrix Eq. 6.35 (see Ex. 6.28).

6.5.2 Load–Displacement Law

Example 6.16 showed, from successive integration with the Macaulay method, that the total displacement of a beam is composed of the sum of terms each linear in its loading: F, M and w.

Now let any combination of loading be represented by p_j where $j = 1, 2, 3, ... n$ identifies each load. It follows that the displacement (Δ_a) at any point a, has a common form for all Hookean structures:

$$\Delta_a = F_{a1}p_1 + F_{a2}p_2 + F_{a3}p_3 + F_{a4}p_4 + .. F_{aj}p_j + .. F_{an}p_n \qquad (6.36a)$$

($\Delta_a = F_{aj}p_j$). The *flexibility coefficients* F_{a1}, F_{a2}, F_{a3}, ... F_{an} depend only upon the position of point a, i.e. they are functions of z, as can be seen from Eq. i in Example 6.16. For a second point b, these coefficients will be different. That is, the displacement Δ, at b under the same loading p_j will become:

$$\Delta_b = F_{b1}p_1 + F_{b2}p_2 + F_{b3}p_3 + F_{b4}p_4 + .. F_{bj}p_j + ... F_{bn}p_n \qquad (6.36b)$$

($\Delta_b = F_{bj}p_j$). The double subscript notation for the coefficients F_{aj} and F_{bj}, within Eqs 6.36a,b is used to refer flexibility across all the loads ($j = 1, 2, ... n$) to the chosen position (a and b). Specifically, if the positions a and b are chosen to correspond to the load points then the displacements beneath the loads are given by:

$$\Delta_i = F_{i1}p_1 + F_{i2}p_2 + F_{i3}p_3 + F_{i4}p_4 + ... F_{in}p_n \qquad (6.37a)$$

within which the superposition principle applies. That is, the total displacement at point i is the sum of the displacements that apply to i when each load acts at its respective position in isolation. It follows that all the load-point displacements: Δ_1, $\Delta_2 ... \Delta_n$ must lie within the simultaneous equations:

$$\Delta_1 = F_{11}p_1 + F_{12}p_2 + F_{13}p_3 + F_{14}p_4 + ... F_{1n}p_n$$
$$\Delta_2 = F_{21}p_1 + F_{22}p_2 + F_{23}p_3 + F_{24}p_4 + ... F_{2n}p_n$$

$$\Delta_i = F_{i1}p_1 + F_{i2}p_2 + F_{i3}p_3 + F_{i4}p_4 + ... F_{in}p_n \qquad (6.37b)$$

$$\Delta_n = F_{n1}p_1 + F_{n2}p_2 + F_{n3}p_3 + F_{n4}p_4 + ... F_{nn}p_n$$

For brevity, the general system of Eqs 6.37a,b may be written either in indicial (subscript) or matrix notations as, respectively:

$$\Delta_i = F_{ij}p_j \quad \text{or} \quad \boldsymbol{\Delta} = \mathbf{F}\,\mathbf{p} \qquad (6.38a,b)$$

Equation 6.38a adopts the *summation convention* in which a repeated subscript j within the term $F_{ij}p_j$ requires its sum to be taken as we set $j = 1, 2, ... n$. Summing over j for each i (= 1, 2, ... n) in turn will recover Eq. 6.37b. Alternatively, Eq. 6.38b adopts the symbolic matrix notation for a matrix multiplication that provides all the equations within Eq. 6.37b:

$$
\begin{bmatrix}
\Delta_1 \\
\Delta_2 \\
\cdot \\
\Delta_i \\
\cdot \\
\cdot \\
\cdot \\
\Delta_n
\end{bmatrix}
=
\begin{bmatrix}
F_{11} & F_{12} & F_{13} & F_{14} & \cdot & \cdot & \cdot & F_{1n} \\
F_{21} & F_{22} & F_{23} & F_{24} & \cdot & \cdot & \cdot & F_{2n} \\
\cdot & \cdot & \cdot & \cdot & \cdot & \cdot & \cdot & \cdot \\
F_{i1} & F_{i2} & F_{i3} & F_{i4} & \cdot & \cdot & \cdot & F_{in} \\
\cdot & \cdot & \cdot & \cdot & \cdot & \cdot & \cdot & \cdot \\
\cdot & \cdot & \cdot & \cdot & \cdot & \cdot & \cdot & \cdot \\
\cdot & \cdot & \cdot & \cdot & \cdot & \cdot & \cdot & \cdot \\
F_{n1} & F_{n2} & F_{n3} & F_{n4} & \cdot & \cdot & \cdot & F_{nn}
\end{bmatrix}
\begin{bmatrix}
p_1 \\
p_2 \\
p_3 \\
p_4 \\
\cdot \\
\cdot \\
\cdot \\
p_n
\end{bmatrix}
\qquad (6.38c)
$$

Equations 6.38b,c show that $\mathbf{F}$ is a square $(n \times n)$ matrix whose elements are the flexibility coefficients F_{ij}. The loads p_j and load-point displacements Δ_i appear as $(n \times 1)$ *column matrices*: $\mathbf{p} = [p_1\ p_2\ p_3\ p_4 \dots p_n]^T$ and $\Delta = [\Delta_1\ \Delta_2\ \Delta_3\ \Delta_4 \dots \Delta_n]^T$. Inverting Eqs 6.38a,b expresses the dependence of the loads upon their displacements:

$$ p_j = K_{ji}\Delta_i \quad \text{or} \quad \mathbf{p} = \mathbf{F}^{-1}\Delta = \mathbf{K}\,\Delta \qquad (6.39a,b) $$

In Eqs 6.39a,b, $\mathbf{K} = \mathbf{F}^{-1}$ is an $(n \times n)$ matrix whose elements become the *stiffness coefficients* K_{ji}. Expanding Eq. 6.38a with a summation over $i = 1, 2, \dots n$:

$$ p_j = K_{j1}\Delta_1 + K_{j2}\Delta_2 + K_{j3}\Delta_3 + K_{j4}\Delta_4 + \dots K_{jn}\Delta_n \qquad (6.39c) $$

Then setting $j = 1, 2, 3 \dots n$, Eq. 6.39c reveals the dependence of each individual load p_j upon all the displacements Δ_i $(i = 1, 2, \dots n)$.

6.5.3 Maxwell's Reciprocal Theorem

An important property of the matrices $\mathbf{F}$ and $\mathbf{K}$ within Eqs 6.38b and 6.39b is that they are symmetrical about their leading diagonals. That is, their elements conform to $F_{ij} = F_{ji}$ and $K_{ij} = K_{ji}$, which we may prove from Maxwell's reciprocal theorem applied to a beam in Figs 6.24ab.

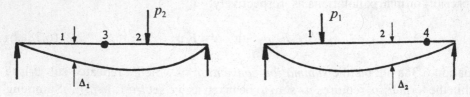

(a)   (b)

Figure 6.24 Maxwell's reciprocal theorem

Given 1, 2, 3 and 4 are different points along the *same* beam, Maxwell's theorem states that the deflection Δ_1 at point 1, caused from applying a load p_2 to point 2, will equal the deflection Δ_2 at point 2 when a load p_1 acts at point 1, provided $p_1 = p_2$. From Eq. 6.37a:

$$\Delta_1 = F_{12}p_2, \quad \Delta_2 = F_{21}p_1 \tag{6.40a,b}$$

Similarly, when the same load ($p_3 = p_4$) is interchanged between points 3 and 4 in Figs 6.24a,b then the displacements $\Delta_3 = \Delta_4$. Equations 6.40a,b appear to show that $F_{12} = F_{21}$ when $p_1 = p_2$ but the latter is the only requirement for $\Delta_1 = \Delta_2$. In fact, $F_{12} = F_{21}$ when $p_1 \neq p_2$ and $F_{34} = F_{43}$ when $p_3 \neq p_4$ etc. since the flexibility coefficients are inherent properties of the structure and not its particular loading. Applying the theorem to the remaining load points reveals the symmetry within the flexibility and stiffness matrices, as in Eq. 6.38c
:

$$F_{ij} = F_{ji}, \quad K_{ij} = K_{ji}$$

In summary, each stiffness coefficient identifies with the spring-like behaviour at each point in a structure, i.e. K is the spring constant of proportionality between the load applied and the displacement it produces ($p = K\Delta$). The spring constant varies with the load position so that when it is referred to the displacement at a given point 1 we have K_{11} for when the load is applied to point 1, K_{12} for when the load is applied to point 2, K_{13} for when the load is applied to point 3, etc. When all the loads act together the displacement at point 1 becomes the sum of displacements from each load contribution. For this it is more convenient to use flexibility coefficients which invert each stiffness [i.e. $\Delta_i = (K_{ij})^{-1}p_j = F_{ij}p_j$], this giving: $\Delta_1 = F_{11}p_1 + F_{12}p_2 + F_{13}p_3$ etc. The system of Eqs 6.37b provide for the displacements at each load point when all the loads act together.

Example 6.17 Use the reciprocal law and the superposition principle to find an expression for the equivalent end load p_1 that would give the same end displacement Δ_1 for the cantilever's loads p_2 and p_3, as shown in Fig. 6.25.

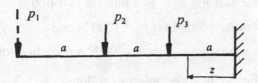

Figure 6.25 Reciprocal theorem for cantilever beam

The superposition principle gives the end displacement under the combined loads p_2 and p_3, as a sum of end displacements when p_2 and p_3 act separately upon the cantilever:

$$\Delta_1 = F_{12}p_2 + F_{13}p_3 \tag{i}$$

With only the end load p_1 applied, the displacements at the three load positions are:

$$\Delta_1 = F_{11}p_1, \quad \Delta_2 = F_{21}p_1 \text{ and } \Delta_3 = F_{31}p_1 \tag{ii}$$

Referring to Example 6.2, the displacements at points 1, 2 and 3 under p_1 are found from putting $z = 3a$ and $z = 2a$ and $z = a$, respectively, in the deflection equation. Combining these with Eq. ii gives the flexibility coefficients:

$$\Delta_1 = 9\,p_1 a^3/EI = F_{11}p_1, \qquad \therefore F_{11} = 9a^3/EI \tag{iii}$$

$$\Delta_2 = 14\,p_1 a^3/3EI = F_{21}p_1, \qquad \therefore F_{21} = 14a^3/3EI \tag{iv}$$

$$\Delta_3 = 4\,p_1 a^3/3EI = F_{31}p_1, \qquad \therefore F_{31} = 4a^3/3EI \tag{v}$$

Applying the reciprocal law $F_{12} = F_{21}$ and $F_{13} = F_{31}$ to Eq. i:

$$\Delta_1 = F_{21}p_2 + F_{31}p_3 \tag{vi}$$

Substituting from Eqs iv and v:

$$\Delta_1 = (14a^3/3EI)\,p_2 + (4a^3/3EI)\,p_3 \tag{vii}$$

when from Eqs ii and iii:

$$\Delta_1 = (9a^3/EI)\,p_1 \tag{viii}$$

Equating vii and viii provides the equivalent end load p_1:

$$p_1 = (2/27)(7p_2 + 2p_3)$$

Example 6.18 The displacements at three points 1, 2 and 3 in a plate of Hookean material were measured when different loads were applied individually to each point. The deflections under an initial load $p_1 = 800\text{N}$ were: $\Delta_1 = 1.5$ mm, $\Delta_2 = -0.5$ mm and $\Delta_3 = -2.5$mm. Find:

(a) Δ_1 given $\Delta_2 = 2$ mm and $\Delta_3 = 1.75$ mm for $p_2 = 600$ N,

(b) Δ_1 and Δ_2 given $\Delta_3 = 5$ mm for $p_3 = 1000$ N and

(c) Δ_1, Δ_2 and Δ_3 when $p_1 = p_2 = p_3 = 1200$ N are applied together.

Equation 6.37a applies to the three points as follows:

$$\Delta_1 = F_{11}p_1 + F_{12}p_2 + F_{13}p_3 \tag{i}$$

$$\Delta_2 = F_{21}p_1 + F_{22}p_2 + F_{23}p_3 \tag{ii}$$

$$\Delta_3 = F_{31}p_1 + F_{32}p_2 + F_{33}p_3 \tag{iii}$$

Substituting the given initial conditions into Eqs i–iii, gives the flexibility coefficients:

$$1.5 = 800 \, F_{11}, \quad \therefore F_{11} = 1.5/800 \tag{iv}$$

$$-0.5 = 800 \, F_{12}, \quad \therefore F_{21} = -0.5/800 \tag{v}$$

$$-2.5 = 800 \, F_{31}, \quad \therefore F_{31} = -2.5/800 \tag{vi}$$

Substituting the information given in (a) into Eqs i–iii and applying the reciprocal law to the flexibility coefficients in Eq. v:

$$\Delta_1 = F_{12}p_2 = F_{21}p_2 = -(0.5/800)600 = -0.375 \text{ mm}$$

$$2 = 600F_{22} \quad \therefore F_{22} = (2/600) \tag{vii}$$

$$1.75 = 600F_{32}, \quad \therefore F_{32} = (1.75/600) \tag{viii}$$

Substituting the information in (b) and Eqs vi–viii into Eqs i–iii gives:

$$\Delta_1 = F_{13}p_3 = F_{31}p_3 = -(2.5/800)1000 = -3.125 \text{ mm}$$

$$\Delta_2 = F_{23}p_3 = F_{32}p_3 = (1.75/600)1000 = 2.917 \text{ mm}$$

$$5 = 1000F_{33}, \quad \therefore F_{33} = (5/1000) \tag{ix}$$

Finally, the coefficients in Eqs iv–ix enable the displacements to be solved under condition (c). From Eqs i–iii, with $p_1 = p_2 = p_3 = 1200$ N:

$$\Delta_1 = 1200 \left[(1.5/800) - (0.5/800) - (2.5/800) \right] = -2.25 \text{ mm}$$

$$\Delta_2 = 1200 \left[-(0.5/800) + (2/600) + (1.75/600) \right] = 6.75 \text{ mm}$$

$$\Delta_3 = 1200 \left[-(2.5/800) + (1.75/600) + (5/1000) \right] = 5.75 \text{ mm}$$

The symmetry in the flexibility and stiffness matrices revealed within Maxwell's theorem is important to the study of finite elements (FE) in Chapter 13. The FE method provides the displacements numerically from inverting the stiffness matrix K_{ij} in a procedure that is simplified by that matrix's symmetry.

Exercises

Flexure Equation

6.1 Derive from the flexure Eq. 6.4 general expressions for the slope and deflection for a beam carrying uniformly distributed loading, w, over its length, L, when it is: (a) simply supported and (b) a cantilever. Give in each case expressions for the maximum slope and deflection.
[Ans: (a) $wL^2/(24EI)$, $5wL^4/(384EI;)$ (b) $wL^3/(6EI)$, $wL^4/(8EI)$]

6.2 A steel beam 15 m long is simply supported at its ends and carries a total uniformly distributed load of 60 kN. Calculate the slope and deflection at a point 6 m from the left support. Take $I = 400 \times 10^6$ mm^4 and $E = 200$ GPa.
(Ans: 0.12°, 31.4 mm)

6.3 If a cantilever instead of the simply supported beam in Exercise 6.2 is used to carry the distributed loading over a similar length, calculate the slope and deflection at a point 12m from the fixed end. What is the maximum slope and deflection?
(Ans: 1.6°, 232.5 mm; 1.61°, 316.4 mm)

6.4 Derive, from Eq. 6.4, expressions for the slope and deflection at the tip of a cantilever carrying a linearly varying distributed loading, from zero at its free-end to a maximum of w_o at the fixed end.
[Ans: $\theta = w_o L^3/(24EI)$, $v = w_o L^4/(30EI)$]

6.5 Find the slope and deflection at the free-end of a cantilever when loaded as shown in Fig. 6.26. Hint: Take the origin for z where shown.
[Ans: $\theta = Fa^2/(2EI)$, $v = F(3a^2L - a^3)/(6EI)$]

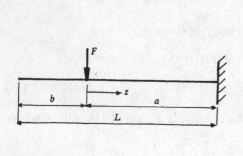

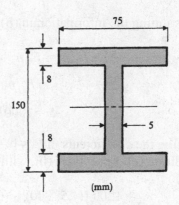

Figure 6.26   **Figure 6.27**

6.6 A steel bin of weight 2 kN is to be supported along two parallel sides of its 1.5 m square base with steel cantilevers of the I-section given in Fig. 6.27. If the maximum stress in the cantilever is not to exceed 85 MPa, calculate the safe load that can be carried in the bin and the maximum deflection of its support under this load.

6.7 A simply supported beam carries a concentrated load W, distance a from the left-hand support and b from the right-hand support, where $a + b = L$ and $a < b$. Derive expressions for the for the slope and deflection at the load point.
[Ans: $Wab(a - b)/6L$, $Wa^2b^2/(3EIL)$]

Mohr's Theorems

6.8 Use Mohr's theorems to confirm the maximum slope and deflection expressions given as answers for the two beams referred to in Exercise 6.1.

6.9 Find the maximum slope and deflection for a 4 m-long cantilever which carries a uniformly distributed load of 15 kN/m from its mid-span position to its free end. Take $I = 85 \times 10^6$ mm^4 and $E = 210$ GPa.
(Ans: 0.45°, 23 mm)

6.10 Using Mohr's theorems find the slope and deflection at mid-length for the cantilever beam loaded uniformly, as shown in Fig. 6.5.

6.11 The symmetrical cross section of a cantilever is 250 mm deep with a uniform $I = 120 \times 10^6$ mm^4 over its 3 m length. Calculate, from the moment-area method, the magnitude of the total loading and the deflection at the free end in each of the following cases when the maximum bending stress in the material is limited to 75 MPa and $E = 207$ GPa:
    (a) load concentrated at the free end
    (b) load uniformly distributed throughout
    (c) load uniformly distributed between the mid-point and the free end
    (d) load concentrated at a point 1 m from the free end.
[Ans: (a) 24 kN, 8.7 mm, (b) 8 kN/m, 3.25 mm, (d) 24 kN, 8.4 mm]

6.12 Show that a prop, when used to eliminate the deflection at the centre of a cantilever, carries 2½ times its concentrated end load. Show, that the free-end deflection of the propped cantilever is given by $7WL^3/96EI$. By what fraction has the central prop reduced the fee-end deflection? (Ans: 7/32)

6.13 A steel cantilever, 4.5 m long, carries a uniformly distributed load of 1.65 t/m over a length of 3.65 m from its built-in end. A free-end prop is to support a load of 2 t when the level at the free end is 7.5 mm above the level at the fixed end. Find the value of I for the cross section and the deflection at the free end given $E = 208$ GPa.
(Ans: $I = 93 \times 10^6$ mm^4, 0.94 mm upwards)

6.14 Calculate the end-prop load P to raise the level of the beam in Fig. 6.28 by 50 mm. Where in the span is the deflection zero? Take $E = 207$ GPa and $I = 60 \times 10^3$ mm^4.

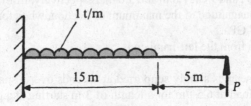

Figure 6.28

6.15 Calculate the prop load P, the maximum slope and deflection when P maintains the level at the fixed end for each of the cantilevers shown in Figs 6.29a and b. Take for both beams: $E = 200$ GPa and $I = 50 \times 10^3$ mm^4.

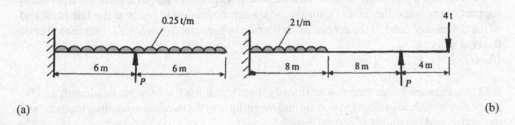

(a)   (b)

Figure 6.29

6.16 Determine the force within a levelling prop support and the beam slope at the prop for a cantilever, loaded as shown in Fig. 6.30a and having the cross section given in Fig. 6.30b.

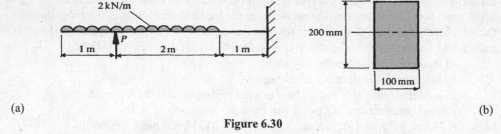

(a)   (b)

Figure 6.30

6.17 An encastré beam carries a uniformly distributed load, w, in addition to a concentrated load W at its centre. Determine the greatest slope and deflection.
[Ans: $(2W + wL)L^2/(128EI)$, $(2W + wL)L^3/(384EI)$]

6.18 A prop is introduced at the centre of an encastré beam to eliminate the central deflection when carrying an offset concentrated load of 10 kN with $a = 1$m and $b = 2$ m (see Fig. 6.13). Find the prop reaction and the position and magnitude of the maximum deflection for this beam. Take $EI = 125$ MNm^2.

Macaulay's Method

6.19 A steel beam is simply-supported at its ends over a span of 6 m. It is loaded with concentrated loads of 5 t and 9 t at 1.8 m and 3.6 m, respectively, from the left-hand support. Find the position and magnitude of the maximum deflection, when for this beam: $I = 450 \times 10^6$ mm^4, and $E = 207$ GPa.
(Ans: 6.3 mm at 3.1 m from the left-hand end)

6.20 A uniform steel beam is simply supported at its ends over a span of 8 m. It carries a uniformly distributed load of 6.5 t/m for a length of 3 m starting at a point 2.75 m from the left-hand support and a concentrated load of 8 t acting at a point 1.5 m from the left-hand support. If $I = 500 \times 10^6$ mm^4, calculate the slope at the left-hand support and the deflection at mid-span. Take $E = 207$ GPa.
(Ans: 0.0062 rad, 30.2 mm)

6.21 A horizontal beam, with a uniform cross section over length L rests on supports at its ends. It carries a uniformly distributed load w/unit length over length l, from the right-hand support. Determine the ratio l/L for the maximum deflection to occur at the left-hand end of the distributed load. If the maximum deflection is expressed as wL^4/kEI, find the value of the coefficient k.
(Ans: $l/L = 0.453$, $k = 179$)

6.22 An encastré beam carries a uniformly distributed load w over its full length L. Use Macaulay's method to find the position and magnitude of the maximum bending moment and deflection and the points of contraflexure.
{Ans: $wL^2/12$, $wL^4/(384EI)$, $[1 \pm (1/\sqrt{3})](L/2)$}

6.23 Find, for each of the two beams, loaded as shown in Figs 6.31a,b, the deflection beneath each of the concentrated loads and the position and magnitude of the maximum deflection. Take $E = 207$ GPa and $I = 10^{10}$ mm^4.

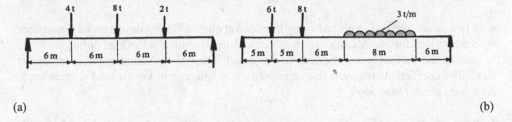

(a)   (b)

Figure 6.31

6.24 A beam with uniform cross section is loaded as shown in Fig. 6.32. Determine the slope and deflection at the left-hand end. Sketch the deflected shape and construct the bending moment diagram inserting principal values. Take $EI = 16.65$ MNm^2.
(Ans: 0.047°, 69 mm, 59.25 kNm)

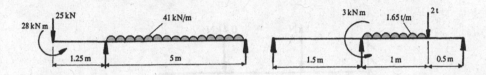

Figure 6.32   Figure 6.33

6.25 Find the deflection of the beam carrying combined loading shown in Fig. 6.33 at the position 1.22 m from the left-hand end. Take $I = 415 \times 10^6$ mm^4, $E = 206$ GPa.
(Ans: 0.15 mm)

6.26 Find the position and magnitude of the maximum deflection for the encastré beam shown in Fig. 6.34. Take $I = 333 \times 10^6$ mm^4 and $E = 208$ GPa.
(Ans: 2.6 mm, 4.86 m from left-hand end)

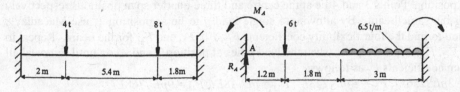

Figure 6.34   Figure 6.35

6.27 Determine the left-hand support reactions R_A and M_A and the deflection at mid-span for the encastré beam shown in Fig. 6.35. Take $I = 415 \times 10^6$ mm^4 and $E = 208$ GPa.
(Ans: $R_A = 91$ kN, $M_A = 110$ kNm, 1.65 mm)

Principle of Superposition

6.28 Show that the inverse matrix Eq. 6.35 gives equivalent loading for Fig. 6.23c as: $F_1 =$ ¼$(A + B + C + D)$, $F_2 =$ ¼$(A + B - C - D)$, $F_3 =$ ¼$(A - B + C - D)$, $F_1 =$ ¼$(A + B + C + D)$.

6.29 In a laboratory test an aircraft wing is loaded at eight different points on its top surface by concentrated forces. Measurements of vertical displacement at the load points are made to enable the flexibility coefficients to be found. Deduce the number of independent flexibility coefficients required from applying the displacement versus load expression to each load point. (Ans: 36)

6.30 A uniform beam of length $3a$ is simply supported in bearings at its ends. Use Maxwell's reciprocal law to find: (i) the vertical deflection at $z = a$ when a concentrated clockwise moment M is applied at $z = 2a$ and (ii) the slope at $z = 2a$ when a vertical force P is applied at $z = a$. Hint: Apply the Macaulay method to find the flexibility coefficient for each case. [Ans: $5a^2(M$, or $P)/(18EI)$]

6.31 Using an equivalent end load, find the end deflection of a cantilever when loaded in the manner shown in Fig. 6.36. Take $E = 70$ GPa and $I = 3.5 \times 10^{-8}$ mm⁴. (Ans: 2.55 mm)

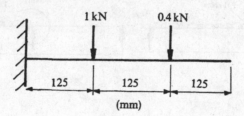

(mm)

Figure 6.36

6.32 In Fig. 6.24a point 3 is at the mid-span position and in Fig 6.24b point 4 is at the three-quarter span position. Show that the superposition principle applies giving equal deflections $\Delta_3 = \Delta_4$ when a load is applied at point 3 to give Δ_4 and then moved to point 4 to give Δ_3.

6.33 In Fig. 6.24a, point 1 is at the quarter-span position and point 2 refers to the two-thirds span position. Points 3 and 4 lie at the centre and three-quarter span positions respectively along the same beam. By allowing a single load p to lie at position 1, use Macaulay's method to find the four flexibility coefficients F_{11}, F_{21}, F_{31} and F_{41} for the beam. Repeat to give F_{12}, F_{22}, F_{32} and F_{42} for when the load p lies at position 2 and so on until to reveal all sixteen coefficients F_{ij} as follows:

$F_{11} = 3pL^3/256EI$, $F_{22} = 4pL^3/243EI$, $F_{33} = pL^3/48EI$, $F_{44} = 9pL^3/768EI$
$F_{12} = F_{21} = 119pL^3/10368EI$, $F_{13} = F_{31} = 11pL^3/768EI$, $F_{23} = F_{32} = 23pL^3/1296EI$
$F_{14} = F_{41} = 7pL^3/768EI$, $F_{24} = F_{42} = 71pL^3/5184EI$, $F_{34} = F_{43} = 11pL^3/768EI$

6.34 Loads p_1, p_2 p_3 and p_4 act together at their respective positions described in Exercise 6.33 above. Use the flexibility coefficients therein to write down the expression for the central deflection. Show that the single equivalent central load p to provide a similar central deflection is given as: $p = 11p_1/16 + 23p_2/27 + p_3 + 11p_4/16$.

CHAPTER 7

THEORIES OF TORSION

Summary: The shear mode of deformation in a bar under axial torsion is distinct, theoretically, from the shear mode in a bar or beam under transverse forces. The latter was considered in Chapter 5 where an examination of the distribution in shear stress was made for solid sections and thin-walled sections in open and closed forms. Here we shall examine the nature of the shear stress when similar sections bear axial torsion. At an introductory level it is sufficient to restrict the analysis to solid and hollow circular sections, with particular applications to stepped, tapered and composite bars. Thereafter, torsion of thin-walled tubes of any shape and open sections shaped from thin strip extent the complexity of the theory.

7.1 Torsion of Circular Bars

The engineering theory of torsion connects the shear stress τ and angular twist θ (in radians), that are produced in a solid circular shaft or a tube, to an applied axial torque T (see Fig. 7.1a). Let the shaft have length L, outer radius R (see Fig. 7.1b) and polar second moment of area J, in a material with rigidity modulus G. The radius r shown in Fig. 7.1a applies to a cylindrical core element lying within the solid shaft with section of outer radius R (Fig. 7.1b).

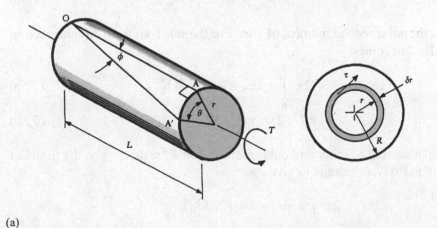

(a)

(b)

Figure 7.1 Solid circular section shaft under torsion

When the far end of the shaft shown is fixed, point A rotates to A′ at the free end as shown in Fig. 7.1a. The angular distortion φ for the core radius, defines the shear strain γ at radius r as:

$$\gamma = \tau/G = \tan \varphi \qquad (7.1a)$$

where $\tan\varphi = $ AA′/OA. Provided the deformation remains elastic, φ is small and Eq. 7.1a approximates closely to:

$$\tan \varphi \approx \varphi \approx \text{AA′}/L \text{ (rad)} \qquad (7.1b)$$

Since AA′ = $r\theta$, it follows from Eqs 7.1a,b that:

$$\gamma = \tau/G = r\theta/L \qquad (7.2)$$

The cross section resists the applied torque by becoming stressed in shear. Figure 7.1b shows the shear stress τ, acting on an annular strip area of thickness δr at radius r. The torque equilibrium condition for this strip is:

$$\delta T = 2\pi r \delta r \times \tau \times r \qquad (7.3a)$$

Substituting Eq. 7.2 into Eq. 7.3a:

$$\delta T = 2\pi \ (G\theta/L) \ r^3 \ \delta r \qquad (7.3b)$$

Equation 7.3b allows an integration over the circular cross section for both solid shafts and tubes. This gives T as:

$$T = \frac{2\pi G\theta}{L} \int_r r^3 \, dr \qquad (7.4a)$$

Equation 7.4a is written as:

$$T/J = G\theta/L \qquad (7.4b)$$

where J is the polar second moment of area. For the solid, uniform circular section in Fig. 7.1b, J becomes:

$$J = 2\pi \int_0^R r^3 dr = 2\pi \left| r^4/4 \right|_0^R \qquad (7.5a)$$

$$J = \pi R^4/2 = \pi D^4/32 \qquad (7.5b)$$

For a tubular section of inner and outer radii, R_i and R_o respectively, the limits of integration in Eq. 7.5a change to give J as:

$$J = 2\pi \int_{R_i}^{R_o} r^3 dr = 2\pi \left| r^4/4 \right|_{R_i}^{R_o}$$

$$= \pi(R_o^4 - R_i^4)/2 = \pi(D_o^4 - D_i^4)/32 \qquad (7.5c)$$

Finally, combining Eqs 7.2 and 7.4b gives the *engineer's torsion theory* outlined above in a single, three-part expression:

$$\frac{T}{J} = \frac{G\theta}{L} = \frac{\tau}{r} \tag{7.6}$$

Examples will follow to show how Eq. 7.6 may be applied to tapered, stepped and composite shafts. Included among them are shafts used to transmit power and those subjected to non-uniform torque.

7.1.1 Power Transmission

The power P (in Watt, W) transmitted by a rotating shaft under a torque T (in Nm) is found from the rate of its work in angular terms:

$$Power \text{ (W)} = Torque \text{ (Nm)} \times Angular\ Speed \text{ (rad/s)} \tag{7.7a}$$

Usually the speed is given as N rev/min (rpm), which is converts to $2\pi N/60$ rad/s. Substituting into Eq. 7.7a:

$$P = \frac{2\pi NT}{60} \tag{7.7b}$$

When required Eq. 7.7b is coupled to Eq. 7.6 given that the torque is common between them. The following example shows how T can be found from Eq. 7.7b for a given power transmitted by a drive shaft at speed. Equation 7.6 will then provide a check that the stress and strain in the shaft remain elastic to the required degree of safety.

Example 7.1 A solid steel drive shaft 50 mm diameter and 1.5 m long is to be used to transmit 20 kW when rotating at 250 rev/min. Determine the angular twist between the ends and the factor of safety that the design ensures given that the ultimate shear stress for the shaft material is $USS = 310$ MPa and that its shear modulus is $G = 85$ GPa.

The solid shaft's polar second moment of area, is from Eq. 7.5b:

$$J = \frac{\pi \times 50^4}{32} = 613.59 \times 10^3 \text{ mm}^4$$

Equation 7.7b shows that at the rated power and speed the shaft is subjected to the following torque:

$$T = \frac{60P}{2\pi N} = \frac{60 \times (20 \times 10^3)}{2\pi \times 250} = 763.94 \text{ Nm}$$

For which the maximum operating shear stress occurs at the outer radius r_o, being given by Eq. 7.6 (in units of $\underline{N}$ and $\underline{mm}$) as:

$$\tau = \frac{Tr_o}{J} = \frac{(763.94 \times 10^3) \times 25}{(613.59 \times 10^3)} = 31.13 \text{ MPa}$$

Alternatively, making this stress calculation in units of $\underline{N}$ and $\underline{m}$:

$$\tau = \frac{Tr_o}{J} = \frac{763.94 \times (25 \times 10^{-3})}{(613.59 \times 10^{-9})} = 31.13 \times 10^6 \ \frac{N}{m^2} = 31.13 \text{ MPa}$$

which reveals a design based upon a tenfold margin of safety:

$$S = \frac{USS}{\tau_{max}} = \frac{310}{31.33} \approx 10$$

The twist in radians (°) is found from the first two terms in Eq. 7.6. Knowing that the radian is a non-dimensional angular measure, a cancellation between the units in the twist expression is required. In $\underline{N}$ and $\underline{mm}$:

$$\theta = \frac{TL}{GJ} = \frac{(763.94 \times 10^3) \times (1.5 \times 10^3)}{(85 \times 10^3) \times (613.59 \times 10^3)} = 0.022^c$$

Alternatively, in $\underline{N}$ and $\underline{m}$

$$\theta = \frac{TL}{GJ} = \frac{763.94 \times 1.5}{(85 \times 10^9) \times (613.59 \times 10^{-9})} = 0.022^c$$

Each calculation confirms an angular twist in degrees, $0.022 \times 180/\pi = 1.26°$, between the ends is small, as would be expected of a safe elastic design.

Example 7.2 A 10 m long hollow steel cylinder, 200 mm o.d. and 100 mm i.d., is used as a drive shaft to transmit power when rotating at 90 rev/min. If the maximum shear stress is limited to 50 MPa, determine the power and the angular twist. If a solid steel shaft was used to transmit the power at this speed with a similar length and limiting stress, what is its % increase in weight? Take $G = 80$ GPa.

Hollow Shaft
From Eq. 7.5c, the polar moment of area for the hollow shaft is:

$$J = \pi (200^4 - 100^4)/32 = 147.26 \times 10^6 \text{ mm}^4$$

Since τ is proportional to r, the maximum shear stress applies to the outer radius r_o. Hence, when the limiting shear stress is reached at the outer radius, the torque is given by Eq. 7.6:

$$T = J\tau/r_o = 147.26 \times 10^6 \times 50 /100$$

$$= 73.63 \times 10^6 \text{ Nmm} = 73.63 \times 10^3 \text{ Nm}$$

The power transmitted follows from Eq. 7.7b:

$$P = \frac{2\pi \times 90 \times 73.63 \times 10^3}{60} = 694 \times 10^3 \text{ W} = 694 \text{ kW}$$

The angular twist (in radians) is found from either of two expressions from Eq. 7.6. Applying the limiting τ to the outer radius r_o:

In degrees:

$$\theta = \frac{\tau L}{Gr_o} = \frac{50 \times 10 \times 10^3}{80 \times 10^3 \times 100} = 0.0625^c$$

$$\theta = 0.0625 \times \frac{360}{2\pi} = 3.58°$$

Solid Shaft

Where a solid shaft of outer radius R is used, Eqs 7.5b and 7.6 connect the torque and limiting shear stress as follows:

$$\tau = \frac{TR}{J} = \frac{TR}{\pi R^4/2} = \frac{2T}{\pi R^3}$$

from which the shaft's outer radius is:

$$R = \left(\frac{2T}{\pi \tau}\right)^{1/3} = \left(\frac{2 \times 73.63 \times 10^6}{\pi \times 50}\right)^{1/3} = 97.87 \text{ mm}$$

For similar shaft lengths, weight is proportional to their section areas. Hence the % weight increase of solid to hollow becomes:

$$\left[\frac{\pi R^2}{\pi(R_o^2 - R_i^2)} - 1\right] \times 100 = \left[\frac{97.87^2}{(100^2 - 50^2)} - 1\right] \times 100 = 27.7\%$$

Despite its reduced diameter, the significant weight increase of the solid shaft may not be tolerable, say, in aerospace applications.

7.1.2 Tapered Bar

To derive an expression for the angular twist in a solid circular shaft that tapers from radius a to b over length L in Fig. 7.2, that part of our torsion theory within Eq. 7.4 may be applied to an element of length δl for which the angular twist is $\delta\theta$.

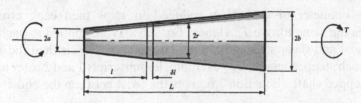

Figure 7.2 Tapered shaft under torsion

It is seen that δl becomes the thickness of a disc of radius r at position l in the length. Applying Eqs 7.5a and 7.6 gives the angular twist in the disc:

$$\delta\theta = \frac{T \times \delta l}{JG} = \frac{2T}{\pi G} \times \frac{\delta l}{r^4} \tag{7.8a}$$

Equation 7.8a may be expressed in terms of the single variable r by substituting the relationship for a tapered shaft: $\delta r / \delta l = (b - a)/L$. This gives:

$$\delta\theta = \frac{2TL}{\pi G(b-a)} \times \frac{\delta r}{r^4} \tag{7.8b}$$

Hence the twist θ over shaft length L is found from the integrating Eq. 7.8b between limits that equal the end radii:

$$\theta = \frac{2TL}{\pi G(b-a)} \int_a^b r^{-4}\, dr = - \frac{2TL}{3\pi G(b-a)} \left| r^{-3} \right|_a^b$$

$$\theta = \frac{2TL}{3\pi G(b-a)}\left(\frac{1}{a^3} - \frac{1}{b^3} \right) = \frac{2TL}{3\pi G} \frac{(a^2 + ab + b^2)}{a^3 b^3}$$

The complete analysis requires that the internal shear stress across each section remains in equilibrium with the applied axial torque. Because of the taper the shear stress distribution varies from section to section in contrast to the uniform distribution which applies in the absence of a taper. We might assume a linear variation in shear stress with both radius and length in which the surface shear stresses at the ends are, given from Eq. 7.6:

$$(\tau)_{r=a} = \frac{2T}{\pi a^3} \quad \text{and} \quad (\tau)_{r=b} = \frac{2T}{\pi b^3}$$

Strictly, a more advanced torsion theory is required to find the shear stresses in tapered bars (see A. S. Saada, *Elasticity: theory and applications*, Pergamon, 1974). This shows that the maximum shear stress at the outer smaller end is far less:

$$(\tau)_{r=a} = \frac{3T}{4\pi a^3}$$

7.1.3 Stepped Shaft

Where the diameter of a shaft is reduced in steps then each cross section experiences the same torque T. Hence, Eq. 7.6 may be applied to each length to give the twist and shear stress within it. The total twist θ_t is then the sum of the twists for each stepped length. For example, let subscripts 1 and 2 refer to each step in a two-stepped shaft. Equation 7.6 gives the twist between the ends:

$$\theta_t = (TL/JG)_1 + (TL/JG)_2 \text{ (radians)}$$

$$\theta_t = (T/G)[(L/J)_1 + (L/J)_2](180/\pi) \text{ (degrees)} \tag{7.9}$$

The maximum shear stresses occurs at each outer radius r_1 and r_2:

$$\tau_1 = T r_1 / J_1 = 2T / \pi r_1^3$$

and

$$\tau_2 = T r_2 / J_2 = 2T / \pi r_2^3$$

Example 7.3 Solid and hollow mild steel shafts are welded together at their ends (see Fig. 7.3). Determine the length L_2 of the solid bar required to limit the total angular twist to 3° under an axial torque of 25 kNm. Take $G = 80$ GPa.

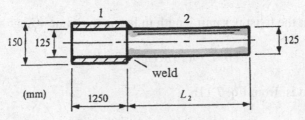

Figure 7.3 Welded shaft

Identifying 1 with the hollow shaft and 2 with the solid shaft, Eq. 7.9 is transposed to give the solid length of bar required (in units of <u>N</u> and <u>mm</u>):

$$L_2 = \left[\frac{\pi \theta_t G}{180 T} - \left(\frac{L}{J} \right)_1 \right] J_2 = \frac{\pi \theta_t G J_2}{180 T} - \frac{L_1 J_2}{J_1}$$

$$= \frac{\pi \times 3 \times 80 \times 10^3 \times \pi (125)^4}{180 \times 25 \times 10^6 \times 32} - \frac{1250 \times 125^4}{(150^4 - 125^4)}$$

$$= 4015.95 - 1164.31 = 2851.6 \text{ mm}$$

7.1.4 Composite Bar

Figure 7.4 shows the cross section of a solid circular bar press-fitted or keyed within an outer tube. When manufactured from bars in two different materials, 1 and 2, the resulting solid circular section becomes a composite torsion bar.

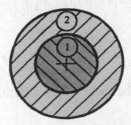

Figure 7.4 Composite shaft section

Provided that there is no slipping at the common diameter, the total torque carried by the composite bar is the sum of the torques that each supports:

$$T = T_1 + T_2 = T_1(1 + T_2/T_1) \tag{7.10}$$

Equation 7.6 refers the torques to the outer radius of each component:

$$T_1 = \tau_1 J_1/r_1, \, T_2 = \tau_2 J_2/r_2 \tag{7.11a}$$

and ensures that the twist per unit length in bar is the same when:

$$\theta/L = (T/JG)_1 = (T/JG)_2 \tag{7.11b}$$

The torque ratio is, from Eq. 7.11b

$$T_2/T_1 = G_2 J_2/G_1 J_1 \tag{7.11c}$$

Substituting Eqs 7.11a,c into Eq. 7.10:

$$T = \frac{2\tau_1}{d_1}\left[J_1 + \frac{G_2}{G_1}J_2\right] = \frac{2\tau_2}{d_2}\left[J_2 + \frac{G_1}{G_2}J_1\right] \tag{7.12}$$

where J_1 and J_2 are defined by Eqs 7.5a,b, respectively.

Example 7.4 A 4 m length of brass tube is shrunk-fit onto a 75 mm steel bar. If each material contributes equally to the total torque carried, determine the outer diameter of the brass tube. Find the maximum shear stress in each material and the angular twist under a total applied torque of 20 kNm. Take $G = 82$ GPa for steel and $G = 41$ GPa for brass.

Identify 1 with steel and 2 with brass so that d_2 is the diameter required and d_1 is the interface diameter. Since $T_1 = T_2$, Eqs 7.10 and 7.11b give:

$$\frac{T_2}{T_1} = \frac{G_2 J_2}{G_1 J_1} = \frac{G_2}{G_1}\left[\left(\frac{d_2}{d_1}\right)^4 - 1\right] = 1$$

from which

$$d_2 = d_1\left(G_1/G_2 + 1\right)^{1/4} = 75(2+1)^{1/4} = 98.7 \text{ mm}$$

Here, with $T_1 = T_2 = 10$ kNm, the shear stresses (units: $\underline{N}$ and $\underline{mm}$) are found from Eq. 7.11a:

$$\tau_1 = \frac{T_1 r_1}{J_1} = \frac{(10 \times 10^6) \times 37.5}{(\pi \times 75^4/32)} = 120.72 \text{ MPa}$$

$$\tau_2 = \frac{T_2 r_2}{J_2} = \frac{(10 \times 10^6) \times 49.35}{\pi(98.7^4 - 75^4)/32} = 79.46 \text{ MPa}$$

The twist in the steel bar is:

$$\theta_1 = \left(\frac{TL}{JG}\right)_1 = \frac{(10 \times 10^6) \times (4 \times 10^3)}{(\pi \times 75^4/32) \times (82 \times 10^3)} = 0.157^c = 9°$$

which is confirmed within the brass tube:

$$\theta_2 = \left(\frac{TL}{JG}\right)_2 = \frac{(10 \times 10^6) \times (4 \times 10^3)}{[\pi(98.7^4 - 75^4)/32] \times (41 \times 10^3)} = 0.157^c = 9°$$

7.1.5 Varying Torque

Figure 7.5 shows a multi-stepped, circular-section shaft with fixed ends. Concentrated torques T_2, T_3 and T_4 are applied at positions in the length where the steps occur, within which θ_1, θ_2, θ_3 and θ_4 are the angular twists.

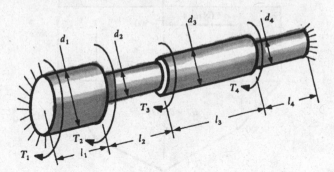

Figure 7.5 Stepped shaft under concentrated torques

Since there can be no relative twist between the ends it must follow that:

$$\theta_1 + \theta_2 + \theta_3 + \theta_4 + \ldots = 0 \qquad (7.13a)$$

The angular twist over each parallel length is found from Eq. 7.6. With the reaction torque T_1 assumed clockwise at the left end, these twists are:

$$\theta_1 = \frac{T_1 l_1}{G_1 J_1}, \quad \theta_2 = \frac{(T_1 + T_2) l_2}{G_2 J_2},$$

$$\theta_3 = \frac{(T_1 + T_2 + T_3) l_3}{G_3 J_3}, \quad \theta_4 = \frac{(T_1 + T_2 + T_3 + T_4) l_4}{G_4 J_4} \qquad (7.13b)$$

It follows that when a shaft of uniform diameter in the same material is subjected to concentrated torques between its fixed ends, similar to Fig. 7.5, Eqs 7.13a,b combine to give a zero net twist condition:

$$T_1 l_1 + (T_1 + T_2) l_2 + (T_1 + T_2 + T_3) l_3 + (T_1 + T_2 + T_3 + T_4) l_4 = 0 \qquad (7.13c)$$

where l_1, l_2, l_3 and l_4 are the length positions at which the torques are applied.

Example 7.5 Construct the torque and twist diagrams for the shaft in Fig. 7.6a. From these find the position and magnitude of the maximum shear stress and the greatest angular twist in the shaft. Take $G = 80$ GPa.

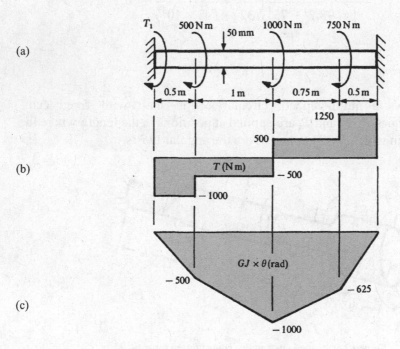

Figure 7.6 Encastré shaft with concentrated torques

Substituting into Eq. 7.13c:

$$0.5T_1 + (T_1 + 500) + (T_1 + 500 + 1000)0.75 + (T_1 + 500 + 1000 + 750)0.5 = 0$$

$$\therefore 2.75T_1 + 2750 = 0$$

This gives $T_1 = -1000$ Nm, indicating that the direction of T_1 in Fig. 7.6a is reversed. Figure 7.6b shows the variation in torque with length. This reveals that a maximum torque $T_{max} = 1250$ Nm is applied over a length of 0.5 m which connects to the right fixing. The maximum shear stress occurs at the outer diameter of this length:

$$\tau_{max} = \frac{T_{max} r_o}{J} = \frac{(1250 \times 10^3) \times 25}{\pi(50)^4/32} = 50.93 \text{ MPa}$$

Equation 7.13b gives relative twists that occur for lengths between the concentrated torques:

$$\theta_1 = (-1000) \times 0.5/GJ = -500/GJ$$
$$\theta_2 = (-1000 + 500) \times 1.0/GJ = -500/GJ$$
$$\theta_3 = (-1000 + 500 + 1000) \times 0.75/GJ = +375/GJ$$
$$\theta_4 = (-1000 + 500 + 1000 + 1250) \times 0.5/GJ = +625/GJ$$

These twists are summed in the construction of the twist diagram in Fig. 7.6c, which confirms $\sum\theta = 0$ between the fixed ends. The maximum ordinate in Fig. 7.6c provides the greatest twist in the shaft:

$$\theta_{max} = -\frac{1000}{GJ} = -\frac{(1000 \times 10^3 \times 10^3)}{(80 \times 10^3) \times (\pi \times 50^4/32)} = 0.0204^c = 1.167°$$

Alternatively, from Eq. 7.8:

$$\theta_{max} = \frac{1}{GJ} \int T dl = \frac{1}{GJ}\left[\int_0^{0.5} (-1000) dl + \int_{0.5}^{1.5} (-500) dl \right]$$

$$= -[(1000 \times 0.5) + 500(1.5 - 0.5)]/(GJ) = -1000/(GJ)$$

7.2 Torsion of Thin Strips

Consider the twist θ, in a thin strip of section $b \times t$, when a torque T is applied about a centroidal axis z, as shown in Fig. 7.7a. Unlike the circular cross section, the plane rectangular section of a thin strip in torsion does not remain in its original plane. As the strip rotates with the torque a warping of the cross section occurs in which the four corners displace axially in opposite directions while its edges remain straight. Figure 7.7b shows the distortion produced in the length and cross section of a re-orientated element $\delta z \times \delta x$, lying at a distance $+y$ from the x-axis.

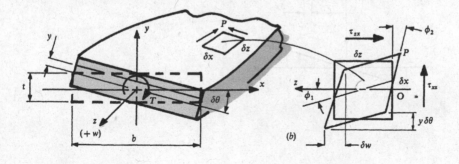

(a)   (b)

Figure 7.7 Torsion of a thin strip

Corresponding to an incremental angular twist $\delta\theta$ (see Fig. 7.7a), two shear displacements appear. Adjacent corners of side δx warp in the direction of z by an amount δw. Also, a tangentially displacement $y\delta\theta$, aligned with direction x, appears between the corners of side δz. The total change in the angle between originally perpendicular sides meeting at P in (Fig. 7.7b) defines the shear strain γ as:

$$\gamma = \tan\varphi_1 + \tan\varphi_2 \qquad (7.14)$$

where φ_1 is the angular rotation in a z-chord and φ_2 is the rotation in the x-span. These shear angles are equal due to the complementary, *horizontal* shear stresses acting on the sides, i.e. $\tau_{zx} = \tau_{xz} (= \tau)$, as shown in Fig. 7.7b. For small elastic distortions: $\tan\varphi_1 \simeq \varphi_1 = \varphi_2$ (rad). It follows that Eq. 7.14 may be written as:

$$\gamma = y\frac{\delta\theta}{\delta z} + \frac{\delta w}{\delta x} = 2y\frac{\delta\theta}{\delta z} \qquad (7.15)$$

where $\delta\theta/\delta z$ is the 'rate' of angular twist with respect to length. Note that the strip's mid-plane retains its original rectangular shape, since $\gamma = 0$ for $y = 0$. Shear displacements within surfaces lying at equal distances $\pm y$ above and below this plane are equal but occur in opposing directions. Equation 7.15 gives the horizontal shear stress at position $+y$ in the cross section (see Fig. 7.8a) as:

$$\tau = G\gamma = 2Gy\frac{\delta\theta}{\delta z} \qquad (7.16)$$

Equation 7.16 shows that the horizontal shear stress τ, lying in the plane x, z, varies linearly across the strip's thickness t. This has its maximum values τ_{max} at the edges (see Fig. 7.8b).

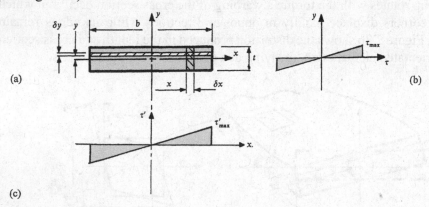

(a)

(b)

(c)

Figure 7.8 Shear stress distributions in a thin strip

The lower diagram in Fig. 7.8c shows that a further linear distribution applies to the *vertical* shear stress τ' lying in the plane y–z. The maximum shear stresses within this distribution is taken as:

$$\tau'_{max} = \frac{t}{b}\tau_{max} \qquad (7.17a)$$

Thus for any point x, y in the cross section τ and τ' are related to their respective maxima:

$$\tau' = \left(\frac{2x}{b} \right) \tau_{max}' \quad \text{and} \quad \tau = \left(\frac{2y}{t} \right) \tau_{max} \qquad (7.17b,c)$$

Combining Eqs 7.16b and c with Eq. 7.17a connects τ and τ':

$$\tau' = \frac{x}{y} \left(\frac{t}{b} \right)^2 \tau \qquad (7.17d)$$

Substituting Eq. 7.16 into Eq. 7.17d gives an expression for the vertical shear stress distribution:

$$\tau' = 2Gx \left(\frac{t}{b} \right)^2 \left(\frac{\delta\theta}{\delta z} \right) \qquad (7.18)$$

With $\varphi_1 = \varphi_2$, the warping displacement w follows from Eq. 7.15 as:

$$\frac{\delta w}{\delta x} = y \frac{\delta\theta}{\delta z} \qquad (7.19a)$$

$$w = yx \frac{\delta\theta}{\delta z} \qquad (7.19b)$$

An integration constant is unnecessary in Eq. 7.19b since there is no warping within the central plane, i.e. $w = 0$ for all x at $y = 0$.

The applied torque must equal the resultant torque from both the horizontal and vertical shear stress distributions. Firstly, let T_1 arise from τ in Eq. 7.16. When τ acts over an elemental strip of thickness δy, distance y from x (see Fig. 7.8a):

$$T_1 = \int_{-t/2}^{t/2} (\tau b \, dy) y = 2Gb \frac{\delta\theta}{\delta z} \int_{-t/2}^{t/2} y^2 dy = 2Gb \left(\frac{\delta\theta}{\delta z} \right) \left(\frac{t^3}{12} \right)$$

Now let a torque T_2 be due to vertical shear stress τ' in Eq. 7.18. For an elemental strip thickness δx, distance x from y (see Fig. 7.8a):

$$T_2 = \int_{-b/2}^{b/2} (\tau' t \, dx) x = 2Gt \left(\frac{t}{b} \right)^2 \left(\frac{\delta\theta}{\delta z} \right) \int_{-b/2}^{b/2} x^2 dx = 2Gb \left(\frac{\delta\theta}{\delta z} \right) \left(\frac{t^3}{12} \right) = T_1$$

Adding T_1 and T_2 gives the net torque

$$T = T_1 + T_2 = 4Gb \left(\frac{\delta\theta}{\delta z} \right) \left(\frac{t^3}{12} \right) = GJ \left(\frac{\delta\theta}{\delta z} \right) \qquad (7.20)$$

Identified within Eq. 7.20 is *St Venant's torsion constant* $J = bt^3/3$ for a rectangular section. Combining Eqs 7.16 and 7.20 presents this theory conveniently as a three-part equation:

$$\frac{T}{J} = G\frac{\delta\theta}{\delta z} = \frac{\tau}{2y} \tag{7.21}$$

Note the similarity between Eq. 7.6 for torsion of circular shafts and Eq. 7.21 for torsion of a rectangular strip. For a thin, uniform rectangular section of length L the twist rate $\delta\theta/\delta z$ in Eq. 7.21 is constant $(= \theta/L)$ so that the angular twist and maximum shear stress, for $y = t/2$, become

$$\theta = \frac{TL}{JG} \quad \text{and} \quad \tau_{max} = \frac{3T}{bt^2} \tag{7.22a,b}$$

7.2.1 Open Sections

Equation 7.21 may also be applied to uniformly thin-walled open tubes in which the thickness t is small compared to the perimeter length b (see Figs 7.9a,b).

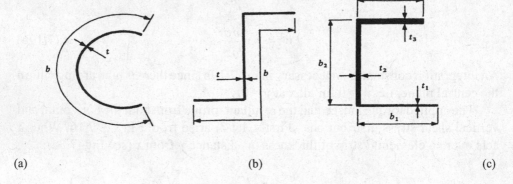

(a)   (b)   (c)

Figure 7.9 Dimensions of thin-walled open sections

In Figs 7.9a and b the torsion constant for a rectangle again applies as $J = bt^3/3$ but in Fig. 7.9c, where the thickness of each rectangular limb differs, J becomes a sum:

$$J = \frac{1}{3}\sum(bt^3) = b_1 t_1^3/3 + b_2 t_2^3/3 + b_3 t_3^3/3 + \ldots$$

Example 7.6 Sheet steel of thickness 5 mm is bent into the semi-circular arch shown in Fig. 7.10. What arch dimensions d and L can support an axial torque of 50 Nm given an allowable shear stress of 40 MPa and a limiting angular twist of 5°? Take $G = 80$ GPa.

The torsion constant applies to a cross section with perimeter b and thickness t

$$J = \frac{bt^3}{3} = \frac{\pi d/2 \times t^3}{3} = \frac{\pi dt^3}{6}$$

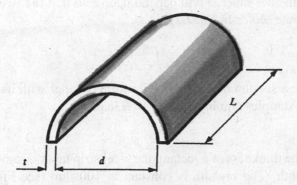

Figure 7.10 Semi-circular arch

From Eq. 7.21, the maximum shear stress applies to $y = t/2$:

$$T = \frac{J\tau_{max}}{2y} = \frac{J\tau_{max}}{t} = \frac{\pi dt^2 \tau_{max}}{6}$$

from which the required diameter is

$$d = \frac{6T}{\pi t^2 \tau_{max}} = \frac{6 \times 50 \times 10^3}{\pi \times 5^2 \times 40} = 95.5 \text{ mm}$$

The limiting angular twist is expressed from Eq. 7.22a as:

$$\theta = \frac{TL}{JG} = \frac{6TL}{\pi dt^3 G}$$

giving the length L of arch required (in Nmm and Nm units, respectively)

$$L = \frac{\pi dt^3 G\theta}{6T} = \frac{\pi \times 95.5 \times 5^3 \times (80 \times 10^3) \times (5 \times \pi/180)}{6 \times (50 \times 10^3)} = 872.73 \text{ mm}$$

$$= \frac{\pi \times 0.0955 \times (5 \times 10^{-3})^3 \times (80 \times 10^9) \times (5 \times \pi/180)}{6 \times 50} = 0.873 \text{ m}$$

7.2.2 Tapered Strip

A strip may taper within its cross section and over its length. The former would apply to Fig. 7.9a, where the thickness t of the arc-section strip tapers gradually around the perimeter b. The thickness variation is expressed as $t = t(s)$ where the perimeter co-ordinate s has its origin at one free end. Provided there is no length taper, the torsion constant J is found from:

$$J = \frac{1}{3} \int_0^b [t(s)]^3 \, ds \qquad (7.23a)$$

Where the length does taper, J will depend upon z so that the twist is found from integrating its 'rate' $\delta\theta/\delta z$ in Eq. 7.21:

$$\theta = \int_z \frac{T\,dz}{J(z)G} \qquad (7.23b)$$

in which the cross section may either be uniform or taper with its thickness. The following two examples employ both Eqs 7.23a and b.

Example 7.7 The thickness of a rectangular steel strip tapers from 5 mm to 15 mm over a 2 m length. The breadth is constant at 100 mm (see Fig. 7.11). If the maximum shear stress in the material is restricted to 100 MPa, calculate the maximum torque, T, the strip can support and the angle through which the strip length twists. Compare with a twist estimate from taking an average thickness value. Take $G = 83$ GPa.

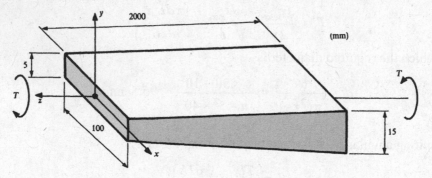

Figure 7.11 Tapered strip

The maximum torque is determined from Eq. 7.21 for $y = t/2$ at the outer edges:

$$T = \frac{J\tau_{max}}{2y} = \frac{(bt^3/3)\,\tau_{max}}{2(t/2)} = \frac{bt^2\tau_{max}}{3}$$

It follows that if τ_{max} is not to be exceeded, then T must be determined from the smaller end where $t = 5$ mm:

$$T = 100 \times 5^2 \times 100/3 = 83.33 \times 10^3 \text{ Nmm} = 83.33 \text{ Nm}$$

The angular twist can be found from Eq. 7.23b. At an intermediate length position $0 \le z \le L$, where the thickness is t and the breadth b, we have:

$$J = b\,[t(z)]^3/3 \qquad (i)$$

and with end thicknesses increasing from t_1 to t_2:

$$t(z) = t_1 + (t_2 - t_1)\,z/L \qquad (ii)$$

Combining Eqs i and ii with Eq. 7.23b:

$$\theta = \frac{3T}{bG} \int_0^L \left[t_1 + (t_2 - t_1)\frac{z}{L} \right]^{-3} dz$$

$$= \frac{3TL}{bG(t_2 - t_1)} \int_0^L \left[t_1 + (t_2 - t_1)\frac{z}{L} \right]^{-3} \frac{(t_2 - t_1)}{L} dz$$

$$= - \frac{3TL}{2bG(t_2 - t_1)} \left| \left[t_1 + (t_2 - t_1)\frac{z}{L} \right]^{-2} \right|_0^L$$

$$= - \frac{3TL}{2bG(t_2 - t_1)} \left(\frac{1}{t_2^2} - \frac{1}{t_1^2} \right) = \frac{3TL(t_1 + t_2)}{2bGt_1^2 t_2^2}$$

$$= \frac{3 \times (83.33 \times 10^3) \times (2 \times 10^3) \times (5 + 15)}{2 \times 100 \times (83 \times 10^3) \times 5^2 \times 15^2} = 0.107^c = 6.14°$$

If the average thickness $t_{av} = 10$ mm is taken for the constant J in Eq. i, the twist is:

$$\theta = \frac{TL}{(bt_{av}^3/3)G} = \frac{3TL}{bt_{av}^3 G}$$

$$= \frac{3 \times (83.33 \times 10^3) \times (2 \times 10^3)}{100 \times 10^3 \times (83 \times 10^3)} = 0.0602^c = 3.45°$$

Example 7.8 An alloy strip tapers in its cross section as shown in Fig. 7.12a. Determine the torsion constant and find the angular twist on a length of 1 m when a torque of 10 Nm is applied about a longitudinal axis passing through its centroid. Find the magnitude and position of the maximum shear stress in the section and the warping displacements at the four corners. Take $G = 30$ GPa.

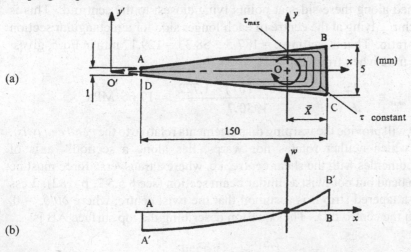

Figure 7.12 Tapered alloy strip

Let an origin O' for co-ordinates x, y' lie at the apex of the section ($= 187.5$ mm from the longer vertical edge). The centroid position $\overline{X}$ is found from taking first moments of area ABCD about the edge BC:

$$\left(5 \times \frac{187.5}{2}\right) \frac{187.5}{3} - \left(1 \times \frac{37.5}{2}\right)\left(150 + \frac{37.5}{3}\right) = \left[\left(5 \times \frac{187.5}{2}\right) - \left(1 \times \frac{3.5}{2}\right)\right]\overline{X}$$

$$\overline{X} = \frac{26250}{450} = 58.33 \text{ mm}$$

With origin at O', the thickness t depends upon x as:

$$t(x) = 2y' = \frac{5}{187.5}x \tag{i}$$

Substituting Eq. i into Eq. 7.23a with the appropriate limits for x:

$$J = \frac{1}{3} \int_{37.5}^{187.5} (5x/187.5)^3 \, \mathrm{d}x = \frac{5^3}{3(187.5)^3} \left.\frac{x^4}{4}\right|_{37.5}^{187.5} = 1950.7 \text{ mm}^4$$

Equation 7.22a provides the angular twist:

$$\theta = \frac{TL}{JG} = \frac{(10 \times 10^3) \times (1 \times 10^3)}{1950.7 \times (30 \times 10^3)} = 0.1709^c = 9.79°$$

When, for a gently tapered thin strip, $b/t > 10$, the maximum shear stress is sensibly constant along the longer edge AB as with a rectangular section. When $b/t < 10$, contours of constant shear stress break at the tapered sides AB and DC, as shown in Fig. 7.12a, indicating that the shear stress is not constant. The maximum shear stress is reached along these sides at points lying closest to the centroid. This is consistent with τ_{max} lying at the centre of each longer side for a rectangular section with a low b/t ratio. That is, setting $x = 187.5 - 58.33 = 129.17$ mm in Eq. i, gives: $y' = y = 1.722$ mm, when from Eq. 7.21:

$$\tau_{max} = \frac{2Ty}{J} = \frac{2 \times (10 \times 10^3) \times 1.722}{1950.7} = 17.66 \text{ MPa}$$

Equation 7.19 will provide the warping displacements relative to the *centre of twist*. This centre, which neither rotates nor warps, lies along a section's axis of symmetry. It coincides with the shear centre, i.e. where a transverse force must be applied if it to bend but not twist a similar beam section (see § 5.5.2, p. 182). Less precisely, for a tapered strip it is assumed that the twist centre, where $\delta\theta/\delta z = 0$, coincides with the centroid O. The equation describing the top surface AB is:

$$y = \frac{2.5}{187.5}x + 1.7223$$

Now: $\delta\theta/\delta z = \theta/L = 0.1709 \times 10^{-3}$ rad/mm, when Eq. 7.19a gives:

$$w = \int y \, (\mathrm{d}\theta/\mathrm{d}z) = (0.1709 \times 10^{-3}) \int_x (0.0133 \, x + 1.7223) \, \mathrm{d}x$$

$$w = (0.1709 \times 10^{-3})(0.00665 x^2 + 1.7223 x) \qquad\qquad \text{(iii)}$$

The parabolic dependence of w on x is a consequence of the tapered section. The warping displacements at A and B, in the length direction z, are found from Eq. iii:

$$\text{At A: } x = -91.67 \text{ mm}, \quad w_A = -0.0174 \text{ mm}$$

$$\text{At B: } x = +58.33 \text{ mm}, \quad w_B = +0.02104 \text{ mm}$$

Here positive w follows positive z (see Fig. 7.7). Hence the top edge AB warps as shown in Fig. 7.12b. From symmetry, the warping displacements at corners C and D are: $w_C = -0.02104$ mm and $w_D = +0.0174$ mm. These four displacements reveal the manner in which the cross section ABCD warps from its original alignment with the x, y plane.

7.2.3 Thick Rectangular Section

For a thicker rectangular strip section, where t is not small compared to b in Fig. 7.9, Eqs 7.22a,b are modified to:

$$\theta = \frac{TL}{\beta b t^3 G} \quad \text{and} \quad \tau_{max} = \frac{T}{\alpha b t^2} \qquad\qquad \text{(7.24a,b)}$$

Table 7.1 shows that the coefficients α and β are dependent upon the b/t ratio but both α and β approach ⅓ as b/t increases beyond 10, thereby returning to the thin-strip theory.

Table 7.1 Coefficients α and β in Eqs 7.24a,b

b/t	1	2	4	6	8	10	> 10
α	0.208	0.246	0.282	0.299	0.307	0.313	0.333
β	0.141	0.299	0.281	0.299	0.307	0.313	0.333

For standard, extruded I and U sections and those fabricated from n-plates, as in Fig. 9.9c, where each thickness, t, is not small compared to breadth, b, Eqs 7.24a,b become:

$$\theta = \frac{TL}{G \sum_{i=1}^{n} (\beta b t^3)_i} \qquad\qquad \text{(7.25a)}$$

For the I- and U-sections in particular:

$$\sum(\beta bt^3) = (\beta bt^3)_1 + (\beta bt^3)_2 + (\beta bt^3)_3 \qquad (7.25b)$$

in which β depends upon b/t, as in Table 7.1. A similar degree of angular twist is experienced by each rectangular limb but the torque is divided between them. This gives:

$$\theta = \frac{T_1 L}{G(\beta bt^3)_1} = \frac{T_2 L}{G(\beta bt^3)_2} = \frac{T_3 L}{G(\beta bt^3)_3} \qquad (7.26a,b,c)$$

where $T = T_1 + T_2 + T_3$. The maximum shear stress, τ_{max} is the greatest of:

$$\tau_1 = \frac{T_1}{(\alpha bt^2)_1}, \quad \tau_2 = \frac{T_2}{(\alpha bt^2)_2}, \quad \tau_3 = \frac{T_3}{(\alpha bt^2)_3} \qquad (7.27a,b,c)$$

in which α depends upon b/t, as in Table 7.1.

Example 7.9 Figure 7.13 gives the cross-sectional dimensions of an extruded aluminium alloy bar that is to withstand a torque of 50 Nm applied about its longitudinal axis. Calculate the position and magnitude of the maximum shear stress in each rectangle and the angular twist over a length of 1 m. Take $G = 30$ GPa.

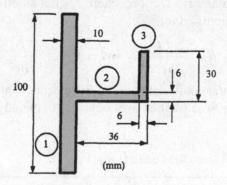

Figure 7.13 Extruded section

Taking the respective b/t ratios for limbs 1, 2 and 3 as 10, 6 and 4, we read from Table 7.1: $\beta_1 = 0.313$, $\beta_2 = 0.299$ and $\beta_3 = 0.281$. Equation 7.25b gives:

$$\sum(\beta bt^3) = (0.313 \times 100 \times 10^3) + (0.299 \times 36 \times 6^3) + (0.281 \times 24 \times 6^3) = 35082 \text{ mm}^4$$

The twist follows from Eq. 7.25a as

$$\theta = \frac{(50 \times 10^3) \times (1 \times 10^3)}{(30 \times 10^3) \times 35082} = 0.0475^c = 2.72°$$

Equations 7.26a-c give the torque contributions from each rectangle:

$$T_1 = \theta\,(\beta bt^3)_1\,G/L = 0.0475(0.313 \times 100 \times 10^3)(30 \times 10^3)/(1 \times 10^3)$$
$$= 44602.5 \text{ Nmm}$$

$$T_2 = \theta\,(\beta bt^3)_2\,G/L = 0.0475(0.299 \times 36 \times 6^3)(30 \times 10^3)/(1 \times 10^3)$$
$$= 3313.16 \text{ Nmm}$$

$$T_3 = \theta\,(\beta bt^3)_3\,G/L = 0.0475(0.281 \times 24 \times 6^3)(30 \times 10^3)/(1 \times 10^3)$$
$$= 2075.8 \text{ Nmm}$$

Correspondingly, the shear stresses follow from Eqs 7.27a–c and Table 7.1:

$$\tau_1 = 44602.5/(0.313 \times 100 \times 10^2) = 14.25 \text{ MPa}$$

$$\tau_2 = 3313.16/(0.299 \times 36 \times 6^2) = 8.55 \text{ MPa}$$

$$\tau_3 = 2075.8/(0.282 \times 24 \times 6^2) = 8.52 \text{ MPa}$$

which apply to the outer, longer side of each rectangle.

It is instructive to compare θ and τ_2 with the approximate solutions from the thin strip theory. Firstly, J is approximated to:

$$J = \frac{1}{3}\left[(100 \times 10^3) + (36 \times 6^3) + (24 \times 6^3)\right] = 37.65 \times 10^3 \text{ mm}^4$$

when Eq. 7.22a gives:

$$\theta = \frac{(50 \times 10^3) \times (1 \times 10^3)}{(37.65 \times 10^3) \times (30 \times 10^3)} = 0.0443^c = 2.48°$$

Taking b from the thicker vertical limb, Eq. 7.22b gives:

$$\tau_{max} = \frac{3\,T}{bt^2} = \frac{3 \times (50 \times 10^3)}{100 \times 10^2} = 15 \text{ MPa}$$

These approximate calculations provide a reasonable estimate of twist and maximum stress, which would befit the use of a suitable safety factor.

7.3 Bredt-Batho Torsion Theory

This useful theory deals with torsion of thin-walled closed tubes of any shape and varying thickness. A single-cell closed tube is able to resist torque by means of a constant shear flow in the wall. The introduction of partitions to form multi-cells will serve to both stiffen and strengthen the tube against torsional failure.

7.3.1 Single Cell Tube

Let any shape of a thin-walled, closed tube have an area A enclosed by its mean wall co-ordinate s in which the wall thickness t varies (see Fig. 7.14a). When the tube is subjected to an axial torque T this produces shear stress τ around the wall and along the length. The variation in shear stress and thickness for an element of wall at A is re-orientated in Fig. 7.14b:

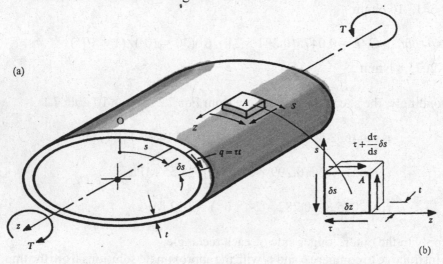

Figure 7.14 Torsion of a closed, single-cell tube with varying thickness

This element appears in co-ordinates of perimeter length s versus axial length z. It shows that as the thickness t increases by δt the shear stress τ increases by an amount $(\mathrm{d}\tau/\mathrm{d}s)\delta s$ over a distance δs. The force equilibrium equation for the element's z-direction becomes:

$$\tau(\delta z \times t) = \left(\tau + \frac{\mathrm{d}\tau}{\mathrm{d}s}\,\delta s \right)(t + \delta t)\delta z \qquad (7.28a)$$

Expanding Eq. 7.28a and dividing through by $\delta s \times \delta z$ leads to:

$$\tau\,\frac{\delta t}{\delta s} + t\,\frac{\mathrm{d}\tau}{\mathrm{d}s} + \frac{\mathrm{d}\tau}{\mathrm{d}s}\,\delta t = 0 \qquad (7.28b)$$

In the limit, as $\delta s \to 0$, we may neglect the product of two small quantities within the third term. Equation 7.28b then becomes a differential product:

$$\frac{\mathrm{d}(\tau t)}{\mathrm{d}s} = 0 \qquad (7.28c)$$

The solution to Eq. 7.28c reveals an important property of a closed tube:

$$q = \tau t = \text{constant}$$

in which the shear flow q is constant within the wall regardless of the tube shape. Note that shear stress $\tau = q/t$ is constant only when t is constant. Figure 7.15a shows how the thin-walled section resists the torque through its shear flow.

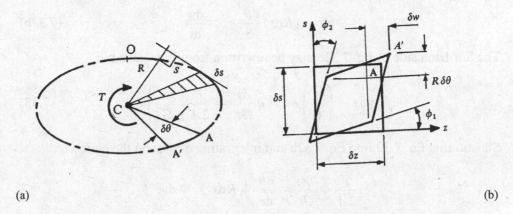

(a)   (b)

Figure 7.15 Torque balance and distortion for a single cell tube

Within the elemental length δs shown, a tangential force $q\delta s$ acts at a perpendicular distance R from the torque axis. Integration around the perimeter provides the torque equilibrium equation:

$$T = \oint_s (q\,ds)R = q\oint_s R\,ds \qquad (7.29a)$$

Now $R \times ds$ in Eq. 7.29a is twice the area of the shaded segment shown. It follows, therefore, that the path integral in Eq. 7.29a is twice the area A enclosed by the mid-wall perimeter. Hence, Eq. 7.29a becomes:

$$T = 2Aq = 2A\tau t \qquad (7.29b)$$

The rate of twist $\delta\theta/\delta z$ is found from assuming that the tube is free to warp. Thus, in the absence of axial strain, chords, initially parallel to the tube axis, are free to rotate without changing their lengths. That is, the point A, on the outer surface is free to rotate (twist) and displace axially (warp) to A', as shown in Fig. 7.15b. The shear strain defines the change to the right angle at A within the distorted shape of the surface element shown:

$$\gamma = \tan\varphi_1 + \tan\varphi_2 \approx \varphi_1 + \varphi_2 \text{ (rad)} \qquad (7.30)$$

in which the approximation $\tan\varphi \approx \varphi$ (radian) holds for elasticity in metals. The tangential and warping components of displacement from A to A' are $R\delta\theta$ and δw, respectively. Hence, the two shear angles in Eq. 7.30 are:

$$\varphi_1 = R\frac{\delta\theta}{\delta z} \quad \text{and} \quad \varphi_2 = \frac{\delta w}{\delta s} \text{ (rad)} \qquad (7.31a,b)$$

Substituting Eqs 7.31a,b into Eq. 7.30, we have γ, as δs and δz approach zero:

$$\gamma = R\frac{d\theta}{dz} + \frac{dw}{ds} \tag{7.32a}$$

from which

$$\gamma\delta s = (R\delta s)\frac{d\theta}{dz} + \delta s\frac{dw}{ds} \tag{7.32b}$$

The left-hand side of Eq. 7.32b may be rewritten from Eqs 7.29a,b:

$$\gamma\delta s = \left(\frac{\tau}{G}\right)\delta s = q\frac{\delta s}{Gt} = \frac{T}{2A}\left(\frac{\delta s}{Gt}\right) \tag{7.33}$$

Substituting Eq. 7.33 into Eq. 7.32b and integrating δs around the perimeter:

$$\frac{T}{2A}\oint\frac{ds}{Gt} = \frac{d\theta}{dz}\oint Rds + \oint dw$$

where $\oint Rds = 2A$. Also, the path integration must start and finish at the same point so that there is no relative warping displacement. Hence $\oint dw = 0$, giving two expressions for the rate of twist:

$$\frac{d\theta}{dz} = \frac{T}{4A^2}\oint\frac{ds}{Gt} = \frac{q}{2A}\oint\frac{ds}{Gt} \tag{7.34a}$$

Combining Eqs 7.29b and 7.34a completes the general theory for a non-uniform tube section consisting of more than one material:

$$T = \frac{4A^2\left(\dfrac{d\theta}{dz}\right)}{\oint\dfrac{ds}{Gt}} = 2Aq \tag{7.34b}$$

There are many types of tube for which Eqs 7.34a,b apply. For a uniform tube in a single material, G is constant, and Eq. 7.34b becomes:

$$T = GJ\frac{d\theta}{dz} = 2Aq \tag{7.35a}$$

where J is the *St Venant Torsion Constant*, defined as:

$$J = \frac{4A^2}{\oint\dfrac{ds}{t}} \tag{7.35b}$$

Where this tube's cross section remains uniform with respect to z, then Eq. 7.34a integrates to provide the angular twist θ over the tube length L:

$$\theta = \frac{TL}{4A^2G}\oint\frac{ds}{t} = \frac{qL}{2AG}\oint\frac{ds}{t} \tag{7.36a}$$

When the cross section of this tube is an n-sided polygon with each flat of different length and thickness:

$$\oint \frac{ds}{t} \equiv \sum_{i=1}^{n} \left(\frac{s}{t} \right)_i \qquad (7.36b)$$

In a tapered tube both J and A in Eq. 7.35a will depend upon z and so too will the twist rate. Examples 7.11 and 7.12 that follow show further how to apply Eq. 7.34a when the torque varies with z.

7.3.2 Warping

The following warping derivative follows from Eq. 7.32a

$$\frac{dw}{ds} = \gamma - R \frac{d\theta}{dz} \qquad (7.37a)$$

and substituting the twist rate from Eq. 7.34a into Eq. 7.37a

$$\frac{dw}{ds} = \frac{q}{Gt} - \frac{qR}{2A} \oint_s \frac{ds}{Gt}$$

when, from Eq. 7.35a

$$\frac{dw}{ds} = \frac{T}{2AGt} - \frac{TR}{4A^2} \oint_s \frac{ds}{Gt} \qquad (7.37b)$$

Integrating Eq. 7.37b from an initial warping displacement w_o at $s = 0$:

$$\int_{w_o}^{w} dw = \frac{T}{2A} \int_0^s \frac{ds}{Gt} - \frac{T}{4A^2} \oint \frac{ds}{Gt} \int_0^s R\,ds \qquad (7.38a)$$

If G is constant, and the origin for s lies on an axis of symmetry, then $w_o = 0$ and Eq. 7.38a gives w directly

$$w = \frac{T}{2AG} \int_0^s \frac{ds}{t} - \frac{T}{4A^2 G} \oint \frac{ds}{t} \int_0^s R\,ds \qquad (7.38b)$$

Equation 7.38b is expressed more conveniently as:

$$w = \frac{Ti}{2AG} \left(\frac{i_{os}}{i} - \frac{A_{os}}{A} \right) \qquad (7.38c)$$

where $i = \oint ds/t$ and $A = \frac{1}{2} \oint R\,ds$ is the total area enclosed by the section. Referring to Fig. 7.15a, $i_{os} = \int ds/t$ is the line integral between the origin O for s and a point A along the mean wall perimeter, distance s from O, at which w is required. A_{os} is the segment of area enclosed between O, A and the *centre of twist* C. The latter lies at a point in the enclosed section that does not suffer rotation or warping displacement. The centre of twist lies at the intersection between axes of symmetry and can normally be found by inspection. If C in Fig. 7.15a displaces because it does not lie at the centre of twist then Eq. 7.37b will supply warping displacements relative to C. Equation 7.38c shows that a section will not warp when:

$$\frac{i_{os}}{i} = \frac{A_{os}}{A} \tag{7.39}$$

For example, Eq. 7.39 is satisfied by a circular tube of mean radius R and uniform thickness t since the LH and RH sides are:

$$\frac{i_{os}}{i} = \frac{s/t}{2\pi R/t} = 2\pi R$$

$$\frac{A_{os}}{A} = \frac{Rs/2}{\pi R^2} = 2\pi R$$

Equation 7.39 is also satisfied by a triangular tube of uniform thickness and a rectangular tube section where the ratio of the side lengths equals the ratio of their thicknesses. These are all known as Neuber tubes.

Example 7.10 The cross section of a 2 m-long aluminium alloy tube is shown in Fig. 7.16a. If the maximum shear stress in the wall is limited to 27.5 MPa find the axial torque which can be applied to the tube and its angular twist. What are the warping displacements at points A, B, C and D? Take $G = 27$ GPa.

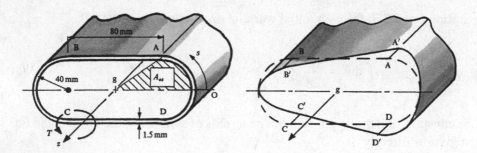

(a)   (b)

Figure 7.16 Closed, non-circular alloy tube

The area properties required are:

$$A = \pi(40)^2 + (80)^2 = 11426.55 \text{ mm}^2$$

$$i = \oint ds/t = (2 \times 80/1.5) + (\pi \times 80/1.5) = 274.22$$

$$J = 4A^2/i = 4 \times (11426.55)^2/274.22 = 1.905 \times 10^6 \text{ mm}^4$$

The torque is found from Eq. 7.29b:

$$T = 2At\tau = 2 \times 11426.55 \times 1.5 \times 27.5 = 942.7 \text{ Nm}$$

The angular twist follows from Eq. 7.36a:

$$\theta = \frac{qLi}{2AG} = \frac{(\tau t)Li}{2AG}$$

$$= \frac{(27.5 \times 1.5) \times (2 \times 10^3) \times 274.22}{2 \times 11426.55 \times (27 \times 10^3)} = 0.0367^c = 2.1°$$

To find the warping displacements at points A, B, C and D, we know that the centre of twist lies at the intersection between axes of symmetry in this doubly symmetric section. A suitable origin O for s is at the intersection between the wall and one symmetry axis where $w_o = 0$. Let the anticlockwise direction of s correspond with that of T so that positive w coincides with positive z. The constant coefficient in Eq. 7.38c is:

$$\frac{Ti}{2AG} = \frac{(942.7 \times 10^3) \times 274.22}{2 \times 11426.55 \times (27 \times 10^3)} = 0.419$$

At point A, we have within Eq. 7.38c:

$$A_{os} = (40 \times 40/2) + \pi(40)^2/4 = 2056.64 \text{ mm}^2$$

$$i_{os} = OA/t = (2\pi \times 40)/(4 \times 1.5) = 13.33\pi$$

$$\frac{i_{os}}{i} - \frac{A_{os}}{A} = \frac{13.33\pi}{274.22} - \frac{2056.64}{11426.55} = -0.0272$$

$$w_A = (0.419)(-0.02727) = -0.0114 \text{ mm}$$

At point B:

$$A_{os} = 2056.64 + (80 \times 40/2) = 3656.64 \text{ mm}^2$$

$$i_{os} = 13.33\pi + 80/15 = 95.22$$

$$\frac{i_{os}}{i} - \frac{A_{os}}{A} = \frac{95.22}{274.22} - \frac{3656.64}{11426.55} = 0.0272$$

$$w_B = (0.419)(0.0272) = 0.0114 \text{ mm}$$

At point C:

$$A_{os} = 3656.64 + (2 \times 2056.64) = 7769.92 \text{ mm}^2$$

$$i_{os} = 95.22 + (80\pi/2 \times 1.5) = 178.996$$

$$\frac{i_{os}}{i} - \frac{A_{os}}{A} = \frac{178.996}{274.22} - \frac{7769.92}{11426.55} = -0.0272$$

$$w_C = (0.419)(-0.0272) = -0.0114 \text{ mm}$$

At D, $w_D = +0.0114$ mm by inspection. These displacements reveal that the section ABCD warps to A'B'C'D' in the manner of Fig. 7.16b.

Example 7.11 A cantilever with the rectangular tube section shown in Fig. 7.17 is subjected to a uniformly distributed anticlockwise torque of 20 kNm per metre of its 2.5 m length. The vertical and horizontal sides are of different materials for which $G = 26$ GPa and 18 GPa, respectively. Calculate the maximum shear stress in the tube and the manner in which the angular twist varies with the length.

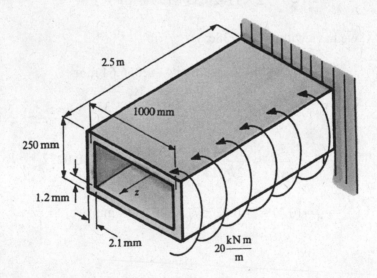

Figure 7.17 Cantilever tube under distributed torque

Since $T_{max} = 20 \times 2.5 = 50$ kNm at the fixed end, it follows from Eq. 7.29b that the maximum shear stress is given by:

$$\tau_{max} = \frac{T_{max}}{2At_{min}} = \frac{50 \times 10^6}{2(1000 \times 250)1.2} = 83.33 \text{ MPa}$$

When $T = T(z)$ and G is not constant, the angular twist is found from integrating Eq. 7.34a:

$$\theta = \frac{1}{4A^2} \oint \frac{ds}{Gt} \int_z T(z)dz \qquad (i)$$

where, in units of $\underline{N}$ and $\underline{mm}$:

$$\oint \frac{ds}{Gt} = \frac{2 \times 1000}{(18 \times 10^3) \times 1.2} + \frac{2 \times 250}{(26 \times 10^3) \times 2.1} = 101.75 \times 10^{-3} \frac{mm^2}{N}$$

$$A^2 = (1000 \times 250)^2 = 625 \times 10^8 \, mm^4$$

Taking the origin for z at the fixed end, the torque (unit of $\underline{Nmm}$) varies as:

$$T(z) = (2500 - z) \times (20 \times 10^3)$$

i.e. a linear variation from zero at the free end to a maximum value at the fixed end. Substituting into Eq. i:

$$\theta = \frac{101.75 \times 10^{-3}}{4(625 \times 10^8)} \times (20 \times 10^3) \int_z (2500 - z)dz$$

$$= (0.814 \times 10^{-8})(2500z - z^2/2) \qquad (ii)$$

Equation ii shows that the twist θ increases non-linearly with length z, from its zero value at the fixing ($z = 0$) to attain its maximum value 0.0254^c ($1.45°$) at the free end ($z = 2.5$ m).

Example 7.12 The idealised section of an aluminium aircraft wing is shown in Fig. 7.18. The vertical sides have uniform thicknesses but those for the sloping sides vary linearly between the ends as shown. Assuming that the fuselage end is fully built-in, calculate the maximum shear stress, the torsion constant and the angular twist at the free end when in flight the wing is subjected to a uniformly distributed torque of 25 kNm per metre of its 6 m length. Take $G = 28.5$ GPa.

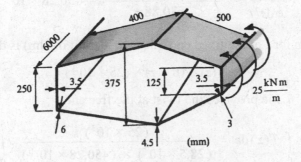

Figure 7.18 Aircraft wing under distributed torque

The area enclosed within the wall is:

$$A = \frac{1}{2}(250 + 375)\sqrt{400^2 - 62.5^2} + \frac{1}{2}(375 + 125)\sqrt{500^2 - 125^2}$$

$$= (12.346 \times 10^4) + (12.1 \times 10^4) = 24.446 \times 10^4 \text{ mm}^2$$

The maximum torque applies to the fixed end:

$$T_{max} = 6 \times 25 = 150 \text{ kNm}$$

where, from Eq. 7.29b, the shear stress is at its maximum where the wall thickness is a minimum:

$$\tau_{max} = \frac{T_{max}}{2At_{min}} = \frac{(150 \times 10^6)}{2 \times (24.446 \times 10^4) \times 3} = 102.3 \text{ MPa}$$

For the bottom (and top)-right sloping sides:

$$\frac{\delta t}{\delta s} = \frac{(4.5 - 3)}{500} \quad \therefore \delta s = 333.33\,\delta t$$

For the top (and bottom)-left sloping sides:

$$\frac{\delta t}{\delta s} = \frac{(6 - 4.5)}{400} \quad \therefore \delta s = 266.67\,\delta t$$

From Eq. 7.36a:

$$\oint \frac{ds}{t} = \frac{125}{3.5} + 2\int_3^{4.5} 333.33\,\frac{dt}{t} + 2\int_{4.5}^{6.0} 266.67\,\frac{dt}{t} + \frac{250}{3.5}$$

$$= 107.14 + 666.67\ln(4.5/3.0) + 533.34\ln(6.0/4.5) = 530.88$$

Equation 7.35b gives the torsion constant as:

$$J = \frac{4A^2}{\oint ds/t} = \frac{4 \times (24.446 \times 10^4)^2}{530.88} = 450.28 \times 10^6 \text{ mm}^4$$

With the origin for z at the fixed end, the torque (unit of <u>Nm</u>) is distributed as:

$$T(z) = (6 - z) \times (25 \times 10^3)$$

Integrating Eq. 7.34a provides the twist at the free end:

$$\theta = \frac{1}{GJ}\int_0^L T(z)\,dz = \frac{(25 \times 10^3)}{(28.5 \times 10^9) \times (450.28 \times 10^{-6})}\int_0^6 (6 - z)\,dz$$

$$= (1.948 \times 10^{-3})\left|6z - z^2/2\right|_0^6 = 0.03507^c = 2.01°$$

7.3.2 Two-Cell Tube

Providing a closed path for shear flow enhances the torque-carrying capacity of a thin-walled tubular section. A similar property applies to a two-cell tube where there is an interaction between the separate shear flow in each tube at their common partition. Let q_1 and q_2 be the respective shear flows within each cell in Fig. 7.19. In general, the thicknesses t_1, t_2 and t_{12} and the shear moduli G_1, G_2 and G_{12} can differ between the cells and the partition.

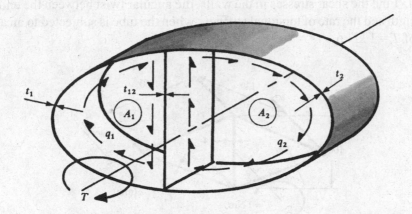

Figure 7.19 Shear flow in a two-cell tube

The applied torque T is the sum of the torques carried by each cell. Thus a torque equilibrium condition follows from Eq. 7.29b:

$$T = 2A_1 q_1 + 2A_2 q_2 \qquad (7.40)$$

With two unknowns, q_1 and q_2, in the equilibrium Eq. 7.40, this structure is *statically indeterminate*. The solution must therefore satisfy both equilibrium and compatibility. The latter requires that the rate of twist within each cell is the same. Now as q_1 and q_2 act in opposing directions at the common wall Eq. 7.34a provides the twist rate in each cell as:

$$\left(\frac{d\theta}{dz} \right)_1 = \left(\frac{q_1}{2A_1} \right) \oint_1 \frac{ds}{Gt} - \left(\frac{q_2}{2A_1} \right) \int_{12} \frac{ds}{Gt} \qquad (7.41a)$$

$$\left(\frac{d\theta}{dz} \right)_2 = \left(\frac{q_2}{2A_2} \right) \oint_2 \frac{ds}{Gt} - \left(\frac{q_1}{2A_2} \right) \int_{12} \frac{ds}{Gt} \qquad (7.41b)$$

Equating 7.41a,b provides the required compatibility condition:

$$q_1 \oint_1 \frac{ds}{Gt} - q_2 \int_{12} \frac{ds}{Gt} = \frac{A_1}{A_2} \left(q_2 \oint_2 \frac{ds}{Gt} - q_1 \int_{12} \frac{ds}{Gt} \right) \qquad (7.42)$$

Once q_1 and q_2 are solved simultaneously from Eqs 7.40 and 7.42, the shear stresses in the three walls are given by:

$$\tau_1 = \frac{q_1}{t_1}, \quad \tau_2 = \frac{q_2}{t_2} \quad \text{and} \quad \tau_{12} = \frac{(q_1 - q_2)}{t_{12}} \qquad (7.43a,b,c)$$

Example 7.13 The two cells in the uniform multi-cell tube shown in Fig.7.20 are of identical dimension and material, for which the shear modulus is $G = 30$ GPa = constant. Find the shear stresses in the walls, the angular twist between the ends of a 2 m length and the rate of torsional stiffness when the tube is subjected to an axial torque of $T = 1$ kNm.

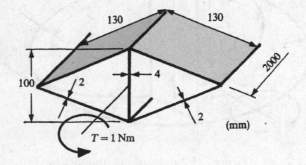

Figure 7.20 Two-cell tube with identical cells

Since $A_1 = A_2$ it follows from Eq. 7.42 that:

$$q_1 \left(\oint_1 \frac{ds}{Gt} + \int_{12} \frac{ds}{Gt} \right) = q_2 \left(\oint_2 \frac{ds}{Gt} + \int_{12} \frac{ds}{Gt} \right)$$

Now $\oint_1 ds/Gt = \oint_2 ds/Gt$ and therefore $q_1 = q_2$, which will always apply to identical cells. For the cells in question:

$$A_1 = A_2 = 100 \times 120/2 = 6000 \ \text{mm}^2$$

$$\oint_1 \frac{ds}{t} = \oint_2 \frac{ds}{t} = \left(2 \times \frac{130}{2} \right) + \frac{100}{4} = 155, \quad \int_{12} \frac{ds}{t} = \frac{100}{4} = 25$$

Equation 7.40 gives the torque as $T = 4Aq_1$. Hence:

$$q_1 = \frac{T}{4A} = \frac{(1 \times 10^6)}{4 \times 6000} = 41.67 \ \frac{\text{N}}{\text{mm}} = q_2$$

The shear stresses in the walls follow from Eqs 7.43a,b:

$$\tau_1 = 41.67/2 = 20.83 \ \text{MPa} = \tau_2$$

Since q_1 and q_2 are equal and opposite in the web, it follows from Eq. 7.43c that the wall is unstressed ($\tau_{12} = 0$). The rate of twist is found from either of Eqs 7.41a or b, which simplify to:

$$\frac{d\theta}{dz} = \frac{q_1\left(\oint_1 \frac{ds}{t} - \int_{12} \frac{ds}{t}\right)}{2A_1 G}$$

$$= \frac{41.67(155 - 25)}{2 \times 6000 \times (30 \times 10^3)} = 15.05 \times 10^{-6} \frac{c}{mm}$$

Hence the twist on a 2 m length is

$$\theta = 15.05 \times 10^{-6} \times 2 \times 10^3 = 0.0301^c = 1.78°$$

The *torsional stiffness rate* is found from the ratio (in N and m):

$$\frac{T}{d\theta/dz} = \frac{(1 \times 10^3)}{(15.05 \times 10^{-3})} = 66.45 \times 10^3 \frac{Nm}{°/m} = 66.45 \frac{kNm^2}{c}$$

Example 7.14 The two-cell, aluminium tube section shown in Fig. 7.21 is uniform for its 2 m length. Find the shear stresses in the walls, the angular twist, the rate of twist and the torsional stiffness when a torque of 1 kNm is applied along its axis as shown. Take $G = 30$ GPa.

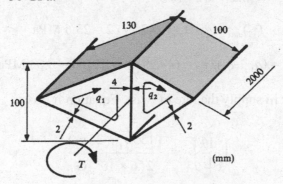

Figure 7.21 Two-cell tube with dissimilar cells

The cell areas are:

$$A_1 = 50 \sqrt{(130^2 - 50^2)} = 6000 \text{ mm}^2, \quad A_2 = 50 \sqrt{(100^2 - 50^2)} = 4330 \text{ mm}^2$$

Applying the torque equilibrium Eq. 7.40:

$$1 \times 10^6 = 2 (6 \times 10^3) q_1 + 2 (4.33 \times 10^3) q_2$$

$$100 = 1.2q_1 + 0.866q_2 \qquad (i)$$

and from Eqs 7.41a,b, the cell twist rates are:

$$2 \times (6 \times 10^3) G \left(\frac{d\theta}{dz} \right)_1 = q_1 \left(2 \times \frac{130}{2} + \frac{100}{4} \right) - q_2 \left(\frac{100}{4} \right)$$

$$= 155 q_1 - 25 q_2 \qquad \qquad \text{(ii)}$$

$$2 \times (4.33 \times 10^3) G \left(\frac{d\theta}{dz} \right)_2 = q_2 \left(2 \times \frac{100}{2} + \frac{100}{4} \right) - q_1 \left(\frac{100}{4} \right)$$

$$= 125 q_2 - 25 q_1 \qquad \qquad \text{(iii)}$$

Solving Eqs ii and iii when $(d\theta / dz)_1 = (d\theta / dz)_2$:

$$111.858 q_1 - 18.042 q_2 = 125 q_2 - 25 q_1$$

$$\therefore q_1 = 1.045 q_2$$

From Eq. i:

$$100 = 1.254 q_2 + 0.866 q_2$$

$$q_2 = 47.17 \text{ N/mm and } q_1 = 49.29 \text{ N/mm}$$

Equations 7.43a,b,c give the wall shear stresses:

$$(\tau_1)_{max} = q_1 / t_{min} = 49.29/2 = 24.7 \text{ MPa}$$

$$(\tau_2)_{max} = q_2 / t_{min} = 47.17/2 = 23.6 \text{ MPa}$$

$$\tau_{12} = (q_1 - q_2)/t_{12} = (49.29 - 47.17)/4 = 0.53 \text{ MPa}$$

Either of Eqs ii or iii supply the rate of twist. Equation ii gives:

$$\left(\frac{d\theta}{dz} \right)_1 = \frac{(155 q_1 - 25 q_2)}{2(6 \times 10^3)G}$$

$$= \frac{(155 \times 49.29) - (25 \times 47.17)}{2(6 \times 10^3) \times (30 \times 10^3)} = 17.95 \times 10^{-6} \ \frac{^c}{mm} = 0.01795 \ \frac{^c}{m}$$

Thus, for a length of 2 m, the twist is $0.0359^c = 2.06°$ and the torsional stiffness is $T/\theta = 1/2.06 = 0.485$ kNm/°.

Example 7.15 The uniform, two-cell, aluminium alloy tube section in Fig. 7.22 is subjected to a pure axial torque of 12 kNm. Find the rate of twist and the multiplying factor t to be applied to the dimensions of cell 2 in order that the vertical web remains unstressed. Take $G = 25$ GPa.

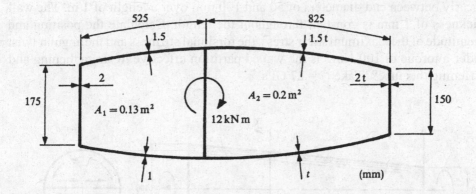

Figure 7.22 Uniform two-cell tube

Since the shear stress in the web is to be zero, Eq. 7.43c shows that $q_1 = q_2$. The shear flow is found from the torque T (Nmm) in Eq. 7.40:

$$q_1 = \frac{T}{2(A_1 + A_2)} = \frac{(12 \times 10^6)}{2(0.13 + 0.20) \times 10^6} = 18.182 \text{ Nmm}$$

Equation 7.42 applies to the equal twist rates for each cell, within which both $q_1 = q_2$ and G will cancel. This gives:

$$\frac{A_2}{A_1} \left(\oint_1 \frac{ds}{t} - \int_{12} \frac{ds}{t} \right) = \oint_2 \frac{ds}{t} - \int_{12} \frac{ds}{t}$$

$$\frac{0.20}{0.13} \left(\frac{525}{1.5} + \frac{525}{1} + \frac{175}{2} \right) = \frac{825}{1.5t} + \frac{825}{t} + \frac{150}{2t}$$

$$1480.77 = 1450/t$$

from which

$$t = 0.98$$

The rate of twist may be calculated from cell 1. Equation 7.41a gives:

$$\frac{d\theta}{dz} = \frac{q_1}{2A_1 G} \left(\oint_1 \frac{ds}{t} - \int_{12} \frac{ds}{t} \right)$$

$$= \frac{18.182}{2 \times (0.13 \times 10^6) \times (25 \times 10^3)} \left(\frac{525}{1.5} + \frac{525}{1} + \frac{175}{2} \right)$$

$$= 2.69 \times 10^{-6} \, {}^{\circ}/\text{mm} = 0.00269 \, {}^{\circ}/\text{m} = 0.154 \, {}^{\circ}/\text{m}$$

Example 7.16 The circular tube with two semi-circular cells in Fig. 7.23 tapers linearly between end diameters of 50 and 100 mm over a length of 1 m. The wall thickness of 1 mm is constant throughout the length. Calculate the position and magnitude of the maximum shear stress, the torsional stiffness and the angular twist under a torque of 100 Nm. Is the walled partition effective in strengthening and stiffening this tube? Take $G = 27$ GPa.

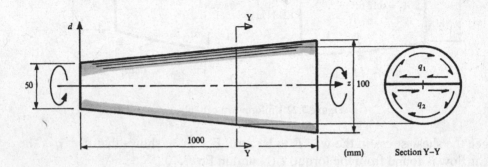

Figure 7.23 Tapered two-cell circular tube

Partitioned Tube

For identical cells $q_1 = q_2$ and the partition wall remains unstressed (see Example 7.13). The shear flow within each cell attains its greatest magnitude at the smaller end diameter, where Eq. 7.40 gives:

$$q_1 = \frac{T}{4A_1} = \frac{(100 \times 10^3)}{4 \times (\pi/8)(50)^2} = 25.47 \text{ N/mm}$$

Correspondingly, Eq. 7.43a gives the maximum shear stress in the tube:

$$\tau_{max} = q_1/t = 25.47/1 = 25.47 \text{ MPa}$$

Equation 7.41a provides the rate of twist for cell 1 which is also the twist rate for the tube:

$$\frac{d\theta}{dz} = \frac{q_1}{2A_1 G}\left(\oint_1 \frac{ds}{t} - \int_{12} \frac{ds}{t}\right) \tag{i}$$

Taking an origin for tube length z at the smaller end in Fig. 7.23, the diameter d (mm) and cell area A_1 (mm²) vary with z (mm) as follows

$$d = 50 + \frac{50}{1000}z$$

$$A_1(z) = \frac{\pi}{8}(50 + 0.05z)^2 \tag{ii}$$

The bracketed quantity in Eq. i indicates that the path integral need only be evaluated for the semi-circular part of the cell:

$$\left(\oint_1 \frac{ds}{t} - \int_{12} \frac{ds}{t} \right) = \frac{\pi d}{2t} = \frac{\pi}{2t}(50 + 0.05z) \tag{iii}$$

The shear flow varies with the cell area as:

$$q_1 = \frac{T}{4A_1(z)} = \frac{(100 \times 10^3)}{4 \times (\pi/8)(50 + 0.05z)^2} = \frac{(200 \times 10^3)}{\pi(50 + 0.05z)^2} \tag{iv}$$

Substituting Eqs ii–iv into Eq. i, gives the twist rate:

$$\frac{d\theta}{dz} = \frac{4 \times (100 \times 10^3)}{\pi Gt(50 + 0.05z)^3} = \frac{4.716}{(50 + 0.05z)^3} \frac{^c}{mm} \tag{v}$$

The torsional stiffness rate, in N and mm, follows from Eq. v as:

$$\frac{T}{d\theta/dz} = \frac{(100 \times 10^3)}{4.716/(50 + 0.05z)^3} = 21.2 \times 10^3 (50 + 0.05z)^3 \frac{Nmm}{^c/mm}$$

Integrating Eq. v gives the twist over the 1000 mm tube length:

$$\theta = 4.716 \int_0^{1000} (50 + 0.05z)^{-3} dz = -47.16 \left| (50 + 0.05z)^{-2} \right|_0^{1000}$$

$$= -47.16 \left(\frac{1}{100^2} - \frac{1}{50^2} \right) = 0.01415^c = 0.811°$$

for which the tube's overall torsional stiffness is:

$$T/\theta = 100/0.811 = 123.3 \ Nm/°$$

Plain Tube

Removal of the partition does not alter the shear flow or the maximum shear stress at the smaller end:

$$q = \frac{T}{2A} = \frac{(100 \times 10^3)}{2 \times (\pi/4)(50)^2} = 25.47 \ N/mm$$

$$\tau_{max} = q/t = 25.47/1 = 25.47 \ MPa$$

Equation 7.34a provides the rate of twist for this tube:

$$\frac{d\theta}{dz} = \frac{T}{4A^2G} \oint \frac{ds}{t} \tag{vi}$$

where

$$A(z) = \frac{\pi}{4} (50 + 0.05z)^2 \qquad\qquad \text{(vii)}$$

$$\oint \frac{ds}{t} = \frac{\pi d}{t} = \frac{\pi}{t}(50 + 0.05z) \qquad\qquad \text{(viii)}$$

Substituting Eqs vii and viii into Eq. vi, gives a twist rate identical to that of the partitioned tube:

$$\frac{d\theta}{dz} = \frac{4 \times (100 \times 10^3)}{\pi Gt(50 + 0.05z)^3} = \frac{4.716}{(50 + 0.05z)^3} \frac{\text{c}}{\text{mm}} \qquad\qquad \text{(v)}$$

Clearly, an unstressed partition plays no role in strengthening or stiffening the tube. However, by introducing an off-centre partition with a different thickness it will support shear stress to become effective in each regard (see Exercise 7.43). This is provided the panel does not buckle under shear, one of a number of topics on instability covered in the following chapter.

Exercises

Torsion of Circular Bars

7.1 A solid steel shaft, 50 mm diameter, can transmit a torque of 2.17 kNm. What diameter steel shaft can transmit 45 kW when rotating at 240 rev/min?
(Ans: 46.88 mm)

7.2 A wrought-iron shaft has a diameter of 65 mm and is 6.5 m long. What will be the twist in the bar when the maximum shear stress reaches 80 MPa? Take $G = 72.5$ GPa.
(Ans: 12.65°)

7.3 A tubular extension, 38 mm o.d., 32 mm i.d. and 2.5 m long, is fitted over the end of a 32 mm diameter solid spindle. If, when transmitting torque, the maximum shear stress in the spindle is 31 MPa, calculate the maximum shear stress and the angle of twist for the tube. Take $G = 82.7$ GPa.
(Ans: 34 MPa, 3.04°)

7.4 A hollow drive shaft with a diameter ratio 3:5 is to transmit 448 kW at 120 rev/min. If the shear stress must not exceed 62 MPa and the angle of twist is restricted to 1° over a 2.5 m length, calculate the external diameter of the shaft. Take $G = 82.7$ GPa.
(Ans: 163 mm)

7.5 The constraints imposed upon a solid circular shaft transmitting 932.5 kW at 240 rev/min are: (i) that it should not twist more than 1° on 15 diameters and (ii) that the maximum shear stress should not exceed 54 MPa. Calculate the shaft diameter and the working shear stress. Take $G = 82.7$ GPa.
(Ans: 158 mm, 48 MPa)

7.6 A hollow drive shaft 203 mm o.d. with a 25 mm wall thickness, is to transmit 933 kW when rotating at 125 rev/min. Calculate the safe working shear stress for the shaft material when based upon a safety factor of 3.
(Ans: 190 MPa)

7.7 A solid steel shaft, 60 mm diameter and 1.5 m long, is subjected to a torque of 2 kNm. Find the angle of twist and the greatest shear stress. Take $G = 80$ GPa.
(Ans: 1.69°, 47.17 MPa)

7.8 A hollow steel shaft is to transmit a torque of 30 kNm without the twist exceeding 3° over a length of 4 m. If the maximum shear stress is restricted to 80 MPa, find the shaft diameters that will minimise its weight. Take $G = 80$ GPa.
(Ans: 159 mm, 135 mm)

7.9 A hollow steel shaft, 90 mm o.d. and 6 mm wall thickness, is to transmit a power of 240 kW at a rotational speed of 300 rev/min. Calculate the shear stresses at the inner and outer diameters and the twist over a length of 2 m. Take $G = 80$ GPa.
(Ans: 106.1 MPa, 122.4 MPa, 3.9°)

7.10 One degree of twist is measured for a steel propeller shaft, 305 mm in diameter and 4.5 m in length. Determine the power transmitted by the steel shaft and the maximum shear stress within it when the shaft rotates at 300 rev/min. If a hollow shaft, with diameter ratio 1:4, is used to transmit this power under similar conditions, determine the % saving in weight. Take $G = 80.2$ GPa.
(Ans: 8.13 MW, 46.6 MPa, 58%)

7.11 A solid steel shaft, 255 mm in diameter, rotates at at 180 rev/min. If the maximum torque exceeds the mean torque by 10% what power can be transmitted when the maximum shear stress is restricted to 62 MPa? Find the length of this shaft over which the angle of twist is 1.5°. Take $G = 82.7$ GPa.
(Ans: 34 kW, 4.45 m)

7.12 Determine the torque that can be transmitted by a 100 mm diameter solid shaft when the maximum shear stress is restricted to 48 MPa. Through what angle would this shaft twist in a length of 3.65 m given $G = 82.7$ GPa? What is the % saving in weight for a hollow shaft of equal strength with a diameter ratio of 1.5?
(Ans: 10 kNm, 2.4°)

7.13 A 100 mm diameter solid steel shaft transmits 180 kW at 800 rev/min with the maximum torque exceeding the mean torque by 50%. What is the maximum shear stress in the shaft? If this shaft contains a coupling, with six bolts equi-spaced on a pitch circle diameter of 180 mm, calculate the required minimum diameter of the bolts when their allowable shear stress value is 27.5 MPa.
(Ans: 15.5 MPa, 165 mm)

7.14 A hollow steel shaft, 205 mm i.d. and 305 mm o.d., is to be replaced by a solid alloy shaft. If their polar second moments of area J are the same, calculate the solid shaft's diameter and the ratio between their stiffness. What would the ratio between their J-values become when the stiffness of each shaft is the same? Take G (steel) $= 2.4G$ (alloy).
(Ans: 283 mm, 2.58, 0.49)

7.15 A hollow steel shaft, 150 mm o.d. with 12.5 mm wall thickness, is coupled to a solid 125 mm diameter steel shaft. If the length of the hollow shaft is 1.25 m, find the length of solid shaft which limits the total twist to 3° under a torque of 25 kNm. Take $G = 69$ GPa.

7.16 One end of a 1 m long aluminium alloy rod is joined to the end of a 0.75 m long nickel alloy rod. The allowable shear stresses for each material are 35 MPa and 75 MPa respectively and their shear moduli are 30 GPa and 80 GPa respectively. If their extreme ends are rigidly fixed, find the torque required and where in the length it must be applied to stress both materials fully.

7.17 The same two materials and their allowable shear stresses, referred to in Exercise 7.16, are to form a composite shaft consisting of two concentric hollow cylinders with diameters: 10 mm i.d. and 20 mm o.d. for nickel alloy; 20 mm i.d. and 25 mm o.d. for aluminium alloy. What is the maximum torque that can be carried?

7.18 A 20 mm steel bar is surrounded firmly by an aluminium alloy tube to prevent slip. Respective shear stresses of 100 and 50 MPa are allowed under an axial torque of 1 kNm. Determine the composite shaft outer diameter and the angle of twist over a 1.5 m length. Take $G = 28$ GPa for aluminium and $G = 80$ GPa for steel.

7.19 The diameter of a 0.5 m long steel shaft increases linearly from 50 mm to 75 mm at its ends. If the maximum shear stress in the shaft is limited to 95 MPa, find the torque which may be transmitted and the angle of twist between the shaft's ends. Take $G = 80$ GPa.

7.20 A solid steel shaft tapers for a length of 1.5 m between end diameters in which one is twice the other. The shear stress and angular twist are limited to 60 MPa and 4°, respectively, under a torque of 10 kNm. Determine the end diameters. Take $G = 80$ GPa.

7.21 A 2 m long steel shaft, 100 mm outer diameter, is solid for 1 m and hollow at 60 mm inner diameter for its remaining 1 m length. Find the greatest torque the shaft can withstand when the maximum shear stress is limited to 100 MPa. What is the angle of twist between the ends? Take $G = 80$ GPa. (Ans: 17.1 kNm, 2.68°)

7.22 The hollow stepped shaft in Fig.7.24 is subjected to clockwise torques of 12.65 kNm at sections A and B and is built-in at section C. Determine the maximum shear stress and the angle of twist between A and C given $G = 75$ GPa. (Ans: 65 MPa, 4.8°)

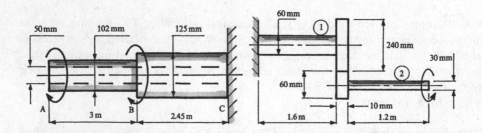

Figure 7.24   **Figure 7.25**

7.23 The ends of the two steel shafts in Fig. 7.25 are connected by gears. The other end of shaft 1 is fixed while that for shaft 2 is free. Determine the angle of twist at the free end of shaft 2 when a torque of 400 Nm is applied at this position. Take $G = 80$ GPa.
(Ans: 10.1°)

7.24 The shaft in Fig.7.26 is supported in bearings when equal opposing torques of 200 Nm are applied through pulleys at positions A and B as shown. Plot diagrams showing the torque and twist variations with the length and determine the position and magnitude of the maximum shear stress.
(Ans: 65.2 MPa)

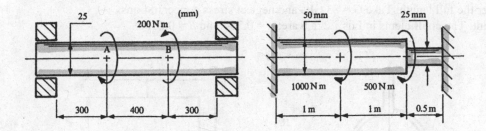

Figure 7.26   **Figure 7.27**

7.25 The stepped shaft in Fig.7.27 is fixed at its ends and is subjected to concentrated torques at the positions shown. Plot the torque and twist diagrams and from them determine the maximum shear stress in the shaft and its maximum angular twist. Take $G = 80$ GPa.

Torsion of Thin Strips

7.26 A thin rectangular strip, 50 mm wide × 2 mm thick, is subjected to a pure axial torque of 10 Nm. Determine the maximum shear stress in the strip and its angle of twist over a strip length of 0.5 m.

7.27 A 5 mm thick rectangular steel sheet is formed into a 270° circular arc. It is to support a torque of 50 Nm without the maximum shear stress exceeding 55 MPa whilst restricting the angular twist to 3°. Find the dimensions required for the rectangular sheet.

7.28 A right-angled section, 5 mm thick with outer dimensions 100 mm × 75 mm, carries a torque of 15 Nm. Calculate the rate of twist and the position and magnitude of the maximum shear stress for the section. Take $G = 82$ GPa.

7.29 A long, straight uniform bar of rectangular section is to act as a door by twisting $\pi/2$ radians along its length. If the maximum shear stress is to be 100 MPa and the modulus of rigidity is 80 GPa, find the torque and the length required. The width of the bar is 20 mm and its thickness is 1.5 mm, for which $\alpha = \beta = 0.32$ from Table 7.1. If this bar is replaced by a circular bar of equal length and section area, find the maximum shear stress and torque required to twist it through $\pi/2$ radians.

7.30 A thin-walled channel section is made from an aluminium alloy with $G = 25$ GPa. The flanges are both 100 mm wide with thickness t and the web is 200 mm long and of thickness $2t$. If the section is to transmit an axial torque of 75 Nm without the twist in a metre length and the shear stress exceeding 25° and 40 MPa, respectively, find the least value of thickness t, to satisfy both of these requirements.

7.31 A torsion member used for stirring in a chemical process is made from a circular tube to which is welded four rectangular strips as shown in Fig.7.28. The tube has inner and outer diameters of 94 and 100 mm, respectively; each strip is 50 mm by 18 mm; and the stirrer is 3 m in length. If the maximum shear stress everywhere in the cross section is limited to 56 MPa, calculate the maximum torque which can be carried by the stirrer and the resulting twist over the full length. Take $G = 83$ GPa and neglect stress concentrations.
(Hint: The coefficients in Eqs 7.22a,b are: $\alpha = 0.264$ and $\beta = 0.258$)

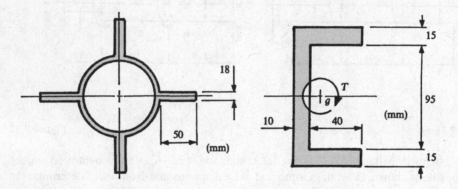

Figure 7.28   **Figure 7.29**

7.32 Find the torque T that can be withstood by the steel channel section in Fig. 7.29 when the greatest shear stress is limited to 120 MPa. What angle would this section twist through in a length of 3 m? Take $G = 83$ GPa with the appropriate coefficients given in Table 7.1.

Bredt-Batho Torsion of Thin-Walled Tubes

7.33 Find the maximum torque that can be carried by a tube having an equilateral triangular section 5 mm thick with 0.2 m mean side length, when the allowable shear stress is restricted to 50 MPa.

7.34 A steel tube, having a 30 mm mean diameter with 1 mm wall thickness, is required to sustain a torque of 80 Nm. Calculate the maximum shear stress and the safety factor that applies when the working stress for the tube is limited to 124 MPa.
(Ans: 61 MPa, 2)

7.35 The walls of a rectangular box section are: 100 mm deep × 2 mm thick and 250 mm wide × 0.5 mm thick. Determine the warping displacements at the four corners of the section when it is subjected to an axial torque of 100 Nm. Take $G = 27$ GPa.
(Ans: ± 0.00837 mm)

7.36 Steel plates 100 mm × 10 mm in cross section are welded together to form: (i) a closed tube of triangular section and (ii) an I-section. Calculate the ratio between their maximum shear stresses and between their torsional stiffness rates when each section is subjected to a similar torque.

7.37 Determine the torsion constant for the thin-walled tubular section shown in Fig. 7.30. If the maximum shear stress is limited to 30 MPa, determine the torque that may be applied and the angular twist over a length of 1.5 m. What is the effect of inserting a thin vertical partition to connect the vertices upon torque and twist? Take $G = 30$ GPa.
(Ans: 78.92×10^3 mm^4, 133.02 Nm, 6.44°)

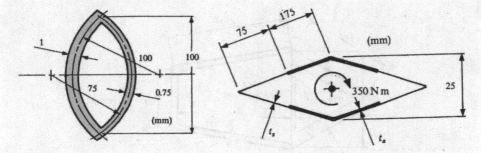

Figure 7.30   **Figure 7.31**

7.38 The symmetric double-wedge tubular section in Fig. 7.31 is made from different thicknesses, t_s and t_a, in steel and aluminium, respectively. The tube, having one end fixed and the other end free, is to support an axial torque of 350 Nm without the twist exceeding 0.035 radian. If the maximum shear stress in the steel is also restricted to 18 MPa determine t_a and t_s. Take $G_s = 3G_a = 81$ GPa.
(Ans: $t_a = 2.56$ mm, $t_s = 1.55$ mm)

7.39 Figure 7.32 shows the cross section of an aircraft fuselage. Calculate the shear stress for each thickness within the section and its angular twist over a 25 m length under an applied torque of 2 MNm. How would the addition of a luggage hold (dotted), to make a two-cell tube, affect the shear stresses and the twist? Take $G = 30$ GPa.
(Ans: 197.8 MPa, 98.9 MPa, 49.5 MPa, 3.34°; with hold: 165 MPa, 82.5 MPa, 7.4 MPa, 67.7 MPa, 2.51°)

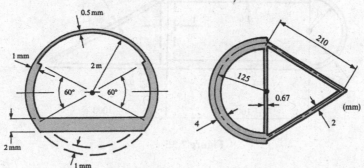

Figure 7.32   **Figure 7.33**

7.40 The two-cell tube in Fig. 7.33 supports a torque of 2 kNm. Calculate the position and magnitude of the maximum shear stress and the rate of twist. What ratio of cell areas would ensure zero shear stress? Compare torsional stiffness/m length for sections with and without the web. Take $G = 79$ GPa.
(Ans: 10.08 MPa, 0.0522 °/m, 2.14, 38.31 kNm/°, 37.27 kNm/°)

7.41 Determine the position and magnitude of the maximum shear stress and the rate of twist when the two-cell tube in Fig. 7.34 is subjected to an axial torque of 900 Nm. Take $G = 81$ GPa for the vertical web, and $G = 27$ GPa for the remaining walls.
(Ans: 12.025 MPa, 0.26°/m)

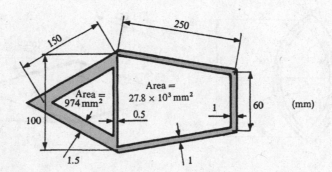

Figure 7.34

7.42 The aluminium alloy wing section in Fig. 7.35 is subjected to an axial torque of 100 kNm over its 10 m length. Calculate the shear stress in each thickness of this section and its angular twist assuming that it is free to warp. Examine the effect on the stress and angular twist when a further 1 mm thick web of the same material is placed in the position of the dashed lines as shown. Take $G = 30$ GPa.
(Ans: single cell: 10°, 188.5 MPa, 94.3 MPa; two cells: 8.2°, 230 MPa, 115 MPa, 93 MPa, 54.8 MPa)

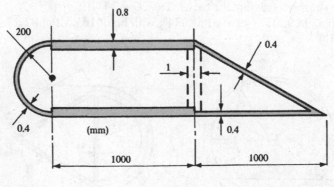

Figure 7.35

7.43 Examine the effects upon strength and stiffness for the tapered, two-cell tube in Fig. 7.23 by: (i) raising the horizontal 1 mm thick partition to give enclosed cell areas in the ratio 2/1 and (ii) increasing the thickness of the central horizontal partition to 2 mm.

CHAPTER 8

BUCKLING OF STRUTS

Summary: There are many forms of structural buckling. The most common form considered here is the flexural lateral buckling which occurs in slender struts under axial compression. The Euler mathematical theory of elastic buckling provides the buckling load but is idealised in that it does not limit the material's stress. Strictly, this theory can only be applied reliably to long thin members that are prone to buckling under low elastic stress levels. The stress increases for shorter struts to a theoretical cut off at the yield stress where direct compression influences the flexural buckling load. Here a number of empirical formulae are available to estimate buckling loads for so-called imperfect struts, the bases for these being a stress summation from these two causes. The Euler theory may be also be modified in a similar manner to provide useful accounts of initial imperfections including eccentric loading, lack of initial straightness and the presence of lateral loading in addition to axial compression. Various standards provide formulae restricted to shorter struts or for applications extending over a wider range from short to medium-length struts. A degree of safety is usually admitted in which Euler's theory is recognised as providing a reliable upper limit to the buckling load for long struts.

8.1 Euler Buckling Theory

Leonhard Euler (1707–1783) derived the axial compressive force required to initiate bucking in a long thin strut. Ideally, a perfectly straight strut, when subjected to a purely compressive load, would compress and not buckle. Buckling is therefore the most likely result of an eccentricity in axial loading applied to a slender strut with imperfect straightness. The combined effects of these minuscule imperfections upon overall buckling behaviour is predictable where long struts behave elastically. Euler based his instability criterion upon the lateral deflection that occurs. This will be applied here to long struts with the following end fixings.

8.1.1 Pinned or Hinged Ends

The flexure Eq. 6.4 is applied to a point upon the deflected strut with initial length L under axial compression P in Fig. 8.1. The co-ordinates (z, v) are arranged, with v positive in the direction shown, to display the strut with sagging curvature.

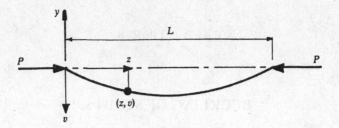

Figure 8.1 Deflection of a pinned-end strut

Applying Eq. 6.4 to the left of point (z, v) shown:

$$EI \, d^2 v/dz^2 = -Pv$$

which is written as:

$$d^2 v/dz^2 + \alpha^2 v = 0 \qquad (8.1)$$

where $\alpha^2 = P/EI$. The solution to Eq. 8.1 is:

$$v = A \sin \alpha z + B \cos \alpha z \qquad (8.2)$$

which must satisfy the following boundary conditions:

(i) $v = 0$ at $z = 0$, $\therefore$ $B = 0$

(ii) $v = 0$ at $z = L$, $\therefore$ $A \sin \alpha L = 0$

We cannot take $A = 0$ as a solution to condition (ii) because this gives $v = 0$ for all z in Eq. 8.2. The solution is therefore $\sin \alpha L = 0$ from which: $\alpha L = \pi, 2\pi, 3\pi.....n\pi$. Buckling of the strut will commence when the least πL value is achieved. Thus:

$$L\alpha = \pi, \quad \therefore L \times \sqrt{(P/EI)} = \pi$$

from which the Euler critical buckling load is:

$$P_E = \pi^2 EI/L^2 \qquad (8.3)$$

Equation 8.3 shows that this load depends upon the strut length and the elastic modulus of the strut material. The second moment of area of the section must apply to a principal axis of bending which offers the least resistance to buckling, i.e. the axis with the least I-value for the cross section. For example, in a rectangular section I in Eq. 8.3 applies to an axis passing through the centroid parallel to the longer side. Another form of Eq. 8.3 employs *the radius of gyration, k,* of a section. Since $I = Ak^2$ where A is the section area:

$$P_E = \frac{\pi^2 EA}{(L/k)^2} \tag{8.4}$$

Within Eq. 8.4 the *slenderness ratio* L/k becomes the determining factor in validating an elastic buckling analysis. The net section stress for an Euler strut is:

$$\sigma_E = \frac{P_E}{A} = \frac{\pi^2 E}{(L/k)^2} \tag{8.5}$$

The Euler theory will remain valid provided σ_E does not exceed the compressive yield stress σ_c of the strut material

$$\frac{\pi^2 E}{(L/k)^2} < \sigma_c \tag{8.6a}$$

For Eq. 8.6a to apply, it follows that the slenderness ratio must exceed a theoretical minimum value:

$$\frac{L}{k} > \sqrt{\frac{\pi^2 E}{\sigma_c}} \tag{8.6b}$$

The critical slenderness ratio appears as a material constant, e.g. for mild steel with $\sigma_c = 325$ MPa and $E = 207$ GPa:

$$\frac{L}{k} > \sqrt{\frac{\pi^2 \times 207 \times 10^3}{325}} \simeq 80$$

When L/k is less than this minimum value the Euler theory is invalid. Instead, one of the semi-empirical formulae given in § 8.3 should be used.

Table 8.1 Euler Buckling Loads for Various End Fixings

End condition	Fixed-fixed	Pinned-fixed	Fixed-free	Fixed-fixed with axial misalignment	Fixed-pinned with axial misalignment
S T R U T					
L_e	$L/2$	$L/\sqrt{2}$	$2L$	L	$2L$
P_E	$4\pi^2 EI/L^2$	$2\pi^2 EI/L^2$	$\pi^2 EI/4L^2$	$\pi^2 EI/L^2$	$\pi^2 EI/4L^2$

8.1.2 Other End Fixings

The effect that different end fixings have on the buckling load may be deduced without detailed derivation. Here the strut length L in Eq. 8.3 is replaced by an effective length L_e, over which the buckling analysis for strut with pinned ends would apply. Table 8.1 gives the effective buckling length and critical buckling load for each of the five end conditions shown. In addition, when both ends of a strut are fixed imperfectly an effective length of $0.6L$–$0.8L$ is used to account for the fixings' loss of rotational restraint. To find the smallest slenderness ratio for which the Euler theory remains valid in each of the struts in Table 8.1, L in Eq. 8.6b must be replaced by the appropriate *effective length* L_e, as will be shown in the examples that follow. A similar approach applies to the semi-empirical formulae for strut buckling given later.

Example 8.1 A strut 2 m long has a tubular cross section, 50 mm outside diameter and 44 mm inside diameter. Determine the Euler critical load when the ends are pinned. If the tube material has a compressive yield stress of 310 MPa, find the shortest length of this tube for which the Euler theory applies. Repeat the calculation when the same strut has fixed ends. Take $E = 207$ GPa.

The cross-sectional area properties for this strut are:

$$A = \pi(d_o^2 - d_i^2)/4 = \pi(50^2 - 44^2)/4 = 442.97 \text{ mm}^2$$

$$I = \pi(d_o^4 - d_i^4)/64 = \pi(50^4 - 44^4)/64 = 122.7 \times 10^3 \text{ mm}^4$$

Pinned Ends

From Eq. 8.3 the pinned-end buckling load is, in units of N and mm:

$$P_E = \pi^2 EI/L^2 = (\pi^2 \times 207 \times 10^3 \times 122.7 \times 10^3)/2000^2$$

$$= 62.7 \times 10^3 \text{ N} = 62.7 \text{ kN}$$

and from Eq. 8.6b the shortest permissible length of an Euler strut is:

$$L = \sqrt{(\pi^2 E k^2/\sigma_c)} = \sqrt{(\pi^2 EI/A\sigma_c)}$$

$$= \sqrt{[\pi^2 \times (207 \times 10^3) \times (122.7 \times 10^3)/(442.97 \times 310)]}$$

$$= 1351 \text{ mm} = 1.351 \text{ m}$$

This shows that the Euler theory is valid for this 2 m-long strut.

Fixed Ends

Table 8.1 gives the fixed-end buckling load as:

$$P_E = 4\pi^2 EI/L^2 = 4 \times 62.7 = 250.8 \text{ kN}$$

The shortest permissible length of an Euler strut becomes:

$$L = \sqrt{(4^2 EI/\sigma_c)} = 2\sqrt{(\pi^2 EI/A\sigma_c)}$$

$$= 2 \times 1.351 = 2.702 \text{ m}$$

Hence, the Euler theory is invalid for this 2 m-long, fixed-end strut.

Example 8.2 The unsymmetrical channel section given in Fig. 1.13 (see Chapter 1, p. 23) is to be used as a strut with one end pinned and the other end fixed. Calculate the buckling load for the minimum permissible length of an Euler strut given that $E = 207$ GPa and $\sigma_c = 310$ MPa for the strut material.

To find the *least value* of the second moment of area for the section (I_u or I_v), we may use any one of the methods outlined in Chapter 1. Both the analytical and graphical methods (see Example 1.9, p. 23) showed that the lowest *I*-value for this section is $I_v = 4.5 \times 10^4 \text{ mm}^4$. Hence from Table 8.1 and Eq. 8.6, the minimum permissible length of a valid Euler strut is in N and mm units:

$$L = \sqrt{2}L_e = \sqrt{(2\pi^2 EI/A\sigma_c)}$$

$$= \sqrt{[2\pi^2 \times (207 \times 10^3) \times (4.5 \times 10^4)/(900 \times 310)]} = 814.75 \text{ mm}$$

The corresponding buckling load is:

$$P_E = 2\pi^2 EI/L^2 = 2\pi^2 \times (207 \times 10^3) \times (4.5 \times 10^4)/(814.75)^2$$

$$= 277 \times 10^3 \text{ N} = 277 \text{ kN}$$

8.2 Imperfect Euler Struts

Practical struts may differ from perfect Euler struts because of their additional lateral loadings, eccentricities and initial curvature. The buckling behaviour of these struts is associated with the attainment of a critical compressive stress at the section subjected to the greatest bending moment. This stress may be taken from applying a reduction (safety) factor to the compressive yield stress of the strut material or identifying it with the critical Euler buckling stress for a perfect strut.

The two stress limits are equivalent for long struts but a short column can bear a greater stress before becoming unstable. Hence the need for a reduction factor is offset in shorter struts and columns where yielding accompanies buckling, i.e. the critical stress becomes the yield stress.

8.2.1 Eccentric Load to Pinned Ends

Let the line of action of the compressive force P be eccentric by an amount e, to the axis of a pinned-end strut, as shown in Fig. 8.2. The strut's z-axis passes through the centroid of the section.

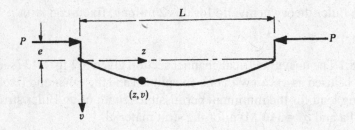

Figure 8.2 Deflection of an eccentrically loaded, pinned-end strut

The bending moment at point (z, v) upon the deflected strut will be modified within the governing flexure Eq. 6.4:

$$EI \frac{d^2v}{dz^2} = - P(e + v)$$

This equation now takes the form:

$$\frac{d^2v}{dz^2} + \alpha^2 v = - \alpha^2 e$$

for which the solution is:

$$v = A\cos\alpha z + B\sin\alpha z - e \tag{8.7}$$

Equation 8.7 is employed to find the safe axial load corresponding to a given allowable stress σ for the strut material. This stress is the sum of a direct compressive stress and a bending stress (see Eq. 5.6a). The latter changes sign across the section giving the net stress as:

$$\sigma = - P/A \pm My/I \tag{8.8}$$

From Eq. 8.8, the net stress is greatest on the compressive side of the section at a position in the length where the bending moment is a maximum:

$$\sigma = -P/A \pm M_{max}y/I \qquad (8.9)$$

where, within the cross section, y is the perpendicular distance from the buckling axis to the furthest compressive edge. Figure 8.2 shows that a maximum deflection v_{max} occurs at the mid-length position $z = L/2$, giving

$$M_{max} = -P(e + v_{max}) \qquad (8.10)$$

The following boundary conditions apply to Eq. 8.7:

(i) $v = 0$ at $z = 0$,

(ii) $dv/dz = 0$ at $z = L/2$

Condition (i) shows that $A = e$, when from condition (ii):

$$-e\alpha \sin(\alpha L/2) + B\alpha \cos(\alpha L/2) = 0, \; \therefore \; B = e \tan(\alpha L/2)$$

Substituting A and B into Eq. 8.7 provides the maximum deflection:

$$v_{max} = e \cos(\alpha L/2) + e \tan(\alpha L/2) \sin(\alpha L/2) - e$$

$$v_{max} = e[\sec(\alpha L/2) - 1] \qquad (8.11)$$

Then, Eqs 8.9 and 8.10 give the maximum moment and compressive stress:

$$M_{max} = -Pe \sec(\alpha L/2) \; \text{(sagging)}$$

$$\sigma = -P/A - (Pey/I) \sec(\alpha L/2) \qquad (8.12)$$

Equation 8.12 does not provide P directly because $\alpha = \sqrt{(P/EI)}$. However, an equation in P follows from Webb's approximation to the secant:

$$\sec\theta \approx \frac{(\pi/2)^2 + (\pi^2/8 - 1)\theta^2}{(\pi/2)^2 - \theta^2} \qquad (8.13a)$$

Putting $\theta = \alpha L/2$ and introducing, from Eq. 8.3, Euler's perfect critical buckling load $P_E = \pi^2 EI/L^2$, Eq. 8.13a gives:

$$\sec\frac{\alpha L}{2} \approx \frac{P_E + P(\pi^2/8 - 1)}{P_E - P} \qquad (8.13b)$$

Substituting Eq. 8.13b into Eq. 8.12 leads to a quadratic in the buckling load for this imperfect strut:

$$aP^2 + bP + c = 0 \qquad (8.14a)$$

The buckling load P identifies with the positive root of this quadratic when σ is taken as a negative value within the coefficients b and c:

$$a = eyA\left(\pi^2/8 - 1\right) - I, \quad b = eyAP_E + IP_E - I\sigma A, \quad c = I\sigma AP_E \qquad (8.14\text{b-d})$$

Conversely, as the following example shows, P is positive within Eq. 8.12 when the maximum compressive stress is to be found. If σ in Eqs 8.14c,d is taken as the critical buckling stress for a perfect strut of similar length and cross section then the substitution $\sigma = P_E/A$ simplifies these two quadratic coefficients:

$$b = eyAP_E, \quad c = IP_E^2$$

Example 8.3 A pinned-end strut has a tubular cross section 75 mm o.d. and 65 mm i.d. Calculate the maximum stress and deflection in a length of 3 m when the strut is to carry an axial compressive force of 50 kN offset by 3 mm from its centroid. Take $E = 200$ GPa.

The cross-sectional area properties of this strut are:

$$A = (\pi/4)(75^2 - 65^2) = 1099.56 \text{ mm}^2$$

$$I = (\pi/64)(75^4 - 65^4) = 67.69 \times 10^4 \text{ mm}^4$$

$$\alpha = \sqrt{\frac{P}{EI}} = \sqrt{\frac{(50 \times 10^3)}{(200 \times 10^3) \times (67.69 \times 10^4)}} = 6.0773 \times 10^{-4} \text{ mm}^{-1}$$

Substituting these, together with $P = 50 \times 10^3$ N, into Eq. 8.12 gives the maximum compressive stress:

$$\sigma = -\frac{(50 \times 10^3)}{1099.56} - \frac{(50 \times 10^3) \times 3 \times 37.5}{(67.69 \times 10^4)} \sec\left[(6.0773 \times 10^{-4}) \times \frac{3000}{2} \times \frac{360}{2\pi}\right]$$

$$= -45.473 - 13.567 = -59.04 \text{ MPa}$$

From Eq. 8.11, the maximum deflection becomes:

$$v_{max} = 3\left[\sec\left((6.0772 \times 10^{-4}) \times \frac{3000}{2} \times \frac{360}{2\pi}\right) - 1\right]$$

$$= 3(1.6327 - 1) = 1.898 \text{ mm}$$

Note that Eq. 8.3 gives the Euler buckling load for an ideal strut of similar dimension as $P_E = 148.46$ kN, at which the critical elastic buckling stress is $P_E/A = -135$ MPa. From this it appears that an eccentric load greater than 50 kN could be supported.

8.2.2 Eccentric Loading at Free End

Figure 8.3 shows a cantilever strut with one end fixed, carrying an eccentric compressive load at its free end. Let the origin of co-ordinates (z, v) lie at the fixed end so that v_L $(z = L)$ becomes the free-end deflection.

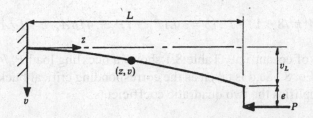

Figure 8.3 Deflection of a cantilever strut eccentrically loaded at its free end

For any point (z, v) on the deflected strut axis, Eq. 6.4 becomes:

$$EI\, d^2v/dz^2 = P\,(e + v_L - v) \quad \text{(hogging)}$$

$$d^2v/dz^2 + \alpha^2 v = \alpha^2 (e + v_L)$$

For which the solution is:

$$v = A \cos \alpha z + B \sin \alpha z + (v_L + e)$$

Applying the boundary conditions gives the constants A and B:

  (i) $dv/dz = 0$ at $z = 0$, $\therefore B = 0$

  (ii) $v = 0$ at $z = 0$, $\therefore A = - (v_L + e)$

$$\therefore v = (v_L + e)(1 - \cos \alpha z)$$

Setting $v = v_L$ at $z = L$,

$$v_L = e\,(\sec \alpha L - 1) \tag{8.15}$$

The bending moment is a maximum at the fixed end:

$$M_{max} = P\,(e + v_L) = Pe \sec \alpha L \tag{8.16}$$

Adopting Webb's approximation to the sec αL (see Eq. 8.13a):

$$\sec \alpha L \approx \frac{P_E + 4P(\pi^2/8 - 1)}{P_E - 4P} \tag{8.17}$$

Substituting Eqs 8.16 and 8.17 into Eq. 8.9, P becomes the positive root of the quadratic for a given maximum compressive (negative) stress:

$$a P^2 + b P + c = 0 \qquad (8.18a)$$

where the coefficients are:

$$a = 4 \left[eyA\left(\pi^2/8 - 1\right) - I \right], \quad b = eyAP_E + IP_E - 4I\sigma A, \quad c = I\sigma AP_E \qquad (8.18b\text{-}d)$$

In the absence of eccentricity Table 8.1 shows a buckling load $P_E/4$ for this strut. Hence, if σ in Eq. 8.18c,d is taken as the corresponding critical buckling stress $\sigma = P_E/4A$ this simplifies the two quadratic coefficients:

$$b = eyAP_E, \quad c = IP_E^2/4$$

Example 8.4 A 3 m-long strut of tubular cross section, 50 mm o.d. and 25 mm i.d., is loaded at its free end with a compressive force eccentric by 75 mm from its centroidal axis. Find the maximum deflection and the safe axial load when, at the fixed end: (i) the maximum compressive stress is limited to 35 MPa and (ii) the net stress on the tensile side is to be zero. Take $E = 210$ GPa.

Part (i) employs Eqs 8.15 and 8.18a where:

$$A = (\pi/4)(50^2 - 25^2) = 1472.62 \text{ mm}^2$$

$$I = (\pi/64)(50^4 - 25^4) = 28.76 \times 10^4 \text{ mm}^4$$

$$P_E = \frac{\pi^2 EI}{L^2} = \frac{\pi^2 \times (210 \times 10^3) \times (28.76 \times 10^4)}{(3 \times 10^3)^2} = 66236.23 \text{ N} = 66.24 \text{ kN}$$

With $\sigma = -35$ MPa, the coefficients in Eqs 8.18b-d become (in N and mm):

$$a = 4[(75 \times 25 \times 1472.62 \times 0.2337) - 28.762 \times 10^4] = 1.431 \times 10^6 \text{ mm}^4$$

$$b = 2.6123 \times 10^{11} \text{ Nmm}^4, \quad c = -9.819 \times 10^{14} \text{ N}^2\text{mm}^4$$

These give the quadratic equation from Eq. 18.18a:

$$P^2 + (182.55 \times 10^3)P - (686.16 \times 10^6) = 0$$

for which the positive root $P = 3.7 \times 10^3$ N becomes the compressive load. Then:

$$\alpha = \sqrt{\frac{P}{EI}} = \sqrt{\frac{(3.7 \times 10^3)}{(210 \times 10^3) \times (28.76 \times 10^4)}} = 2.475 \times 10^{-4} \text{ mm}^{-1}$$

whereupon Eq. 8.15 provides the maximum displacement:

$$v_L = 75[\sec(2.475 \times 10^{-4} \times 3000 \times 180/\pi) - 1] = 26.8 \text{ mm}$$

For part (ii) a zero stress on the tensile side of the fixed end can be achieved by substituting $\sigma = 0$ and $M = M_{max}$ (from Eq. 8.16) into Eq 8.8:

$$\sigma = -\frac{P}{A} + Pe\sec(\alpha L) \times \frac{y}{I} = 0 \tag{i}$$

Substituting $\alpha^2 = P/EI$ into Eq. i leads to the compressive load:

$$P = \frac{EI}{L^2} \left[\sec^{-1} \left(\frac{I}{Aye} \right) \right]^2$$

$$= \frac{(210 \times 10^3) \times (28.76 \times 10^4)}{(3 \times 10^3)^2} \left[\frac{360}{2\pi} \sec^{-1} \left(\frac{28.76 \times 10^4}{1472.62 \times 25 \times 75} \right) \right]^2$$

$$= 6710.7 \left[\frac{360}{2\pi} \times \frac{1}{\cos^{-1} 0.1042} \right]^2 = 3120.8 \text{ N} = 3.121 \text{ kN}$$

This gives

$$\alpha = \sqrt{\frac{P}{EI}} = \sqrt{\frac{(3.121 \times 10^3)}{(210 \times 10^3) \times (28.76 \times 10^4)}} = 2.273 \times 10^{-4} \text{ mm}^{-1}$$

when from Eq. 8.15, the maximum deflection is:

$$v_L = 75\{\sec[(2.273 \times 10^{-4}) \times (3 \times 10^3) \times 180/\pi] - 1\} = 21.6 \text{ mm}$$

8.2.3 Lateral Loading with Pinned Ends

Figure 8.4 shows a pinned-end strut, carrying a lateral load W at mid-span in addition to an axial compressive load P. The net moment to the left side of point (z, v) on the deflected strut shown, consists of two sagging components: Pv and $Wz/2$.

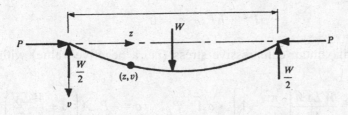

Figure 8.4 Pinned-end strut with axial and lateral loading

Hence, the flexure Eq. 6.4 becomes:

$$EI\,d^2v/dz^2 = -Pv - Wz/2 \quad \text{(sagging)}$$

$$d^2v/dz^2 + \alpha^2 v = -Wz/(2EI)$$

The corresponding solution is:

$$v = A\cos\alpha z + B\sin\alpha z - (Wz)/(2P)$$

Applying the boundary conditions:

(i) $v = 0$ at $z = 0$, $\therefore A = 0$

(ii) $dv/dz = 0$ at $z = L/2$, $\therefore B = W/[2P\alpha\cos(\alpha L/2)]$

This leads to a maximum displacement at $z = L/2$ of:

$$v_{max} = \frac{W}{2P}\left[\frac{1}{\alpha}\tan\frac{\alpha L}{2} - \frac{L}{2}\right] \tag{8.19}$$

and a maximum sagging moment at $z = L/2$ of:

$$M_{max} = -Pv_{max} - \frac{WL}{4} = -\frac{W}{2\alpha}\tan\frac{\alpha L}{2} \tag{8.20}$$

Equation 8.9 gives the maximum compressive stress in the central section:

$$\sigma = -\frac{P}{A} - \frac{Wy}{2I\alpha}\tan\frac{\alpha L}{2} \tag{8.21}$$

from which P may be solved from Webb's tangent approximation:

$$\tan\theta \approx \frac{(\pi/2)^2\theta - (\pi^2/12 - 1)\theta^3}{(\pi/2)^2 - \theta^2}$$

Setting $\theta = \alpha L/2$ and $P_E = \pi^2 EI/L^2$, this becomes:

$$\tan\frac{\alpha L}{2} = \frac{\alpha L}{2} \times \frac{[P_E - P(\pi^2/12 - 1)]}{P_E - P}$$

This substitution converts Eq. 8.21 into a quadratic in P

$$aP^2 + bP + c = 0$$

where, σ is the maximum compressive stress (i.e. a negative value) within its coefficients:

$$a = 1, \quad b = \frac{WLyA}{4I}\left(\frac{\pi^2}{12} - 1\right) + \sigma A - P_E, \quad c = -P_E A\left(\sigma + \frac{WLy}{4I}\right)$$

As with Eqs 8.18a-d, these coefficients are simplified when $\sigma = P_E/A$.

Example 8.5 A pinned-end strut, 2 m long and 30 mm diameter, supports an axial compressive force of 20 kN. If the maximum compressive stress is 200 MPa, calculate the additional load that may be applied laterally at mid-span. What is the maximum deflection and the maximum bending moment? Take $E = 207$ GPa.

$$I = \pi d^4/64 = \pi(30)^4/64 = 39760.8 \text{ mm}^4$$

$$A = \pi d^2/4 = \pi(30)^2/4 = 706.86 \text{ mm}^2$$

$$\alpha = \sqrt{\frac{P}{EI}} = \sqrt{\frac{(20 \times 10^3)}{(207 \times 10^3) \times 39760.8}} = 1.559 \times 10^{-3} \text{ mm}^{-1}$$

Substituting into Eq. 8.21:

$$-200 = -\frac{20 \times 10^3}{706.86} - \frac{W \times 15}{2 \times 39760.8 \times (1.559 \times 10^{-3})} \times \tan\left(1.559 \times \frac{2}{2} \times \frac{360}{2\pi}\right)$$

$$= -28.494 - 10.256\,W$$

From which $W = 16.72$ N. This value is low because, by the Euler theory, the strut is almost at the point of buckling under the critical axial load (in $\underline{N}$ and $\underline{mm}$):

$$P_E = \frac{\pi^2 EI}{L^2} = \frac{\pi^2 \times (207 \times 10^3) \times 39760.8}{(2 \times 10^3)^2} = (20.31 \times 10^3) \text{ N} = 20.31 \text{ kN}$$

From Eq. 8.19, the maximum deflection is:

$$v_{max} = \frac{16.72}{2 \times (20 \times 10^3)}\left[\frac{1}{1.559 \times 10^{-3}} \tan\left(1.559 \times \frac{2}{2} \times \frac{180}{\pi}\right) - \frac{(2 \times 10^3)}{2}\right]$$

$$= 0.418 \times 10^{-3}[641.44 \tan 89.32° - 1000] = 22.17 \text{ mm}$$

and from Eq. 8.20, the maximum central moment is:

$$M_{max} = -\frac{16.72}{2 \times (1.559 \times 10^{-3})} \tan\left(1.559 \times \frac{2}{2} \times \frac{180}{\pi}\right)$$

$$= -454.56 \times 10^3 \text{ Nmm} = 454.56 \text{ Nm (sagging)}$$

8.2.4 Distributed Loading with Pinned Ends

Figure 8.5 shows a pinned-end strut carrying a uniformly distributed lateral loading w/unit length in addition to an axial compressive force P. The bending moment to the left of point (z, v) is composed of two sagging components (Pv) and $(wLz/2)$ and a hogging component $(wz^2/2)$.

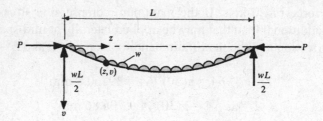

Figure 8.5 Pinned-end strut with laterally distributed and axial loading

Hence, Eq. 6.4 becomes:

$$EI \, d^2 v/dz^2 = - Pv - wLz/2 + wz^2/2$$

$$d^2 v/dz^2 + \alpha^2 v = - (w/2EI)(Lz - z^2)$$

for which the solution is:

$$v = A \cos \alpha z + B \sin \alpha z + (w/2P)(z^2 - Lz - 2EI/P)$$

Applying the boundary conditions:

(i) $v = 0$ at $z = 0$, $\therefore A = EIw/P^2$

(ii) $v = 0$ at $z = L$, $\therefore B = (EIw/P^2) \tan(\alpha L/2)$

The maximum displacement at $z = L/2$ becomes:

$$v_{max} = \frac{EIw}{P^2} \cos\left(\frac{\alpha L}{2}\right) + \frac{EIw}{P^2} \tan\left(\frac{\alpha L}{2}\right) \sin\left(\frac{\alpha L}{2}\right) - \frac{w}{2P}\left(\frac{L^2}{4} + \frac{2EI}{P}\right)$$

$$= \frac{EIw}{P^2}\left[\sec\left(\frac{\alpha L}{2}\right) - 1\right] - \frac{wL^2}{8P} \qquad (8.22a)$$

Correspondingly, the maximum sagging bending moment at the centre is:

$$M_{max} = - Pv_{max} - \frac{wL^2}{4} + \frac{wL^2}{8}$$

and with v_{max} from Eq. 8.22a

$$M_{max} = - \frac{EIw}{P}\left[\sec\left(\frac{\alpha L}{2}\right) - 1\right] \qquad (8.22b)$$

Equation 8.9 gives the maximum compressive stress at the central section:

$$\sigma = - \frac{P}{A} - \frac{Ewy}{P}\left[\sec\left(\frac{\alpha L}{2}\right) - 1\right] \qquad (8.23)$$

Webb's approximation to $\sec(\alpha L/2)$ may again be employed to solve for P. Thus, combining Eq. 8.13b with Eq. 8.23 leads to a quadratic in P:

$$a P^2 + b P + c = 0$$

where the coefficients are:

$$a = 1, \quad b = -(P_E - A\sigma), \quad c = -A(\sigma P_E + Ewy\pi^2/8)$$

In a simplification afforded by setting $\sigma = P_E/A$, the quadratic provides a closed solution to P:

$$P = \sqrt{P_E^2 + AEwy\pi^2/8} = P_E \sqrt{1 + (wy/8)(L/k)^2}$$

If P in Fig. 8.5 is tensile, the flexure equation, Eq. 6.4, becomes:

$$EI\, d^2 v/dz^2 = M = Pv - wLz/2 + wz^2/2$$

The solution gives the maximum central moment:

$$M_{max} = -\frac{EIw}{P}\left[1 - \operatorname{sech}\left(\frac{\alpha L}{2}\right)\right]$$

Consequently, the stress expression becomes

$$\sigma = -\frac{P}{A} - \frac{Ewy}{P}\left[1 - \operatorname{sech}\left(\frac{\alpha L}{2}\right)\right]$$

for which P is found by trial.

Example 8.6 The I-section of a 2.5 m-long strut has flange dimensions 50×20 mm and a web of 60×15 mm. The strut loading is shown in Fig. 8.5, where $P = 125$ kN and $w = 5$ kN/m. Find the values of the maximum bending moment, compressive and tensile stresses. Take $E = 210$ GPa.

$$A = 2(50 \times 20) + (60 \times 15) = 2900 \text{ mm}^2$$

$$I = 50 \times 100^3/12 - 35 \times 60^3/12 = 3.5367 \times 10^6 \text{ mm}^4$$

$$\alpha = \sqrt{\frac{P}{EI}} = \sqrt{\frac{(125 \times 10^3)}{(210 \times 10^3) \times (3.5367 \times 10^6)}} = 4.103 \times 10^{-4} \text{ mm}^{-1}$$

From Eq. 8.22b, the maximum moment is (in units of N and mm):

$$M_{max} = -\frac{(210 \times 10^3) \times (3.5367 \times 10^6) \times 5}{(125 \times 10^3)}\left[\sec \frac{(4.103 \times 10^{-4}) \times (2.5 \times 10^3)}{2} \times \frac{360}{2\pi} - 1\right]$$

$$= -4.3856 \times 10^6 \text{ Nmm} = -4.3856 \text{ kNm}$$

Equation 8.23 provides the stress on the compressive side of the central section, which may also be written as:

$$\sigma = -\frac{P}{A} - \frac{M_{max}y}{I} = -\frac{(125 \times 10^3)}{2900} - \frac{(4.3856 \times 10^6) \times 50}{(3.5367 \times 10^6)}$$

$$= -43.1 - 62.01 = -105.11 \text{ MPa}$$

On the tensile side of the central section, Eq. 8.9 gives:

$$\sigma = -\frac{P}{A} + \frac{M_{max}y}{I}$$

$$= -43.1 + 62.01 = +18.91 \text{ MPa}$$

8.2.5 Encastré Strut with Lateral Loading

Figure 8.6 shows similar lateral strut loadings to the previous case covered in § 8.2.3, except that each end is prevented from rotating within guides as it displaces under P. Consequently, $L_e = L/2$ and a fixed-end moment M_o accompanies the end-force reactions to the lateral loading.

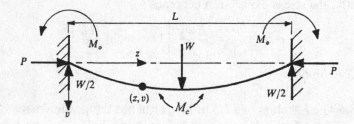

Figure 8.6 Encastré strut with central, concentrated lateral load

At point (z, v) on the deflected strut in Fig. 8.6, its left-side moment expression is modified to:

$$EI\, d^2 v/dz^2 = M = -Pv - Wz/2 + M_o$$

The solution to this equation leads to a central sagging moment M_c of equal magnitude to the fixed-end hogging moment:

$$M_c = -M_o = -(W/2\alpha)[\text{cosec}(\alpha L_e/2) - \cot(\alpha L_e/2)]$$

It follows from Eq. 8.9 that the net tensile and compressive stress exist at the surface of both the central and end sections but they are in opposition, i.e. maximum tension at the bottom central section gives maximum compression at the bottom end section and vice versa. Moreover, unstressed sections exist symmetrically in the length at contraflexure positions between each end and the centre where the moment is zero. As no contraflexure points apply to the simply supported strut the rotational restraint provided by guided ends will serve to offset buckling

Figure 8.7 shows similar lateral strut loadings to the case covered in § 8.2.4, except that the ends are restrained from rotating as they displace axially. The flexure equation is applied left of point (z, v), in which it is seen that there are two hogging moments (positive) and two sagging moments (negative):

$$EI\, d^2 v/dz^2 = M = -Pv - wLz/2 + wz^2/2 + M_o$$

Here the solution shows different sagging and hogging bending moments at their respective central and end positions:

$$M_c = -(wL/2\alpha)[\text{cosec}(\alpha L_e/2) - 2/\alpha L_e] \quad \text{(sagging)}$$

$$M_o = (wL/2\alpha)[\cot(\alpha L_e/2) - 2/\alpha L_e] \quad \text{(hogging)}$$

At each position the maximum tensile and compressive surface stresses are provided by taking the appropriate sign in Eq. 8.9 with the conversion of each bending moment into stress.

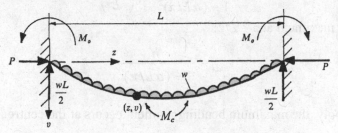

Figure 8.7 Encastré strut with uniformly distributed lateral loading

8.2.6 Initial Curvature within Pinned Ends

Figure 8.8 shows a pinned-end strut with an initial curvature in which h refers to the value of the maximum central deviation from the horizontal axis. Elsewhere, for any point (z, v_o) in the length the central deviation for the unloaded strut is described by the sinusoidal function $v_o = h \sin(\pi z/L)$.

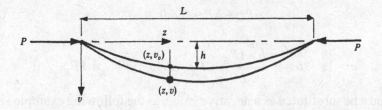

Figure 8.8 Deflection of a pinned-end strut with initial curvature

With a total lateral deflection v, arising from a compressive axial force P, the flexure Eq. 6.4 becomes:

$$EI\frac{d^2(v - v_o)}{dz^2} = -Pv$$

$$EI\frac{d^2v}{dz^2} + Pv = EI\frac{d^2v_o}{dz^2} = EIh\frac{d^2}{dz^2}\sin\left(\frac{\pi z}{L}\right)$$

$$\frac{d^2v}{dz^2} + \alpha^2 v = -\frac{h\pi^2}{L^2}\sin\left(\frac{\pi z}{L}\right)$$

where $\alpha^2 = P/EI$. The solution is:

$$v = A\cos\alpha z + B\sin\alpha z + \frac{h}{1 - (\alpha L/\pi)^2}\sin\left(\frac{\pi z}{L}\right)$$

Applying the boundary conditions: $v = 0$ at $z = 0$ and $z = L$, leads to:

$$v = \frac{h}{1 - (\alpha L/\pi)^2}\sin\left(\frac{\pi z}{L}\right)$$

which has its maximum at $z = L/2$:

$$v_{max} = \frac{h}{1 - (\alpha L/\pi)^2} \tag{8.24}$$

Correspondingly, the maximum bending moment occurs at the centre:

$$M_{max} = -Pv_{max} = -\frac{Ph}{1 - (\alpha L/\pi)^2}$$

Since $(\alpha L/\pi)^2 = P/P_E$, the maximum compressive stress is, from Eq. 8.9:

$$\sigma = -\frac{P}{A} - \frac{Phy}{I[1 - (\alpha L/\pi)^2]} = -\frac{P}{A} - \frac{Phy}{I(1 - P/P_E)}$$

This equation leads to a quadratic in P:

$$aP^2 + bP + c = 0 \tag{8.25}$$

where:

$$a = 1, \quad b = -\frac{hyAP_E}{I} + P_E - \sigma A, \quad c = -\sigma A P_E$$

in which σ must be substituted as a negative value, as the following example shows. Setting $\sigma = P_E/A$ gives: $b = -hyAP_E/I$ and $c = -P_E^2$ where y is the distance within the cross section from the neutral axis to the furthest compressive edge.

Example 8.7 A 2 m-long pinned-end strut, with a 25 mm-diameter solid circular section, has a 5 mm maximum amplitude of initial curvature (see Fig. 8.8). Find the compressive end load P that would just produce yielding at its mid-span given that the compressive yield stress is 300 MPa. What is the additional central deflection under this load? Take $E = 210$ GPa.

$$A = \pi d^2/4 = (\pi/4)25^2 = 490.87 \text{ mm}^2$$

$$I = \pi d^4/64 = (\pi/64)25^4 = 19.17 \times 10^3 \text{ mm}^4$$

$$P_E = \pi^2 EI/L^2 = \pi^2 \times 210000 \times 19.17 \times 10^3/2000^2 = 9935.1 \text{ N}$$

Working in units of N and mm, $\sigma = -300$ N/mm^2, $h = 5$ mm and $y = 12.5$ mm, the coefficients b and c in Eq. 8.25 become:

$$b = -\left[\frac{5 \times 12.5 \times 490.87 \times 9935.1}{19.17 \times 10^3} + 9935.1 - (-300 \times 490.87) \right] = -173.2 \times 10^3 \text{ N}$$

$$c = -(-300 \times 490.87 \times 9935.1) = 1.463 \times 10^9 \text{ N}^2$$

Hence the quadratic Eq. 8.25 in P becomes:

$$P^2 - (173.2 \times 10^3)P + (1.463 \times 10^9) = 0$$

from which $P = 8.95 \times 10^3$ N = 8.95 kN. Equation 8.24, provides the maximum deflection:

$$v_{max} = \frac{h}{1 - (\alpha L/\pi)^2} = \frac{h}{1 - P/P_E} = \frac{5}{1 - 8.95/9.935} = 51.1 \text{ mm}$$

8.2.6 Initial Curvature, Cantilever Strut

Figure 8.9 shows a cantilever strut with an initial curvature described by the cosine expression: $v_o = h_o[1 - \cos(\pi z/2L)]$, for which h_o is the maximum end deviation from the straight, horizontal z-axis.

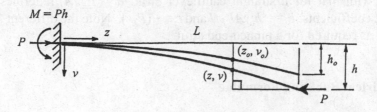

Figure 8.9 Deflection of an initially curved, cantilever strut

For any point (z, v_o) in the length let an additional deflection $v - v_o$ arise from the compressive end force P as shown. Correspondingly, the end deviation has increased to h. Applying the flexure Eq. 6.4 to the right of point (z, v):

$$EI\frac{d^2}{dz^2}(v - v_o) = P(h - v)$$

$$EI\frac{d^2v}{dz^2} + Pv = Ph + EI\frac{d^2v_o}{dz^2} = Ph + EIh_o\frac{d^2}{dz^2}\left[1 - \cos\left(\frac{\pi z}{2L}\right)\right]$$

$$\frac{d^2v}{dz^2} + \alpha^2 v = \alpha^2 h - h_o\left(\frac{\pi}{2L}\right)^2 \cos\frac{\pi z}{2L}$$

where $\alpha^2 = P/EI$. The solution shows that the initial deviation is increased to a total deviation from the horizontal axis:

$$v = h_o\left(1 - \cos\frac{\pi z}{2L}\right)\left(1 + \frac{P}{P_E - P}\right)$$

in which $P_E = \pi^2 EI/4L^2$ is the Euler buckling load for a straight cantilever strut (see Table 8.1, p. 289). When $z = L$, the maximum displacement is: $h = h_o/(1 - P/P_E)$, from which the maximum bending moment at the fixed end is shown as Ph. Hence, the maximum compressive stress at this end becomes:

$$\sigma = -\frac{P}{A} - \frac{Ph_o y}{I(1 - P/P_E)}$$

where y is the distance within the cross section from the neutral axis to the furthest compressive edge. This equation leads to a quadratic in P similar to Eq. 8.25:

$$aP^2 + bP + c = 0$$

where:

$$a = 1, \quad b = -\frac{h_o y A P_E}{I} + P_E - \sigma A, \quad c = -\sigma A P_E$$

in which σ must be substituted as a negative value. Identifying the stress at buckling with that for a straight cantilever strut: $\sigma = P_E/A$, redefines the two quadratic coefficients: $b = -h_o y A P_E/I$ and $c = -(P_E^2)$. Note the different P_E within Eq. 8.25, as required for a pinned-end strut.

8.3 Empirical Buckling Formulae

As the slenderness ratio (l/k) decreases for shorter struts so the Euler theory diverges from the observed behaviour. Semi-empirical solutions to the buckling load for short- to medium-length struts are used to account for the combined effects of direct compression and bending as the net stress in the cross-section approaches the yield stress. Figure 8.10 shows three empirical forms that apply to a region below Euler's limiting slenderness ratio (see Eq. 8.6b). None allows the net stress to exceed the compressive yield stress within a region of decreasing L/k below this limiting ratio. The Euler theory is invalid for this region (broken line) since it does not impose a similar limit on the compressive stress, i.e. the Euler stress exceeds the limiting stress of the strut material.

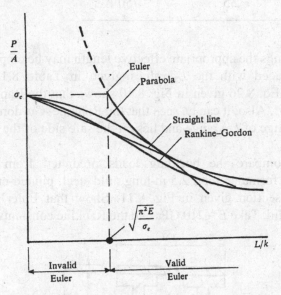

Figure 8.10 Comparison between Euler's and empirical predictions

8.3.1 Rankine–Gordon

For shorter, pinned-end struts the Euler's ideal buckling load P_E is combined with the compressive yield load P_c for a short column as a sum of their reciprocals:

$$\frac{1}{P_R} = \frac{1}{P_E} + \frac{1}{P_c}$$

from which the Rankine buckling load P_R is:

$$P_R = \frac{P_c}{1 + P_c/P_E}$$

Substituting $P_c = A\sigma_c$ and P_E from Eq. 8.3:

$$P_R = \frac{A\sigma_c}{1 + a(L/k)^2} \tag{8.26}$$

where σ_c is the compressive yield stress and a replaces $\sigma_c/\pi^2 E$. The empirical constant a is found from fitting Eq. 8.26 to test results. Table 8.2 gives typical values for σ_c and a for the pinned-end struts in four materials.

Table 8.2 Rankine–Gordon constants in Eq. 8.26

Material	σ_c, MPa	$1/a$
Mild steel	325	7500
Wrought iron	247	9000
Cast iron	557	1600
Timber	35	750

For other end fixings the appropriate effective length may be employed. Thus L in Eq. 8.26 is replaced with the L_e values given in Table 8.1. The graphical representation of Eq. 8.26 given in Fig. 8.10 shows that the Rankine stress P_R/A does not exceed σ_c. Also, it can be seen that the Rankine–Gordon formula applies over the whole range of L/k range and lies on the safe side of the Euler curve.

Example 8.8 Compare the buckling loads predicted from the Euler and Rankine–Gordon formulae for a 2.5 m-long mild steel, pinned-end strut with the fabricated cross section given in Fig. 8.11. Show that Euler's buckling load prediction is invalid. Take $E = 210$ GPa and the Rankine constants from Table 8.2.

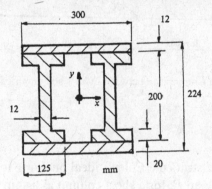

Figure 8.11 Strut section

$$A = 2[(300 \times 12) + 2(125 \times 20) + (160 \times 12)] = 21.04 \times 10^3 \text{ mm}^2$$

Buckling occurs about the axis x or y with the lesser I-value:

$$I_x = \frac{300 \times 224^3}{12} - \frac{300 \times 200^3}{12} + 2\left(\frac{125 \times 200^3}{12} - \frac{113 \times 160^3}{12}\right) = 170.5 \times 10^6 \text{ mm}^4$$

$$I_y = 2\left(\frac{20 \times 300^3}{12} - \frac{20 \times 50^3}{12}\right) + \left(\frac{160 \times 187^3}{12} - \frac{160 \times 163^3}{12}\right) = 173.06 \times 10^6 \text{ mm}^4$$

$$\therefore \ k = \sqrt{\frac{I_x}{A}} = \sqrt{\frac{170.5 \times 10^6}{21.04 \times 10^3}} = 90.02 \text{ mm}$$

The slenderness ratio is

$$\frac{L}{k} = \frac{2500}{90.02} = 27.77$$

The buckling load prediction from Euler's Eq. 8.4 is:

$$P_E = \frac{\pi^2 EA}{(L/k)^2} = \frac{\pi^2 \times (207 \times 10^3) \times (21.04 \times 10^3)}{27.77^2} = 56.21 \times 10^6 \text{ N} = 56.21 \text{ MN}$$

The Rankine–Gordon Eq. 8.26 gives a much reduced buckling load:

$$P_R = \frac{(21.04 \times 10^3) \times 325}{1 + (27.77)^2 / 7500} = 6.2 \times 10^6 \text{ N} = 6.2 \text{ MN}$$

Only the Rankine load is valid because Eq. 8.6b shows that the Euler theory ceases to apply for lengths below:

$$L = k \sqrt{\frac{\pi^2 E}{\sigma_c}} = 90.02 \sqrt{\frac{\pi^2 \times (207 \times 10^3)}{325}} = 7137 \text{ mm} = 7.137 \text{ m}$$

Example 8.9 The equal-angle section in Fig. 8.12 is used for a 2 m-long mild steel strut, having one end fixed and the other end pinned. Calculate the safe axial load from the Rankine–Gordon theory using a safety factor of 4 and the constants given in Table 8.2.

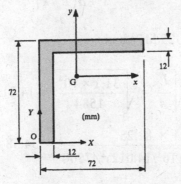

Figure 8.12 Equal-angle strut section

Buckling occurs about the axis with the least *principal second moment of area* from Eqs 1.14a,b (see p. 16). That is, either I_u or I_v applies, inclined at 45° and 135° to the centroidal axes x and y shown. The section area is:

$$A = (72 \times 12) + (60 \times 12) = 1584 \text{ mm}^2$$

The centroid position $\bar{y}$ is found from: $\sum(A_i y_i) = A\,\bar{y}$

$$(72 \times 12 \times 6) + (60 \times 12 \times 42) = [(72 \times 12) + (60 \times 12)]\,\bar{y}$$

This gives $\bar{y} = 22.4$ mm $(= \bar{x})$. To calculate I_x and I_y in Eqs 1.14a,b the convenient form $I = \sum(bd^3/3)$ applies to rectangles whose bases lie on the x-axis:

$$I_x = \frac{72 \times 22.4^3}{3} - \frac{60 \times 10.4^3}{3} + \frac{12 \times 49.6^3}{3} = 73.5 \times 10^4 \text{ mm}^4 \ (= I_y)$$

The product moment of area $I_{xy} = \sum(A_i x_i y_i)$ is:

$$I_{xy} = [(72 \times 12)(+13.6)(+16.4)] + [(60 \times 12)(-19.6)(-16.4)] = 42.4 \times 10^4 \text{ mm}^4$$

Substituting into Eqs 1.14a,b the principal moments are

$$I_{u,v} = \frac{1}{2}\left[(I_x + I_y) \pm \sqrt{(I_x - I_y)^2 + 4I_{xy}^2}\right]$$

$$= \frac{1}{2}\left[(73.5 + 73.5) \pm \sqrt{(73.5 - 73.5)^2 + (4 \times 42.4^2)}\right] \times 10^4$$

giving

$$I_u = 31.1 \times 10^4 \text{ mm}^4, \quad I_v = 115.9 \times 10^4 \text{ mm}^4$$

with respective orientations $\theta = 45°$ and $135°$ to x (from Eq. 1.13).

For a strut with these end fixings Table 8.1 provides an equivalent length $L_e = L/\sqrt{2} = 2\sqrt{2} = 1.414$ m. The Rankine–Gordon buckling load follows from Eq. 8.26:

$$P_R = \frac{A\sigma_w}{1 + a(L_e/k)^2}$$

in which $\sigma_w = 325/4 = 81.25$ MPa and

$$k = \sqrt{\frac{I_u}{A}} = \sqrt{\frac{31.1 \times 10^4}{1584}} = 14.012 \text{ mm}'$$

$$\therefore \quad P_R = \frac{1584 \times 81.25}{1 + (1.414 \times 10^3/14.012)^2/7500} = 54.58 \times 10^3 \text{ N} = 54.58 \text{ kN}$$

8.3.2 Straight-Line Formula

A straight-line equation may be used to provide the buckling load for a particular L/k range. The straight line shown in Fig. 8.10 takes the form:

$$P = A\sigma_c\left[1 - n\left(\frac{L}{k}\right)\right] \tag{8.27}$$

where n and σ_c are constants determined from two points through which the line passes, corresponding to: (i) the compressive yield stress σ_c ($L/k = 0$) and (ii) the stress at the point of tangency between the straight line and the Euler curve for a given L/k value. The latter restricts the L/k range over which Eq. 8.27 applies. When conditions (i) and (ii) above are applied to a pinned-end strut in this manner the constants are those given in Table 8.3.

Table 8.3 Constants in the straight-line Eq. 8.27

Material	n	σ_c, MPa	L/k
Mild steel	0.004	258	≤ 150
Cast iron	0.004	234	≤ 100
Al-alloy	0.009	302	≤ 85
Oak	0.005	37	≤ 65

8.3.3 Johnson's Parabola

Figure 8.10 shows further how the following parabolic equation may be applied to struts of smaller slenderness ratios lying within the invalid Euler region:

$$P = A\sigma_c\left[1 - b\left(\frac{L}{k}\right)^2\right] \qquad (8.28)$$

where b and σ_c are constants. For mild steel $\sigma_c = 275$ MPa with typical values for b lying between 23×10^{-6} to 30×10^{-6} for $L/k \leq 150$. Equation 8.28 may take σ_c to be less than the compressive yield stress of the strut material to offset the need for a safety factor. One particular form of Eq. 8.28 is:

$$P = A\sigma_c\left[1 - \frac{1}{2C_c^2}\left(\frac{L}{k}\right)^2\right] \qquad (8.29a)$$

where C_c is the slenderness ratio value that applies to the intersection between Eq. 8.29a and the Euler curve at $\sigma_E = P_E/A = \sigma_c/2$. Identifying this buckling stress condition with an Euler stress ordinate from Eq. 8.5 reveals that C_c is a material constant for a pinned-end strut:

$$C_c = \sqrt{\frac{2\pi^2 E}{\sigma_c}} \qquad (8.29b)$$

Alternatively, b in Eq. 8.28 may be found from the condition that a point of tangency between the parabola and the Euler curve applies to a given L/k (typically 120). The following examples show how to determine the constants in Eqs 8.27 and 8.28 at each cut-off condition.

Example 8.10 Determine the constants in Eqs 8.27 and 8.28 for a pinned-end strut given that each is to intersect with an Euler buckling condition: (a) at a slenderness ratio $L/k = 120$ and (b) for a stress $\sigma_c/2$, as shown in Figs 8.13a,b. Take $E = 210$ GPa and $\sigma_c = 320$ MPa.

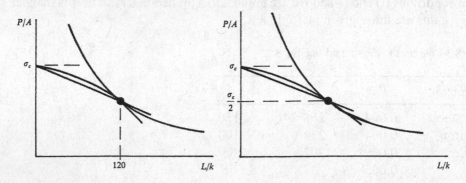

(a)   (b)

Figure 8.13 Euler and Johnson curves

(a) Referring to Fig. 8.13a for the common intersection stress P/A between Eq. 8.5 and the straight-line Eq. 8.27 gives:

$$\sigma_c[1 - n\,(L/k)\,] = \pi^2 E/(L/k)^2$$

from which

$$n = (L/k)^{-1}[1 - \pi^2 E/\sigma_c(L/k)^2]$$

$$n = (1/120)[1 - (\pi^2 \times 210 \times 10^3)/(320 \times 120^2)] = 4.59 \times 10^{-3}$$

Also shown in Fig. 8.13a is the intersection between Eq. 8.5 and Johnson's parabola in Eq. 8.28. The common stress ordinate P/A gives:

$$\sigma_c[1 - b(L/k)^2] = \pi^2 E/(L/k)^2$$

when

$$b = (L/k)^{-2}[1 - \pi^2 E/\sigma_c(L/k)^2]$$

$$b = (1/120)^2[1 - (\pi^2 \times 210 \times 10^3)/(320 \times 120^2)] = 38.21 \times 10^{-6}$$

(b) Figure 8.13b refers the intersection stress $\sigma_c/2$ to Euler's equation, Eq. 8.5:

$$\sigma_c/2 = \pi^2 E/(L/k)^2$$

from which the slenderness ratio is:

$$\frac{L}{k} = \sqrt{\frac{2\pi^2 E}{\sigma_c}} = \sqrt{\frac{2 \times \pi^2 \times 210 \times 10^3}{320}} = 113.82$$

Substituting $P/A = 160$ MPa, $\sigma_c = 320$ MPa and $L/k = 113.82$ into the straight-line Eq. 8.27:

$$160 = 320(1 - 113.82\, n), \qquad \therefore n = 0.0044$$

Similar substitutions in the parabola Eq. 8.28 lead to:

$$160 = 320[1 - (113.82)^2 b], \quad \therefore b = 38.6 \times 10^{-6}$$

Alternatively, the parabola may be expressed directly from Eqs 8.29a,b with:

$$C_c = L/k = 113.82$$

Example 8.11 Determine the constants in the straight line and parabolic strut formulae when their cut-off slenderness ratios become tangents with the Euler curve for $L/k = 120$, as shown in Fig. 8.14. Take $E = 207$ GPa.

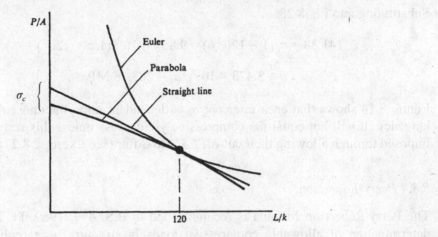

Figure 8.14 Strut buckling curves

The common stress ordinate is found from the Euler equation, Eq. 8.5:

$$\sigma = \frac{\pi^2 E}{(L/k)^2} = \frac{\pi^2 \times (207 \times 10^3)}{120^2} = 141.88 \text{ MPa}$$

The tangents to the curves given by each formula will have the same gradients at $L/k = 120$ (see Fig. 8.14). In general, the gradients from Eqs 8.5, 8.27 and 8.28 are given in Table 8.4.

Table 8.4 Gradients $\dfrac{\mathrm{d}\sigma}{\mathrm{d}(L/k)}$ to strut buckling curves in Fig. 8.14

(a) Euler	(b) Straight-line	(c) Johnson
$\dfrac{-2\pi^2 E}{(L/k)^3}$	$-n\sigma_c$	$-2\sigma_c b(L/k)$

Equating (a) and (b) in Table 8.4:

$$- n\sigma_c = - (2 \times \pi^2 \times 207 \times 10^3)/120^3 = - 2.365$$

Substituting into Eq. 8.27:

$$141.88 = \sigma_c(1 - 120n) = 2.365(1/n - 120)$$

$$\therefore n = 5.556 \times 10^{-3}, \ \sigma_c = 425.68 \text{ MPa}$$

Equating (a) and (c) in Table 8.4:

$$-2\sigma_c b \times 120 = - 2.365$$

$$\therefore \sigma_c = (9.854 \times 10^{-3})/b$$

Substituting into Eq. 8.28:

$$141.88 = \sigma_c(1 - 120^2 b) = 9.854 \times 10^{-3}(1/b - 120^2)$$

$$\therefore b = 3.473 \times 10^{-5}, \ \sigma_c = 283.78 \text{ MPa}$$

Figure 8.14 shows that each intercept σ_c is dependent upon the chosen point of tangency. It will not equal the compressive yield stress unless this restriction is imposed through allowing their cut-off L/k's to differ (see Exercise 8.27).

8.3.4 Perry-Robertson

The Perry-Robertson formula is recommended in B.S. 449, 1969, Pt. 2 for the determination of allowable compressive loads in structural steel columns. In common with the Rankine–Gordon method, the Perry–Robertson formula identifies the transition between pinned-end strut buckling at high L/k to short column compressive yielding at low L/k. The said formula applies to slenderness ratios in the range: $80 \le L/k \le 350$.

$$P = \frac{A}{2K}\left\{[\sigma_c + (\mu + 1)\sigma_E] - \sqrt{[\sigma_c + (\mu + 1)\sigma_E]^2 - 4\sigma_E\sigma_c]}\right\} \quad (8.30a)$$

where σ_c is the minimum yield strength, and K is a loading safety factor, taken as 1.7 - 2.0. The Standard accounts for two imperfections identified earlier. Thus, an initial sinusoidal curvature and eccentricity of loading appear indirectly in a deformation factor:

$$\mu = \alpha(L/k) \quad (8.30b)$$

where α lies between 0.001 and 0.003 based upon Robertson's experiments. An alternative measure of μ is given by the *Dutheil factor*:

$$\mu = \frac{0.3}{\pi^2}\left(\frac{\sigma_c}{E}\right)\left(\frac{L}{k}\right)^2 \quad (8.30c)$$

Equation 8.30c was found to provide a match with the lowest buckling loads observed in pinned-end struts sensitive to each imperfection.

8.3.5 Fidler

Fidler employed a simplified form of Eq. 8.30a for the design of columns used in bridge construction. The particular combination of the Euler buckling stress σ_E and the compressive yield stress σ_c was given as:

$$P = \frac{A}{q}\left[(\sigma_c + \sigma_E) - \sqrt{(\sigma_c + \sigma_E)^2 - 2q\sigma_E\sigma_c}\right] \tag{8.31}$$

where A is the section area and q is a load factor with an average value of 1.2. The strut's end-condition is accounted for within σ_E appearing in Eqs 8.30a–c and 8.31. This allows the effective length L_e for an Euler strut, giving σ_E as:

$$\sigma_E = \frac{\pi^2 E}{(L_e/k)^2} \tag{8.32}$$

where L_e may be read from Table 8.1. As they are written, Eqs 8.30a and 8.31 provide safe, axial compressive loading for $K > 1$ and $q > 1$. Their buckling loads apply when each load factor is taken to be unity (see Example 8.12).

8.3.6 Engesser

Engesser employed a simple modification to the Euler theory to account for inelastic buckling. He replaced the elastic modulus with the tangent modulus E_t, thereby modifying Eq. 8.32 to:

$$\sigma = \frac{\pi^2 E_t}{(cL/k)^2} \tag{8.33}$$

where c accounts for the effect of end fixing upon the plastic buckling stress. Here, it is known that the effect of end constraint upon struts in the range $50 \le L/k \le 100$ is much less severe than the corresponding elastic constraint given in Table 8.1. The tangent modulus E_t is the gradient of the tangent to a uniaxial compressive stress–strain curve (as shown in Fig. 8.15). Hence $E_t = d\sigma/d\varepsilon$ from which the buckling stress in Eq. 8.33 must satisfy the following condition:

$$\sigma = \left(\frac{\pi}{c}\right)^2\left(\frac{L}{k}\right)^{-2}\left(\frac{d\sigma}{d\varepsilon}\right) \tag{8.34}$$

Equation 8.34 may be satisfied graphically or by employing a suitable empirical representation to the σ versus ε curve. The simplest of these is the Hollomon law:

$$\sigma = B\varepsilon^n \tag{8.35a}$$

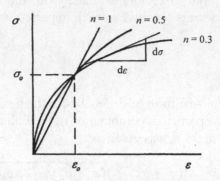

Figure 8.15 Tangent modulus $E_t = d\sigma/d\varepsilon$ with hardening exponents n

where material constants are: B the strength coefficient and $n < 1$ the hardening exponent. From Eq. 8.35a:

$$\frac{d\sigma}{d\varepsilon} = nB\varepsilon^{n-1} = nB\left(\frac{\sigma}{B}\right)^{\frac{(n-1)}{n}} \tag{8.35b}$$

Substituting Eq. 8.35b into Eq. 8.34 enables σ to be found from:

$$\sigma = Bn^n\left(\frac{\pi}{cL/k}\right)^{2n} \tag{8.36a}$$

Applying Considére's tensile instability condition to Eq. 8.35a:

$$d\sigma/d\varepsilon = \sigma$$

shows that, at the inception of necking, the ultimate stress and strain are $\sigma_u = Bn^n$ and $\varepsilon_u = n$. Hence Eq. 8.36a gives the plastic buckling stress as a fraction of its ultimate tensile strength:

$$\frac{\sigma}{\sigma_u} = \left(\frac{\pi}{cL/k}\right)^{2n} \tag{8.36b}$$

Equation 8.36b allows for the determination of constraint factor c experimentally given the material properties n and σ_u for a strut whose ends are not pinned.

Example 8.12 A pinned-end steel strut 1.5 m long has the cross section shown in Fig. 8.16. Find Engesser's prediction to the plastic buckling load given that the Hollomon law $\sigma/\sigma_c = 5\varepsilon^{1/3}$ describes the compressive stress–strain curve in the region beyond the yield stress $\sigma_c = 300$ MPa. Compare this with the elastic buckling loads predicted from the Perry-Robertson and Fidler formulae. Take $E = 207$ GPa.

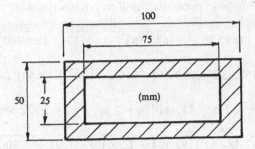

Figure 8.16 Strut's rectangular tube section

The geometric properties for the pinned strut are:

$$A = (100 \times 50) - (75 \times 25) = 3125 \text{ mm}^2$$

$$I = 100 \times 50^3/12 - 75 \times 25^3/12 = 11.328 \times 10^6 \text{ mm}^4$$

$$k = \sqrt{(I/A)} = \sqrt{(11.328 \times 10^6/3125)} = 60.21 \text{ mm}$$

$$L/k = 1500/60.21 = 24.91$$

Hollomon and Engesser's Eqs 8.35a and 8.36a are applied for $c = 1$ and $B = 5\sigma_c$:

$$\sigma = 5\sigma_c n^n \left(\frac{\pi}{L/k} \right)^{2n} = 5 \times 300 \left(\frac{1}{3} \right)^{\frac{1}{3}} \left(\frac{\pi}{24.91} \right)^{\frac{2}{3}} = 261.56 \text{ MPa}$$

from which Engesser's buckling load is predicted as:

$$P = \sigma A = (261.56 \times 3125) = 817.2 \text{ kN}$$

In applying the Perry-Robertson equation, Eq. 8.30a, firstly Euler's reference stress is found from Eq. 8.32:

$$\sigma_E = \pi^2 E/(L/k)^2 = \pi^2 \times 207000/(24.91)^2 = 3292.48 \text{ MPa}$$

Then the Dutheil deformation factor is found from Eq. 8.30c:

$$\mu = \frac{0.3}{\pi^2} \left(\frac{\sigma_c}{E} \right) \left(\frac{L}{k} \right)^2 = \frac{0.3}{\pi^2} \left(\frac{300}{207 \times 10^3} \right) \times 24.91^2 = 0.0273$$

Within Eq. 8.30a, $K = 1.0$ and

$$[\sigma_c + (\mu + 1)\sigma_E] = [300 + (1.0273 \times 3292.48)] = 3682.36$$

Hence, the buckling load predicted by Perry-Robertson is:

$$P = \frac{3125}{(2 \times 1.0)} \left[3682.36 - \sqrt{(3682.36)^2 - (4 \times 3292.48 \times 300)} \right]$$

$$= 1562.5(3682.36 - 3099.81) = 910.24 \times 10^3 \text{ N} = 910.24 \text{ kN}$$

In Fidler's equation, Eq. 8.31, with $q = 1$, $\sigma_c + \sigma_E = 3592.48$, the buckling load is:

$$P = 3125 \left[3592.48 - \sqrt{(3592.48)^2 - (2 \times 1.0 \times 3292.48 \times 300)} \right]$$

$$= 3125(3592.48 - 3306.12) = 894.88 \times 10^3 \text{ N} = 894.88 \text{ kN}$$

Engesser's buckling load prediction is the most conservative. It appears from the Perry-Robertson and Fidler predictions that, unless load factors $K > 1$ and $q > 1$ are introduced into Eqs 8.30a and 8.31, the elastic buckling range is extended unrealistically. The effect is similar to applying Euler's theory to struts that are insufficiently long.

8.4 Inelastic Buckling of Struts

8.4.1 Tangent Modulus

We have seen how Engesser's modification to the Euler's buckling theory accounts for inelastic buckling simply by replacing the elastic modulus with a plastic tangent modulus. With this reduction in stiffness, the section stress is given by Eq. 8.33:

$$\sigma = \frac{P}{A} = \frac{\pi^2 E_t}{\left(L_e / k \right)^2} \qquad (8.37a)$$

where L_e is an effective length that accounts for the particular rotational restraint exerted by the end fixing. Within the range of slenderness ratios $50 \leq L_e / k \leq 100$, the effect of end constraint on the plastic buckling load P is less than that of an elastic strut with similar end fixings. For example, we should not assume that the plastic buckling load of a pinned-end strut, lying in the same range, will be doubled by fixing its ends. However, for $L_e / k \leq 50$ an elastic constraint is assumed and an equivalent elastic length $L_e = cL$ may be employed, particularly when a safety factor is used. This gives: $L_e = L$ for pinned ends; $L_e = L/2$ for fixed ends; $L_e = L/\sqrt{2}$ for pinned-encastré end fixings; and $L_e = 2L$ for fixed-free ends fixings. The tangent modulus $E_t = d\sigma/d\varepsilon$ in Eq. 8.37a is the gradient of the tangent to the uniaxial compressive stress–strain curve within its plastic range (see Fig. 8.15). It follows from this definition of E_t that the buckling stress σ in Eq. 8.37a must satisfy the following condition

$$\sigma = \frac{\pi^2 (d\sigma/d\varepsilon)}{\left(L_e / k \right)^2} \qquad (8.37b)$$

Since σ appears on both sides of Eq. 8.37b it may be solved by trial. That is, a plastic buckling stress σ, is selected to be greater than the yield stress σ_o. Assuming that a stress–strain curve is available, $E_t = d\sigma/d\varepsilon$ is found and Eq. 8.37b is solved for L_e/k. The solution is correct only when L_e/k matches that for the given strut. The procedure is aided when all such solutions to Eq. 8.37b appear as points on a plot of σ versus L_e/k. Alternatively, a suitable empirical representation to the σ versus ε curve may be employed, the simplest being the Hollomon equation, Eq. 8.35, which is represented graphically in Fig. 8.15 for various hardening exponents n. Here, the stress–strain curves are described more conveniently in an alternative form of the Hollomon law, Eq. 8.35a:

$$\frac{\sigma}{\sigma_o} = \left(\frac{\varepsilon}{\varepsilon_o} \right)^n \tag{8.38a}$$

The gradient of its tangent is, from Eq. 8.38a:

$$E_t = \frac{d\sigma}{d\varepsilon} = n \left(\frac{\sigma_o}{\varepsilon_o} \right) \left(\frac{\sigma}{\sigma_o} \right) \left(\frac{\sigma_o}{\sigma} \right)^{1/n} \tag{8.38b}$$

where σ_o, ε_o and n are the material constants shown. Substituting Eq. 8.38b into Eq. 8.37b results in an equation that is soluble in σ:

$$\frac{\sigma}{\sigma_o} = \left(\frac{n}{\varepsilon_o} \right)^n \left(\frac{\pi}{L_e/k} \right)^{2n} \tag{8.39}$$

Equation 8.39 defines the Engesser's curve 1 in Fig. 8.17, this curve being valid for a net section stress $\sigma > \sigma_o$.

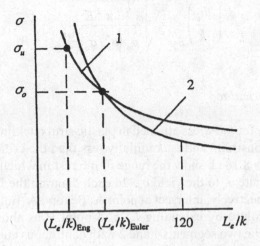

Figure 8.17 Buckling curves (Key: 1 – Engesser; 2 – Euler)

In contrast, Euler's elastic curve 2 in this figure applies to section stresses $\sigma < \sigma_o$ and is expressed as:

$$\sigma = \frac{P}{A} = \frac{\pi^2 E}{\left(L_e/k\right)^2} \tag{8.40}$$

To ensure an intersection between the two curves at $\sigma = \sigma_o$, we first find Euler's critical slenderness ratio from Eq. 8.40

$$\left(\frac{L_e}{k}\right)_{Euler} = \sqrt{\frac{\pi^2 E}{\sigma_o}} \tag{8.41}$$

Since $(L_e/k)_{Euler}$ defines an intersection co-ordinate, substitution of Eq. 8.41 into Eq. 8.39 leads to the condition

$$n\sigma_o = E\varepsilon_o \tag{8.42}$$

We then see that Euler becomes a special case of Eq. 8.39 when $n = 1$. Thus, a curve of any $0 < n < 1$ value will pass through this common point $[(L_e/k)_{Euler}, \sigma_o]$. Correspondingly, the intersection occurs at the common yield point $(\varepsilon_o, \sigma_o)$ for Hollomon curves with different n-values (see Fig. 8.15).

Contrary to the Hollomon law (8.38a) in which the true stress increased monotonically with the true strain (also called natural or logarithmic strain), the stress in a plastic strut cannot be increased indefinitely. The cut-off ordinate in Fig. 8.17 occurs at the ultimate compressive strength σ_u as shown. Thus, the critical value of L_e/k for an Engesser strut follows from Eqs 8.39 and 8.42, in terms of the yield and ultimate strengths σ_o and σ_u respectively, as

$$\left(\frac{L_e}{k}\right)_{Eng} = \sqrt{\frac{\pi^2 E}{\sigma_o\left(\sigma_u/\sigma_o\right)^{1/n}}} \tag{8.43}$$

8.4.2 Empirical Formulae

Empirical formulae for the stress attained in plastic strut buckling cover short struts with a given end-constraint and L_e/k initially less than the critical Engesser value in Eq. 8.43. Figures 8.18a,b show the range of interest in which section stress may vary from the ultimate σ_u to the yield σ_o. In each diagram the Engesser and Euler curves, 1 and 2 respectively, intersect at point A. Beyond A, Euler's curve 2 allows for elastic stress failure by extending L_e/k to point B as shown. The empirical approach replaces the two segments 1 and 2 with continuous curves (3 - 5) that plot to their safe side and terminate at σ_u on the σ–axis. The following strut formulae are available for this, the choice between them depending upon the strut material.

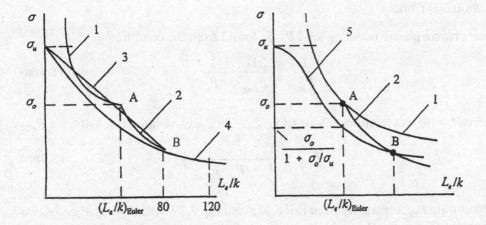

Figure 8.18 Approximations to Engesser (1) and Euler (2) from straight line (3) and parabola (4)

(a) Straight Line

The straight line 3 in Fig. 8.18a is simply written as

$$\sigma = \sigma_u \left[1 - q \left(\frac{L_e}{k} \right) \right] \tag{8.44}$$

where q is a material constant. For aluminium alloy struts, Eq. 8.44 is arranged to intersect the Euler curve at $L_e/k = 80$ as shown, while for other non-ferrous metals an intersection at $L_e/k = 120$ may be preferred. The Euler stress ordinate at each intersection is found from Eq. 8.40. Substituting this into Eq. 8.44, q is found. Typically, $E = 75$ GPa, $\sigma_u = 320$ MPa and $q = 8 \times 10^{-3}$ for an aluminium alloy.

(b) Parabola

The parabola 4 in Fig. 8.18a provides a safer prediction for steel struts with effective slenderness ratios lying in the range $L_e/k \leq 120$. This is written as

$$\sigma = \sigma_u \left[1 - b \left(\frac{L_e}{k} \right)^2 \right] \tag{8.45}$$

where b is a material constant, found from its intersection co-ordinates ($L_e/k, \sigma$) with Euler's curve. Values of b lie in the range $(40–50) \times 10^{-6}$ for steels with ultimate strengths between 400 and 500 MPa and a modulus $E = 210$ GPa.

(c) Rankine–Gordon

Their critical plastic buckling load P_{RG} is found from the condition

$$\frac{1}{P_{RG}} = \frac{1}{P_{Eng}} + \frac{1}{P_u} \tag{8.46a}$$

where $P_u = A\sigma_u$ is the ultimate load and $P_{Eng} = A\sigma$. Rearranging Eq. 8.46a for P_{RG}:

$$P_{RG} = \frac{P_u}{1 + P_u / P_{Eng}} \tag{8.46b}$$

Substituting P_{Eng} separately from Eqs 8.37a and Eq. 8.39, with $P_u = \sigma_u A$, gives two alternative forms for P_{RG}:

$$P_{RG} = \frac{A\sigma_u}{1 + \dfrac{\sigma_u}{\pi^2 E_t}\left(\dfrac{L_e}{k}\right)^2} = \frac{A\sigma_u}{1 + \dfrac{\sigma_u}{\sigma_o}\left(\dfrac{\sigma_o}{\pi^2 E}\right)^n \left(\dfrac{L_e}{k}\right)^{2n}} \tag{8.47a,b}$$

Equation 8.47b is a more useful for combining with Eq. 8.42. This reveals material constants σ_o, σ_u, E and n within the section stress

$$\sigma = \frac{\sigma_u}{1 + \dfrac{\sigma_u}{\sigma_o}\left(\dfrac{\sigma_o}{\pi^2 E}\right)^n \left(\dfrac{L_e}{k}\right)^{2n}} \tag{8.48}$$

Equation 8.48 appears as curve 5 in Fig. 8.18b for which the stress corresponding to $(L_e/k)_{Euler}$ is: $\sigma = \sigma_o/(1 + \sigma_o/\sigma_u)$, lying on the safe side of Engesser and Euler's curves 1 and 2 respectively. The slenderness ratio at an intersection point B with curve 2, is found from equating Eqs 8.40 and 8.48:

$$rq\left(\frac{L_e}{k}\right)^2 = 1 + rq^n\left(\frac{L_e}{k}\right)^{2n} \tag{8.49a}$$

where, for simplicity:

$$q = \frac{\sigma_o}{\pi^2 E}, \qquad r = \frac{\sigma_u}{\sigma_o} \tag{8.49b,c}$$

It is seen from Eqs 8.49a–c that the solution to L_e/k depends solely upon the material constants for elasticity E, yield σ_o and plasticity n and σ_u.

Exercises

Euler Theory

8.1 Find the shortest length of tube, 50 mm o.d. with 3 mm wall thickness, for which the Euler theory would be applicable when it is to act as a strut made from (i) mild steel with σ_c = 310 MPa and having pinned ends and (ii) high tensile steel with σ_c = 618 MPa and having fixed ends. Take E = 207 GPa. (Ans: 1.37 m, 1.94 m)

8.2 A 1 m-long connecting rod of rectangular cross-section may be considered pinned at its ends across the stronger section axis but rigidly fixed about the weaker axis. If the rod is to be equally resistant to buckling when the breadth of section is limited to 20 mm, determine its depth using a factor of safety of 4 when the axial compressive working load is 80 kN. Take E = 208 GPa.

8.3 A pinned-end strut is formed from two mild steel T-sections, as shown in Fig. 8.19. The section properties are: A = 2430 mm^2, I_x = 1.05 × 10^6 mm^4, I_y = 2.09 × 10^6 mm^4 and $\bar{y}$ = 18.82 mm. Using a safety factor of 6, determine the length of an Euler strut with this cross section when it is to carry an end load of 135 kN. Take E = 208 GPa. (Ans: 3.1 m)

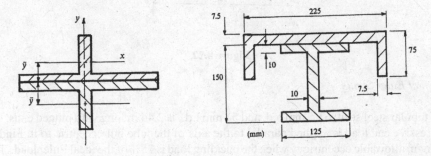

Figure 8.19   **Figure 8.20**

8.4 Find the buckling axis and the Euler buckling load for the strut cross section shown in Fig. 8.20. The 12 m-long, pinned-end strut is loaded in compression along its centroidal axis. Take E = 207 GPa. (Ans: 378 kN)

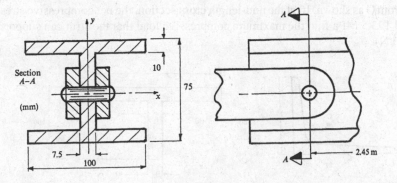

(a)   (b)

Figure 8.21

8.5 An *I*-section steel strut is pinned at its ends to a shackle that fits closely to the web thickness about its centre, as shown between the two views in Figs 8.21a,b. A centre-to-centre distance of 2.45 m defines the strut length. Find the Euler buckling load considering the end condition which applies to buckling about both axes x and y. Take $E = 207$ GPa.
(Ans: 768 kN)

8.6 The plane pin-jointed framework in Fig. 8.22 is designed to lift heavy loads at B and C. It takes the form of an isosceles triangle, with the angle at the support A being 90°. The strut member at the base is 3m long and is made from a uniform solid circular steel rod. Each of the joints B and C supports a vertical load of 10 kN. Find the smallest diameter of the strut BC to avoid buckling, and hence find the compressive stress therein. Take $E = 208$ GPa.
(Ans: 40.93 mm, 7.6 MPa)

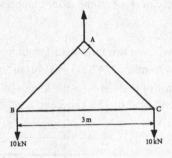

Figure 8.22

Imperfect Euler Struts

8.7 A tubular steel strut, 63.5 mm o.d. and 51 mm i.d., is 2.44 m long with pinned ends. The compressive end load is applied parallel to the axis of the tube but eccentric to it. Find the maximum allowable eccentricity when the buckling load is 75% of the ideal Euler load. Take the compressive yield stress as 310 MPa and $E = 207$ GPa.
(Ans: 5.2 mm)

8.8 The 75 mm equal-angle section shown in Fig. 8.23 has an area 1775 mm² and a least radius of gyration of 14.5 mm about a 45° axis passing through the centroid G and the corner. When used as a 1.5 m-long strut the section supports a compressive load through a point (*O*), 6.5 mm from G as shown. If, at the mid-length cross section, the net compressive stress is not to exceed 123.5 MPa find the maximum compressive load that the strut can support.
(Ans: 90 kN)

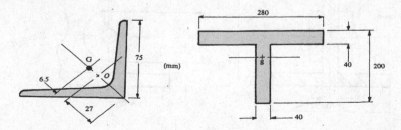

Figure 8.23   Figure 8.24

8.9 A 3 m long mild steel strut, with the T-section given in Fig. 8.24, is built-in at one end and free at the other end. A maximum eccentricity of 30 mm vertically from the centroidal axis is permissible when supporting an axial compressive load of 200 kN. Calculate the additional eccentricity allowed at the free end due to the strut not being initially straight. The maximum stress produced by combined compression and bending is limited to 80 MPa.

8.10 Determine the maximum compressive stress in a strut of solid circular section, with diameter 150 mm and length 5 m, when an axial compressive force of 15 kN is applied at a distance of 50 mm from the centroid of the section. Take $E = 207$ GPa.

8.11 A pinned, steel strut, 25 mm in diameter and 1.5 m long, supports an axial compression of 16 kN. Find the magnitude of an additional, concentrated force that would cause the strut to buckle if applied laterally at mid-span. Take $\sigma_c = 278$ GPa, $E = 207$ GPa. (Ans 133 N)

8.12 A 25 mm diameter steel tie is 2 m long with pinned ends. It carries an axial tensile load of 1.5 kN together with a concentrated lateral load of 0.3 kN at mid-length. Working from first principles, use the flexure equation to calculate the maximum bending moment, the maximum tensile and compressive stresses in the bar. What magnitude of lateral load would cause this strut to buckle under a similar compressive stress level σ_E to that of an ideal Euler strut of the same length and cross section? Take $\sigma_E = \pi^2 E/(L/k)^2$ with $E = 200$ GPa.

8.13 A coupling rod, with rectangular section 100×31.75 mm and pinned ends, supports an axial compressive force of 120 kN together with a uniformly distributed lateral load of 4.2 kN/m applied over its 2.5 m length. Find the bending moment and the maximum tensile and compressive stresses at mid-span. Estimate the effect that fixed ends would have upon the maximum stress What magnitude of axial compressive load would cause the pinned-end strut to buckle under a similar stress level σ_E for an ideal Euler strut of the same length and cross section? Take $\sigma_E = \pi^2 E/(L/k)^2$ with $E = 208$ GPa. (Ans: 3.94 kNm, 109.46 MPa)

8.14 A horizontal I-strut, of depth 150 mm, area 2280 mm^2 and $I = 8.75 \times 10^6$ mm^4, is simply supported over a span of 6.1 m. The beam carries axial compression 44.5 kN together with a uniformly distributed lateral load 1.46 kN/m over its length. Find the greatest bending moment and the maximum stress in the strut. Take $E = 207$ GPa. (Ans: 86 MPa)

8.15 A vertical cantilever strut of length 2.5 m is made from tubular section 150 mm outside diameter and 100 mm inside diameter. When z is measured from the fixed end, the initial deformed shape is given by $v_o = C(1 - \cos \alpha z/2L)$ where C is a constant. If the strut carries a vertical force P at the centroid of the free end show that v_o increases by $CP/(P_E - P)$ where P_E is the Euler load for a straight vertical cantilever strut. Hence determine the allowable, initial tip displacement C under a vertical force of 500 kN if the maximum compressive stress is limited to 100 MPa. Take $E = 210$ GPa.

Empirical Strut Formulae

8.16 Compare graphically the Euler and Rankine–Gordon predictions to the buckling load for a mild steel pipe: 4 m long with pinned ends, outer diameter 175 mm and wall thickness 12 mm. Use a safety factor of 3 and take $a = 7500^{-1}$, $\sigma_c = 320$ MPa and $E = 207$ GPa. (Ans: $P_E = 0.88$ MN, $P_R = 0.88$ MN)

8.17 An axially loaded steel strut has both ends built-in. If the length is 160 times the least radius of gyration, calculate the Euler buckling stress. Is the Rankine prediction more realistic in this case? Take $E = 207$ GPa, $\sigma_c = 278$ MPa and $a = 7500^{-1}$ for pinned ends.
(Ans: 150 MPa)

8.18 A 5 m long rectangular box section, with outer dimensions 250 mm × 200 mm and respective wall thicknesses of 30 and 15 mm, is to act as a pinned-end strut. Compare predictions upon the Euler and Rankine–Gordon buckling loads and show why one value is invalid. Take $E = 200$ GPa, $\sigma_c = 330$ MPa, $a = 7500^{-1}$.
(Ans: $P_E = 9.2$ MN, $P_R = 4.09$ MN)

8.19 A circular steel tube, 38 mm i.d. and 6.35 mm thick, is bolted rigidly between two casings 1 m apart. The tube is then subjected to a temperature change of 50°F without change in length. Find the factor of safety against buckling from the Rankine–Gordon theory. Take $E = 207$ GPa, $\alpha = 6.36 \times 10^{-6}/°F$, $\sigma_c = 324$ MPa and $a = 7500^{-1}$. (Ans: 4.35)

8.20 A 6 m length of cast-iron tube, with 50 mm outer and 38 mm inner diameters, buckled between fixed ends under 136 kN. A much shorter length of tube material failed in compression under a load of 420 kN. Determine the Rankine–Gordon constants and hence find the buckling load for a 1.5 m length of tube when used as a strut with pinned ends. Would the Euler buckling load prediction be valid in this case? Take $E = 92.6$ GPa.
(Ans: $\sigma_c = 472$ MPa, $a = 1590^{-1}$, 61.5 kN)

8.21 A 7.6 m long steel strut has a section area of 9675 mm². It is required to support an axial compressive load of 900 kN with fixed ends. Find the least possible value of the radius of gyration according to the Rankine–Gordon theory with the constants $\sigma_c = 324$ MPa and $a = 7500^{-1}$. (Ans: 27.8 mm)

8.22 Compare values of the Euler and Rankine–Gordon buckling loads for 5 m long fixed end struts with the cross sections given in Figs 8.25a and b. Take $E = 210$ GPa with the constants $\sigma_c = 325$ MPa and $a = 7500^{-1}$.
[Ans: (a) $P_E = 13.68$ MN, $P_R = 3.14$ MN; (b) $P_E = 553.5$ kN, $P_R = 355.6$ kN]

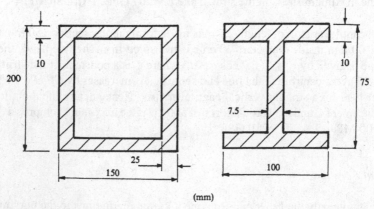

(mm)

(a)   (b)

Figure 8.25

8.23 Find the axis about which buckling takes place for the strut section shown in Fig. 8.26. What is the safe axial load that may be applied to an 11 m length of this strut when the ends are fixed? Decide between Euler and Rankine–Gordon buckling loads given the following area properties for a single channel: $I_x = 34.5 \times 10^6 \text{ mm}^4$, $I_y = 1.66 \times 10^4 \text{ mm}^4$, $A = 3660 \text{ mm}^2$, $\bar{x} = 18.8 \text{ mm}$. Take $E = 207$ GPa, $a = 7500^{-1}$, $\sigma_c = 324$ MPa. (Ans: 9880 kN, 3645 kN)

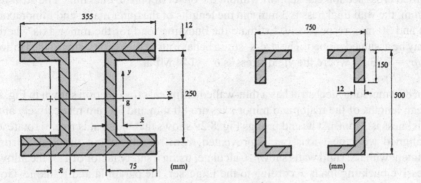

Figure 8.26   **Figure 8.27**

8.24 A steel strut 10 m long consists of four equal angles 150 mm side length and 12 mm thick, bolted together as shown in Fig. 8.27. Find the safe axial load that may be applied for a load factor of 6 when the ends are pinned and when they are fixed. The Rankine–Gordon constants are: $\sigma_c = 320$ MPa and $a = 7500^{-1}$.

8.25 A tubular box section, with outer dimensions 150×200 mm and 10 mm wall thickness, is to act as a steel strut of length 6 m with fixed ends. Find the safe axial compressive load using a load factor of 4 based upon the Rankine–Gordon, straight line, parabolic and Perry–Robertson formulae. Where appropriate take $E = 207$ GPa, $\sigma_c = 325$ MPa, $a = 7500^{-1}$, $\mu = 0.003L/k$. The straight line and parabola are to pass through σ_c and $L/k = 120$. (Ans: $P_R = 1.63$ MN)

8.26 A 7.5 m long steel column is constructed by joining two channel sections together as shown in Fig. 8.28. If, for one section, the area is 6950 mm² and the greatest and least I-values are 91.1×10^6 and $5.74 \times 10^6 \text{ mm}^4$, compare the axial loads that the column may carry according to the straight line, parabola and Perry-Robertson predictions. Assume that these all pass through a common yield point of 310 MPa and that the straight line and parabola two make tangents with the Euler curve at $L/k = 120$.

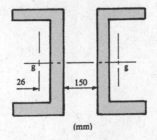

Figure 8.28

8.27 If, in Example 8.11 (see p. 313), a common intercept, equal to the compressive yield stress, together with points of tangency with the Euler curve are imposed upon the straight-line and parabola buckling formulae, over what range of L/k do they apply?

8.28 It is required to find the maximum compressive load that a brass strut with an elliptical, thin-walled cross section can support without the onset of plastic buckling. The strut length is 825 mm, the wall thickness is 3 mm and the lengths of the outer major and minor axes are 80 mm and 30 mm, respectively. Estimate the buckling load for the minor axis where the ends may be assumed pinned. The stress–strain behaviour of brass conforms to the Hollomon law: $\sigma/\sigma_c = 7.83\varepsilon^{1/3}$ where the yield stress is $\sigma_c = 144$ MPa.

8.29 An 800 mm long steel strut has a thin-walled elliptical cross section shown in Fig. 8.29. The mean lengths of the major and minor axes are 80 mm and 30 mm respectively and the wall thickness is 3 mm. At its end fixings Fig. 8.29 shows that the strut is free to rotate about a pin aligned with the y-axis but is prevented from rotating about the x-axis from its containment with the rigid walls shown. Calculate, using a safety factor of 1.5, the allowable compressive buckling loads according to the Engesser, the parabola and Rankine–Gordon formulae. For steel take the Euler reference stress $\sigma_E = 300$ MPa the ultimate compressive stress $\sigma_u = 450$ MPa and Hollomon's hardening exponent $n = \frac{1}{3}$.
(Ans: 555.35 kN, 613.58 kN, 409.27 kN)

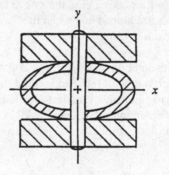

Figure 8.29

CHAPTER 9

BUCKLING OF PLATES AND TUBES

Summary: Flat plates supporting a one-dimensional axial compression can buckle laterally in a similar manner to struts considered in the previous chapter. The buckling mode for plates becomes more complex when all four sides are supported with an in-plane loading that is two-dimensional. The latter includes special cases of shear buckling arising in the thin, vertical walls of a beam section under transverse shear or within the individual plates of both open and closed thin-walled sections placed under axial torsion. Torsional buckling of thin-walled circular tubes is treated separately with its sensitivity to a number of geometrical factors including the ratios between tube length, radius and thickness. Local buckling of the cross section can also occur alongside the global, flexural buckling of a plate strut. For example, the limb tips of an equal-angle section rotate as the strut bows laterally from end to end when supporting an axial compression. The following buckling modes are described from their relevance to engineering design.

9.1 Buckling Modes

Primary buckling of a plate refers to the lateral shift in transverse sections away from the longitudinal axis [1] of compression. Whether the cross section is that of a wide strut, a shear web, a beam or a torsion bar it is susceptible to the primary buckling mode where flexure promotes a combined translation and rotation of the cross section. Both effects accompany the lateral deflection of a strut or the bending of a beam subjected to their respective axial and transverse loadings. The cross sections concerned lie away from the ends such that rotation and translation occur by varying amounts, depending upon their position in the length.

Secondary buckling refers to any other mode of buckling that is not primarily flexural. Both closed and open thin-walled sections are prone to local compressive buckling within individual 'flats' that form its web, flange and sides. As the straight, thin walls of a short column bearing compression these flats distort in the section plane without translation or rotation. Localised stress concentrations at corners can exceed the yield stress and cause a local crippling failure. A similar distortion to a strut's cross section can also reduce its resistance to buckling in a primary mode. Thus, the local buckling mode, considered here refers to the accompanying distortion within the cross section itself as the flats bow and rotate under the action of the compression and shear they are expected to support [1].

(a)   (b)

Figure 9.1 Local buckling in closed and open and cross sections

For example, local buckling modes for slender struts are shown within the walls of a box section in Fig. 9.1a and within both flanges and the web of an I-section in Fig. 9.1b. Similar bucking modes can occur within the thin walls of wide strut sections. Typical examples are shown for thin aerofoils in Figs 9.2a,b. The integrally machined and Z-stiffened sections are shown having regular pitch (b) wherein the local buckling modes appear.

(a)   (b)

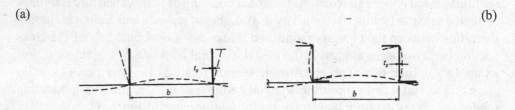

Figure 9.2 Local buckling in closed and open and cross-sections

In the former design a buckling within the foil simply rotates the web stiffening but for the latter an additional local buckling of the Z-section's plates occurs. Where a stiffener is riveted or spot welded to a wide plate strut or a shear web the local mode prevails when the individual flats lying between the attachment points are placed either under compression or under shear. Here an accompanying primary compressive buckling of the sheet may also occur between rivets along the stiffener's rivet line. Such inter-rivet buckling [2, 3] occurs with a wavelength that differs from the rivet pitch and is influenced by the nearness of the stiffener's webs and flanges. When the wavelength is less than the rivet pitch the undulations resemble the wrinkling that occurs in thin unsupported sheet under compression.

9.2 Unsupported Plate Under Axial Compression

Consider the axial compression of a wide, thin plate, of thickness t, with similar length and width dimensions, a and b respectively (see Fig. 9.3a). Having simply supported (pinned) ends and unsupported sides, the strut deflects laterally with uniform curvature R. The cross section $b \times t$ remains rectangular during bending when biaxial stresses σ_1 and σ_2 arise in the plane of the plate's top surface as shown.

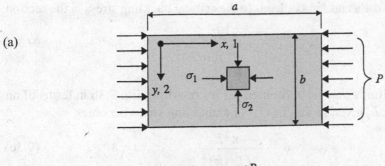

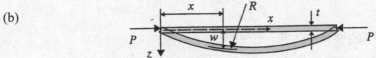

Figure 9.3 Buckling of a wide plate strut

A lateral stress σ_2 arises with there being no strain in the 2-direction:

$$\varepsilon_2 = 0 = \frac{1}{E}(\sigma_2 - v\sigma_1), \quad \therefore \quad \sigma_2 = v\sigma_1$$

$$\varepsilon_1 = \frac{1}{E}(\sigma_1 - v\sigma_2) = \frac{1}{E}(1 - v^2)(\pm\sigma_1) \tag{9.1}$$

In a slender strut (see Chapter 8), where b is small compared to a, the presence of lateral strains ε_2 results in *anticlastic curvature*, i.e. the cross-section does not remain rectangular. Such distortion arises from an opposing lateral strain ε_2 within the tensile and compressive surfaces, known as the *Poisson effect*. Here, as $\sigma_2 = 0$, the axial and lateral surface strains are: $\varepsilon_1 = (\pm\sigma_1)/E$ and $\varepsilon_2 = -v(\pm\sigma_1)/E$. Comparing ε_1 with Eq. 9.1 reveals a difference in ε_1 between slender and wide struts. Consequently, it becomes necessary to modify the flexure equation for a wide strut by the factor $(1 - v^2)$. Firstly, the curvature R in Fig. 9.3b, follows from Eq. 6.3 as:

$$\frac{d^2w}{dx^2} = \frac{1}{R} = \frac{\varepsilon_1}{z} \tag{9.2a}$$

where z is the section's thickness position, measured from the neutral axis, having as its limits $z = \pm t/2$. Substituting into Eq. 9.2a, ε_1 from Eq. 9.1, with $M/I = \sigma_1/z$:

$$\frac{d^2w}{dx^2} = \frac{(1 - v^2)}{E}\left(\frac{\sigma_1}{z}\right) = \frac{(1 - v^2)M}{EI} \tag{9.2b}$$

in which the sign of σ_1 will depend upon sign of z. With a moment $M = -Pw$, the solution to the critical buckling load for a pinned-end plate is from Eq. 9.2b:

$$P_{cr} = \frac{\pi^2 EI}{(1 - v^2)a^2} \tag{9.3a}$$

Substituting $P_{cr} = \sigma_{cr} bt$ and $I = Ak^2$, leads to the critical buckling stress in the section

$$\sigma_{cr} = \frac{\pi^2 E}{(1 - v^2)(a/k)^2} \qquad (9.3b)$$

Other edge conditions may be accounted for by rewriting Eq. 9.3b in terms of an equivalent length L_e given in Table 9.1. The buckling stress becomes

$$\sigma_{cr} = \frac{\pi^2 E}{(L_e/k)^2} \qquad (9.3c)$$

Table 9.1 Equivalent length for plates with various end conditions

Edge Condition	Equivalent Length, L_e	Bucking Coefficient, K
a) Pinned	$a(1 - v^2)^{1/2}$	$(\pi b/a)^2 / 12(1 - v^2)$
b) Fixed	$(a/2)(1 - v^2)^{1/2}$	$(\pi b/a)^2 / 3(1 - v^2)$
c) Pinned-Fixed	$(a/\sqrt{2})(1 - v^2)^{1/2}$	$(\pi b/a)^2 / 6(1 - v^2)$
d) Fixed-Free	$2a(1 - v^2)^{1/2}$	$(\pi b/a)^2 / 48(1 - v^2)$

Moreover, an inelastic buckling stress may be estimated from Eq. 9.3c in using the equivalent lengths of wide-plate struts provided E in Eq. 9.3c is replaced by the tangent modulus E_T. In this chapter we shall express the elastic plate buckling stress formulae by setting $k^2 = I/A = t^2/12$ in Eq. 9.3c to give:

$$\sigma_{cr} = \frac{E}{12}\left(\frac{\pi b}{L_e}\right)^2 \left(\frac{t}{b}\right)^2 = KE\left(\frac{t}{b}\right)^2 \qquad (9.4)$$

where K is the buckling coefficient which, from Table 9.1, is seen to depend upon the plate's side lengths a and b, Poisson's ratio v and the manner in (a)–(d) by which its ends are supported.

9.3 Supported Plate Under Axial Compression

More generally, with the uni-directional compression of thin, wide plates the long sides may be free, simply supported or clamped. Free or unsupported sides, returns to the wide strut problem considered in § 9.2 above. Compressive loading of thin, wide plates, as opposed to slender struts, refers to plates whose side lengths a and b (see Fig. 9.3) do not differ by more than an order of magnitude. The theory outlined in § 9.6 accounts for buckling of thin plates with this geometry under biaxial compression when its four sides have various types of edge support. Here the following reductions to an axial compression are given.

9.3.1 Simple Supports

Under a uni-directional compression, the critical buckling stress σ_1, for a plate with all four sides simply supported and free to rotate, follows from setting $\sigma_2 = 0$ (i.e. $\beta = 0$) in Eq. 9.14. This gives [4]:

$$\sigma_{cr} = \left(\frac{D\pi^2}{t}\right)\left[\left(\frac{m}{a}\right)^2 + \left(\frac{n}{b}\right)^2\right]^2 \left(\frac{a}{m}\right)^2 = \left(\frac{D\pi^2}{tb^2}\right)\left(\frac{mb}{a} + \frac{n^2 a}{mb}\right)^2 \quad (9.5a)$$

where $D = Et^3/[12(1 - v^2)]$ is the *flexural stiffness*, t, a and b are the plate thickness, length and breadth, respectively. The number of half-waves of buckling in the principal stress directions 1 and 2 (coincident with x and y) are denoted by m and n, respectively. For example, the simply supported plate shown in Fig. 9.4 has buckled with a single half-wave ($n = 1$) in the 2-direction and three half-waves ($m = 3$) in the 1- direction.

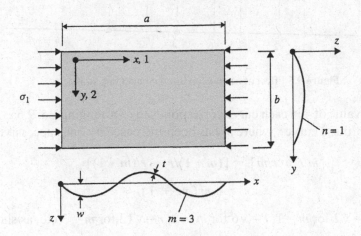

Figure 9.4 Buckling of a thin, simply supported plate under uniaxial compression

When $n = 1$ in Eq. 9.5a, m will correspond to a minimum in the buckling stress expression:

$$\sigma_{cr} = \left(\frac{D\pi^2}{tb^2}\right)\left(\frac{m}{r} + \frac{r}{m}\right)^2 \quad (9.5b)$$

where $r = a/b$. Differentiating for the m-values that minimise Eq. 9.5b with integral values for r:

$$d\sigma_{cr}/dm = 2(m/r + r/m)(1/r - r/m^2) = 0$$

$$m/r^2 + 1/m - 1/m - r^2/m^3 = 0$$

$$m^4 - r^4 = 0$$

$$(m - r)(m + r)(m^2 + r^2) = 0$$

The condition $m = r$ implies that the plate will buckle into an integral number of square cells $b \times b$ each under the same stress. That is, from Eq. 9.5b

$$\sigma_{cr} = \frac{4D\pi^2}{tb^2} = \frac{\pi^2 E}{3(1 - v^2)} \left(\frac{t}{b} \right)^2 \qquad (9.6a,b)$$

which has a similar form to the wide strut buckling stress in Eq. 9.4. However, K will now take a different value depending upon r and m. Figure 9.5 provides the buckling stress when both integer and non-integer values of r are taken with values of $m = 1, 2, 3$ and 4 in Eq. 9.5b.

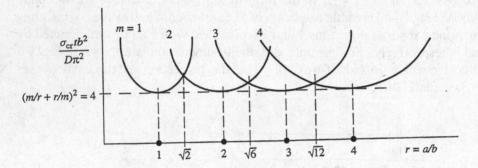

Figure 9.5 Effect of r and m on uniaxial buckling stress

The minimum value of 4 in each curve corresponds to re-arranging Eq. 9.6a. At the intersections of these curves, where m has been increased by one successively:

$$[m/r + r/m] = [(m + 1)/r + r/(m + 1)]$$

$$\therefore \ r = \sqrt{m(m + 1)}$$

from which: $r = \sqrt{2}$ for $m = 1$; $r = \sqrt{6}$ for $m = 2$; $r = \sqrt{12}$ for $m = 3$ etc, as shown.

9.3.2 Other Edge Fixings

Various methods account for the effect of constraining the plate's edges. The simplest of these [5] generalises Eq. 9.6a by introducing an elastic constraint coefficient K_r as follows:

$$\sigma_{cr} = \frac{K_r D\pi^2}{tb^2} = K_e E \left(\frac{t}{b} \right)^2 \qquad (9.7a,b)$$

where Eq. 9.7a applies to any edge fixing when

$$K_r = \left(\frac{m}{r} \right)^2 + p + q \left(\frac{r}{m} \right)^2 \qquad (9.7c)$$

Taking edge rotational restraint factors $p = 2$ and $q = 1$, with $m = r$, Eq. 9.7c contains Eq. 9.6a in the special case of simple supports. The dependence of K_r upon the edge restraint factors (p and q), the plate aspect ratio ($r = a/b$) and the buckling mode m, has been established experimentally in certain cases [4–6]. For example, Table 9.2 applies to a plate with fixed edges.

Table 9.2 Restraint coefficients for a plate with fixed edges

$r = a/b$	0.75	1.0	1.5	2.0	2.5	3.0
K_r	11.69	10.07	8.33	7.88	7.57	7.37

As r ($= a/b$) increases, the effect of edge restraint lessens and K_r approaches the minimum value of 4 as found from Eq. 9.6a for a plate with simply supported edges. A more convenient graphical approach employs Eq. 9.7a,b within design curves that provide the restraint coefficient $K_e = K_r \times \pi^2/[12(1 - v^2)]$ for both uniaxial and biaxial loadings [6]. These curves present the ratio between the critical elastic buckling stress $(\sigma_{cr})_e$ for a plate with a particular edge fixings to the buckling stress σ_{cr} for a simply supported plate (i.e. Eq. 9.6b) with similar aspect ratio $r = a/b$. Figure 9.6 shows how this stress ratio varies with r for clamped and various mixed-edge fixings for plates under uniaxial compression.

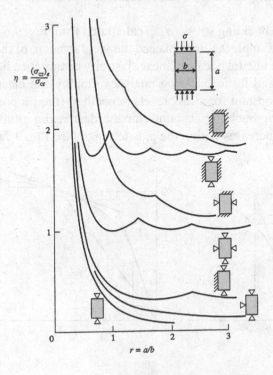

Figure 9.6 Plate buckling under compression

Example 9.1 Using Fig. 9.6, find the critical, elastic buckling stress for a 5 mm-thick aluminium alloy plate, with $a = 400$ mm and $b = 200$ mm, with compression applied through a rigid clamp to the edges of its shorter sides while its longer sides are simply supported. Take $E = 70$ GPa and $v = 0.33$.

Firstly, the reference stress, used to normalise the ordinate in Fig. 9.6, is given by Eq. 9.6b:

$$\sigma_{cr} = \frac{\pi^2 E}{3(1 - v^2)} \left(\frac{t}{b} \right)^2 = \frac{\pi^2 \times 70000}{3(1 - 0.33^2)} \left(\frac{5}{200} \right)^2 = 25.9 \text{ MPa}$$

Here the elastic buckling stress is required for sides clamped and simply-supported. Reading from the appropriate graph in Fig. 9.6 gives the elastic buckling coefficient for $a/b = 2$:

$$\eta = (\sigma_{cr})_e / \sigma_{cr} = 1.22$$

The critical, elastic buckling stress follows with no plasticity correction required (see § 9.4)

$$(\sigma_{cr})_e = \eta \sigma_{cr} = 1.22 \times 25.9 = 31.6$$

9.4 Inelastic Buckling of Plates Under Axial Compression

In thicker plates the buckling stress $(\sigma_{cr})_e$, calculated from Fig. 9.6 and Eq. 9.6b, in the manner of the Example 9.1, may exceed the yield stress σ_y of the plate material. The solution will be invalid because linear elasticity cannot then be assumed at the critical stress level. Elastic buckling analyses employ the elastic compressive modulus which is constant for stress levels beneath σ_y (the proportional limit). A correction for plastic buckling accounts for the decreasing gradient to the flow curve as the stress increases within the plastic range (see Fig. 9.7a).

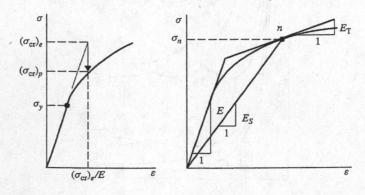

(a)

(b)

Figure 9.7 Tangent and secant moduli

In a deflected slender strut a stress gradient arises where the sum of direct and bending stresses amounts to a greater compression on its concave side. Where the elastic limit is exceeded a tangent modulus E_T (see Fig. 9.7b) is often employed with Euler's strut theory to provide an average, compressive stress at buckling. Equally, E_T may replace E for an account of plastic buckling of thin plates, particularly where through-thickness stress gradients are less severe. Figure 9.7a shows that where plasticity has occurred it reduces the buckling stress to a lower level $(\sigma_{cr})_p$. Figure 9.7b shows how both the tangent and secant moduli, E_T and E_S respectively, are used to account for a plastic stress level. The secant modulus gives the total strain at a reference point n as

$$\varepsilon_n = \frac{\sigma_n}{E_S} \tag{9.8a}$$

The Ramberg–Osgood description [7] to a stress–strain curve gives the total strain under a plastic stress level as:

$$\varepsilon = \frac{\sigma}{E} + \alpha \left(\frac{\sigma}{E} \right)^m \tag{9.8b}$$

Combining Eqs 9.8a,b at the reference stress σ_n gives

$$\frac{E}{E_S} - 1 = \alpha \left(\frac{\sigma_n}{E} \right)^{m-1} \tag{9.9a}$$

Differentiating Eq. 9.8b admits the gradient $E_T = d\sigma/d\varepsilon$ at point n in Fig. 9.7

$$\frac{1}{m} \left(\frac{E}{E_T} - 1 \right) = \alpha \left(\frac{\sigma}{E} \right)^{m-1} \tag{9.9b}$$

Now let σ_n be the stress level at which $E_T = E/2$. Equation 9.9b gives

$$\alpha = \frac{1}{m} \left(\frac{\sigma_n}{E} \right)^{1-m} \tag{9.9c}$$

Substituting Eq. 9.9c into Eq. 9.8b leads to a normalised stress–total strain relationship

$$\varepsilon = \frac{\sigma_n}{E} \left[\frac{\sigma}{\sigma_n} + \frac{1}{m} \left(\frac{\sigma}{\sigma_n} \right)^m \right] \tag{9.10}$$

where σ_n and m are material properties found from fitting Eq. 9.10 to a stress-strain curve. Table 9.3 gives typical m-values that lie in a range 16–29 for heat treated aluminium alloys and in a range 5–105 for heat-treated steels [8]. Correspondingly, the σ_n values, at which $E_T = \frac{1}{2}E$, vary from 180–440 MPa for aluminium alloys (L and DTD series), from 300–650 MPa for corrosion resistant steels (S500 series) and from 700–1200 MPa for alloy steels (S96 and S134). Table 9.3 shows that the m-value also depends upon whether the material is sheet (S), plate (P) or bar (B).

Table 9.3 Hardening exponent values for aluminium (L and DTD - grades) and steel (S - grades)

Matl	L71	L73	DTD687	L65	S520	S514	S515	S134	S96
m	16	17	15	29	4.9	17	8.4	7	105
Form	S	S	P	B	S	S	S	B	B

Figure 9.7a shows that the elastic strain under $(\sigma_{cr})_e$ is $\varepsilon = (\sigma_{cr})_e/E$. Substituting this value for ε in Eq. 9.10 allows $\sigma = (\sigma_{cr})_p$ to be found. Design data [9] adopts a graphical solution to $(\sigma_{cr})_p$ by employing a *plasticity reduction factor* $\mu < 1$ within:

$$\mu = (\sigma_{cr})_p/(\sigma_{cr})_e \tag{9.11}$$

Both m and σ_n influence μ in the manner shown in Fig. 9.8.

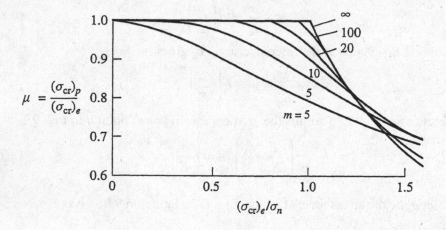

Figure 9.8 Plasticity reduction factor for a simply supported plate

When applying Fig. 9.8, firstly we find from Fig. 9.6 the critical elastic buckling stress $(\sigma_{cr})_e$. This determines the ratio $(\sigma_{cr})_e/\sigma_n$ from which a μ value is found from Fig. 9.8. Equation 9.11 is then employed to find $(\sigma_{cr})_p$ as in the following example.

Example 9.2 A 5 mm-thick S520 steel plate with side lengths $a = 600$ mm and $b = 200$ mm is subjected to uniaxial compression across its shorter sides. Find the critical plastic buckling stress when all four sides are: (a) simply supported and (b) clamped. Take this material's properties as: $E = 210$ GPa, $v = 0.27$, $m = 4.9$, $\sigma_y = 300$ MPa and $\sigma_n = 450$ MPa..

Firstly, we find from Eq. 9.6b the critical elastic buckling stress $\sigma_{cr} = (\sigma_{cr})_e$:

$$(\sigma_{cr})_e = \frac{\pi^2 E}{3(1-v^2)}\left(\frac{t}{b}\right)^2 = \frac{\pi^2(210 \times 10^3)}{3(1-0.27^2)}\left(\frac{5}{200}\right)^2 = 465.75 \text{ MPa}$$

As this stress exceeds $\sigma_y = 300$ MPa, a correction for plasticity it required. The abscissa in Fig. 9.8 is 465.75/450 = 1.035, from which its ordinate $\mu = 0.83$ applies to $m = 4.9$. Hence, from Eq. 9.11, the critical plastic buckling stress becomes:

$$(\sigma_{cr})_p = \mu\,(\sigma_{cr})_e = 0.83 \times 465.75 = 386.57 \text{ MPa}$$

(a) Ends Simply Supported

It should be noted here that Fig. 9.8 is constructed for a plate having all four sides simply supported in material with Poisson's ratios 0.3 and 0.5 within its elastic and plastic ranges, respectively. Fortunately, the greatest possible deviation in v (from 0.25 to 0.35) can have only a negligible influence upon the reduction factor and therefore this effect may be ignored. However, the elastic Poisson's ratio effect within Fig. 9.6 should be accounted for. The influence of the plate supports upon the reduction factor μ has been discussed further in [10, 11]. This amounts to a vertical shift in the curves within Fig. 9.8 when the sides are supported differently.

(b) Ends Clamped

Where the plate has clamped edges at both ends and along both sides, reading from the upper curve in Fig. 9.6, the following critical stress ratio, corresponds to an integral number of buckled panels (i.e. $r = a/b = 2$) in the sheet:

$$\mu = (\sigma_{cr})_e / \sigma_{cr} = 2$$

As the denominator has been identified with 465.75 MPa for a simply supported plate, the required elastic buckling stress for the clamped plate follows as: $(\sigma_{cr})_e = 2 \times 465.75 = 931.5$ MPa. Strictly, three further corrections should be applied to this stress magnitude:

(i) Firstly, $(\sigma_{cr})_e$, as calculated, should be multiplied by a factor $(1 - 0.3^2)/(1 - v^2)$ where Poisson's ratio differs from 0.3 (the basis of Fig. 9.6). This gives:

$$(\sigma_{cr})_e = \frac{(1 - 0.30^2)}{(1 - 0.27^2)} \times 931.5 = 914.3 \text{ MPa}$$

(ii) Secondly, $(\sigma_{cr})_e$ in (i) above should then be corrected for plasticity by extending Fig. 9.8 to an abscissa value of 914.3/450 = 2.03. This gives $\mu \approx 0.62$, at $m = 4.9$, for a plate with simply supported edges.

(iii) Finally, with clamped plate edges, a downward shift in the respective curve in Fig. 9.8 has shown [6] that $\mu \approx 0.58$, i.e. a multiplying factor of 0.58/0.62 = 0.935 applies to the plasticity reduction factor found in (ii). Hence, the critical plastic buckling stress is estimated as:

$$(\sigma_{cr})_p = \mu\,(\sigma_{cr})_e = (0.935 \times 0.62) \times 914.3 = 530.3 \text{ MPa}$$

In the event of a near ultimate strength value being found from this procedure the implication is that, in the manner of its fixing, this plate cannot buckle without a significant amount of accompanying plasticity. Elastic-plastic buckling at a lesser critical stress would be more likely to arise in a plate of reduced thickness.

9.5 Post-Buckling of Plates Under Axial Compression

When a plate has buckled the load may be increased further as the axial compressive stress increases mostly within two strips of material adjacent to the side supports (see Fig. 9.9).

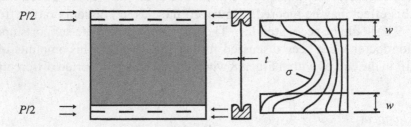

Figure 9.9 Stress distribution in a buckled plate

As only a slight increase in axial stress can occur in the central buckled material, it may be assumed that the whole load P is carried by the two edge strips of effective width $2w$, over which σ is assumed constant. The load supported becomes

$$P = 2wt\sigma \qquad (9.12a)$$

With all the edges of our equivalent elastic plate being simply supported, the critical buckling stress becomes [5]

$$\sigma_{cr} = \frac{\pi^2 E}{3(1 - v^2)} \left(\frac{t}{2w}\right)^2 \qquad (9.12b)$$

from which w may be found once σ_{cr} attains the yield stress σ_y:

$$w = \frac{\pi t}{2} \sqrt{\frac{E}{3\sigma_y(1 - v^2)}} \qquad (9.12c)$$

Taking $v = 0.3$ in Eq. 9.12c gives the effective half-width

$$w = 0.95\, t \sqrt{\frac{E}{\sigma_y}} \qquad (9.13a)$$

When $b = 2w$ the full plate width is used most effectively in post-buckling. The load that an be supported across this width follows from Eq. 9.12a as:

$$P = 1.9 \, t^2 \sqrt{E\sigma_y} \qquad (9.13b)$$

Experiment [12] has shown that the coefficient in Eq. 9.13a is nearer 0.85. Moreover, when the 'plate' forms the compressive surface of a box section in bending the coefficient becomes 1.14.

With alternative supports along the longer sides, the asymptotic values of the critical elastic buckling stress ratio applies. These appear in Fig. 9.6 in which Eq. 9.12b defines the denominator for the normalised ordinate $\eta = (\sigma_{cr})_e / \sigma_{cr}$ for a plate with a given combination of side supports. The following example shows how the calculation for post buckling load is made.

Example 9.3 Find the semi-effective width expression and load-carrying capacity for a thin-aluminium plate in post-buckling when one long side is simply supported and the other parallel side is free, assuming Poisson's ratio is ⅓.

The second lowest plot in Fig. 9.6 shows that the critical asymptotic stress ratio is $\eta = 0.106$. That is:

$$\eta = \frac{(\sigma_{cr})_e}{\sigma_{cr}} = 0.106$$

and substituting σ_{cr} from Eq. 9.12b:

$$(\sigma_{cr})_e = 0.106 \times \frac{\pi^2 E}{3(1 - v^2)} \left(\frac{t}{2w} \right)^2$$

from which the half-width w may be found for $v = \frac{1}{3}$, when $\sigma_{cr} = \sigma_y$

$$w = 0.313 \, t \sqrt{\frac{E}{\sigma_y}}$$

This show a reduction in effective width of ⅓ compared to Eq. 9.13a for a plate with simple guided supports. Correspondingly, the reduced load that can be supported across this width follows from Eq. 9.12a as:

$$P = 0.616 \, t^2 \sqrt{E\sigma_y}$$

This load would only remain unchanged in the absence of work-hardening during post-buckling.

9.6 Plate Under Biaxial Compression

Uniaxial compressive buckling of wide plates considered above and their shear buckling behaviour (to follow in § 9.7) are each special cases of plate buckling under biaxial compression. An outline of the general theory is given here. The appropriate reductions to uniaxial compression have been given in § 9.3 and those for shear are considered in § 9.7. Consider a simply supported, thin rectangular plate $a \times b$ with thickness t. Let uniform compressive stresses σ_1 act normal to $b \times t$ and σ_2 act normal to $a \times t$, as shown in Fig. 9.10.

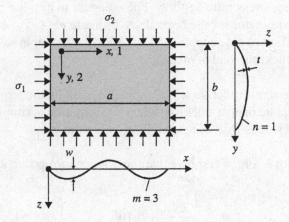

Figure 9.10 Buckling of a thin plate under biaxial compression

Let the applied stresses increase proportionately in the constant ratio $\beta = \sigma_2 / \sigma_1$. With $a > b$, the actual number of half-waves of buckling is that which minimise σ_1. In general, this principal stress is expressed as [4]

$$(\sigma_1)_{cr} = \frac{D\pi^2 [(m/a)^2 + (n/b)^2]^2}{t[(m/a)^2 + \beta(n/b)^2]} \qquad (9.14)$$

where $D = Et^3/[12(1 - v^2)]$ is the *flexural stiffness* and m and n are the respective number of half waves for buckling in the x- and y- directions. For example, in Fig. 9.10 a buckled plate is shown where $m = 3$ and $n = 1$. For a square plate, where $a = b$, buckling occurs with one half-wave in each direction. That is, $m = n = 1$ and Eq. 9.14 reduces to

$$(\sigma_1)_{cr} = \frac{\pi^2 E}{3(1 - v^2)(1 + \beta)} \left(\frac{t}{a}\right)^2 \qquad (9.15a)$$

In the case of an equi-biaxial compression, where $\beta = 1$, Eq. 9.15a gives

$$(\sigma_1)_{cr} = \frac{\pi^2 E}{6(1 - v^2)} \left(\frac{t}{a}\right)^2 \qquad (9.15b)$$

A useful buckling stress formula $(\sigma_1)_{cr}$ applies to a slender plate with aspect ratio $r = a/b$ in which $b << a$. Firstly, we must find the number of half-wavelengths for the buckled shape for the x-direction. Taking $n = 1$ for the much smaller b - dimension aligned with y, Eq. 9.14 becomes

$$(\sigma_1)_{cr} = \frac{D\pi^2(m^2 + r^2)^2}{a^2t(m^2 + \beta r^2)} = \frac{D\pi^2(m/r + r/m)^2}{b^2t[1 + \beta(r/m)^2]} \qquad (9.16)$$

and m is found from the condition that $(\sigma_1)_{cr}$ in Eq. 9.16 is a minimum. This gives

$$d(\sigma_1)_{cr}/dm = [1 + \beta(r/m)^2] \times 2(m/r + r/m)(1/r - r/m^2)$$
$$- (m/r + r/m)^2 \times 2\beta(r/m)(-r/m^2) = 0$$

$$[1 - (r/m)^2][1 + \beta(r/m)^2] + \beta(r/m)^2[1+(r/m)^2] = 0$$

$$\therefore \qquad \frac{r}{m} = \frac{1}{(1 - 2\beta)^{1/2}} \qquad (9.17a)$$

whereupon, the number of half-wavelengths in the x-direction becomes

$$m = r(1 - 2\beta)^{1/2} \qquad (9.17b)$$

The corresponding half-wavelength is

$$\frac{a}{m} = b\left(\frac{r}{m}\right) = \frac{b}{(1 - 2\beta)^{1/2}} \qquad (9.17c)$$

Substituting Eq. 9.17a into Eq. 9.16

$$(\sigma_1)_{cr} = \frac{D\pi^2[(1 - 2\beta) + 2 + 1/(1 - 2\beta)]}{bt^2[1 + \beta/(1 - 2\beta)]} = \frac{4D\pi^2}{b^2t}(1 - \beta) \qquad (9.18a)$$

It is apparent from Eq. 9.17c that as $\beta \rightarrow \frac{1}{2}$ the half-wavelength approaches infinity. The corresponding buckling stress is, from Eq. 9.18a:

$$(\sigma_1)_{cr} \rightarrow \frac{2D\pi^2}{b^2t} \qquad (9.18b)$$

For each solution to $(\sigma_1)_{cr}$ the critical lateral stress follows as $(\sigma_2)_{cr} = \beta(\sigma_1)_{cr}$. The two critical stresses remain elastic provided they satisfy a von Mises yield criterion (see § 12.1.5 p. 492):

$$\sigma_1^2 - \sigma_1\sigma_2 + \sigma_2^2 \le \sigma_y^2 \qquad (9.19a)$$

where σ_y is the uniaxial yield stress. For a given β-ratio Eq. 9.19a limits σ_1 to

$$\sigma_1 \le \sigma_y / (1 - \beta + \beta^2)^{1/2} \qquad (9.19b)$$

If $(\sigma_1)_{cr}$ from Eq. 9.18a exceeds the limiting elastic stress value in Eq. 9.19b it must be corrected for plasticity. For this, we can use the uniaxial plasticity reduction factor given in Fig. 9.8. However, when applying this figure to biaxial compression it becomes necessary to redefine the abscissa as $(\sigma_1)_{cr}/\sigma_n'$. Here σ_n' follows from Eq. 9.19b as: $\sigma_n' = \sigma_n/(1 - \beta - \beta^2)^{1/2}$ where σ_n refers to the plastic stress value corresponding to a secant modulus $E_s = E/2$ under uniaxial stressing.

9.7 Plate Under In-Plane Shear

When the sides of a thin plate $a \times b \times t$ are subjected to shear stress τ, the principal stress state (σ_1, σ_2) within the plate becomes one of diagonal tension and compression. The principal stresses are equal in magnitude but have opposite sense in compression and tension giving $\sigma_1 = -\tau$ and $\sigma_2 = \tau$ respectively, as shown in Fig. 9.11a. Hence, we have a principal, biaxial stress ratio $\beta = -1$ under pure shear for which a buckling condition will arise. Shear buckling of flat plates occurs with parallel wrinkles lying perpendicular to the compressive stress σ_1 (see Fig. 9.11b).

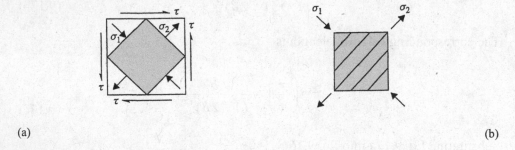

(a)   (b)

Figure 9.11 Plate in shear showing principal stress directions and wrinkling normal to σ_1

9.7.1 Elastic Buckling Stress

To achieve the pure shear state within Fig. 9.10 the direction of σ_2 is reversed to tension and its magnitude is made equal to that of σ_1. Upon setting $\beta = -1$, $n = 1$ and $r = a/b \gg 1$ in Eq. 9.14, the critical, elastic compressive stress σ_1 for shear is:

$$(\sigma_1)_{cr} = \frac{D\pi^2}{a^2 t} \frac{(m^2 + r^2)^2}{(m^2 - r^2)} \qquad (9.20a)$$

where m, the number of half-waves in the 1-direction, is found when $(\sigma_1)_{cr}$ in Eq. 9.20a is a minimum. That is:

$$\frac{d(\sigma_1)_{cr}}{dm} = (m^2 - r^2) \times 2(m^2 + r^2) \times 2m - (m^2 + r^2)^2 \times 2m = 0 \qquad (9.20b)$$

which gives $m = \sqrt{3}r$. Taking integral half-wave values $m = 1, 2, 3, \ldots$ within the length a, the corresponding values of the ratio $r = 1/\sqrt{3}, 2/\sqrt{3}, 3/\sqrt{3}, \ldots$ apply to these minima. From Eq. 9.20a, each (m, r) combination gives the common buckling stress expression:

$$(\sigma_1)_{cr} = \frac{8\pi^2 D}{tb^2} = \frac{2\pi^2 E}{3(1-v^2)}\left(\frac{t}{b}\right)^2 \qquad (9.21a)$$

Given that $\tau_{cr} = (\sigma_1)_{cr}$, the critical elastic shear stress for any (non-minimal) r-value takes a similar form to Eq. 9.21a:

$$\tau_{cr} = K_e E\left(\frac{t}{b}\right)^2 \qquad (9.21b)$$

where b is the shorter side and K_e depends upon the manner in which the sides are supported. Figure 9.12 provides the elastic shear buckling coefficient K_e for various edge fixings when $v = 0.3$ [13]. If $v \neq 0.3$ the K_e value should be corrected with the multiplication factor: $(1 - 0.3^2)/(1 - v^2)$.

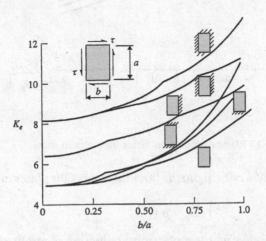

Figure 9.12 Dependence of shear buckling coefficient on plate geometry and edge-fixing ($v = 0.3$)
Key: shaded sides - fixed; clear sides - simply supported

Pure shear also arises when thin strips and plates are subjected to axial torsion. Thus, for the thin strip considered earlier in Fig. 7.7 (see p. 253), the length L replaces a so that we may find the elastic buckling coefficient where b/L has replaced the plate aspect ratio $b/a \leq 1$. Figure 9.12 is used extensively to provide optimum designs arising from the coincidence of critical shear stress levels within appropriate failure criteria including local plate buckling.

9.7.2 Inelastic Shear Buckling

Equation 9.21b supplies a critical *elastic* buckling stress $\tau_{cr} = (\tau_{cr})_e$ but it will need to be corrected if this stress is found to exceed the shear yield stress τ_y of the plate material [14]. The correction is similar to that described in § 9.4 for the effect of plasticity upon the compressive buckling stress. Again, the Ramberg–Osgood description [7, 8] of the material's uniaxial flow curve is used (see Eq. 9.10) to derive a plasticity reduction factor μ for shear. This factor is provided graphically in Fig. 9.13 [13, 14].

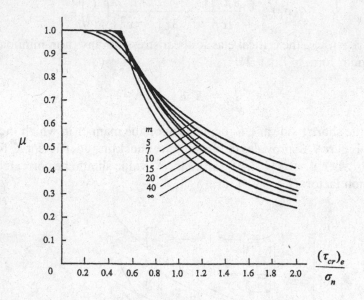

Figure 9.13 Plasticity reduction factor for plates in shear

Thus, for the calculation of a critical *plastic* buckling stress for plates either in shear or in torsion:

$$(\tau_{cr})_p = \mu \, (\tau_{cr})_e \qquad (9.22)$$

The following example shows where a plasticity reduction is required.

Example 9.4 A 4.5 mm-thick DTD 687 aluminium alloy plate, with an aspect ratio: $b/a = 200$ mm / 267 mm $= 0.75$, has both its longer sides clamped and its shorter sides simply supported. Show that inelastic shear buckling will occur and find the critical buckling stress. Take the tensile properties for this material $E = 73$ GPa, $\sigma_n = 340$ MPa and from Table 9.3, $m = 15$.

Firstly, Fig. 9.12 gives the elastic shear buckling coefficient $K_e = 10$ for a plate supported in this manner. The critical, elastic buckling stress follows from Eq. 9.21b as:

$$(\tau_{cr})_e = 10 \times 73 \times 10^3 \, (4.5/200)^2 = 369.6 \text{ MPa}$$

This is well in excess of the limiting elastic shear stress, which is estimated from $\sigma_n = 340$ MPa and the von Mises yield criterion (see Eq. 12.12c) to be less than $340/\sqrt{3} = 196.3$ MPa. Hence, the abscissa in Fig. 9.13 becomes:

$$\frac{(\tau_{cr})_e}{\sigma_n} = \frac{\tau_e}{\sigma_n} = \frac{369.6}{340} = 1.087$$

Correspondingly, with $m = 15$ from Table 9.3, the plasticity reduction factor is read from Fig. 9.13 as $\mu = 0.56$. The plastic buckling stress follows from Eq. 9.22 as:

$$(\tau_{cr})_p = 0.56 \times 369.6 = 207 \text{ MPa}$$

Note that Fig. 9.13 shows $\mu = 1$ when $(\tau_{cr})_e / \sigma_n \leq 0.4$ and therefore no correction would be required.

9.8 Tension Field Beams

Once buckled in shear, a thin plate cannot sustain a further increase in diagonal compression though an increase in diagonal tension is possible. Hence any increase in load beyond the critical buckling load is carried solely by an increase in the tensile stress σ_t. Experiment has shown that the thin webs of beams buckle under a diagonal compressive stress σ_c from loads much lower than the working load. For very thin webs $\sigma_c \ll \sigma_t$, this allowing the assumption that $\sigma_c = 0$ for when the web acts as a purely tension field beam. Longer tension field beams would normally be stiffened with the attachment of vertical stiffeners to the web at regular intervals b, as shown in Fig. 9.14.

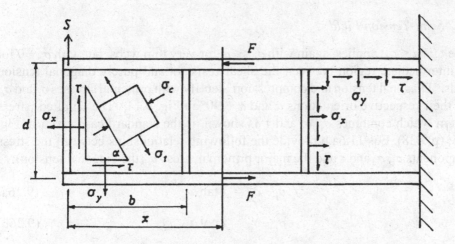

Figure 9.14 Web-stiffening in a tension field beam

To attain a pure tension field the shear load S must pass firstly through shear resistant and semi-tension stages as it is increased from zero to its ultimate value for the beam. In their respective analyses that follow, it is taken that the web supports only shear and the flange supports only bending. The boom refers to a concentration of material placed in line with and separated by the web of depth d. Typically, the boom area A_b would be made equal to each flange area within uniform, asymmetric, thin-walled sections (e.g. I, T, Z and $\sqsubset$ sections).

9.8.1 Shear Resistant Beam

The web shear stress and shear flow are assumed constant:

$$\tau = \frac{S}{dt}, \qquad q = \tau t = \frac{S}{d} \tag{9.23a,b}$$

In this condition normal stresses are absent along horizontal and vertical directions within the web, i.e. $\sigma_x = \sigma_y = 0$. Also the vertical stiffeners carry no end load. However, the booms carry an end-loading F shown which constitutes a couple with applied bending moment M for all positions in the length. Thus at x:

$$M = Fd = Sx, \qquad F = \frac{Sx}{d} \tag{9.24a,b}$$

Consequently, Eq. 9.24b shows that the boom's direct stresses are equal in magnitude in tension and compression:

$$\sigma = \pm \frac{F}{A_b} = \pm \frac{Sx}{dA_b} \tag{9.25a,b}$$

No buckling of the web occurs when $\sigma < \tau < \tau_{cr}$ in which τ_{cr} is the critical shear stress for the web as given by Eq. 9.21b.

9.8.2 Semi-Tension Field

Here $\tau_{cr} > \tau < \tau_{ult}$ applies, again with $\sigma_c << \sigma_t$ for very thin webs, but with $\sigma_c \neq 0$ for the intermediate region between the shear resistant and purely diagonal tension fields. Diagonal tension and compression identify with principal stresses σ_t and σ_c, and their respective orientations α and $\alpha + 90°$ in Fig. 9.14. The 'applied' stress system which combines σ_x, σ_y and τ as shown, is the standard case given in Fig. 11.5, (p. 438). Eqs 11.6a,b provide the following relationships between the stress components σ_x, σ_y and τ and the major principal stress σ_t (the diagonal tension):

$$\sigma_t - \sigma_x = \tau \tan\alpha \tag{9.26a}$$

$$\sigma_t - \sigma_y = \tau \cot\alpha \tag{9.26b}$$

For the perpendicular plane setting $\alpha \rightarrow \alpha + 90°$ and Eqs 9.26a,b give the corresponding relationships for diagonal compression:

$$\sigma_c - \sigma_x = \tau \tan(\alpha + 90°) \tag{9.27a}$$

$$\sigma_c - \sigma_y = \tau \cot(\alpha + 90°) \tag{9.27b}$$

Multiplying Eqs 9.26a,b and 9.27a,b leads to the usual expression for the principal stresses, combined as follows:

$$\sigma_{t,c} = \frac{1}{2}(\sigma_x + \sigma_y) \pm \frac{1}{2}\sqrt{(\sigma_x - \sigma_y)^2 + 4\tau^2} \tag{9.28a,b}$$

In the graphical solution to Eqs 9.28a,b it is convenient to introduce a k-factor based upon the component stress ratios as follows:

$$k = 1 + \frac{\sigma_c}{\sigma_t} = \frac{1 + \sigma_y/\sigma_x}{\frac{1}{2}(1 + \sigma_y/\sigma_x) + \frac{1}{2}\sqrt{(1 - \sigma_y/\sigma_x)^2 + 4(\tau/\sigma_x)^2}} \qquad (9.28c)$$

where k depends upon the known ratio τ/τ_{cr} between Eqs 9.23a and 9.21b. This factor lies between $k = 0$ ($\sigma_c/\sigma_t = -1$) for the shear resistant field and $k = 1$ ($\sigma_c/\sigma_t = 0$) for the purely tension field, in the manner of Fig. 9.15a [15].

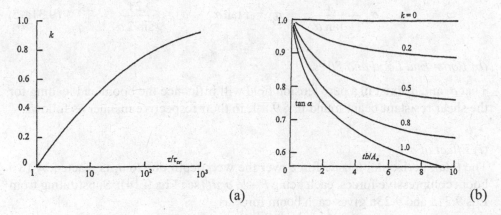

(a)   (b)

Figure 9.15 Diagonal tension factor and its orientation

Equation 9.26a provides the inclination of the diagonal tension:

$$\tan \alpha = \frac{\sigma_t/\sigma_x - 1}{\tau/\sigma_x} \qquad (9.29)$$

Once k is found from Fig. 9.15a then α follows from the graphical solution to Eq. 9.29, provided in Fig. 9.15b. This figure requires the calculation of the geometric ratio tb/A_e where tb is the web area associated with one stiffener of section area A in the case where stiffeners are attached, with spacing b, to both sides of the web. Where they are attached to one side of the web an effective section area must be used to allow for the eccentricity e its centroid to the mid-plane of the web:

$$A_e = \frac{A}{1 + (e/\rho)^2}$$

9.8.3 Pure Tension Field

When $\tau > \tau_{cr}$ and $\sigma_c \approx 0$ the web acts as series of parallel strips with inclination α each under a uniaxial tension σ_t. Here the normal stress components σ_x and σ_y must accompany the shear stress so that the stresses upon the wedge element of material shown in Fig. 9.14 remain in static equilibrium when $\sigma_c = 0$.

(a) Stress State

Resolving forces in the direction of diagonal tension confirms Eqs 9.26a,b (see Exercise 9.13). For the perpendicular plane $\alpha + 90°$ the minor principal stress (the diagonal compression) σ_c is now absent and so Eqs 9.27a,b become:

$$0 - \sigma_x = \tau \tan(\alpha + 90°) \tag{9.30a}$$

$$0 - \sigma_y = \tau \cot(\alpha + 90°) \tag{9.30b}$$

in which τ is given by Eq. 9.23a. Solving Eqs 9.26a,b with Eqs 9.30a,b leads to:

$$\sigma_x = \frac{\tau}{\tan \alpha}, \quad \sigma_y = \tau \tan\alpha, \quad \sigma_t = \frac{2\tau}{\sin 2\alpha} \tag{9.31a-c}$$

(b) Boom End Load and Stress

That σ_x and σ_y exist in a pure tension field will influence the boom end loading for the shear resistant beam found in § 9.8.1, in their respective manner as follows.

(i) Effect of σ_x

The total, horizontal tensile force over the web depth due to σ_x is reacted by two boom compressive forces, each being $F = \frac{1}{2}\,\sigma_x dt$ (see Fig. 9.14). Substituting from Eqs 9.31a and 9.23a gives each boom force as

$$\frac{\sigma_x dt}{2} = \frac{\tau dt}{2\tan\alpha} = \frac{S}{2\tan\alpha} \tag{9.32a-c}$$

Accounting for this compression the combined stress in the booms follows from Eqs 9.25b and 9.32c as:

$$\sigma = \pm\frac{Sx}{dA_b} - \frac{S}{2A_b\tan\alpha} \tag{9.33}$$

(ii) Effect of σ_y

The total vertical force upon the width of each interior panel is $\sigma_y bt$, which is distributed uniformly upon the boom per unit of its length as $w = \sigma_y t$ in Fig. 9.16a.

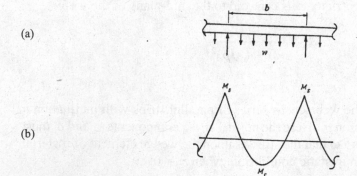

Figure 9.16 Vertical force distribution along boom

This boom distribution is supported by a compressive end load in each vertical stiffener (see § 9.8.3c). Hence the boom may be regarded as a continuous beam with its distributed loading w simply supported at regular intervals b by the panel stiffeners. Figure 9.16b shows the bending moment diagram for an interior panel in which the hogging moment at each vertical stiffener position is given by:

$$M_s = \frac{wb^2}{12} = \frac{\sigma_y t b^2}{12} = \frac{\tau t b^2 \tan \alpha}{12} \qquad (9.34a)$$

The sagging moment at each mid-span position is:

$$M_c = \frac{wb^2}{24} = \frac{\sigma_y t b^2}{24} = \frac{\tau t b^2 \tan \alpha}{24} \qquad (9.34b)$$

These moments are converted to a bending stress distribution for the boom using Eq. 5.4: $\sigma = M_{s,c} y / I_B$ where y defines the position in the boom section from its centre and I_B is its second moment of the boom area. This is added to Eq. 9.33 to give the net stress in the boom at position y:

$$\sigma_{B_{s,c}} = \pm \frac{Sx}{dA_B} - \frac{S}{2A_B \tan \alpha} \pm \frac{M_{s,c} y}{I_B} \qquad (9.35)$$

where S is referred to x in the length and $M_{s,c}$ are taken from Eqs 9.34a,b at the respective stiffener (subscript s) and centre (subscript c) positions within the length b. The following example shows that the greatest influence of this effect occurs at the top and bottom surfaces of each boom where y equals one half the boom depth.

Example 9.5 Find the maximum, net boom stresses for a thin-walled I-section cantilever beam with flange width c, web depth d of uniform thickness t when stiffened at b intervals throughout its length L.

Here the booms replace the two flanges with equal section area. Setting the following relationships:

$$A_B = ct, \quad I_B = \frac{ct^3}{12}, \quad y = \frac{t}{2}$$

we have from Eq. 9.35, the net, top and bottom surface stresses at the boom's stiffener positions, as defined by x:

$$\sigma_{B_s} = \frac{S}{cdt}\left(x - \frac{d}{2\tan\alpha} \pm \frac{b^2\tan\alpha}{2t} \right)$$

in which the orientation α remains to be determined (see § 9.8.3f). When the end force S acts as shown in Fig. 9.14 the third term is positive for a hogging moment at the each boom's top surface and negative for its bottom surface. At the centre of each stiffened panel, where, again, x must be specified, Eq. 9.35 provides the net surface stresses as:

$$\sigma_{B_c} = \frac{S}{cdt}\left(x - \frac{d}{2\tan\alpha} \pm \frac{b^2\tan\alpha}{4t} \right)$$

Under a sagging bending moment, the third term is positive for the each boom's bottom surface and negative for its top surface.

(c) Stiffener End Load and Stress

In a shear resistant beam (see § 9.8.1) there is no axial loading placed upon the stiffeners. It can be seen from Eq. 9.21b, that with the stiffener spacing b, their role is to control the critical shear stress at which buckling occurs. In the presence of normal stress components σ_x and σ_y for the pure tension field each component introduces end and lateral loading to the stiffener in the following manner.

(i) Effect of σ_y
In the presence of σ_y, a compressive end load wb acts upon each stiffener from adjacent spans. The end-load follows from Eq. 9.31b as

$$wb = \sigma_y tb = \tau t b \tan\alpha \tag{9.36a}$$

where α is to be defined (see § 9.8.3e). Hence, the uniform compressive stress in a stiffener, with section area A_s, follows from Eq. 9.36a

$$\sigma_s = \frac{wb}{A_s} = \frac{\tau t b \tan\alpha}{A_s} \tag{9.36b}$$

(ii) Effect of σ_x
The horizontal normal stress places a distributed lateral load $\sigma_x t$ along the depth d of the vertical stiffener. Only the free-end stiffener may be treated as a beam with this distributed loading. For interior stiffeners, the distribution applies to each side of the stiffener from its adjacent panels. The opposing distributions ensure that the stiffener does not bend and therefore, unlike the boom, Eq. 9.36b provides the net axial stress in the stiffener from a single source.

(d) Web–Boom Attachments

The stress state upon a horizontal plane in Fig. 9.14, shows that σ_y and τ act upon the attachment between web and boom across each panel width. The former stress places a similar normal load distribution upon the web–boom connection to that acting upon the boom, i.e. $w = \sigma_y t$. The shear stress τ produces a shear load distribution, known as the shear flow $q = \tau t$. Hence the resultant load Q acting, with inclination α, upon each panel between the stiffeners becomes:

$$Q = \sqrt{(\tau tb)^2 + (\sigma_y tb)^2} \tag{9.37a}$$

Substituting Eq. 9.31b into Eq.9.37a, this resultant force Q is distributed along the horizontal attachment as follows:

$$\frac{Q}{b} = \tau t \sqrt{1 + \tan^2\alpha} = \tau t \sec\alpha \tag{9.37b}$$

(e) Web–Stiffener Attachments

For the vertical plane within Fig. 9.14, a stress state σ_x and τ act upon the attachment between web and stiffener over the panel depth d. The former stress places a normal load distribution upon the web–stiffener connection, i.e. $w = \sigma_x t$. The shear stress τ again produces a shear flow $q = \tau t$ similar to that in § 9.8.3d, i.e. q is constant around the panel perimeter. Hence, the resultant load R acting with an inclination $\alpha + 90°$ upon each panel between its stiffeners becomes:

$$R = \sqrt{(\tau t d)^2 + (\sigma_x t d)^2} \qquad (9.38a)$$

Substituting Eq. 9.31a into Eq. 9.38a, shows that this resultant is distributed along the vertical attachment as follows:

$$\frac{R}{d} = \tau t \sqrt{1 + \cot^2 \alpha} = \tau t \csc \alpha \qquad (9.38b)$$

(f) Orientation of Diagonal Tension

When it is arranged that the boom and stiffener have equal stiffness, or that they are infinitely rigid, then $\alpha = 45°$. In the former case the deflection profile for the two components may be matched only at the free end when their ends are taken to be simply supported. Using the standard expression (see Example 6.8, p. 205) for the central deflection Δ of a beam under uniformly distributed loading, the match in central stiffness w/Δ occurs when:

$$\frac{(EI)_s}{d^4} = \frac{(EI)_B}{b^4} \qquad (9.39)$$

Thus, given a uniform material the end-panel design, Eq. 9.39 shows that its geometry should conform to $I_s/d^4 = I_B/b^4$. Equation 9.39 does not apply to the interior panels whose stiffeners do not bend. Here, it has been shown, from Eqs 9.35 and 9.36b, how the character of the axial stress within the boom and stiffener differ. Under a uniform compression the latter would be stiffer than a boom in bending and therefore it would be expected that $\sigma_x > \sigma_y$. With the exception of the end panels, experiment has shown that α is usually found to be less than 45°, though it is rarely less than 38° for thin aerofoil panels [15]. A more rigorous analysis of the orientation α lies within its correspondence to a minimum in the total strain energy of the web–boom–stiffener assembly [16].

9.9 Secondary Buckling of Plate Elements

Where a cross section, formed from plate elements, is subjected to an axial compression an interconnected plate may bow at its centre or one with a free edge may rotate. Secondary buckling can appear in both slender and wide struts in the manner of Figs 9.1a,b and 9.2a,b. Here, a local buckling refers to where the half-wavelength of bowing is generally of the same order as the flat width. The local buckling stress is independent of an axial length that is at least three times the width of the largest flat. Stresses at the plate junctions grow more rapidly than the average section stress and can become sufficiently high to cause a local collapse. In the determination of the local buckling stress due consideration must be given to the manner of the flat support. At the junction between two flats a rotational restraint is exerted by one flat upon the other. For example, each plate within the box section shown in Fig. 9.1a has both its long edges supported by its connecting plates. Each flange within the I-section in Fig. 9.1b has its centre supported and both edges free. The web has both its edges supported. The degree of restraint lies between that provided by simple supports and a full clamping. Open sections with equal width flats (see Fig. 9.17a–c) may be taken as having their edges simply supported along the join since neither flat can exert a restraint upon the other.

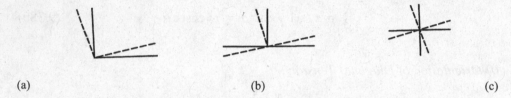

(a)   (b)   (c)

Figure 9.17 Local buckling in sections with equal flats

Equal width flats will buckle simultaneously in uniform thickness cross sections so that the critical stress for the section becomes that for a single flat. A similar, coincident buckling behaviour can be arranged in sections with unequal flats by adjusting their thicknesses. It follows that the analyses of plate buckling in thin-walled sections must consider the geometry of the section's individual flats.

9.9.1 Local Compressive Buckling

(a) Equal Flats
Conveniently, that form of uniaxial plate buckling expression, given in Eq. 9.7b, may be applied to each limb as K_e provides for the appropriate restraint exerted upon it by neighbouring limbs. We may apply Fig. 9.6 to each limb when it is taken to be a plate under uniform compression with specified edge supports. A similar approach is taken to find the buckling coefficient for the horizontal plates in thin-walled beam sections placed under bending. However, the vertical plate of a beam-section experiences a stress gradient, from tension to compression, this providing a greater resistance to local buckling.

Use is made of Fig. 9.6 to find the critical, elastic stress for local buckling of each flat within a thin-walled section placed under uniform compression. The ordinate in this figure is the scaling factor η, for each plate's length/width ratio (a/b) given its support condition. The critical compressive stress follows as

$$(\sigma_{cr})_e = \eta\,(\sigma_{cr}) = \frac{\eta\,\pi^2 E}{3(1-v^2)}\left(\frac{t}{b}\right)^2 = K_e E\left(\frac{t}{b}\right)^2 \qquad (9.40a)$$

in which Eq. 9.6b provides the reference stress σ_{cr} for a simply supported plate with identical loading and geometry (a/b and t). In the case of simple supports it is seen that $\eta = 1$ and when taken with $v = 0.3$, Eq. 9.40a gives $K_e = 3.62$. For other edge supports, η accounts for the change to the elastic restraint within which the buckling coefficient follows from Eq. 9.40a and Fig. 9.6 as:

$$K_e = \frac{\eta \pi^2}{3(1 - v^2)} \qquad (9.40b)$$

The buckling stress, so modified for the given supports, may need scaling down further should plasticity arise. Here $(\sigma_{cr})_e$, found from Eqs 9.40a,b, would have exceeded the stress at the material's proportional limit. Figure 9.8 supplies the plasticity reduction factor μ, based upon a Ramberg–Osgood hardening equation, Eq. 9.10. This reduction factor is applied to Eq. 9.40a as follows:

$$(\sigma_{cr})_p = \mu \, (\sigma_{cr})_e = \mu \eta \, \sigma_{cr} \qquad (9.41a)$$

Substituting Eq. 9.40a into Eq. 9.41a gives the plastic buckling stress:

$$(\sigma_{cr})_p = \frac{\mu \eta \pi^2 E}{3(1 - v^2)} \left(\frac{t}{b} \right)^2 = \mu K_e E \left(\frac{t}{b} \right)^2 \qquad (9.41b)$$

An example of the application of Eq. 9.41a to an optimum design now follows.

Example 9.6 Each side plate within a thin-walled, square steel tube is 50 mm wide and 1 mm thick. Find the critical, flexural and local plastic buckling stress when the tube acts as a 1.5-m long strut subjected to axial compression. What should the width and thickness be for an optimally designed cross section? Take the material properties as: $E = 210$ GPa, $v = 0.27$, $m = 5$, $\sigma_y = 250$ MPa and $\sigma_n = 350$ MPa.

Flexural Buckling
The following approximation applies to the second moment of area about the neutral axis for a thin-walled square tube:

$$I = 2 \left[\frac{bt^3}{12} + bt \left(\frac{b}{2} \right)^2 + \frac{tb^3}{12} \right] \approx \frac{2tb^3}{3}$$

and with the section's wall area $A = 4bt$:

$$k^2 = \frac{I}{A} = \frac{2tb^3/3}{4bt} = \frac{b^2}{6}$$

The critical elastic flexural buckling stress follows from the Euler theory as:

$$\sigma = \frac{\pi^2 E}{(L/k)^2} = \frac{\pi^2 E b^2}{6L^2} \qquad (i)$$

$$\sigma = \frac{\pi^2 \times (207 \times 10^3) \times 50^2}{6 \times (1.5 \times 10^3)^2} = 378.3 \text{ MPa}$$

As this stress exceeds the yield stress $\sigma_y = 250$ MPa it is necessary to examine whether the strut's integrity is pre-determined by a local buckling mode.

Local Buckling
All four sides may be assumed to be simply supported when they bow in secondary buckling. Equation 9.6b provides the critical, local elastic buckling stress:

$$\left(\sigma_{cr}\right)_e = \frac{\pi^2 E}{3(1 - v^2)} \left(\frac{t}{b}\right)^2 = \frac{\pi^2(210 \times 10^3)}{3(1 - 0.27^2)} \left(\frac{1}{50}\right)^2 = 298.1 \text{ MPa}$$

This lesser stress indicates a more likely local buckling mode for this strut but as the critical value exceeds the yield stress, it requires a correction for plasticity. The abscissa in Fig. 9.8 is $298.1/350 = 0.85$, from which the ordinate $\mu = 0.9$ applies to $m = 5$. Hence, Eq. 9.41a provides the critical plastic buckling stress:

$$(\sigma_{cr})_p = \mu(\sigma_{cr})_e = 0.9 \times 298.1 = 268.3 \text{ MPa}$$

Optimum Design
The separate analyses given of flexural and local buckling stress, point to one method of optimising a strut's geometry. Thus, Eqs i and 9.6b ensure that local and flexural buckling occur simultaneously at the yield stress when:

$$\sigma_y = \frac{\pi^2 E b^2}{6L^2} = \frac{\pi^2 E}{3(1 - v^2)} \left(\frac{t}{b}\right)^2$$

providing an optimum breadth:

$$b = \sqrt{\frac{6L^2 \sigma_y}{\pi^2 E}} = \sqrt{\frac{6 \times (1.5 \times 10^3)^2 \times 250}{\pi^2 \times 210 \times 10^3}} = 40.65 \text{ mm}$$

and thickness:

$$t = \frac{3\sigma_y L}{\pi^2 E} \sqrt{2(1 - v^2)} = \frac{3 \times 250 \times (1.5 \times 10^3)}{\pi^2 \times (207 \times 10^3)} \sqrt{2(1 - 0.27^2)} = 0.74 \text{ mm}$$

(b) Unequal Flats
Many thin-walled strut sections have straight, unequal sides. Irregular polygon closed tubes– triangular, rectangular, etc.–have sides with different lengths. The edges of the plates (flats) within closed, regular polygon tubes may be assumed to be simply supported (see Fig. 9.1a). However, in closed, irregular tubes the restraint which exists between the bowing of unequal flats disclaims an assumption of simple edge supports. In open sections, typically I, ⊤, ⌐, ⌊ and ⊔, the flange and

web lengths often differ. The 'flats' within open sections are prone to a local buckling mode between their junctions, as shown in Fig. 9.1b. The webs may be taken to have their ends simply supported at the flange centre but the flanges have their ends unsupported. Referring to Fig. 9.18, it is seen how the local buckling coefficient K_e depends upon the d/h ratio and the neighbouring restraints from plates within four of the most common thin-walled cross sections for struts [17, 18]. The plate which defines each section depth h, may be taken to have simple edge supports from the neighbouring flats when the strut length is at least four greater than h. Equation 9.40a takes a common form for the plate's local elastic compressive buckling stress

$$(\sigma_{cr})_e = K_e E \left(\frac{t}{h} \right)^2 \tag{9.42}$$

The elastic coefficient K_e applies to wall thickness $t < d/5$ and Poisson's ratio $v = 0.3$. When $v \neq 0.3$, Eq. 9.40a shows that K_e should be corrected by a multiplying factor: $(1 - 0.3^2)/(1 - v^2)$. If $(\sigma_{cr})_e$, from Eq. 9.42, is found to exceed the yield stress of the strut material it becomes necessary to employ the plasticity reduction factor μ in Eq. 9.41a, as was shown in Example 9.6.

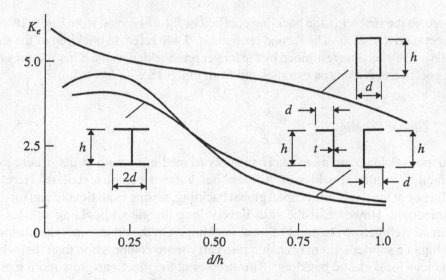

Figure 9.18 Local buckling coefficient for uniform thin-walled sections

The flats that are open flanges require an account of the degree to which a supported side restrains an unsupported side. We may take $\eta = 0.14$ from the second lowest plot in Fig. 9.6 given that the strut length is more than three times greater than dimension d. Equation 9.40b gives the buckling coefficient as $K_e \approx 0.5$ for $v = 0.3$ within which the critical buckling stress for each flange becomes:

$$(\sigma_{cr})_e = \frac{E}{2} \left(\frac{t}{d} \right)^2 \tag{9.43}$$

Local buckling begins at the lesser of the two critical stresses given by Eqs 9.42 and 9.43. The onset of local instability results in a loss of stiffness but not necessarily in complete failure. After buckling the section may continue to carry more load before the remaining flats buckle. Ignoring work hardening, the total load to produce local buckling across each flat can be estimated from adding the buckling loads for individual flats. Hence, for the open sections shown in Fig. 9.18, the full local buckling load is estimated from:

$$P_{cr} = E\left[htK_e\left(\frac{t}{h}\right)^2 + dt\frac{n}{2}\left(\frac{t}{d}\right)^2\right] \tag{9.44a}$$

where n refers to the number of flanges with unsupported edges: the channel and Z-section have 2, an I-section has 4.

In the case of a uniform rectangular tube, the upper plot in Fig. 9.18 applies to local buckling of its longer flats. The shorter flats might be assumed to post-buckle given that the load can be raised further to attain the magnitude:

$$P_{cr} = 2Et\left[dK_e\left(\frac{h}{d}\right)\left(\frac{t}{h}\right)^2 + hK_e\left(\frac{d}{h}\right)\left(\frac{t}{d}\right)^2\right] \tag{9.44b}$$

where, in the first term, the buckling coefficient $K_e(d/h)$ is read from Fig. 9.18 at an aspect ratio $d/h < 1.0$. The second term in Eq. 9.44b refers to buckling of the short walls as they become restrained by the longer pre-buckled walls. This requires that the coefficient $K_e(h/d)$ be extrapolated from Fig. 9.18 with $h/d > 1$.

9.9.2 Global Buckling

Equations 9.44a,b are more likely to apply to medium-length struts, where local instability results in a loss of stiffness but without complete failure. Here the influence of local instability upon global buckling, arising from flexure and torsion, is important. However, if the strut is very long the global buckling stresses are attained well before the onset of local buckling. On the other hand, the plasticity arising in a shorter strut may enable it to carry more compression than that which produces local elastic buckling. The stresses at the junctions grow more rapidly than the average section stress and can become sufficiently high to cause a local plastic collapse known as crippling. Here, the final failure load is estimated from a crippling stress σ_c, which exceeds that at the proportional limit [17]:

$$\sigma_c = \sqrt{\sigma_{cr} \times \sigma_{0.1}} \tag{9.45}$$

where σ_{cr} is taken to be the local buckling stress, either elastic or plastic, estimated from Eqs 9.40a,b–9.42 and $\sigma_{0.1}$ refers to the 0.1% compressive proof stress for the strut material. Test data shows that Eq. 9.45 is accurate to within 10% as the following example shows.

Example 9.7. Estimate the local buckling and crippling stresses and loads for medium- and short-length struts with the Z-section, shown in Fig. 9.18. Take dimensions $h = 125$ mm, $d = 25$ mm and $t = 4$ mm, for an aluminium section with $E = 74$ GPa, $\sigma_n = 280$ MPa, $m = 10$ and $\sigma_{0.1} = 325$ MPa.

Elastic Analysis
From Fig. 9.18, we find $K_e = 4$ for $d/h = 0.2$. Hence, Eq. 9.42 gives the local elastic compressive buckling stress for the web:

$$(\sigma_{cr})_e = 4\,(74 \times 10^3)\left(\frac{4}{125}\right)^2 = 303.1 \text{ MPa}$$

with a corresponding local buckling load:

$$P_{cr} = th(\sigma_{cr})_e = 4 \times 125 \times 303.1 = 151.6 \text{ kN}$$

In post-buckling the two flanges support an additional loading, when according to Eq. 9.44a their buckling loads are:

$$P_{cr} = Eth\left[K_e\left(\frac{t}{h}\right)^2 + \left(\frac{d}{h}\right)\left(\frac{t}{d}\right)^2\right]$$

$$= (74 \times 10^3) \times 4 \times 125\left[4\left(\frac{4}{125}\right)^2 + \left(\frac{25}{125}\right)\left(\frac{4}{25}\right)^2\right] = 341 \text{ kN}$$

For a shorter strut of similar section the crippling stress is estimated from Eq. 9.45:

$$\sigma_c = \sqrt{303.1 \times 325} = 313.9 \text{ MPa}$$

from which the buckling load is:

$$P_{cr} = t(h + 2d)\sigma_c = 4(125 + 50) \times 313.9 = 219.7 \text{ kN}$$

A comparison of the three failure loads shows that local buckling of the web is the most likely failure mode. This mode occurs at a critical stress level which requires a further adjustment for plasticity (to follow) given $\sigma_n = 280$ MPa.

Plasticity Allowance
With a flow stress $\sigma_n = 280$ MPa and hardening exponent $m = 10$ for the aluminium alloy, the ratio $(\sigma_{cr})_e/\sigma_n = 303.1/280 = 1.083$ is used with Fig. 9.8 to give a plasticity reduction factor $\mu = 0.877$. Equation 9.41a provides the corrected, local plastic buckling stress

$$(\sigma_{cr})_p = \mu(\sigma_{cr})_e = 0.877 \times 303.1 = 265.82 \text{ MPa}$$

with a modified local buckling load:

$$P_{cr} = th(\sigma_{cr})_e = 4 \times 125 \times 265.82 = 132.9 \text{ kN}$$

The modified crippling stress becomes

$$\sigma_c = \sqrt{265.82 \times 325} = 293.9 \text{ MPa}$$

from which the crippling load is revised:

$$P_{cr} = t(h + 2d)\sigma_c = 4(125 + 50) \times 293.9 = 205.8 \text{ kN}$$

Again, with its low d/h ratio the web is primarily responsible for buckling. The flanges offer some restraint delaying the buckling until a load is reached for which the cross section as a whole becomes unstable by the appropriate measure for the given strut length.

9.9.3 Local Shear Buckling

(a) Shear Web

A local shear buckling is also possible in a rectangular shear web and in the plates of stiffened and corrugated sections where they are to act as shear webs. Normally, a shear web would be mounted vertically carrying opposing shear forces through its top and bottom fixtures. Here the shear buckling formula for a plate (see § 9.7) is applied to the ratio between each individual flat breadth and the web depth (b/a in Fig. 9.12). Given that the plate's sides are simply supported along the folds of a corrugation, say, we may read from Fig. 9.12 a local buckling coefficient for each flat when it is taken to be a plate under uniform shear. Figure 9.12 also supplies the critical shear stress for local buckling of flat plate elements within a section (i.e. a flange, web or thin wall) when placed under torsion. Here the particular edge condition governs the critical shear stress at buckling:

$$\tau_{cr} = K_e E \left(\frac{t}{b} \right)^2 \tag{9.46}$$

where the local, elastic, shear buckling coefficient, K_e, is read directly from the figure for the flat's width/length ratio (b/a). When the widths b of adjoining flats are different, Eq. 9.46 shows that the flat with greater length b buckles first. Thus, the critical stress for the whole section is based upon the weaker flat but the stress does depend upon the restraint exerted by the other flat(s).

Example 9.8 The widest flat in a corrugated-aluminium shear web, when vertically mounted, has a breadth to depth (length) ratio: $b/a = 1/4$ and a thickness to breadth ratio: $t/b = 1/50$. Find the elastic shear stress to promote local buckling when opposing horizontal shear forces are applied to its top and bottom corrugated edges through (i) simply supports and (ii) rigid clamps. Take $E = 70$ GPa.

(i) *Simple Supports*

Taking $b/a = 1/4$, Fig. 9.12 shows $K_e \approx 5$ for a simply supported plate. Hence, from Eq. 9.46, local shear buckling would be expected to occur when the shear stress attains the value:

$$\tau_{cr} = 5 \times (70 \times 10^3) \times (0.02)^2 = 140 \text{ MPa}$$

(ii) *Rigid Clamps*

Taking the two vertical sides to be simply-supported, Figure 9.12 shows $K_e \approx 5.2$ for the fixed-edge plate. Hence from Eq. 9.46 a local shear buckling would be expected when the shear stress attains the value:

$$\tau_{cr} = 5.2 \times (70 \times 10^3) \times (0.02)^2 = 145.6 \text{ MPa}$$

Here both the critical stress estimates are elastic. Further design data in § 9.7.2 is given to account for plasticity effects should τ_{cr}, from Eq. 9.46, be found to exceed the proportional limit.

(b) Beam

The webs within I-, T- and U-section beams may be prone to local shear buckling when subjected to transverse concentrated loading. Example 5.12 (see p. 175) shows how the shear stress in the web attains its maximum at the neutral axis of bending but falls away where it joins the flanges. Example 5.12 also shows that the web carries more than 90% of the shear force applied to the section. Thus, it may be assumed, reasonably, that an average shear stress in the web will controls its local buckling. In reading the buckling coefficient K_e from Fig. 9.12, b now becomes the web depth d and a is that region of the beam's length for which the shear force is constant. For example, referring to the shear force diagram for an end-loaded cantilever, Fig. 2.15c shows that a is the full length, but in a centrally loaded, simply supported beam a is its half-length (see Fig. 2.15a). Where a beam carries many transverse loads a would be identified with that region between loads where the shear force F has attained a constant maximum, e.g. between A and C in Fig. 2.16 and between A and D in Fig. 2.17. These figures also show that a length a cannot be defined for portions of the beam over which the load is distributed.

Taking the average shear stress to be $\tau_{av} = F/dt$, its critical value for local buckling is again that given by Eq. 9.46. This may be made to coincide with the shear yield stress, when taken from a Tresca yield criterion to be half the tensile yield stress ($\tau_y = \sigma_y/2$, see Eq. 12.6c), so the governing equation for a safe elastic design becomes

$$\frac{F}{dt} = K_e E \left(\frac{t}{d} \right)^2 = \frac{\sigma_y}{2} \qquad (9.47a)$$

from which the web depth and thickness may be pre-selected to lie in the ratio:

$$t = \left(\frac{2F^2}{\sigma_y K_e E} \right)^{\frac{1}{4}}, \qquad d = \left(\frac{8F^2 K_e E}{\sigma_y^3} \right)^{\frac{1}{4}}, \qquad \frac{d}{t} = \left(\frac{2 K_e E}{\sigma_y} \right)^{\frac{1}{2}} \qquad (9.47\text{b-d})$$

In applying Eqs 9.47b–d much depends upon the assessment of the side supports in Fig. 9.12 for a plate of dimension $b \times a$. The side b, i.e. the web depth, will be assumed simply supported to the connecting flanges. The support of side a will depend upon whether a defines the whole or a part of the beam length. In the examples mentioned above the full length of a cantilever a has fixed–free ends but the half-length of a beam with central load a has fixed–simply supported ends. The buckling coefficient may be read from Fig. 9.12 assuming that length a has its ends fixed and simply supported. In each case, Fig. 9.12 shows the lower $K_e = 4.84$ for long beams when $b/a \approx 0$ and $K_e = 9.9$ for short beams where $b/a \approx 1$. Thus, contrary to common belief shear as well as bending can be a controlling factor in the design of long beams when their sections are thin-walled. Bending of the web produces an axial stress gradient that varies from maximum tension to maximum compression over its depth. In fact, local compressive buckling of the web is six times less likely than in the connecting flange which experiences a near uniform maximum compression. Their critical elastic stress expressions are [5]:

$$\sigma_w = 21.7 E \left(\frac{t}{d} \right)^2, \qquad \sigma_f = 3.62 E \left(\frac{t}{d} \right)^2 \qquad (9.48\text{a,b})$$

Despite Eq. 9.48b overriding 9.48a as the critical equation, a local buckling of the web is likely to be the result of combined compression and shear. To account for this it is necessary to establish the principal biaxial stress state (σ_1, σ_2) on the compressive side of the web. Where this shows that the assumption (see § 9.8.1) of pure shear across the web $(\sigma_1 = -\sigma_2)$ is an over-simplification, it becomes necessary to analyse buckling where the magnitudes of the principal compressive and tensile stresses differ $(\sigma_1 \neq -\sigma_2)$. For this, we should return to Eq. 9.14 with the particular (negative) stress ratio $\beta = \sigma_2/\sigma_1$ and plate aspect ratio $r = a/b > 1$ that apply. In Fig. 9.10 the compressive stress σ_1 acts upon the area $b \times t$ as shown but now a tensile stress σ_2 acts upon the area $a \times t$. The earlier analysis, where σ_2 was compressive, was restricted to large r values where $b \ll a$ to justify taking $n = 1$ for one half-wave of buckling in alignment with the y-direction. Where σ_2 is tensile buckling with $n > 1$ is unlikely for all r. The critical compressive stress becomes:

$$(\sigma_1)_{cr} = \frac{D \pi^2}{a^2 t} \frac{(m^2 + r^2)^2}{(m^2 + \beta r^2)} \qquad (9.49\text{a})$$

The number of half-waves for the x-direction, m in Eq. 9.49a, corresponds to a minimum $(\sigma_1)_{cr}$:

$$d(\sigma_1)_{cr}/dm = (m^2 + \beta r^2) \times 2(m^2 + r^2) \times 2m - (m^2 + r^2)^2 \times 2m = 0$$

which leads to

$$m = \sqrt{r^2(1 - 2\beta)} \tag{9.49b}$$

Substituting Eq. 9.49b into 9.49a:

$$(\sigma_1)_{cr} = \frac{4D\pi^2}{b^2 t}(1 - \beta) \tag{9.49c}$$

So far, Eqs 9.49a-c are identical to those given earlier (see Eqs 9.17a–c and 9.18a,b) for buckling under a biaxial compression restricted to large r. It was seen that integral values of m waves appeared with the half-wavelength is a/m. Moreover, Eq. 9.17c showed $a/m = \infty$ for $\beta = \frac{1}{2}$. However, this condition cannot arise within the compressive side of the web since $-1 \le \beta < 0$ between its neutral axis and where it joins the flange. Given the stress state (σ, τ) at a depth position y from the neutral axis (see Fig. 9.16a), the principal stress ratio β within Eq. 9.49c becomes:

$$\beta = \frac{\sigma + (\sigma^2 + 4\tau^2)^{1/2}}{\sigma - (\sigma^2 + 4\tau^2)^{1/2}} = \frac{\alpha + (\alpha^2 + 4)^{1/2}}{\alpha - (\alpha^2 + 4)^{1/2}} \tag{9.50a}$$

in which σ is negative. At any depth position y in the web the *negative* ratio between the bending and shear stresses $\alpha = \sigma/\tau$ appears from Eqs 5.4 and 5.26b as:

$$\alpha = -\frac{|My|/I}{(|F|A\bar{y})/(It)} = -\frac{t|My|}{|F|A\bar{y}} \tag{9.50b}$$

Conveniently, Eq. 9.50b refers the magnitudes of F and M (without sign) to the depth position y (without sign) on the compressive side of the neutral axis. For a uniformly thin, symmetrical I-section with thickness t, web depth d and flange breadth b:

$$I \approx \frac{td^3}{12}\left[1 + 6\left(\frac{b}{d}\right)\right] \tag{9.50c}$$

$$A\bar{y} \approx \frac{td^2}{2}\left\{\frac{b}{d} + \frac{1}{4}\left[1 - \left(\frac{y}{d/2}\right)^2\right]\right\} \tag{9.50d}$$

The approximations apply when t/d is small. Equations 9.50a–d show that for the web centre:

$$y = 0: \quad \alpha = 0, \beta = -1 \text{ (pure shear)}$$

For the web top, α (and β) depend upon M, F and b/d:

$$y = d/2: \quad \alpha = -\frac{M}{Fd(b/d + 1/4)}$$

Consequently, the local buckling analysis may only proceed when the beam geometry is specified and the manner of beam loading is known. The following example illustrates this.

Example 9.9 Examine the possibility of local buckling occurring within the web of a thin-walled I-section (see Fig. 9.19a) of uniform thickness t and having $b = d/2$, when it is to be used as a short cantilever beam of length $L = 4d$, loaded at its free end with a concentrated force F.

With $b = d/2$, Eqs 9.50c,d become:

$$I = \frac{td^3}{3}, \quad A\bar{y} \approx t\left(\frac{d}{2}\right)^2\left\{1 + \frac{1}{2}\left[1 - \left(\frac{y}{d/2}\right)^2\right]\right\} \tag{i, ii}$$

Figure 2.15c shows that over the length of this beam the shear force magnitude $|F|$ is constant while the hogging bending moment varies as:

$$|M| = F(L - z) \tag{iii}$$

where z has its origin at the fixed end. Substituting Eqs ii and iii into Eq. 9.50b provides the stress ratio:

$$\alpha = -\frac{t|My|}{|F|A\bar{y}} = \frac{-L\left(1 - \frac{z}{L}\right)y}{\left(\frac{d}{2}\right)^2\left\{1 + \frac{1}{2}\left[1 - \left(\frac{y}{d/2}\right)^2\right]\right\}} \tag{iv}$$

which indicates where in the length and cross section the lower half of the web is at risk of buckling. The free and fixed ends bound Eq. iv as follows:

Free End
At the free end $z/L = 1$, $\alpha = 0$ and $\beta = -1$, indicating that pure shear exists over the depth. At the neutral axis $y = 0$ the maximum shear stress becomes

$$\tau_{max} = \frac{A\bar{y}}{It} = \frac{t(d/2)^2 \times 3/2}{(td^3/3)t} = \frac{9F}{8td} \tag{v}$$

Buckling would occur when Eq. v attains the critical value, from Eq. 9.21b:

$$\tau_{cr} = K_e E\left(\frac{t}{d}\right)^2 \tag{vi}$$

where $K_e = 5.15$ for $d/L (\equiv b/a) = 0.25$, from Fig. 9.12. Equating (v) and (vi) shows how, as design criteria, F may be limited for a given d/t ratio, or, how d/t should be selected for a given F, respectively:

$$F \leq \frac{4.58Ed^2}{(d/t)^3} \quad \text{or} \quad \frac{d}{t} \geq \left(\frac{4.58Ed^2}{F}\right)^{1/3}$$

Fixed End

When $z = 0$ and $L = 4d$, Eq. iv shows the manner in which α varies with y:

$$\alpha = \cfrac{-8y}{(d/2)\left\{1 + \cfrac{1}{2}\left[1 - \left(\cfrac{y}{d/2}\right)^2\right]\right\}} \qquad \text{(vii)}$$

Again, pure shear ($\alpha = 0, \beta = -1$) applies to the neutral axis where $y = 0$. At the web/flange junction, where $y = d/2$, Eq. vii gives $\alpha = -8$. Correspondingly, Eq. 9.50a gives $\beta = -0.0152$. Between these two positions α varies non-linearly according to Eq. vii, without attaining a stationary value. Similarly, Eq. 9.50a shows that there is no stationary β-value within the range of y. For example, at mid-depth where $y = d/4$ and $\alpha = -2.909, \beta = -0.0965$. As β falls away from -1 most rapidly in the material adjacent to the neutral axis it is likely that a pure shear buckling analysis is the most appropriate choice for this narrow region of the depth across the whole beam length. Here buckling is assumed to occur simultaneously over the full length of this zone with a wavelength aligned with the 45° direction of principal compression (see Fig. 9.19b). At this inclination, we may take a principal element to have simply supported sides in the ratio $b/a = 1$, for which Fig. 9.12 gives the buckling coefficient $K_e = 8.4$.

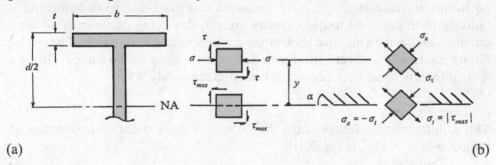

(a)   (b)

Figure 9.19 (a) Web stress state and (b) shear buckling within the length of a narrow shear zone

However, it must be concluded, finally, that the critical elastic stress is realistic only when d in Eq. vi becomes the web thickness and t is a very much smaller web zone depth (giving $t/d \ll 1$) above the NA. In this zone axial bending stress is negligible within interior beam sections.

9.10 Torsional Buckling of Thin Circular Tubes

Previously in Chapter 7 the Bredt-Batho torsion theory (see § 7.3, p. 263) was applied to a thin-walled circular tube with uniform thickness. This theory provides the torque T in terms of the wall shear flow $q = \tau t$ and the cross-sectional area A_o enclosed by the tube wall's mean centre line:

$$T = 2A_o q \qquad (9.51)$$

Firstly, it will be seen that the axial length and radius of a circular tube wall play an important role in its shear buckling behaviour. In fact, a minimal weight design ensures that the shear stress from torsion matches that required to cause shear buckling in the wall [19]. The symbols used for circular tubes, mean radius r_m, wall thickness t and length L, are shown in Fig. 9.20. Torque T is applied about the central, longitudinal axis and this produces a constant shear flow $q = \tau t$ around the wall centre line as shown.

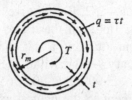

Figure 9.20 Circular tube geometry

9.10.1 Failure Criteria

There are two possible modes of failure: plastic collapse and buckling. In the former it is assumed that once the yield stress has been reached the tube can support no further increase in torque. The assumption implies that work hardening is unlikely to increase the torque capacity greatly, this being particularly so for a circular tube having a minimal shear stress gradient within its thin wall. The second failure mode which affects all thin-walled tubes is torsional buckling. This is a special form of shear buckling which is treated separately here.

(a) Limiting Stress
This failure criterion follows from the maximum shear stress under the applied torque, with $A_o = \pi r_m^2$, in Eq. 9.51:

$$\tau_{max} = \frac{T}{2A_o t} = \frac{T}{2\pi r_m^2 t} \tag{9.52a}$$

where τ_{max} is limited by

$$\tau_{max} \leq \tau_y \tag{9.52b}$$

in which τ_y is a limiting yield (or proof) stress for the tube material. In earlier sections this limiting stress had been placed at k, the stress at the proportional limit, but from now we shall admit an amount of plastic strain (typically 0.1%) with τ_y as a proof stress corresponding to a similar amount of offset strain.

(b) Torsional Buckling
Whilst the shear stress limit from Eq. 9.52b applies to any length of tube, the buckling criterion needs to be divided between the following three length regimes. The respective buckling formulae developed within three categories of tube length are attributed to Gerard [10, 11, 14, 20] and Batdorf et al. [21, 22], who re-

interpreted Donnell's original work [23] on the elastic stability of thin-walled tubes.

9.10.2 Tubes with Short Lengths

A more practical interpretation of the condition $r_m/L \rightarrow \infty$ for a short-length tube is to apply $L < \sqrt{(r_m\, t)}$, in order to identify shear buckling with a known flat plate solution. Thus, when the cylinder's mean circumference $2\pi r_m$ is rolled out flat this length becomes very long compared to the cylinder's short length L. The flat plate buckles under a critical shear stress [14]:

$$\tau_{cr} = \frac{\pi^2 K_s \eta_s E}{12(1 - v_e^2)} \left(\frac{t}{L} \right)^2 \tag{9.53a}$$

where $K_s = 5.35$ for a tube with simply supported ends and $K_s = 8.98$ for clamped ends. For the former K_s value the ends are free to rotate and for the latter the ends are held perpendicular to the twisting action. The plasticity reduction factor η_s is required when the critical shear stress exceeds that at the elastic limit [20]:

$$\eta_s = \left(\frac{1 - v_e^2}{1 - v_p^2} \right)^{\frac{3}{4}} \frac{E_S}{E} \tag{9.53b}$$

where the elastic and plastic Poissons' ratios are taken to be $v_e = 0.3$ and $v_p = 0.5$, respectively, and E_S is the secant modulus at the plastic buckling stress level. Here we may combine Eqs 9.53a,b in a more conservative form using the tangent modulus E_T:

$$\tau_{cr} = KE_T \left(\frac{t}{L} \right)^2 \tag{9.53c}$$

Depending upon the end condition K in Eq. 9.53c takes one of two values, each differing from values of 5.35 and 8.98 (above) by the common multiplying factor:

$$K = \frac{\pi^2 \eta_s K}{12(1 - v_e^2)} \tag{9.53d}$$

9.10.3 Tubes with Medium Lengths

The manner in which torque is applied to the tube ends becomes less important when the tube geometry falls within a range defined by:

$$50 \frac{t}{r} < \left(\frac{L}{r} \right)^2 < 10 \frac{r}{t} \tag{9.54}$$

For a simply supported tube the critical buckling stress is expressed as [10, 11]:

$$\tau_{cr} = \frac{0.7}{(1 - v^2)^{5/8}} E\left(\frac{t}{r}\right)^{5/4}\left(\frac{r}{L}\right)^{1/2} \tag{9.55a}$$

To allow for plasticity and other end constraints we shall modify Eq. 9.55a to:

$$\tau_{cr} = KE_T\left(\frac{t}{r}\right)^{5/4}\left(\frac{r}{L}\right)^{1/2} \tag{9.55b}$$

where K follows from a plasticity reduction factor μ by the usual method (see Example 9.4, p. 346).

9.10.4 Very Long Tubes

In the elastic case the cylinder distorts into an elliptical shape, with two lobe buckling, when the geometry conforms to:

$$\left(\frac{L}{r}\right)^2\left(\frac{t}{r}\right) > a \tag{9.56}$$

where $a = 42$ for simple supports and $a = 60$ for clamped ends, though these and other intermediate end-fixing have a negligible effect on the elastic buckling stress:

$$\tau_{cr} = \frac{0.272}{(1 - v^2)^{3/4}} E\left(\frac{t}{r}\right)^{3/2} \tag{9.57a}$$

With Gerard's allowance for plasticity [20], Eq. 9.53b modifies Eq. 9.57a to:

$$\tau_{cr} = KE_T\left(\frac{t}{r}\right)^{3/2} \tag{9.57b}$$

in which K becomes a 'plastic' buckling coefficient, as with Eq. 9.55b.

9.11 Torsional and Shear Buckling of Non-Circular Tubes

The Bredt-Batho torsion theory (see § 7.3, p. 263) shows that shear flow $q = \tau t$ is constant in a closed thin-walled tube. The derivation allows both τ and t to vary within irregular shaped, non-circular tubes, as in Fig. 7.14. Equation 7.29b shows that in a tube with varying wall thickness, the torque corresponding to the constant shear flow is written as:

$$T = 2A_o q = 2A_o \tau t \tag{9.51}$$

where τ and t apply to any position in the wall and A_o is the area enclosed by the wall within its mean centre path. Figures 9.21a–c give three examples of square and rectangular thin-walled tube sections, with uniform and non-uniform thickness, to which Eq. 9.51 is to be applied.

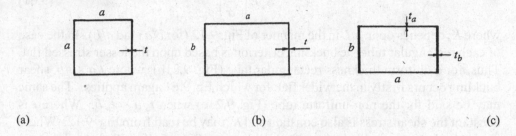

(a)   (b)   (c)

Figure 9.21 Thin-walled torsion tube sections

9.11.1 Limiting Shear Stress

When each tube in Figs 9.21a–c is subjected to an axial torque T, Eq. 9.51 provides the shear stress τ within its wall. This gives, for the two uniform section geometries shown in Figs 9.21a,b, the following constant shear stresses:

$$\tau = \frac{T}{2a^2t}, \qquad \tau = \frac{T}{2abt} \qquad (9.58a,b)$$

For the closed, non uniform rectangular tube (Fig. 9.21c) the shear flow is constant in the wall $q = (\tau t)_a = (\tau t)_b$. Here, the shear stress within each wall thickness is not the same, these being:

$$\tau_a = \frac{T}{2abt_a}, \qquad \tau_b = \frac{T}{2abt_b} \qquad (9.59a,b)$$

where $\tau_a < \tau_b$ if $t_a > t_b$. If this tube is to remain elastic then the design criterion must ensure that the maximum shear stress, i.e. Eq. 9.59b, does not exceed the shear yield stress: $\tau_b < \tau_y$. Since the shear yield stress τ_y is less often published as a material property than the tensile yield stress σ_y, it may be found from the latter using either the Tresca or the von Mises yield criteria, respectively:

$$\tau_y = \frac{\sigma_y}{2} \quad \text{or} \quad \tau_y = \frac{\sigma_y}{\sqrt{3}} \quad (\tau_y = Q\sigma_y) \qquad (9.60a,b,c)$$

Tresca's criterion, Eq. 9.60a, is always the more conservative across the range of Q ($\frac{1}{2} \le Q \le 1/\sqrt{3}$) provided by Eq. 9.60c.

9.11.2 Shear Buckling

The 'plate element' is identified with the individual flats that comprise the sides of each tubular section. Thus, for the square tube in Fig. 9.21a each flat becomes a plate of width a, thickness t and length L. Here we may assume that the sides are simply supported for which the critical buckling stress is given by Eq. 9.21b

$$\tau_{cr} = K_e E \left(\frac{t}{a} \right)^2 \tag{9.61}$$

where K_e depends upon a/L in the manner of Fig. 9.12 (for b/a read a/L). In the case of each rectangular tube the buckling criterion is based upon the lesser stressed flat. Thus, for a uniform thickness rectangular tube (Fig. 9.21b), where $t/a < t/b$, shear buckling occurs firstly in the wider flat, for which Eq. 9.61 again applies. The same may be said for the non-uniform tube (Fig. 9.21c) when $t_a/a < t_b/b$. Where t is constant the shear stress is also constant and K_e may be read from Fig. 9.12. Where t is not constant it is likely that the buckling coefficient K_e will depend upon a degree of restraint that lies within the ratios of the adjoining flat widths and thicknesses. However, if the flats are to buckle simultaneously when $t_a/a = t_b/b$ then K_e may be read directly from Fig. 9.12.

9.11.3 Design Criteria

The most economical tube designs ensure that wall buckling occurs as the shear yield stress of the material is reached. Thus, for a medium-length, thin-walled circular tube with simply supported ends, Eqs 9.52a,b are combined with Eq. 9.55a according to this basis for an elastic design

$$\tau_y = \frac{T}{2\pi r_m^2 t} = \frac{0.7}{(1 - v^2)^{5/8}} E \left(\frac{t}{r_m} \right)^{5/4} \left(\frac{r_m}{L} \right)^{1/2} \tag{9.62a}$$

Taking $v = \frac{1}{3}$ in Eq. 9.62a, with τ_y from Eq. 9.60c, allows the tube's mean radius and thickness to be optimised for a given torque and lengths within the range of Eq. 9.54 as follows:

$$r_m = 0.45 \left[\frac{E^4 T^5}{(Q\sigma_y)^9 L^2} \right]^{\frac{1}{13}}, \quad t = 0.786 \left[\frac{(Q\sigma_y)^5 L^4 T^3}{E^8} \right]^{\frac{1}{13}} \tag{9.62,c}$$

A units check confirms the validity of the indices appearing within Eqs 9.62b,c:

$$r_m \rightarrow \left(\frac{N^4 N^5 m^5 m^{18}}{m^8 N^9 m^2} \right)^{\frac{1}{13}} = m, \quad t \rightarrow \left(\frac{N^5 m^4 N^3 m^3 m^{16}}{m^{10} N^8} \right)^{\frac{1}{13}} = m$$

A similar approach may be adopted to optimise the geometry for the torsional buckling of (i) short and long thin-walled circular tubes (see § 9.10) and (ii) non circular tubes given in Figs 9.21a–c (see Exercises 9.9–9.11).

References

[1]   ESDU 78021, Guide to items on the strength and stability of struts, October 1978
[2]   ESDU 02.01.08, Buckling in compression of sheet between rivets, October 1962.
[3]   ESDU 02.01.09, Inter-rivet buckling curves for specific materials, October 1962.
[4]   Vinson, J. R. *Structural Mechanics*, Wiley, 1974.
[5]   Timoshenko, S. P. and Gere, J. *Theory of Elastic Stability*, Second Edition, McGraw-Hill, 1961.
[6]   ESDU 72019, Buckling of flat isotropic plates under uniaxial and biaxial loading, August 1972.
[7]   Ramberg, W. and Osgood, W. R. Description of stress-strain curves by three parameters, NACA Tech Note 902, April 1943.
[8]   ESDU 76016, Generalisation of smooth, continuous stress-strain curves for metallic materials, May 1985.
[9]   ESDU 83044, Plasticity correction factors for plate buckling, December 1983.
[10] Gerard, G. and Becker, H. *Handbook of Structural Stability*, Part I - Buckling of flat plates, NACA Tech Note 3781, October 1954.
[11] Gerard, G. *Introduction to Structural Stability Theory*, McGraw-Hill, 1962.
[12] Yu, Wei-wen, *Cold Formed Steel Structures*, McGraw-Hill, 1973.
[13] ESDU 71005, Buckling of flat plates in shear, February 1971.
[14] Gerard, G. Critical shear stress of plates above the proportional limit, *Jl Applied Mech*, **15**(1), 7–12, March 1948.
[15] ESDU 02005, Flat Panels in shear - post buckling analysis.
[16] Kuhn, P., Peterson, J. P. and Levin, L. R. A summary of diagonal tension. Part I Method of analysis, NACA TN 2661, Oct 1951; Part II Experimental evidence, NACA TN 2662, January 1952.
[17] ESDU 78020, Local buckling and crippling of I, Z and channel section struts, July 1978.
[18] ESDU 01.01.09, Local buckling and crippling of rectangular tube section struts, November 1978.
[19] Rees, D. W. A. *Mechanics of Optimum Structural Design*, Wiley, 2009.
[20] Gerard, G. Compressive and torsional buckling of thin-walled cylinders in the yield region, NACA Tech Note 3726, 1956.
[21] Batdorf, S. B. A simplified method of elastic-stability analysis for thin cylindrical shells I - Donnell's equations, NACA Tech Note 1341, 1947.
[22] Batdorf, S.B., Stein, M. and Schildercrout, M. Critical stress of thin-walled cylinders in torsion, NACA Tech Note 1345, 1947.
[23] Donnell, L. H. Stability of thin-walled tubes under torsion, NACA Report 479, 1933.

Exercises

9.1 Using Fig. 9.6, find the critical, elastic buckling stress for a 3 mm thick aluminium alloy plate with $a = 500$ mm and $b = 200$ mm when compression is applied through simple supports to its shorter sides with its longer sides being clamped.

9.2 A 5 mm thick S 520 steel plate (see Table 9.3 for the hardening exponent) with side lengths $a = 500$ mm and $b = 250$ mm is subjected to uniaxial compression across its shorter sides. Find the critical plastic buckling stress when all four sides are: (a) simply supported and (b) clamped. The other material properties are: $E = 210$ GPa, $v = 0.28$, $\sigma_y = 310$ MPa and $\sigma_n = 460$ MPa.

9.3 Find the semi-effective width expression and compression carrying capacity for a thin aluminium plate in post-buckling when all four sides are simply supported. Take $v = \frac{1}{3}$.

9.4 A 3.5 mm thick DTD 687 aluminium alloy plate, with $a = 350$ mm and $b = 175$ mm, has both its longer sides clamped and its shorter sides simply supported when placed under shear. Find the critical buckling stress allowing for plasticity. Take $E = 75$ GPa, $\sigma_y = 250$ MPa with the tensile flow stress $\sigma_n = 340$ MPa and $m = 15$ (see Table 9.3).

9.5 Each side within a thin-walled, square steel tube is 40 mm wide and 1 mm thick. Find the critical, flexural stress and the local plastic buckling stress when the tube acts as a 2 m long strut. What should the width and thickness be for an optimally designed cross section? Take the elastic constants as: $E = 210$ GPa and $v = 0.27$ and the material flow properties as: $m = 5$, $\sigma_y = 275$ MPa and $\sigma_n = 360$ MPa.

9.6 Estimate the local buckling and crippling stresses and loads for medium- and short-length struts with the I-section, shown in Fig. 9.18. Take dimensions $h = 125$ mm, $d = 25$ mm and $t = 5$ mm, for an aluminium section with $E = 70$ GPa and $\sigma_{0.1} = 320$ MPa.

9.7 Each flat within a vertically mounted corrugated aluminium shear web has a breadth to depth ratio, $b/a = 1/5$ and a thickness to breadth ratio, $t/b = 1/50$. Find the local, elastic shear buckling stress with shear forces applied to its top and bottom edges through (a) simple supports and (b) rigid clamps. Take $E = 70$ GPa.

9.8 Examine the possibility of local buckling occurring within the web of a thin-walled I-section, of uniform thickness t and flange width $b = d/2$, when used as a short cantilever beam of length $L = 5d$, loaded at its free end with a concentrated force F.

9.9 The geometry of a short circular tube, as defined in § 9.10.2, is to be optimised for torque transmission such that the critical shear stress for torsional buckling (see Eqs 9.53a–c) coincides with the shear yield stress for the tube material. Derive expressions for the tube's thickness and mean radius for a given length and torque.

9.10 Optimise the geometry for the torsional buckling of a long circular tube, as defined in § 9.10.4, by applying Eqs 9.57a,b in a similar manner to the short tube Exercise 9.9.

9.11 Optimise the geometry for torsional buckling of the non-circular tubes given in Figs 9.21a–c in a similar manner to Exercise 9.9.

9.12 Compare the axial buckling load for a 1 m long aluminium strut, with one end fixed and the other end free, with the vertical end load to buckle a 1 m long cantilever beam when the cross sections are (a) solid rectangles 25 mm × 5 mm and (b) 1 mm-thick rectangular tubes with outer dimensions 25 mm × 5 mm. The loads are applied without eccentricity and for each beam in (a) and (b) the depth is 25 mm. Take $E = 75$ GPa.

9.13 Show that Eqs 9.26a,b apply to the wedge element within the first panel in Fig. 9.14. Note, that correspondingly, in Fig. 11.5a, τ_{xy} is reversed and $\theta = 90° - \alpha$ in which α is the orientation of principal tension in Fig. 9.14.

CHAPTER 10

ENERGY METHODS

Summary: In this chapter it is shown how great use is made of the balance between work done by external forces and energy stored within a structure that bears load. This energy balance leads to the two theorems of Castigliano and the principle of virtual work that provide for the deflection of a structure. Theory is presented here between the various approaches following the particular to the general. Thus, from the direct application of the conservation of energy the deflection beneath a single load point is found. The two theorems mentioned apply to many loads. They allow the deflection to be found at each point where a load is applied and vice versa. Virtual work principles are wider ranging, allowing the deflections to be found at points beneath or between loads. These techniques are applied to close- and open-coiled helical springs, leaf springs, curved beams, rings and pin-jointed frames. The potential energy component of energy conservation appears where loads are impacted upon their respective structures. One notable advantage of adopting an energy approach is that being a scalar the energy stored is the sum of its contributions from all sources. The latter refers to all the modes of deformation that apply in mechanics: tension, compression, shear, torsion and bending. The chapter begins with the derivation of the strain energy expressions under each mode applied separately and then for various combined modes.

10.1 Internal Energy and External Work

The first law of thermodynamics states that energy is conserved between the heat and work transfers and the internal energy of a closed system. This law can be expressed in an incremental form:

$$\delta Q - \delta W = \delta U \tag{10.1}$$

where δQ and δW are, respectively, the changes to the heat and work transfers that result in a change δU to a system's store of internal energy. Now consider the closed system to be a solid elastic body in equilibrium with the forces applied to it. With a small increase in these forces, external work $(-\delta W)$ is done on the body and this will result in a positive change to its internal energy $(+\delta U)$. The process may be assumed to be *adiabatic*, i.e. it occurs without heat flow $(\delta Q = 0)$. It then follows from Eq. 10.1 that:

$$0 - (-\delta W) = \delta U$$

$$\therefore \quad \delta W = \delta U \qquad (10.2)$$

which shows that the external work done by these forces is equal to the increase in internal energy of the body. Provided the forces are applied slowly, such that kinetic energy due to the rate of deformation is negligible, then δU is the increase in elastic strain energy stored within the body. It is possible to derive δU for elastic materials that obey Hooke's law. We need only to consider two basic stress systems, direct and shear, to determine δU under any loading combination. External forces associated with the direct strain energy expression will include tension, compression and bending. The external actions appropriate to the shear strain energy expression include shear force and torsion. In practice, we often have combinations of direct and shear energies, as would arise, for example, with the bending and shear of a beam subjected to transverse forces.

10.1.1 Direct Stress

Consider an element $\delta x \times \delta y \times \delta z$ of a rectangular section bar under a tensile loading applied in the z-direction (see Fig. 10.1a). During loading let the axial stress increase from σ to $\sigma + \delta\sigma$ when, correspondingly, the strain increases from ε to $\varepsilon + \delta\varepsilon$, as shown in Figs 10.1b,c.

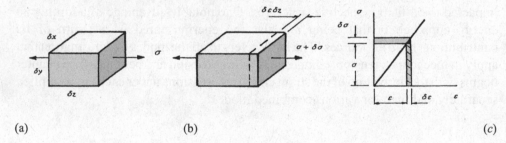

(a)   (b)   (c)

Figure 10.1 Strain energy under direct stress

The change to the external work arises from the net force: $(\sigma + \delta\sigma)(\delta x \times \delta y)$, moving through the displacement $(\delta\varepsilon \times \delta z)$. From Eq. 10.1, their product will correspond to the following change in internal strain energy:

$$\delta U = (\sigma + \delta\sigma)(\delta x \times \delta y)(\delta\varepsilon \times \delta z) \qquad (10.3)$$

Setting the volume $\delta V = \delta x \times \delta y \times \delta z$, we may approximate Eq. 10.3 to:

$$\delta U = \sigma \times \delta\varepsilon \times \delta V \qquad (10.4a)$$

Writing Eq. 10.4a as the direct *strain energy density*

$$\delta U/\delta V = \sigma \, \delta\varepsilon \qquad (10.4b)$$

We see that Eq. 10.4b approximates to the shaded area of the strip in Fig. 10.1c. The neglected triangular area (darkened) is the error from ignoring the second-order term: $\delta\sigma \times \delta\varepsilon \times \delta V$ within Eq. 10.3. Substituting $\delta\varepsilon = \delta\sigma/E$ into Eq. 10.4a and then integrating provides the total strain energy stored:

$$U = \frac{1}{E} \int_V \int_\sigma \sigma \, d\sigma \, dV$$

$$= \frac{1}{2E} \int_V \sigma^2 dV = \frac{1}{2} \int_V \sigma\varepsilon \, dV \qquad (10.5a,b)$$

Equation 10.5b may be applied to uni-directional stressing arising from tension, compression and bending.

10.1.2 Shear Stress

Consider an element $\delta x \times \delta y \times \delta z$ under a shear stress in Fig. 10.2a. Let the shear stress increase from τ to $\tau + \delta\tau$ and the shear strain from γ to $\gamma + \delta\gamma$ along an elastic load path as shown in Figs 10.2b,c.

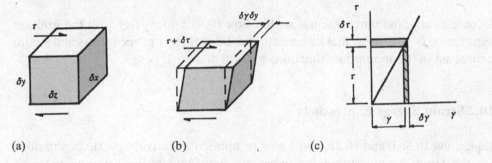

(a)   (b)   (c)

Figure 10.2 Strain energy under shear

A change to the external work occurs as the shear force: $(\tau + \delta\tau)(\delta x \times \delta z)$, moves through a displacement: $(\delta\gamma \times \delta y)$. Their product gives the following change to the element's internal strain energy:

$$\delta U = (\tau + \delta\tau)(\delta x \times \delta z)(\delta\gamma \times \delta y) \qquad (10.6)$$

We may approximate Eq. 10.6 to:

$$\delta U = \tau \times \delta\gamma \times \delta V \qquad (10.7a)$$

The *strain energy density* for shear follows from Eq. 10.7a as:

$$\delta U/\delta V = \tau \, \delta\gamma \qquad (10.7b)$$

Equation 10.7b is a good approximation to the area of the strip in Fig. 10.2c. Setting $\delta\gamma = \delta\tau/G$, and integrating Eq. 10.7a gives the total strain energy stored:

$$U = \frac{1}{G} \int_V \int_\tau \tau \, d\tau \, dV$$

$$= \frac{1}{2G} \int_V \tau^2 \, dV = \frac{1}{2} \int_V \tau\gamma \, dV \qquad (10.8a,b)$$

Note that while the elastic response of most polycrystalline materials is linear, the elastic response of non-metals is often non-linear. For the latter we must write the direct and shear strain energy from the integration of Eqs 10.4a and 10.7a as, respectively:

$$U = \int_V \left[\int_\varepsilon \sigma(\varepsilon) \, d\varepsilon \right] dV \qquad (10.9a)$$

$$U = \int_V \left[\int_\gamma \tau(\gamma) \, d\gamma \right] dV \qquad (10.9b)$$

Geometrically, the inner integral within Eqs 10.9a,b identifies with the limiting summation of strip areas that lie beneath non-linear plots between stress and strain, expressed in the appropriate function: $\sigma = \sigma(\varepsilon)$ and $\tau = \tau(\gamma)$.

10.2 Strain Energy Expressions

Equations 10.5a,b and 10.8a,b will now be applied to derive separate expressions for the strain energy stored for linear elasticity under five common types of loading–tension, compression, bending, shear and torsion–in which the corresponding applied force, moment and torque all appear explicitly.

10.2.1 Tension and Compression

Consider a bar of length L, with uniform section area A under an axial tensile force W, as shown in Fig. 10.3.

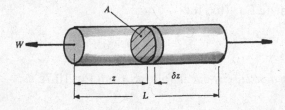

Figure 10.3 Bar in tension

The strain energy stored is found by substituting: $\sigma = W/A$ and $\varepsilon = \sigma/E$ with $\delta V = A \times \delta z$ into Eq. 10.5b. This gives:

$$U = \frac{1}{2} \int_z \left(\frac{W}{A} \right) \left(\frac{W}{AE} \right) (A\,dz) \tag{10.10a}$$

When W, A and E are constants, Eq. 10.10a integrates directly:

$$U = \frac{W^2 L}{2AE} \tag{10.10b}$$

If the cross-section is non-uniform, described by the function $A = A(z)$, the strain energy is found from Eq. 10.10a as:

$$U = \frac{W^2}{2E} \int_0^L \frac{dz}{A(z)} \tag{10.10c}$$

Applications of Eqs 10.10b,c are illustrated in the following examples.

Example 10.1 Steel (S) and brass (B) bars, which form a stepped composite shaft (see Fig. 10.4), are required to extend equally under an axial tensile force of 40 kN. Determine the diameter of the brass bar, the stress in each material and the total strain energy stored. The respective moduli are: $E_S = 207$ GPa and $E_B = 82.7$ GPa.

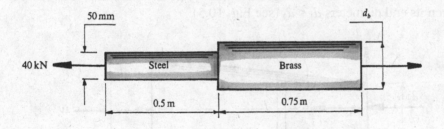

Figure 10.4 Stepped shaft

Since the displacements in the two materials are the same:

$$\left(\frac{WL}{AE} \right)_S = \left(\frac{WL}{AE} \right)_B \quad \rightarrow \quad \left(\frac{WL}{d^2 E} \right)_S = \left(\frac{WL}{d^2 E} \right)_B$$

$$\therefore \quad d_B^2 = \left(\frac{d^2 E}{WL} \right)_S \left(\frac{WL}{E} \right)_B$$

Setting $W_S = W_B$ and with a cancellation between the units for E, W and L:

$$d_B^2 = \frac{50^2 \times 207}{40 \times 0.5} \times \frac{40 \times 0.75}{82.7} = 93.86 \times 10^2 \, \text{mm}^2, \quad \therefore \ d_B = 96.88 \ \text{mm}$$

The stresses are:

$$\sigma_S = \left(\frac{W}{A}\right)_S = \frac{40 \times 10^3}{\pi(25)^2} = 20.37 \ \text{MPa}$$

$$\sigma_B = \left(\frac{W}{A}\right)_S = \frac{40 \times 10^3}{\pi(48.44)^2} = 2.33 \ \text{MPa}$$

From Eq. 10.10b, the strain energy stored becomes:

$$U = \frac{W^2}{2}\left\{\left[\left(\frac{L}{AE}\right)_S + \left(\frac{L}{AE}\right)_B\right]\right\} = \frac{2W^2}{\pi}\left[\left(\frac{L}{d^2 E}\right)_S + \left(\frac{L}{d^2 E}\right)_B\right]$$

$$= \frac{2 \times (40 \times 10^3)^2}{\pi}\left[\frac{(0.5 \times 10^3)}{50^2 \times (207 \times 10^3)} + \frac{(0.75 \times 10^3)}{96.88^2 \times (82.7 \times 10^3)}\right]$$

$$= 1.968 \times 10^3 \ \text{Nmm} = 1.97 \ \text{J}$$

Example 10.2 Derive expressions for the strain energy stored and the axial compression for a tapered shaft of length L when it bears an axial compressive force W between its end diameters $d_1 < d_2$ (see Fig. 10.5).

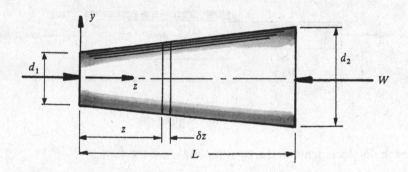

Figure 10.5 Tapered shaft under axial compression

In the z, y co-ordinates shown, the function $A = A(z)$ in Eq. 10.10c becomes:

$$A(z) = \frac{\pi}{4}(2y)^2 = \pi\left[\frac{d_1}{2} + (d_2 - d_1)\frac{z}{2L}\right]^2$$

$$= \frac{\pi d_1^2}{4}\left[1 + \left(\frac{d_2}{d_1} - 1\right)\left(\frac{z}{L}\right)\right]^2$$

Substituting into Eq. 10.10c:

$$U = \frac{2W^2}{\pi E d_1^2} \int_0^L \frac{dz}{\left[1 + (d_2/d_1 - 1)(z/L)\right]^2}$$

Integrating by substitution:

$$U = \frac{-2W^2L}{\pi E d_1 (d_2 - d_1)} \left| \left[1 + \left(\frac{d_2}{d_1} - 1 \right) \frac{z}{L} \right]^{-1} \right|_0^L = \frac{2W^2L}{\pi E d_1 d_2}$$

The axial displacement Δ may be found from equating U to the external work done. In the case of linear elasticity the work done becomes the triangular area beneath the linear load versus displacement plot:

from which

$$\frac{W\Delta}{2} = \frac{2W^2L}{\pi E d_1 d_2}$$

$$\Delta = \frac{4WL}{\pi E d_1 d_2}$$

Example 3.2 (see p. 63) confirms this result from the direct integration of Eq. 3.4.

10.2.2 Biaxial Principal Stress

Figure 4.12 (see p. 121) shows a thin plate under an in-plane biaxial tension. Each principal stress σ_1 and σ_2 contributes to the strain energy stored in loading the plate and therefore Eq. 10.5a is written as

$$\delta U = (\sigma_1 \delta\varepsilon_1 + \sigma_2 \delta\varepsilon_2) \delta V \qquad (10.11)$$

The strain increments within Eq. 10.11 will depend upon increments in both stresses. Using Eqs 4.33a,b:

$$\delta\varepsilon_1 = (\delta\sigma_1 - v\delta\sigma_2)/E \qquad (10.12a)$$

$$\delta\varepsilon_2 = (\delta\sigma_2 - v\delta\sigma_1)/E \qquad (10.12b)$$

in which E and v are elastic constants. Substituting Eqs 10.12a,b into Eq. 10.11 and then integrating:

$$\delta U/\delta V = (\sigma_1 \delta\sigma_1 - v\sigma_1 \delta\sigma_2 + \sigma_2 \delta\sigma_2 - v\sigma_2 \delta\sigma_1)/E$$

$$U = \frac{1}{E} \int_V \int_\sigma [\sigma_1 d\sigma_1 - v\dot{d}(\sigma_1\sigma_2) + \sigma_2 d\sigma_2] dV = \frac{V}{2E} \left(\sigma_1^2 - 2v\sigma_1\sigma_2 + \sigma_2^2 \right) \qquad (10.13)$$

We would not arrive at Eq. 10.13 from summing U due to each stress acting independently. With combined in-plane stressing, Poisson's central term arises from the plate-thinning strain.

10.2.3 Beam in Bending

Consider the uniform section beam, in Fig. 10.6, carrying a bending moment M at position z in its length.

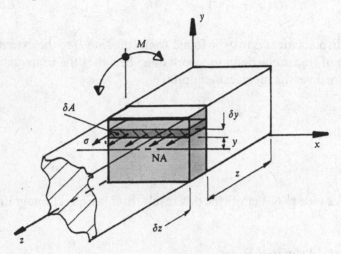

Figure 10.6 Strain energy in bending

The bending stress and strain follow from Eq. 5.4 as: $\sigma = My/I$ and $\varepsilon = \sigma/E = My/IE$. Within the slice of beam shown, an elemental volume $\delta V = \delta A \delta z$, is identified where δA $(= x\delta y)$ lies at a distance y from the neutral axis (NA). Substituting σ, ε and δV into Eq. 10.5b leads to:

$$U = \frac{1}{2} \int_A \int_z \left(\frac{My}{I} \right) \left(\frac{My}{EI} \right) dA\,dz \qquad (10.14a)$$

Both M and I may be regarded as constants where they do not vary with length z. Also, as E is constant for a beam of one material, Eq. 10.14a becomes:

$$U = \frac{M^2}{2EI^2} \int_z \left(\int_A y^2 dA \right) dz \qquad (10.14b)$$

The inner integral $I = \int y^2 dA$ defines the second moment of area for the beam cross section whether it be rectangular, circular, I, T, etc. Thus, for a beam having a uniform cross section over its length L, Eq. 10.14b becomes:

$$U = \frac{M^2 L}{2EI} \qquad (10.15a)$$

Equation 10.15a has limited application to the bending of a spring wire and to a strut or column under an offset axial load. In practice, beams are used to carry lateral loads for which the bending moment diagram shows the manner in which M varies with z (see § 2.5, p. 47), i.e. $M = M(z)$, when Eq. 10.14b is written as:

$$U = \frac{1}{2EI} \int_0^L [M(z)]^2 \, dz \tag{10.15b}$$

For a non-uniform section, $I = I(z)$, where I varies with z and Eq. 10.15b is written more generally as:

$$U = \frac{1}{2E} \int_0^L \frac{[M(z)]^2}{I(z)} \, dz \tag{10.15c}$$

Invariably, any bending moment variation along a beam is accompanied by a variation in shear force. Hence, the total strain energy for a beam is the sum of the contributions from shear and bending. The U expressions for block shearing that follow may be adapted to transverse shear in beams to accompany Eqs 10.15a–c.

10.2.4 Block in Shear

Consider a block with an elemental length δz having a uniform cross-sectional area A subjected to the shear force F in its x-y plane, as shown in Fig. 10.7.

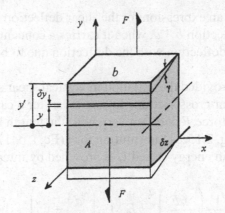

Figure 10.7 Block in shear

Assuming that F is distributed uniformly over A we may substitute $\tau = F/A$, and $\gamma = \tau/G$ with $\delta V = A \times \delta z$ into Eq. 10.8b:

$$U = \frac{1}{2} \int_z \left(\frac{F}{A} \right) \left(\frac{F}{GA} \right) (A \, dz) \tag{10.16a}$$

When the element applies to a shear pin or key for which F, G and A are constants over length L, Eq. 10.16a simplifies to:

$$U = \frac{F^2 L}{2AG} \tag{10.16b}$$

However, it is possible for both F and A to be dependent upon z as with the flexural shearing of a beam with a non-uniform cross section. Correspondingly, the energy dependence appears within Eq. 10.16a as:

$$U = \frac{1}{2G} \int_0^L \frac{[F(z)]^2}{A(z)} \, dz \qquad (10.16c)$$

in which $F(z)$ is defined by the beam's shear force diagram. Because F is squared in Eq. 10.16c the discontinuities and sign changes in the F-diagram amount to a summation of U for those regions in the length where F remains constant or varies linearly (e.g. see Example 2.9, p. 51). Equation 10.16c employs an average shear stress $\tau = F/A$ for the beam cross section. In fact, τ is not distributed uniformly through the section depth but varies parabolically between being zero at the top and bottom surfaces to attain its maximum at the neutral axis (see § 5.4, p.171). Equation 10.8b provides this within the volume element $\delta V = \delta A \times \delta z$ in Fig. 10.7:

$$U = \frac{1}{2G} \int_A \int_z \tau^2 \, dA \; dz \qquad (10.16d)$$

where $dA = b \times dy$ and $\tau = \tau(y)$ is expressed from either Eq. 5.26a or b. It should be noted, however, that for long beams U in shear is normally small and can often be neglected compared to the strain energy stored under bending. The following examples illustrate the relative magnitudes of U under each mode.

Example 10.3 Find an expression for the shear deflection of a cantilever of length L with rectangular section $b \times d$, when it carries a concentrated force F at its free-end. Compare this deflection with the deflection due to bending.

Example 5.11 has provided the variation in vertical shear stress for any beam with a uniform rectangular cross section. In particular, for a cantilever bearing only an end load, the shear force F does not vary with the length but remains equal to the applied load (see Fig. 2.15c). Substituting for τ (Eq. i, p. 174) with $\delta A = b \times \delta y$ into Eq. 10.16d, the strain energy stored U is provided by integrating over the depth:

$$U = \frac{1}{2G} \left(\frac{6F}{bd^3} \right)^2 bL \int_{-d/2}^{d/2} \left(\frac{d^2}{4} - y^2 \right)^2 dy = \frac{3F^2L}{5bdG}$$

The external shear work done is $W = F \times \Delta_s/2$, which, when equated to U (see Eq. 10.2) provides the shear deflection Δ_s within every cross section

$$\Delta_s = \frac{6FL}{5bdG} \qquad (i)$$

Correspondingly, the free-end bending deflection Δ_b is found from equating external bending work $W = F\Delta_b/2$ to the strain energy of bending (Eq. 10.15b):

$$\frac{F\Delta_b}{2} = \frac{1}{2EI} \int_0^L M^2 dz \qquad (ii)$$

Taking the origin for z at the fixed end in Fig. 2.15c, $M = F(L - z)$. Eq. ii gives:

$$\frac{F\Delta_b}{2} = \frac{F^2}{2EI} \int_0^L (L^2 - 2Lz + z^2)\,dz = \frac{F^2 L^3}{6EI}$$

$$\Delta_b = \frac{FL^3}{3EI} \qquad\qquad\qquad\qquad \text{(iii)}$$

The total free-end deflection is given by the sum of Eqs i and iii. However, the expressions show: $\Delta_s \sim L$ and $\Delta_b \sim L^3$, so unless the beam is very short, the former contribution from shear will be negligible.

Example 10.4 Derive the expression for the shear strain energy stored within a circular section of diameter d (see Fig. 10.8a) for cantilever beam of length L, carrying uniformly distributed loading w/unit length.

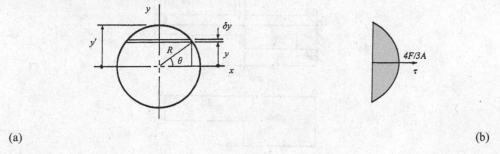

(a) (b)

Figure 10.8 Circular section under shear

Within Eq. 10.16d the shear stress now appears with a dependence upon both the shear force F which varies with the length z as: $F(z) = |w(L - z)|$ (see Fig. 2.15d, p. 48) and the position y in the section depth. With an origin of x, y at the circle's centre in Fig. 10.8a, Eq. 5.26a must be written as:

$$\tau(y) = \frac{F(z)}{I(2x)} \int_y^y y(2x)\,dy \qquad\qquad \text{(i)}$$

Substituting $y = R \sin\theta$, $x = R \cos\theta$ and $y' = R$, the integration leads to τ

$$\tau(y) = \frac{F(z)}{3I} x^2 = \frac{F(z)}{3I}(R^2 - y^2) \qquad\qquad \text{(ii)}$$

as distributed in Fig. 10.8b. Substituting Eq. ii into Eq. 10.16d with $\delta A = 2x \times \delta y$:

$$U = \frac{w^2}{9I^2 G} \int_0^L (L - z)^2\,dz \int_{-R}^R (R^2 - y^2)^2 x\,dy \qquad\qquad \text{(iii)}$$

and, again, with the conversion of the variable y to θ, Eq. iii becomes:

$$U = \frac{w^2 L^3}{27 I^2 G} \int_{-\pi/2}^{\pi/2} \cos^6\theta\,d\theta = \frac{5\pi w^2 L^3}{432 I^2 G} \qquad\qquad \text{(iv)}$$

The integral in Eq. iv has been evaluated as $5\pi/16$ from applying a standard reduction formula. Here, as the beam deflection does not arise from a single force, the external work W must be equated to U within an energy theorem to find the separate displacements due to shear and bending. In particular, Castigliano's second theorem (see § 10.5.2, p. 410) provides displacements under multiple point loads. This problem will be introduced through the following example and developed further in Example 10.14.

Example 10.5 Find the total strain energy stored for the stepped cantilever in Fig. 10.9a due to bending and shear effects. Is it possible to find the free-end deflection in this case? The elastic moduli are $E = 100$ GPa and $G = 38$ GPa respectively.

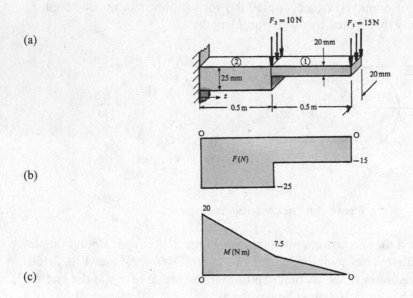

(a)

(b)

(c)

Figure 10.9 Stepped cantilever

The shear force $F(z)$ and bending moment $M(z)$ diagrams are shown in Figs 10.9b,c. The shear strain energy U_s for beam sections 1 and 2 is found from Eq. 10.16b:

$$U_s = \frac{1}{2GA_1} \int_z F_1^2 \, dz + \frac{1}{2GA_2} \int_z F_2^2 \, dz \tag{i}$$

With the origin for z at the fixed end, F_1 and F_2 remain constant in their respective regions (m): $0.5 \le z \le 1$ and $0 \le z \le 0.5$ within Fig. 10.9b. Hence Eq. i gives:

$$U_s = \frac{15^2}{(2 \times 38 \times 10^3 \times 20^2)} \int_{0.5}^{1.0} dz + \frac{25^2}{(2 \times 38 \times 10^3 \times 20 \times 25)} \int_{0}^{0.5} dz \tag{ii}$$

$$= (0.0037 + 0.0039)10^{-3} \, \text{Nm} = 0.0076 \times 10^{-3} \, \text{J}$$

Note how that the mixed units lead directly to Joules (J).

Applying Eq. 10.15b, the strain energy in bending is:

$$U_b = \frac{1}{2EI_1} \int_z M_1^2\, dz + \frac{1}{2EI_2} \int_z M_2^2\, dz$$

where:

$EI_1 = 100 \times 10^3 \times 20^3/12 = 13.33 \times 10^8\ \text{Nmm}^2 = 1333\ \text{Nm}^2$

$EI_2 = 100 \times 10^3 \times 20 \times 25^3/12 = 26.04 \times 10^8\ \text{Nmm}^2 = 2604\ \text{Nm}^2$

Figure 10.9c shows that the moments M_1 and M_2 (Nm) vary in the region of each section according to:

$M_1 = 15(1 - z)$ for $0.5 \le z \le 1$

$M_2 = (1 - z)15 + (0.5 - z)10 = 5(4 - 5z)$ for $0 \le z \le 0.5$

Substituting into Eq. ii in Nm units:

$$U_b = \frac{225}{(2 \times 1333)} \int_{0.5}^{1.0} (1 - z)^2\, dz + \frac{25}{(2 \times 2604)} \int_0^{0.5} (4 - 5z)^2\, dz$$

$$= 0.0844 \left| z - z^2 + z^3/3 \right|_{0.5}^{1.0} + 0.0048 \left| 16z - 20z^2 + 25z^3/3 \right|_0^{0.5}$$

$$= (3.517 + 19.4)10^{-3}\ \text{Nm} = 22.92 \times 10^{-3}\ \text{J}$$

The total strain energy $U_t = U_s + U_b = 22.932 \times 10^{-3}$ J, though clearly the contribution from shear is negligible. Shear influences the design of shorter beams with lengths similar to the cross-section dimensions. In the presence of the two forces in Fig. 10.9a the external work increment is:

$$\delta W = F_1 \delta \Delta_1 + F_2 \delta \Delta_2$$

where Δ_1 and Δ_2 refer to load-line displacements. Equation 10.2 cannot supply these displacements directly since each depends upon both F_1 and F_2. Equation 10.2 could be employed to find the free-end deflection if this were the only point under load. Refer to Example 10.14 to see how Castigliano's second theorem is applied to separate these displacements from within the total energy expression.

10.2.5 Shaft in Torsion

Figure 10.10a shows a uniform, solid circular shaft of diameter d with equal, opposing axial torques T applied over its length L For a uniform shaft the shear stress and shear strain have been derived previously as: $\tau = Tr/J$, $\gamma = \tau/G = Tr/JG$ (see Eq. 7.6, p. 245).

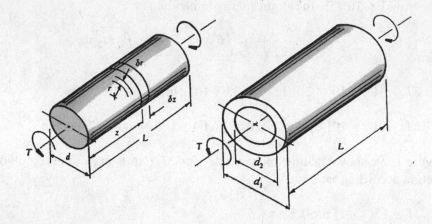

(a)   (b)

Figure 10.10 Solid and hollow circular shafts under torsion

The volume of an annular element lying at mid-length in Fig. 10.10a, with inner radius r, radial depth δr and thickness δz, is: $\delta V = 2\pi r \times \delta r \times \delta z$. Substituting these into Eq. 10.8b:

$$\delta U = \frac{1}{2}\left(\frac{Tr}{J}\right) \times \left(\frac{Tr}{JG}\right) \times 2\pi r \times \delta r \times \delta z \qquad (10.17a)$$

When T, G and J are constants, Eq. 10.17a integrates as follows:

$$U = \frac{T^2}{2J^2G}\int_z\left(\int_r 2\pi r^3\,\mathrm{d}r\right)\mathrm{d}z \qquad (10.17b)$$

Equation 7.5a (see p. 244) shows that the inner integral is the polar second moment of area for a solid shaft ($J = \pi r^4/2 = \pi d^4/32$). Hence Eq. 10.17b reduces to:

$$U = \frac{T^2L}{2JG} \qquad (10.17c)$$

In a more general case the torque $T = T(z)$, may vary with z when applied to a non-uniform circular shaft, $J = J(z)$. Equation 10.17a remains valid but its integration must admit these variations:

$$U = \frac{1}{2G}\int_0^L \frac{[T(z)]^2\,\mathrm{d}z}{J(z)} \qquad (10.17d)$$

Equations 10.17b,c also apply to a hollow shaft (see Fig. 10.10b) when the inner radius appears as the lower limit of integration. This shows that Eq. 10.17c also applies to a uniform tube when J is

$$J = \frac{\pi}{2}\left(r_1^4 - r_2^4\right) = \frac{\pi}{32}\left(d_1^4 - d_2^4\right)$$

Example 10.6 The solid shaft, with diameter $d = 180$ mm, in Fig. 10.10a is to be replaced by the hollow shaft in Fig. 10.10b having the same material, length and weight. Find the diameters of the hollow shaft in order to make its store of strain energy 20% greater than that of the solid shaft when transmitting torque at the same maximum shear stress.

With the same weight and length, the section areas of the two bars become equal:

$$\pi(d_1^2 - d_2^2)/4 = \pi d^2/4$$

$$\therefore \quad d_1^2 - d_2^2 = d^2 \qquad \text{(i)}$$

Applying Eq. 10.17c to the solid shaft, with $\tau_{max} = T(d/2)/J$ and $J = \pi d^4/32$:

$$U_S = \frac{T^2 L}{2JG} = \frac{\tau_{max}^2 LJ}{2G(d/2)^2}$$

from which

$$U_S = \frac{\tau_{max}^2}{4G} \times \frac{\pi}{4} d^2 L = \frac{\tau_{max}^2}{4G} \times V \qquad \text{(ii)}$$

For the hollow shaft, $\tau_{max} = T(d_1/2)/J$ and $J = \pi(d_1^4 - d_2^4)/32$. Eq. 10.17c becomes:

$$U_H = \frac{T^2 L}{2JG} = \frac{\tau_{max}^2 LJ}{2G(d_1/2)^2} = \frac{\tau_{max}^2 \pi L(d_1^4 - d_2^4)}{16 d_1^2 G}$$

$$U_H = \frac{\tau_{max}^2}{4G} \times \frac{(d_1^2 + d_2^2)}{d_1^2} V \qquad \text{(iii)}$$

Now $U_H = 1.2 U_S$ when, from Eqs ii and iii:

$$d_1^2 + d_2^2 = 1.2 d_1^2 \qquad \text{(iv)}$$

Substituting Eq. i, with $d = 180$ mm, into Eq. iv:

$$d_1^2 + (d_1^2 - 180^2) = 1.2 d_1^2$$

$$0.8 d_1^2 = 180^2$$

$$\therefore d_1 = 201.25 \text{ mm and } d_2 = \sqrt{0.2} \times d_1 = 90 \text{ mm}$$

Table 10.1 summarises the strain energy expression for the four types of loading. The integrals in the second column apply when the loading and the section area vary with length, z. The U-expressions given in the third column apply when the loading is constant over the length and where the cross-section is uniform. Note the similarity between all expressions. The fact that each loading W, M, F and T is squared shows that U does not depend upon the sense of loading.

Table 10.1 Summary of strain energy expressions

Loading	U (general)	U
Tension/Compression	$(1/2E) \int [W(z)]^2 \, dz/A(z)$	$W^2L/2AE$
Bending	$(1/2E) \int [M(z)]^2 \, dz/I(z)$	$M^2L/2EI$
Shear Force	$(1/2G) \int [F(z)]^2 \, dz/A(z)$	$F^2L/2AG$
Torsion	$(1/2G) \int [T(z)]^2 \, dz/J(z)$	$T^2L/2JG$

10.3 Application to Springs

Springs play a useful role in maintaining contact between bodies, transmitting forces, storing energy in driving mechanisms and in suspension systems. There are various coil designs which may be flat, conical or cylindrical. Also there are close-coiled and open-coiled springs for which the work-energy interchange facilitates their design most conveniently. In the close-coiled helical spring the coil's helix angle is $\approx 5°$, which simplifies their design for bearing axial load and torque. The design of an open-coiled spring must be based upon the combined stress state under both the bending and torsion of its wire arising from each external action. The leaf spring used in vehicle suspension is fabricated from overlaying plates with contacting ends to form beams under four-point loading. The uniform moment between the supports is converted into stored energy under a desired deflection, this energy being available for dissipation when suspending a fluctuating vehicle weight between its wheels.

10.3.1 Close-Coiled Helical Spring

The energy interchange within Eq. 10.2 provides the elastic deflection of closely wound springs where they are subjected to tension and torsion as shown in Figs 10.11a and b.

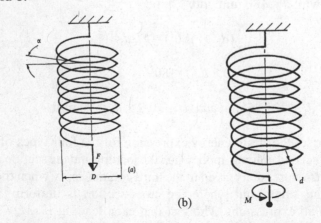

(a)   (b)

Figure 10.11 Tensile and torsional loading of a close-coiled helical spring

Note how the wire reacts to the external action upon the coil in each case. When for (a) the coil is under tension the wire is placed under torsion and for (b) the coil is under torsion the wire is placed under bending. Both the torque and the moment are uniform over the wire length producing their greatest stresses at the wire's outer surface. In a close-coiled spring the helix angle α is normally less than 10° when it may be assumed that there is no interaction between torsion and bending under each loading.

(a) Spring under an Axial Force F

For the spring in Fig. 10.11a, let the wire diameter be d, the mean coil diameter D, the number of turns N, and the wire's coiled length $L = \pi DN$. The applied force F will twist the wire along its coiled length under a constant torque: $T = FD/2$. Let Δ be the deflection beneath F. The external work done upon a linear, elastic spring is $W = F\Delta/2$ and Eq. 10.17b provides the strain energy stored (resilience) from twisting the wire. Applying Eq. 10.2:

$$\frac{F\Delta}{2} = \frac{T^2 L}{2GJ} \tag{10.18a}$$

Substituting for T, L and $J = \pi d^4/32$, Eq. 10.18a leads to:

$$G = \frac{8FD^2 L}{\pi \Delta d^4} = \frac{8FD^3 N}{\Delta d^4} \tag{10.18b}$$

Rearranging Eq. 10.18b, the spring stiffness K is:

$$K = \frac{F}{\Delta} = \frac{\pi G d^4}{8D^2 L} = \frac{G d^4}{8D^3 N} \tag{10.18c}$$

The angular twist (rad) of the wire between its ends is found from Eq. 7.6:

$$\theta = \frac{TL}{GJ} = \frac{16NFD^2}{Gd^4} \tag{10.19a}$$

The maximum shear stress in the wire is attained at its outer diameter remaining constant over the full length. Again, from Eq. 7.6:

$$\tau = \frac{Td}{2J} = \frac{8FD}{\pi d^3} \tag{10.19b}$$

Example 10.7 A close-coiled helical spring is made from steel wire with 12 mm diameter. The mean diameter of the coils is 150 mm and the number of effective coils is 21. Find the elongation and twist of the spring when it carries an axial load of 25 N. What is the maximum shear stress in the wire? Take $G = 85$ GPa.

From Eqs 10.18a–c to 10.19a,b, working in units of N and mm:

$$\Delta = \frac{8FD^3 N}{Gd^4} = \frac{8 \times 25 \times 150^3 \times 21}{85 \times 10^3 \times 12^4} = 8.04 \text{ mm}$$

$$\theta = \frac{16NFD^2}{Gd^4} = \frac{16 \times 21 \times 25 \times 150^2}{85 \times 10^3 \times 12^4} = 0.107^c \text{ (rad)}$$

$$\tau = \frac{8FD}{\pi d^3} = \frac{8 \times 25 \times 150}{\pi \times 12^3} = 5.53 \text{ MPa}$$

Example 10.8 A close-coiled spring is designed to have a stiffness of 1 N/mm when it is to carry a maximum axial force of 40 N, without the shear stress exceeding 125 MPa. With the spring coils touching, the solid length of the compressed spring is 45 mm. Find the diameter of the spring wire, the mean coil diameter and the number of coils required. Take $G = 41.5$ GPa.

As the solid length is $l_s = Nd$, the stiffness Eq. 10.18c becomes:

$$K = \frac{F}{\Delta} = \frac{Gd^4}{8D^3N} = \frac{Gd^5}{8D^3 l_s}$$

Applying the design criteria

$$1 = \frac{41.5 \times 10^3 \times d^5}{8 \times 45 \times D^3}$$

which connects the coil and wire diameters:

$$D^3 = 115.28 d^5 \tag{i}$$

Equation 10.19b places a further constraint upon these two diameters

$$\tau = \frac{8FD}{\pi d^3}$$

from which

$$125 = \frac{8 \times 40 \times D}{\pi d^3}$$

$$D = 1.227 d^3 \tag{ii}$$

Combining Eqs i and ii

$$(1.227 d^3)^3 = 115.28 d^5 \text{ mm}$$

$$d = 2.81 \text{ mm}$$

$$\therefore D = 1.227 (2.81)^3 = 27.24 \text{ mm}$$

$$N = l_s / d = 45/2.81 = 16$$

(b) Spring under an Axial Moment or Torque M

When an axial twisting moment (or torque) M, is applied to a closely coiled helical spring (see Fig. 10.11b) it is reacted by every cross section. Strain energy (i.e. the spring's resilience) is stored in bending within Eq. 10.15a and the external work done is $W = \int M d\varphi$, where φ is the angular twist (in radians) accompanying M. Taking a linear, elastic relationship between M and φ, the area beneath their plot becomes the work done $W = M\varphi/2$ and Eq. 10.2 gives:

$$\frac{M\varphi}{2} = \frac{M^2 L}{2EI}$$

from which

$$\varphi = \frac{ML}{EI} \tag{10.20}$$

Equation 10.20 and the theory of bending (Eq. 5.4) may now be applied according to the shape of the wire cross section:

(i) Circular section of diameter, d

$$I = \pi d^4/64, \quad L = \pi DN$$

when from Eq. 10.20

$$\varphi = \frac{64M(\pi DN)}{E\pi d^4}$$

From Eq. 5.4

$$E = \frac{64MDN}{\pi d^4} \tag{10.21}$$

$$\sigma_{max} = \frac{My}{I} = \frac{M(d/2)}{(\pi d^4/64)} = \frac{32M}{\pi d^3} \tag{10.22}$$

(ii) Square section side length, a

$$I = a^4/12 \quad L = \pi DN$$

From Eq. 10.20:

$$\varphi = \frac{12M(\pi DN)}{Ea^4}$$

From Eq. 5.4:

$$E = \frac{12M(\pi DN)}{\varphi a^4} \tag{10.23}$$

$$\sigma_{max} = \frac{M(a/2)}{(a^4/12)} = \frac{6M}{a^3} \tag{10.24}$$

Example 10.9 A close-coiled helical spring with mean coil diameter 30 mm, is to twist by 60° under an axial torque of 8 Nm when the bending stress in the wire is restricted to 415 MPa. Calculate the wire size and the number of effective coils when the section is: (a) circular and (b) square. Take $E = 207$ GPa.

(*a*) *Circular Section*
The wire diameter is found from rearranging Eq. 10.22:

$$d^3 = \frac{32M}{\pi\sigma_{max}}$$

$$\therefore d = \left[\frac{32 \times (8 \times 10^3)}{\pi \times 415}\right]^{1/3} = 5.8 \text{ mm}$$

The number of coils required, to the nearest whole number, follows from Eq. 10.21:

$$N = \frac{E\varphi d^4}{64MD} = \frac{(207 \times 10^3) \times (60 \times 2\pi/360) \times 5.8^4}{64 \times (8 \times 10^3) \times 30} \approx 16$$

(*b*) *Square Section*
Equation 10.24 provides the square side dimension a:

$$a^3 = \frac{6M}{\sigma_{max}}$$

from which

$$a = \left[\frac{6 \times (8 \times 10^3)}{415}\right]^{1/3} = 4.87 \text{ mm}$$

Equation 10.23 provides the number of coils required for a spring wire in square section:

$$N = \frac{E\varphi a^4}{12M\pi D} = \frac{(207 \times 10^3) \times (60 \times 2\pi/360) \times 4.87^4}{12 \times (8 \times 10^3) \times \pi \times 30} \approx 14$$

10.3.2 Close-Coiled Conical Spring

Let θ be the angle in radians described by the conical winding measured from its smaller end radius R_1 such that with a spring having N turns the large end radius R_2 describes $2\pi N$ radians. Hence, an intermediate radius R ($R_1 < R < R_2$), corresponding to θ ($0 \le \theta \le 2\pi N$), becomes:

$$R = R_1 + \frac{(R_2 - R_1)\theta}{2\pi N} \tag{10.25a}$$

Rearranging Eq. 10.25a, the rotation angle θ at radius R $(R_1 \le R \le R_2)$ becomes:

$$\theta = \frac{2\pi N(R - R_1)}{(R_2 - R_1)} \qquad (10.25b)$$

from which:

$$\frac{d\theta}{dR} = \frac{2\pi N}{(R_2 - R_1)} \qquad (10.25c)$$

The wire's full length L follows from Eq. 10.25c by integrating an element of its length $\delta L = R \times \delta\theta$:

$$L = \int_0^{2\pi N} R\,d\theta = \frac{2\pi N}{R_2 - R_1} \int_{R_1}^{R_2} R\,dR = \pi N(R_1 + R_2) \qquad (10.25d)$$

(a) Axial Force F

The torque T arising from an axial load F applied at the larger end is referred to R and θ through Eq. 10.25a:

$$T = FR = F\left[R_1 + \frac{(R_2 - R_1)\theta}{2\pi N}\right] \qquad (10.26a)$$

Table 10.1 gives the strain energy stored in an element of the spring's helical length δL arising from torsion:

$$\delta U = \frac{T^2 \delta L}{2GJ} = \frac{(FR)^2 (R \times \delta\theta)}{2GJ} \qquad (10.26b)$$

Substituting from Eq. 10.25c allows for an integration of Eq. 10.26b between the end radii:

$$U = \frac{\pi NF^2}{GJ(R_2 - R_1)} \int_{R_1}^{R_2} R^3\,dR = \frac{\pi NF^2}{4GJ}(R_1 + R_2)(R_1^2 + R_2^2) \qquad (10.26c)$$

Equating 10.26c to the external work done $W = F\Delta/2$ provides the extension of a conical spring with wire diameter d:

$$\Delta = \frac{\pi NF}{2GJ}(R_1 + R_2)(R_1^2 + R_2^2) = \frac{16NF}{d^4 G}(R_1 + R_2)(R_1^2 + R_2^2) \qquad (10.26d)$$

Example 10.10 Design a conical spring having a required stiffness of 40 N/mm from a 5 m length of 20 mm-diameter steel wire with the end radii in the ratio 3/1. Take $G = 80$ GPa.

The stiffness is defined from Eq. 10.26d as:

$$K = \frac{F}{\Delta} = \frac{d^4 G}{16N(R_1 + R_2)(R_1^2 + R_2^2)} \qquad (i)$$

where N turns require a wire length L according to Eq. 10.25d:

$$N = \frac{L}{\pi(R_1 + R_2)} \tag{ii}$$

Setting $r = R_2/R_1$, Eqs i and ii combine into

$$K = \frac{\pi d^4 G}{16 L R_1^2 (1 + r^2)}$$

giving end radii and number of turns:

$$R_1 = \sqrt{\frac{\pi d^4 G}{16 L (1 + r^2) K}}, \quad R_2 = r R_1, \quad N = \frac{L}{\pi R_1 (1 + r)} \tag{iii–v}$$

Applying Eqs iii–iv to the design requirements:

$$R_1 = \sqrt{\frac{\pi \times 20^4 \times (80 \times 10^3)}{16 \times (5 \times 10^3) \times (1 + 3^2) \times 40}} = 35.45 \text{ mm}$$

$$R_2 = 3 \times 35.35 = 106.35 \text{ mm}, \quad N = \frac{(5 \times 10^3)}{\pi \times 35.45 \times (1 + 3)} \approx 11$$

(b) Axial Torque M

The bending moment upon every section of the wire's length arising from an axial torque M applied at the larger end is constant. Table 10.1 gives the strain energy stored within the spring length given by Eq. 10.25d:

$$U = \frac{M^2 L}{2EI} = \frac{M^2 \pi N (R_1 + R_2)}{2EI} \tag{10.27a}$$

in which I may apply to solid square $(a \times a)$ or circular section wire of diameter d respectively as: $I = a^4/12$ and $I = \pi d^4/64$. The rotation φ (rad) between the ends is found from equating 10.27a to the external work done: $W = \frac{1}{2} M \varphi$. This gives

$$\varphi = \frac{M \pi N (R_1 + R_2)}{EI} \tag{10.27b}$$

10.3.3 Flat Spiral Spring

Figures 10.12a,b show a flat coil when unwound and fully wound with a couple M applied to the coil's centre spindle. The coil is wound to an involute spiral from a thin flat strip of rectangular cross section with breadth b and thickness t. Within a clock the spring is attached to a central spindle which extends through the casing with a square-end key register.

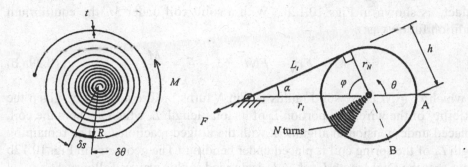

(a)   (b)

Figure 10.12 Flat spiral spring under a central couple and floating-end

(a) Free End (Fig. 10.12a)

Firstly, consider the spring having a hypothetical floating end in Fig. 10.12a. A change in slope occurs between the ends of an elemental length δs of the coil during its winding under a central couple M. This change follows from Eq. 6.2 (p. 196) as $\delta\theta = \delta s/R$, where R is the element's radius of curvature (see Fig. 6.1). Now R is connected to the applied couple M from bending theory as $R = EI/M$. Hence:

$$\delta\theta = \frac{M\delta s}{EI} \qquad\qquad (10.28a)$$

Allowing the end of the coil to float ensures that it will exert only a restoring moment similar to subjecting the close-coiled helical spring to an axial twisting moment (see § 10.3.1b, p. 391). It is the nature of a pure couple that it may be translated from the central axis to every cross section as the spring's equilibrium condition. The moment M is therefore a constant within the integration of Eq. 10.28a to provide the total angle turned through during 'wind-up' of coil length L:

$$\theta = \frac{ML}{EI} \qquad\qquad (10.28b)$$

Alternatively, it can be seen that Eq. 10.28b follows directly from equating the external work done to the strain energy stored:

$$\frac{1}{2}M\theta = \frac{M^2L}{2EI}$$

(b) Fixed End (Fig. 10.12b)

In fact, the couple is not 'pure' because position fixing the far end of the spring as its central coil is would tightly introduces a reaction at the fixing point. Let us assume that M is sufficient to wind the spiral coil fully from its initial radius r_i to a final solid radius r_N, where N turns of the wire, of radial thickness t, become in contact, as shown in Figs 10.12b. With a solid coil under M, the equilibrium condition follows as:

$$M = Fr_N = FNt \quad \therefore \quad F = \frac{M}{Nt} \qquad (10.29a,b)$$

for which $r_N \approx Nt$ is the solid radius within N turns. More accurately, given the flexibility of the wire, that portion L_t of its total length L, which anchors the coil, is placed under tension in alignment with the hinged reaction F. The remaining length L_b of the spring coil is placed under bending. The geometry in Fig. 10.12b shows that the total length $L = L_t + L_b$ is divided in this way as follows:

$$L_b = 2\pi(N-1)t + r_N(\pi - \varphi) \qquad (10.30a)$$

$$L_t = L - L_b = \sqrt{r_i^2 - r_N^2} \qquad (10.30b)$$

where the angle φ (rad), subtended within the final turn is:

$$\varphi = \frac{\pi}{180} \times \cos^{-1}\left(\frac{r_N}{r_i}\right) \qquad (10.30c)$$

Hence the total energy stored within the spring arising from tension and bending becomes:

$$U = \frac{F^2 L_t}{2AE} + \int_0^{L_b} \frac{[M(\theta)]^2 ds}{2EI} \qquad (10.31)$$

It can now be seen how the moment $M(\theta)$, acting upon at each cross section of the solid coil, varies with its perpendicular distance to the line of action of the force F. For example, at point A within the right side of the solid coil in Fig. 10.12b, the perpendicular distance h is:

$$h = (r_N + r_i)\sin\alpha = Nt\left(1 + \frac{r_i}{r_N}\right) \times \frac{r_N}{r_i} \qquad (10.32a,b)$$

giving:

$$M_A = Fh = M\left(1 + \frac{r_N}{r_i}\right) \qquad (10.33a)$$

Of particular interest is the maximum moment that occurs in a bottom section B at the greatest perpendicular distance $2r_N$ from F:

$$M_{max} = 2Fr_N = 2M \qquad (10.33b)$$

and, therefore, with the coil's strip of breadth b, the greatest bending stress in the wire applies to B as:

$$\sigma_{max} = \frac{(2M)(t/2)}{(bt^3/12)} = \frac{12M}{bt^2} \qquad (10.33c)$$

In the straight connecting length L_t of a strip, with breadth b and thickness t, the tensile stress is simply $\sigma = F/bt$. The integral providing the energy from a full wind-up to N coils is the sum of three parts from the following sectors of the coil shown in Fig. 10.13.

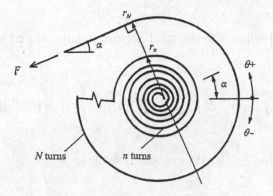

Figure 10.13 Coil sector contributions to strain energy:

(i) $\alpha \le \theta \le \alpha + \pi/2$; $M(\theta) = F[r_N - r_n \sin(\theta - \alpha)]$

(ii) $0 \le \theta \le \alpha$; $M(\theta) = F[r_N + r_n \sin(\alpha - \theta)]$

(iii) $-(\pi/2 - \alpha) \le \theta \le 0$; $M(\theta) = F[r_N + r_n \sin(\theta + \alpha)]$

where $F = M/Nt$. Within each region the contribution to U from $n = 1, 2 \ldots N$ coils is mirrored across a line of symmetry coincident with $\theta = \alpha + \pi/2$. For example, within region (i) the bending energy follows from Eq. 10.31:

$$U_i = 2 \sum_{n=1}^{N} \int_{\alpha}^{\alpha + \pi/2} \frac{M^2(\theta)\,ds}{2EI} = \frac{M^2}{EI(Nt)^2} \sum_{N=1}^{N} \int_{\alpha}^{\alpha + \pi/2} [r_N - r_n \sin(\theta - \alpha)]^2 \, r_n \, d\theta$$

Substituting $r_N = Nt$ and $r_n = nt$ gives the strain energy stored in region (i):

$$U_i = \frac{M^2 t}{EI} \sum_{n=1}^{N} \left\{ \int_{\alpha}^{\alpha + \pi/2} n\left[1 - \frac{n}{N} \sin(\theta - \alpha)\right]^2 d\theta \right\} \qquad (10.34a)$$

Correspondingly, the energy of bending stored in region (ii) becomes:

$$U_{ii} = \frac{M^2 t}{EI} \sum_{n=1}^{N} \left\{ \int_{0}^{\alpha} n\left[1 + \frac{n}{N} \sin(\alpha - \theta)\right]^2 d\theta \right\} \qquad (10.34b)$$

and in region (iii) the bending energy is:

$$U_{iii} = \frac{M^2 t}{EI} \sum_{n=1}^{N} \left\{ \int_{-(\pi/2 - \alpha)}^{0} n \left[1 + \frac{n}{N} \sin(\theta + \alpha) \right]^2 d\theta \right\} \qquad (10.34c)$$

Adding Eqs 10.34a–c leads to the total bending energy stored within the fully-wound spring:

$$U_b = \frac{M^2 t}{EI} \sum_{n=1}^{N} \left\{ \int_{0}^{\alpha + \pi/2} n \left[1 - \frac{n}{N} \sin(\theta - \alpha) \right]^2 d\theta + \int_{-(\pi/2 - \alpha)}^{0} n \left[1 + \frac{n}{N} \sin(\theta + \alpha) \right]^2 d\theta \right\}$$

$$(10.35a)$$

The integration within Eq. 10.35a eventually leads to:

$$U_b = \frac{M^2 t}{EI} \sum_{n=1}^{N} \left\{ n \left[\pi + \alpha \left(\frac{n}{N} \right)^2 \right] - \frac{4n^2}{N} \cos\alpha \left[1 + \left(1 + \frac{n}{4N} \sin\alpha \right) \right] - \frac{n}{4} \left(\frac{n}{N} \right)^2 (\sin 4\alpha + \cos 4\alpha) \right\}$$

$$(10.35b)$$

where $\alpha = \sin^{-1}(r_N/r_i)$. The estimate of U_b given by Eq. 10.35b is judged to be sufficiently accurate but if the tensile energy within the straight length L_t is to be accounted for the above summation is taken to $N - 1$ and the following term added from Eq. 10.31:

$$U_t = \frac{F^2 L_t}{2AE} = \frac{M^2 (1 + r_i^2/r_N^2)}{2(bt)E} \qquad (10.35c)$$

Then, from equating $U_b + U_t$ to the external work done the number of additional turns N_a required to wind the spring fully is found:

$$\frac{1}{2} \times 2\pi N_a \times M = U_b + U_t \quad \therefore \ N_a = \frac{U_b + U_t}{\pi M} \qquad (10.36)$$

Hence, given that the unwound spring has N_i initial turns within its outer radius r_i (see Fig. 10.12b), the total number of turns required for the above summations is $N = N_i + N_a$. If N is not provided in the spring specification then N_a may be assumed in order to compute U_b. Equation 10.36 allows N_a to be re-calculated within an iteration procedure until agreement is found.

10.3.4 Open-Coiled Helical Springs

The helix of a spring is defined by the angle α between the wire axis and a plane perpendicular to the spring axis (see Fig. 10.14). When α equals 10° or more, the spring is said to be open-coiled. Both bending and torsion of an open-coiled wire will occur when the spring carries an axial load F or a twisting moment M.

(a) Axial Load

With the exception of the last quarter turn, where the moment varies linearly from zero at the free ends (coil centre) to a maximum $M = FD/2$ at the coil's outer diameter, this maximum is taken as constant within every section in the body length. Consider a portion AB, of the helically wound wire in this region (see Fig. 10.14). The cross section at A lies in the vertical plane and the moment: $M = FD/2$, applied at A due to the axial load, lies in the horizontal plane.

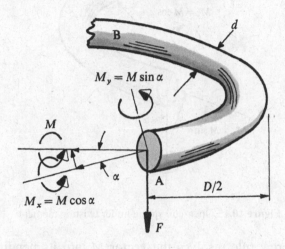

Figure 10.14 Open-coil spring under tension

Taking M to act about a horizontal axis passing through the centroid at A it appears as a vector $\mathbf{M}$ from the right-hand screw rule as shown. Vector $\mathbf{M}$ is resolved into the two components: (i) $M_x = M\cos\alpha$, parallel to the helix axis at A, which twists the wire; and (ii) $M_y = M\sin\alpha$, normal to the helix axis at A, which bends the wire. It is assumed that both (i) and (ii) are constant over the spring's full helical length L. The net strain energy arising from constant bending and torsion effects are given by equating the sum of Eqs 10.15a and 10.17c to the external work done: $F \times \Delta/2$:

$$\frac{F\Delta}{2} = \frac{M_x^2 L}{2GJ} + \frac{M_y^2 L}{2EI} \tag{10.37a}$$

Substituting for M_x and M_y into Eq. 10.37a, the deflection under the load is:

$$\Delta = \frac{FD^2 L}{4}\left(\frac{1}{GJ}\cos^2\alpha + \frac{1}{EI}\sin^2\alpha\right) \tag{10.37b}$$

where $L = \pi DN/\cos\alpha$. Substituting $J = \pi d^4/32$ and $I = \pi d^4/64$ into Eq. 10.37b gives:

$$\Delta = \frac{8FD^3 N}{d^4 \cos\alpha}\left(\frac{1}{G}\cos^2\alpha + \frac{2}{E}\sin^2\alpha\right) \tag{10.37c}$$

(b) Axial Twisting Moment

A moment M applied to the axis of the spring will act upon a vertical axis within the cross-section at A, as shown in Fig. 10.15.

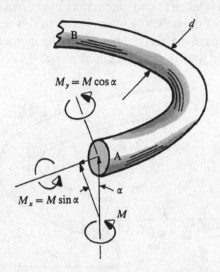

Figure 10.15 Open-coil spring under twisting moment

The right-hand screw rule resolves the vector M into its bending and twisting components; $M_y = M\cos\alpha$ and $M_x = M\sin\alpha$, perpendicular and parallel to the helix axis at A, respectively. The external work done is $M\varphi/2$ where φ is the rotation (in radians) produced by M. From Eq. 10.2:

$$\frac{M\varphi}{2} = \frac{M_x^2 L}{2GJ} + \frac{M_y^2 L}{2EI} \qquad (10.38a)$$

Equation 10.38a provides the rotation under M:

$$\varphi = \frac{32MDN}{d^4\cos\alpha}\left(\frac{1}{G}\sin^2\alpha + \frac{2}{E}\cos^2\alpha\right) \qquad (10.38b)$$

Note that under each loading condition (a) and (b), the shear stress τ and the twist θ (rads) in the wire are found from Eq. 7.6 (see p. 245):

$$\tau = \frac{M_x(d/2)}{J}, \qquad \theta = \frac{M_x L}{JG} \qquad (10.39a,b)$$

In addition the bending stress and the radius of curvature R of the wire are found from Eq. 5.4 (see p. 150):

$$\sigma = \frac{M_y(d/2)}{I}, \qquad R = \frac{EI}{M_y} \qquad (10.40a,b)$$

Example 10.11 An open-coiled helical spring has 10 coils with a 50 mm mean diameter, a wire diameter of 6.5 mm and a helix angle of 30°. Determine the axial force necessary to extend the spring by 12.5 mm and the shear and bending stresses in the wire due to this extension. What magnitude of an axial moment, when applied separately to this spring, would reduce the bending stress in the wire to 55 MPa? Find the rotation of the spring about its axis under this torque. Take $E = 207$ GPa and $G = 82.7$ GPa.

Transposing Eq. 10.37c and substituting what is given in units of N and mm:

$$F = \frac{\Delta \times d^4 \cos\alpha}{8D^3N[(1/G)\cos^2\alpha + (2/E)\sin^2\alpha]}$$

$$= \frac{12.5 \times 6.5^4 \cos 30°}{8 \times 50^3 \times 10[(1/82700)\cos^2 30° + (2/207000)\sin^3 30°]} = 168.26 \text{ N}$$

Equation 10.39a provides the shear stress in the wire:

$$\tau = \frac{(FD/2)\cos\alpha \times (d/2)}{J}$$

$$= \frac{(168.26 \times 50/2)\cos 30° \times 6.5/2}{\pi(6.5)^4/32} = 67.56 \text{ MPa}$$

Equation 10.40a provides the bending stress in the wire:

$$\sigma = \frac{(FD/2) \times \sin\alpha \times (d/2)}{I}$$

$$= \frac{(168.26 \times 50/2)\sin 30° \times 6.5/2}{\pi(6.5)^4/64} = 78 \text{ MPa}$$

The component M_y of the axial moment M, that produces a reduced bending stress value of $\sigma = 55$ MPa, is also found from Eq. 10.40a:

$$M_y = M\cos\alpha = \frac{2\sigma I}{d}$$

giving

$$M = \frac{2\sigma I}{d\cos\alpha} = \frac{\sigma\pi d^3}{32\cos\alpha}$$

$$= \frac{55 \times \pi(6.5)^3}{32\cos 30°} = 1712.3 \text{ Nmm}$$

The accompanying rotation is found from substituting into Eq. 10.38b:

$$\varphi = \frac{32 \times 1712.3 \times 50 \times 10}{6.5^4 \cos 30°}\left(\frac{1}{82.7\times 10^3}\sin^2 30° + \frac{2}{207\times 10^3}\cos^2 30°\right) = 0.182° = 10.43°$$

10.3.5 Open-Coiled Conical Spring

A similar moment resolution into bending and torsion components applies to the helical axis of inclination α for the widely-spaced coils of a conical spring when placed under axial force and torque. In each case we may refer the strain energy to an element of the spring's wire as

$$\delta U = \frac{M_y^2 \, \delta L}{2EI} + \frac{M_x^2 \, \delta L}{2GJ} \tag{10.41}$$

Compared to a close-coil in § 10.3.2 (p. 392), the elemental length is increased for an open-coil to $\delta L = R \delta\theta / \cos\alpha$. Following § 10.3.4a for the components M_x and M_y and then integrating Eq. 10.41 between the end radii R_1 and R_2 $(R_2 > R_1)$, the spring's total store of strain energy under an *axial force F* becomes:

$$U = \frac{\pi N F^2 (R_1 + R_2)(R_1^2 + R_2^2)}{4 \cos\alpha} \left(\frac{\cos^2\alpha}{GJ} + \frac{\sin^2\alpha}{EI} \right) \tag{10.42a}$$

Equation 10.42a may be equated to external work done $\tfrac{1}{2} F \Delta$ to provide the extension Δ for a spring wound from a circular section wire of diameter d:

$$\Delta = \frac{16 N F (R_1 + R_2)(R_1^2 + R_2^2)}{d^4 \cos\alpha} \left(\frac{1}{G} \cos^2\alpha + \frac{2}{E} \sin^2\alpha \right) \tag{10.42b}$$

Following § 10.3.4b for the components M_x and M_y and then integrating Eq. 10.41, the spring's total store of strain energy under an *axial torque T* becomes:

$$U = \frac{\pi N M^2 (R_1 + R_2)}{2 \cos\alpha} \left(\frac{\sin^2\alpha}{GJ} + \frac{\cos^2\alpha}{EI} \right) \tag{10.43a}$$

Equating 10.43a to the external work done $\tfrac{1}{2} M \varphi$ provides the spring's rotation φ for wire of diameter d:

$$\varphi = \frac{32 N M (R_1 + R_2)}{d^4 \cos\alpha} \left(\frac{1}{G} \sin^2\alpha + \frac{2}{E} \cos^2\alpha \right) \tag{10.43b}$$

Setting $R_1 = R_2 = D/2$ in Eqs 10.42b and 10.43b recovers Eqs 10.37c and 10.38b.

10.3.6 The Leaf Spring

The leaf spring is constructed by bending a diamond-shaped steel plate in Fig. 10.16 along its longer diagonal L into the arc of a circle. The bent plate is split into a number n of smaller leaves ($n = 5$ shown) with tapered ends then stacked and strapped into a beam of width b, length L and leaf thickness t (see Fig. 10.17).

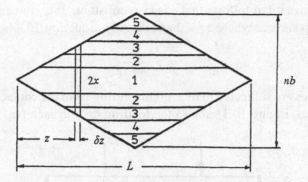

Figure 10.16 Leaf-cutting plan from a single plate

Within the spring's assembled stack of leaves contact is made at their pointed ends. This ensures that when the spring carries the vertical load F each leaf slides and bends over its neighbour with minimal friction. Varying the leaf geometry alters the stiffness of the spring for an application in vehicle suspension, say, where considerable deflection is required under variable loading. The leaf-spring beam shown in Fig. 10.17 is designed to ensure that the maximum bending stress is constant over the length of each plate.

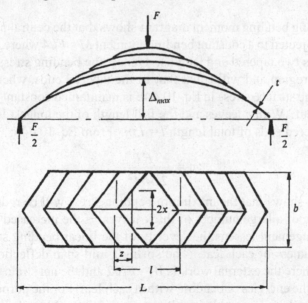

Figure 10.17 Leaf-spring construction

Referring to the top leaf shown in the lower plan view in Fig. 10.17, the bending stress at a section of breadth $2x$, distance z from its left-hand pointed end, is:

$$\sigma = \frac{My}{I} = \frac{(Fz/2)(t/2)}{(2xt^3/12)} = \frac{3Fz}{2xt^2} \qquad (10.44a)$$

Since z/x is constant, it follows that σ is also constant. It is convenient to refer σ at the central top surface where $z = l/2$ and $2x = b$. Equation 10.44a gives:

$$\sigma = \frac{3Fl}{2bt^2} \tag{10.44b}$$

All other leaves are subjected to four-point loading from the contact they make with adjacent leaves. Figure 10.18 shows the loading on one such leaf.

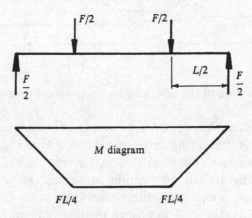

Figure 10.18 Single leaf in four-point loading

The accompanying bending moment diagram shows that the central parallel region of width b is subjected to a constant bending moment $M = Fl/4$ where, for each leaf, l is the sum of its two tapered end lengths. Hence, the bending stress, $\sigma = My/I$, is constant in this region and will equal that in the tapered ends, where Eq. 10.44a applies. Thus, the surface stress in Eq. 10.44b is maintained constant along the full length of each leaf. With n leaves and the full length of the longest leaf as L, each leaf has two tapered ends of total length $l = L/n$. From Eq. 10.44b:

$$\sigma = \frac{3FL}{2nbt^2} \tag{10.44c}$$

Equation 10.44c shows that the maximum bending stress will be reduced within a spring consisting of a large number of thick leaves. If the increased weight is not crucial this arrangement lessens the severity of the linear bending stress gradient through the thickness of each leaf. The spring's mid-span deflection Δ is found from Eq. 10.2 where the external work is $W = F\Delta/2$ and the net strain energy stored U, is the sum of the energies of bending within each leaf. For the purpose of finding U, the leaves may be re-assembled within a single plate of length L and diagonal breadth nb (see Fig. 10.16), that sustains the same surface stress in bending (Eq. 10.44c). The left corner becomes the origin for co-ordinates (z, x) lying in the plane of the equivalent plate. The second moment of area for a plate section with variable breadth $x = z \times bn/L$, for $0 \le z \le L/2$, becomes:

$$I(z) = \frac{(2x)t^3}{12} = \frac{zbnt^3}{6L} \tag{10.45a,b}$$

Taking the left- and right-hand corners to be simply supported and with the force F applied to its centre, the moment at position z in the plate is

$$M(z) = \frac{Fz}{2} \qquad (10.45c)$$

Substituting Eqs 10.45b,c into Eq. 10.15c, gives the energy of bending for whole plate as:

$$U = 2 \times \frac{1}{2E} \int_0^{L/2} \frac{M^2(z)\,dz}{I(z)} = \frac{3F^2 L}{2Ebnt^3} \int_0^{L/2} z\,dz = \frac{3F^2 L^3}{16Ebnt^3} \qquad (10.46a)$$

Equating 10.46a to the external work done $F\Delta/2$ gives the central deflection:

$$\Delta = \frac{3FL^3}{8Ebnt^3} \qquad (10.46b)$$

Equation 10.46b may also be derived from curvature of the deflected arc or from integrating the flexure Eq. 6.4. The spring's maximum *proof load* F_{max} is the force required to flatten the spring to its maximum deflection Δ_{max} (see Fig. 10.17). Equation 10.46b gives this as:

$$F_{max} = \frac{8Ebnt^3 \Delta_{max}}{3L^3} \qquad (10.46c)$$

Example 10.12 A leaf spring is made from seven steel plates each 6.5 mm thick. A deflection of 38 mm is required under a central concentrated force of 5 kN. If the maximum bending stress in the plates is not to exceed 230 MPa, calculate the length of the spring and a suitable common width for its leaves. Take $E = 207$ GPa.

Transposing Eq. 10.44c for the leaf width b and substituting into the proof load Eq. 10.46c:

$$L^3 = \left(\frac{8Et^3 n \Delta_{max}}{3F_{max}} \right) \times b = \left(\frac{8Et^3 n \Delta_{max}}{3F_{max}} \right) \times \left(\frac{3F_{max}L}{2\sigma nt^2} \right)$$

This provides the spring's full length

$$L = \sqrt{\frac{4Et\Delta_{max}}{\sigma}} = \sqrt{\frac{4 \times (207 \times 10^3) \times 6.5 \times 38}{230}} = 943 \text{ mm}$$

and each leaf's breadth:

$$b = \frac{3F_{max}L}{2\sigma nt^2} = \frac{3 \times (5 \times 10^3) \times 943}{2 \times 230 \times 7 \times 6.5^2} = 104 \text{ mm}$$

10.4 Impact Loading

So far we have considered the strain energy stored in a structure when its loading under tension, shear, bending and torsion, has been applied to it slowly. A greater amount of elastic energy will be imparted to a stationary structure under impact loading. Here the load may be accelerating or moving with constant velocity before impact. The stress and strain reached upon impact may be estimated from an energy balance using an *equivalent static force* (P_E). The latter is used to replace the weight P of an impacting body, which having fallen freely through height h, causes an instantaneous deflection Δ of the impacted structure. Equating the loss in potential energy of the fallen weight to the work done by an equivalent static force, that would produce the same displacement of the body:

$$P(h + \Delta) = P_E \times \Delta/2 = U \tag{10.47a}$$

where U is the strain energy stored from applying P_E slowly. Now for an elastic body Δ is a linear in P_E and hence Eq. 10.47a will lead to a quadratic equation in either P_E or Δ. The quadratic may refer to the structure's stiffness $k = P_E/\Delta$ under a static load, when Eq. 10.47a becomes:

$$\Delta^2 - \left(\frac{2P}{k}\right)\Delta + \left(\frac{2P}{k}\right)h = 0 \tag{10.47b}$$

Alternatively, using the structure's compliance (inverse stiffness) $c = \Delta/P_E$, Eq. 10.47a becomes:

$$P_E^2 - 2PP_E - \frac{2Ph}{c} = 0 \tag{10.47c}$$

The respective roots Δ and P_E to Eqs 10.47b,c provide the strain and stress at impact in the manner of the following example.

Example 10.13 Find the instantaneous deflection, the equivalent static load and the maximum stress produced, when a weight $P = 135$ N falls freely through a height $h = 50$ mm to make elastic impact with each of the following structures in (a), (b) and (c) below. Where appropriate take $E = 207$ GPa and $G = 80$ GPa.

(*a*) *Tie bar with circular section: diameter 6.35 mm, length 3 m, impacted axially.*

P_E is found from equating external work to energy stored in Eqs 10.2 and 10.10b:

$$\frac{P_E\Delta}{2} = \frac{P_E^2 L}{2EA}$$

giving

$$P_E = \frac{AE\Delta}{L}$$

which is confirmed within the displacement expression $\Delta = \varepsilon L = \sigma L/E$. Substituting P_E into Eq. 10.47a:

$$P(h + \Delta) = \frac{AE\Delta^2}{2L}$$

which gives the quadratic in Δ:

$$(AE)\Delta^2 - (2LP)\Delta - (2LPh) = 0 \qquad \text{(i)}$$

Substituting for L, P, E, h and $A = 31.67$ mm^2 in Eq. i gives $\Delta = 2.548$ mm.

$$\therefore P_E = AE\Delta/L = 31.67 \times 207 \times 10^3 \times 2.548/3000 = 5568 \text{ N}$$

$$\sigma = P_E/A = 5568/31.67 = 175.8 \text{ MPa}$$

Compare with a static stress under P of $\sigma = 135/31.67 = 4.26$ MPa. Alternatively, a quadratic in P_E follows directly from Eq. 10.47c for $c = \Delta/P_E = L/AE$:

$$P_E^2 - 2PP_E - 2AEPh/L = 0$$

$$P_E^2 - 270P_E - (29.5 \times 10^6) = 0 \qquad \text{(ii)}$$

The solution to Eq. ii confirms the equivalent load value, $P_E = 5568$ N.

(*b*) *Simply supported steel beam, with 25 mm-square section, impacted at the centre of its 1.25 m span*

Figure 2.15a shows $M(z) = Pz/2$, from which the central deflection under the equivalent load is found from Eqs 10.2 and 10.15b

$$\frac{P_E\Delta}{2} = 2 \times \frac{1}{2EI} \int_0^{L/2} \left(\frac{P_E z}{2}\right)^2 dz$$

which provides the well-known central deflection expression

$$\Delta = \frac{P_E L^3}{48EI}$$

Hence, the beam's compliance is $c = \Delta/P_E = L^3/(48EI)$, when from Eq. 10.47c:

$$P_E^2 - (2P)P_E - \frac{96PIEh}{L^3} = 0 \qquad \text{(ii)}$$

Substituting for P, h, E, $L = 1.25$ m and $I = 25^4/12 = 32.55 \times 10^3$ mm^4 into Eq. ii, gives its positive root as $P_E = 1366.2$ N. Correspondingly, the central deflection and maximum bending stress follow as

$$\Delta = \frac{P_E L^3}{48EI} = \frac{1366.2 \times (1.25 \times 10^3)^3}{48 \times (207 \times 10^3) \times (32.55 \times 10^3)} = 8.25 \text{ mm}$$

$$\sigma = \frac{M_{max} y}{I} = \frac{P_E L y}{4I} = \frac{1366.2 \times (1.25 \times 10^3) \times 12.5}{4 \times (32.55 \times 10^3)} = 163.96 \text{ MPa}$$

Note that the collision will be elastic provided the instantaneous stress does not exceed the static yield stress. For low carbon steels, the yield stress lies in the range: 250–350 MPa.

(c) Torsion bar, 20 mm diameter × 500 mm long, impacted at the end of a 200 mm long arm at right angles to its axis, as shown in Fig. 10.19

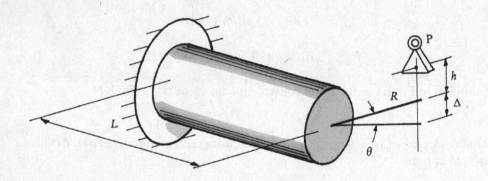

Figure 10.19 Torsion bar

Firstly, by neglecting shear and bending effects we have from Eqs 10.2 and 10.17c:

$$\frac{T\theta}{2} = \frac{T^2 L}{2GJ} \tag{iii}$$

For the equivalent static load the torque is $T = P_E R$, when Eq. iii gives the angular twist and deflection for Fig. 10.19 as:

$$\theta = \frac{(P_E R)L}{GJ}$$

$$\Delta = R\theta = \frac{P_E R^2 L}{GJ}$$

The following quadratic in P_E results from Eq. 10.47a:

$$P_E^2 - (2P)P_E - \frac{2GJPh}{R^2 L} = 0 \tag{iii}$$

Substituting for P, h, G, $R = 200$ mm, $L = 500$ mm and $J = \pi(20)^4/32 = 15708$ mm^4 in Eq. iii gives $P_E = 1065.8$ N. From this the angular twist and maximum shear stress for the bar are found:

$$\theta = \frac{\Delta}{R} = \frac{P_E RL}{GJ} = \frac{1065.8 \times 200 \times 500}{(80 \times 10^3) \times 15708} = 0.085^c \; (= 4.86°)$$

$$\tau_{max} = \frac{Tr}{J} = \frac{(P_E R)r}{J} = \frac{(1065.8 \times 200) \times 10}{15708} = 135.7 \text{ MPa}$$

Accounting for shear, bending and torsion in both arm (subscript a) and bar (subscript b) the equivalent static force P_E is found from equating the external work it does to the total store of strain energy within the assembly:

$$\frac{P_E \Delta}{2} = \frac{P_E^2 R}{2G_a A_a} + \frac{1}{2E_a I_a} \int_0^R (P_E z)^2 \, dz + \frac{P_E^2 L}{2G_b A_b} + \frac{1}{2E_b I_b} \int_0^L (P_E z)^2 \, dz + \frac{(P_E R)^2 L}{2G_b J_b}$$

and for similar arm and bar materials:

$$\frac{P_E \Delta}{2} = \frac{P_E^2 R}{2GA_a} + \frac{P_E^2 R^3}{6EI_a} + \frac{P_E^2 L}{2GA_b} + \frac{P_E^2 R^3}{6EI_b} + \frac{(P_E R)^2 L}{2GJ_b}$$

Hence, in the solution to P_E from Eq. 10.47c, the compliance is written as:

$$c = \frac{\Delta}{P_E} = \frac{R}{GA_a} + \frac{R^3}{3EI_a} + \frac{L}{GA_b} + \frac{R^3}{3EI_b} + \frac{R^2 L}{GJ_b}$$

10.5 Castigliano's Theorems

The young Alberto Castigliano (1847–1884) proposed in his Ph.D. thesis two theorems of general utility, based upon strain energy of elastic structures. His theorems connect the forces applied to the displacements produced and vice versa.

10.5.1 Force Theorem

The force theorem follows from applying Eq. 10.2 to a structure in equilibrium under n externally applied forces, F_i ($i = 1, 2, 3 \ldots n$). Let the displacements Δ_i, in line with the forces F_i, change by the amount $\delta\Delta_i$ to correspond with a change to the forces δF_i. From Eq. 10.2, the work–energy interchange applies as:

$$\delta U = \delta W = F_i \, \delta\Delta_i$$

$$U = \int_0^{\Delta_i} F_i \, d\Delta_i \tag{10.48a}$$

The double subscript appearing within the integral in Eq. 10.48a implies a summation. Written in full:

$$U = \int_0^{\Delta_1} F_1 \, d\Delta_1 + \int_0^{\Delta_2} F_2 \, d\Delta_2 + \int_0^{\Delta_3} F_3 \, d\Delta_3 + \dots \int_0^{\Delta_n} F_n \, d\Delta_n \qquad (10.48b)$$

The following partial derivatives apply to Eq. 10.48b:

$$\frac{\partial U}{\partial \Delta_1} = F_1, \quad \frac{\partial U}{\partial \Delta_2} = F_2, \quad \frac{\partial U}{\partial \Delta_3} = F_3, \quad \dots \quad \frac{\partial U}{\partial \Delta_n} = F_n$$

In general,

$$F_i = \frac{\partial U}{\partial \Delta_i} \qquad (10.49)$$

That is, each force F_i may be obtained by differentiating U partially with respect to its corresponding in-line displacement Δ_i. The theorem is useful only when the forces are to be found for prescribed displacements. More often the displacement is to be found beneath each force as provided by Castigliano's second theorem.

10.5.2 Displacement Theorem

The second, deflection theorem leads to a more useful result than Eq. 10.49. It employs the concepts of complementary energy and work which will now be defined. Firstly, an external work increment $\delta W = F \delta \Delta$ is identified with the strip of area beneath the F versus Δ diagram in Fig. 10.20.

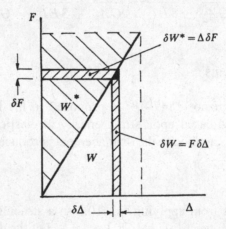

Figure 10.20 External and complementary work

The complementary work increment is the horizontal strip area $\delta W^* = \Delta \delta F$ shown in Fig. 10.20. Hence, the total complementary work W^* becomes the shaded area given by the integral:

$$W^* = \int_0^F \Delta \, dF \qquad (10.50a)$$

The physical interpretation of Eq. 10.50a is that of the work done when a variable force F, undergoes a displacement Δ. Whilst the concept is hard to realise in practice, it nonetheless provides a useful mathematical tool when complementary work is taken to be stored as *complementary energy* U^*. This gives:

$$U^* = W^* = \int_0^F \Delta \, dF \qquad (10.50b)$$

Where a number of variable forces F_i are applied to their corresponding displacements Δ_i, the complementary energy is written from Eq. 10.50b as:

$$U^* = \int_0^{F_i} \Delta_i \, dF_i \qquad (10.51a)$$

Again, Eq. 10.51a adopts the summation convention:

$$U^* = \int_0^{F_1} \Delta_1 \, dF_1 + \int_0^{F_2} \Delta_2 \, dF_2 + \int_0^{F_3} \Delta_3 \, dF_3 + \dots \int_0^{F_n} \Delta_n \, dF_n \qquad (10.51b)$$

The partial derivatives to Eq. 10.51b are:

$$\frac{\partial U^*}{\partial F_1} = \Delta_1, \quad \frac{\partial U^*}{\partial F_2} = \Delta_2, \quad \frac{\partial U^*}{\partial F_3} = \Delta_3, \quad \dots \quad \frac{\partial U^*}{\partial F_n} = \Delta_n$$

or, in general,

$$\Delta_i = \frac{\partial U^*}{\partial F_i} \qquad (10.52a)$$

That is, the deflection Δ_i at each force point i, may be obtained by differentiating U^* partially with respect to the force F_i. Equations 10.49 and 10.52a apply to both linear and non-linear elastic bodies. For a linear (Hookean) solid where $\sigma = E\varepsilon$ applies (see Fig. 10.1c, p. 374), the strain and complementary *energy densities* become:

$$u = \frac{U}{V} = \int_\varepsilon \sigma \, d\varepsilon = E \int_\varepsilon \varepsilon \, d\varepsilon = \frac{E\varepsilon^2}{2} = \frac{\sigma\varepsilon}{2}$$

$$u^* = \frac{U^*}{V} = \int_\sigma \varepsilon \, d\sigma = \frac{1}{E} \int_\sigma \sigma \, d\sigma = \frac{\sigma^2}{2E} = \frac{\sigma\varepsilon}{2}$$

Thus, $u = u^*$, $U = U^*$ and Eq. 10.52a appears in its more common form:

$$\Delta_i = \frac{\partial U}{\partial F_i} \qquad (10.52b)$$

That is, the deflection Δ_i beneath each force F_i may be obtained by differentiating U partially with respect to that force. Note that Eq. 10.52b will only apply to a linear elastic response for which both U and U^* discharge with unloading. Metals and their alloys conform very closely to this material model but non-metals may not. Rubber, for example, is elastic but non-linear and therefore the strain energy and its density do not have an equal complements, i.e. $u \neq u^*$, $U \neq U^*$. Consequently, Eq. 10.52b cannot be applied to non-linear elasticity but Eq. 10.52a remains valid and may be applied to find Δ_i when the function $U^* = U^*(F_i)$ is known.

10.5.3 Application to Structures

The four examples that follow show how Castigliano's second theorem, through the partial derivative given in Eq. 10.52b, provides deflections at load points in beams, arches and frames.

Example 10.14 Determine the load-point displacements for the stepped-cantilever beam in Fig. 10.9. This figure refers to Ex. 10.5 (p. 384), where it was shown that the shear strain energy is negligible compared to the strain energy of bending.

Putting $F_1 = 15$ N and $F_2 = 10$ N the moments in units of Nm, (lengths in m) are:

$$M_1 = F_1(1 - z) \text{ for } 0.5 \le z \le 1 \tag{i}$$

$$M_2 = F_1(1 - z) + F_2(0.5 - z) \text{ for } 0 \le z \le 0.5 \tag{ii}$$

The net strain energy in bending is, from Eq. 10.15b:

$$U = \frac{1}{2EI_1} \int_z M_1^2 \, dz + \frac{1}{2EI_2} \int_z M_2^2 \, dz \tag{iii}$$

Here, and elsewhere where U requires an integration, it is more convenient to apply Eq. 10.52b before integrating. Thus Eq iii gives the deflection Δ_1 beneath F_1 as:

$$\Delta_1 = \frac{\partial U}{\partial F_1} = \frac{1}{EI_1} \int_z M_1 \left(\frac{\partial M_1}{\partial F_1} \right) dz + \frac{1}{EI_2} \int_z M_2 \left(\frac{\partial M_2}{\partial F_1} \right) dz$$

Substituting from Eqs i and ii:

$$\Delta_1 = \frac{1}{EI_1} \int_{0.5}^{1.0} F_1(1 - z)^2 \, dz + \frac{1}{EI_2} \int_0^{0.5} [F_1(1 - z)^2 + F_2(1 - z)(0.5 - z)] \, dz$$

$$= \frac{15}{EI_1} \left| z - z^2 + \frac{z^3}{3} \right|_{0.5}^{1.0} + \frac{1}{EI_2} \left[15 \left| z - z^2 + \frac{z^3}{3} \right|_0^{0.5} + 10 \left| 0.5z - 1.5\frac{z^2}{2} + \frac{z^3}{3} \right|_0^{0.5} \right]$$

$$= \frac{15 \times 0.0417}{1333} + \frac{(15 \times 0.2917) + (10 \times 0.1042)}{2604} = 2.55 \times 10^{-3} \text{ m} = 2.55 \text{ mm}$$

Applying again Eq. 10.52b to Eq iii also gives the deflection Δ_2 beneath F_2 as:

$$\Delta_2 = \frac{\partial U}{\partial F_2} = \frac{1}{EI_1} \int_z M_1 \left(\frac{\partial M_1}{\partial F_2} \right) dz + \frac{1}{EI_2} \int_z M_2 \left(\frac{\partial M_2}{\partial F_2} \right) dz$$

Substituting from Eqs i and ii and noting that $\partial M_1/\partial F_2 = 0$:

$$\Delta_2 = \frac{1}{EI_2} \int_0^{0.5} \left[F_1(1-z)(0.5-z) + F_2(0.5-z)^2 \right] dz$$

$$= \frac{1}{EI_2} \left[15 \left| 0.5z - \frac{1.5z^2}{2} + \frac{z^3}{3} \right|_0^{0.5} + 10 \left| 0.25z - \frac{z^2}{2} + \frac{z^3}{3} \right|_0^{0.5} \right]$$

$$= \frac{(15 \times 0.1042) + (10 \times 0.04167)}{2604} = 0.76 \times 10^{-3} \text{ m} = 0.76 \text{ mm}$$

Example 10.15 The thin semi-circular cantilever beam in Fig. 10.21 supports vertical and horizontal end loads, V and H. Derive expressions for the vertical and horizontal end deflections. What do the deflections become when H is absent?

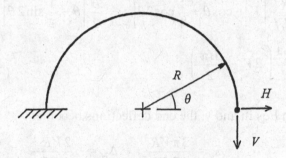

Figure 10.21 Semi-circular cantilever

Taking θ to be measured anticlockwise from the free end, the net positive clockwise moment acting about any section in the region defined by: $0 \le \theta \le \pi$, is:

$$M = VR(1 - \cos\theta) - HR\sin\theta \tag{i}$$

The strain energy stored in bending for the whole beam then becomes:

$$U = \frac{1}{2EI} \int_s M^2 \, ds$$

when, from Eq. 10.52b, the deflection beneath V is:

$$\Delta_V = \frac{\partial U}{\partial V} = \frac{1}{EI} \int_s M \left(\frac{\partial M}{\partial V} \right) ds \tag{ii}$$

From Eqs i and ii, with $ds = R\,d\theta$

$$\Delta_V = \frac{R^3}{EI}\int_0^\pi \left[V(1-\cos\theta) - H\sin\theta\right](1-\cos\theta)\,d\theta$$

$$= \frac{R^3}{EI}\left[V\left|\frac{3\theta}{2} - 2\sin\theta + \frac{1}{4}\sin2\theta\right|_0^\pi - H\left|-\cos\theta + \frac{1}{4}\cos2\theta\right|_0^\pi\right]$$

$$\Delta_V = \frac{R^3}{EI}\left(\frac{3\pi V}{2} - 2H\right) \tag{iii}$$

The horizontal deflection is found from Eqs i and 10.52b, with $ds = R\,d\theta$

$$\Delta_H = \frac{\partial U}{\partial H} = \frac{2}{EI}\int_s M\left(\frac{\partial M}{\partial H}\right)ds \tag{iv}$$

$$\Delta_H = \frac{-R^3}{EI}\int_0^\pi \left[V(1-\cos\theta) - H\sin\theta\right]\sin\theta\,d\theta$$

$$= -\frac{R^3}{EI}\left[V\left|-\cos\theta + \frac{1}{4}\cos2\theta\right|_0^\pi - \frac{H}{2}\left|\theta - \frac{1}{2}\sin2\theta\right|_0^\pi\right]$$

$$\Delta_H = \frac{-R^3}{EI}\left(2V - \frac{H\pi}{2}\right) \tag{v}$$

When $H = 0$ in Eqs iii and v, the end deflections become:

$$\Delta_V = \frac{3\pi VR^3}{2EI}, \qquad \Delta_H = -\frac{2VR^3}{EI}$$

This shows how Castigliano's method may be used to find deflections at positions where loads are not applied. That is, a load is placed at that position aligned with the direction of the required displacement. This enables the partial differential $\partial M/\partial H$ from Eq i to appear within the integration for $\Delta_H = \partial U/\partial H$ in Eq. iv. To simplify the analysis it can be seen that the 'dummy load' H within the integral so formed in Eq. iv could have been set to zero before the integration is carried out.

Example 10.16 The thin semi-circular arch in Fig. 10.22 supports a central vertical load P. Derive expressions for the vertical deflection beneath P when the arch is (a) fixed at A and B and (b) fixed at A and supported on rollers at B. What is the horizontal reaction for (a) the horizontal deflection at B for (b)?

Taking part (a) as the more general case, the net positive clockwise moment acting about any section in the region defined by: $0 \le \theta \le \pi/2$, is:

$$M = HR\sin\theta - (PR/2)(1-\cos\theta) \tag{i}$$

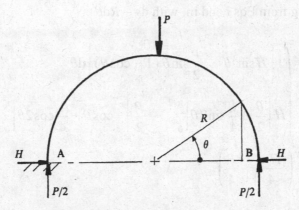

Figure 10.22 Semi-circular arch

Equation i provides the following two partial derivatives to be used in the derivation of the displacement expressions required

$$\frac{\partial M}{\partial P} = -\frac{R}{2}(1 - \cos\theta), \quad \frac{\partial M}{\partial H} = R\sin\theta \qquad \text{(ii, iii)}$$

The strain energy stored in bending for the whole beam then becomes:

$$U = 2 \times \frac{1}{2EI}\int_s M^2\, ds$$

when, from Eq. 10.52b, the deflection beneath P is:

$$\Delta_P = \frac{\partial U}{\partial P} = \frac{2}{EI}\int_s M\left(\frac{\partial M}{\partial P}\right) ds$$

Substituting from Eqs i and ii, with $ds = R\, d\theta$

$$\Delta_P = \frac{R^3}{EI}\int_0^{\pi/2}\left[\frac{P}{2}(1 - \cos\theta)^2 - H\sin\theta(1 - \cos\theta)\right] d\theta$$

$$= \frac{R^3}{EI}\left[\frac{P}{2}\left|\frac{3\theta}{2} - 2\sin\theta + \frac{1}{4}\sin 2\theta\right|_0^{\pi/2} - H\left|-\cos\theta + \frac{1}{4}\cos 2\theta\right|_0^{\pi/2}\right]$$

$$= \frac{R^3}{EI}\left[\frac{P}{2}\left(\frac{3\pi}{4} - 2\right) - \frac{H}{2}\right]$$

$$\Delta_P = \frac{R^3}{2EI}\left[P\left(\frac{3\pi}{4} - 2\right) - H\right] \qquad \text{(iv)}$$

Equation 10.52b also provides the horizontal deflection beneath H as:

$$\Delta_H = \frac{\partial U}{\partial H} = \frac{2}{EI}\int_s M\left(\frac{\partial M}{\partial H}\right) ds$$

Substituting from Eqs i and iii, with $ds = Rd\theta$

$$\Delta_H = \frac{2R^3}{EI} \int_0^{\pi/2} \left[H\sin^2\theta - \frac{P}{2}\sin\theta(1 - \cos\theta) \right] d\theta$$

$$= \frac{2R^3}{EI} \left[H \left| \frac{\theta}{2} - \frac{1}{4}\sin 2\theta \right|_0^{\pi/2} - \frac{P}{2} \left| -\cos\theta + \frac{1}{4}\cos 2\theta \right|_0^{\pi/2} \right]$$

$$= \frac{2R^3}{EI} \left(\frac{H\pi}{4} - \frac{P}{4} \right)$$

$$\Delta_H = \frac{R^3}{2EI}(H\pi - P) \tag{v}$$

(*a*) With B fixed, $\Delta_B = \Delta_H = 0$ and Eqs iv and v give:

$$H = \frac{P}{\pi}, \quad \Delta_P = \frac{PR^3}{EI}\left(\frac{3\pi}{8} - 1 - \frac{1}{2\pi} \right)$$

(*b*) With B on rollers, $H = 0$ and Eqs iv and v give:

$$\Delta_P = \frac{PR^3}{EI}\left(\frac{3\pi}{8} - 1 \right), \quad \Delta_H = -\frac{PR^3}{2EI}$$

Example 10.17 Figure 10.23 shows a davit with vertical and horizontal forces, F_V and F_H, respectively, applied at its free end C. Derive an expression for the vertical displacement at point C. The solution may ignore shear but should account for bending in AB and BC together with compression in AB. Take area A, E, and I to be constants.

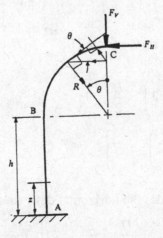

Figure 10.23 Davit with free end loading

With origins for z at A and θ at C, as shown, the bending moments (CW +ve) are:

$$M_{BC} = F_V R \sin\theta - F_H R(1 - \cos\theta), \quad \text{for } 0 \le \theta \le \pi/2 \tag{i}$$

$$M_{AB} = F_V R - F_H(R + h - z), \quad \text{for } 0 \le z \le h \tag{ii}$$

Each section of the davit is subjected to F_V and F_H. These may be resolved in tangential directions to give the compressive forces:

$$F_{BC} = -F_V \sin\theta - F_H \cos\theta, \quad \text{for } 0 \le \theta \le \pi/2 \tag{iii}$$

$$F_{AB} = -F_V, \quad \text{for } 0 \le z \le h \tag{iv}$$

The net strain energy in bending and compression is, from Eqs 10.15b and 10.10a:

$$U = \frac{1}{2EI} \int M^2 \mathrm{d}s + \frac{1}{2AE} \int F^2 \mathrm{d}s$$

From Eq. 10.52b the vertical deflection becomes:

$$\Delta_V = \frac{1}{EI} \int M_{BC} \left(\frac{\partial M_{BC}}{\partial F_V} \right) \mathrm{d}s + \frac{1}{EI} \int M_{AB} \left(\frac{\partial M_{AB}}{\partial F_V} \right) \mathrm{d}s + \frac{1}{AE} \int F_{BC} \left(\frac{\partial F_{BC}}{\partial F_V} \right) \mathrm{d}s + \frac{1}{AE} \int F_{AB} \left(\frac{\partial F_{AB}}{\partial F_V} \right) \mathrm{d}s$$

Substituting from Eqs i–iv:

$$\Delta_V = \frac{R^3}{EI} \int_0^{\pi/2} \left[F_V \sin^2\theta - F_H \sin\theta(1 - \cos\theta) \right] \mathrm{d}\theta + \frac{R}{EI} \int_0^h \left[F_V R - F_H(R + h - z) \right] \mathrm{d}z$$

$$+ \frac{R}{AE} \int_0^{\pi/2} (F_V \sin^2\theta + F_H \sin\theta \cos\theta) \mathrm{d}\theta + \frac{1}{AE} \int_0^h (-F_V)(-1) \mathrm{d}z$$

The integration gives:

$$\Delta_V = \frac{R^3}{EI} \left[\frac{F_V}{2} \left| \theta - \frac{1}{2}\sin 2\theta \right|_0^{\pi/2} - F_H \left| -\cos\theta + \frac{1}{4}\cos 2\theta \right|_0^{\pi/2} \right] + \frac{R}{EI} \left| (F_V R - F_H R - F_H h)z + \frac{F_H z^2}{2} \right|_0^h$$

$$+ \frac{R}{AE} \left[\frac{F_V}{2} \left| \theta - \frac{1}{2}\sin 2\theta \right|_0^{\pi/2} - \frac{F_H}{4} \left| \cos 2\theta \right|_0^{\pi/2} \right] + \frac{F_V}{AE} \left| z \right|_0^h$$

and substituting the limits:

$$\Delta_V = \frac{R^3}{EI} \left(\frac{\pi F_V}{4} - \frac{F_H}{2} \right) + \frac{R}{EI} \left[(F_V - F_H)Rh - \frac{F_H h^2}{2} \right] + \frac{R}{2AE} \left(\frac{\pi F_V}{2} + F_H \right) + \frac{F_V h}{AE}$$

The first two terms are contribution from bending and the final two terms are contributions from compression.

10.5.4 Application to Frameworks

The total strain energy stored in a point-loaded frame is the sum of the strain energies stored in all the bars. We assume that U is due to direct stress only by Eq. 10.10b. Thus, with a different force P_i (tensile or compressive), within each of the structure's n bars of differing length L_i, area A_i and moduli E_i:

$$U = \sum_{i=1}^{n} \frac{P_i^2 L_i}{2 A_i E_i} \tag{10.53}$$

More usually the frame's bars are of a single material with common cross-sectional area. Bar lengths will vary except for frames with equilateral triangular connections. When the bar forces P arise from a single external force F, the deflection Δ beneath F is found from equating Eq. 10.53 to the external work done:

$$\frac{F\Delta}{2} = \sum_{i=1}^{n} \left(\frac{P^2 L}{2AE} \right)_i \tag{10.54a}$$

If a unit force replaces F to produce bar forces k, it follows that $P_i = k_i F$ and Eq. 10.54a becomes:

$$\frac{F\Delta}{2} = \frac{F^2}{2} \sum_{i=1}^{n} \left(\frac{k^2 L}{AE} \right)_i$$

giving

$$\Delta = F \sum_{i=1}^{n} \left(\frac{k^2 L}{AE} \right)_i \tag{10.54b}$$

Castigliano's second theorem provides deflections at the nodes of multiply loaded frames. Let $F_A, F_B, F_C \dots$ be the external forces and $k', k'', k''' \dots$ be the bar forces when unit loads are applied separately at their respective positions A, B, C ... The net bar forces follow as: $P = F_A k' + F_B k'' + F_C k''' + \dots$, when from Eq. 10.53, the energy stored within the frame's strained bars becomes:

$$U = \sum_{i=1}^{n} \left[\left(F_A k' + F_B k'' + F_C k''' + \dots \right)^2 \frac{L}{2AE} \right]_i \tag{10.55}$$

Applying Eq. 10.52b to Eq. 10.55, the second theorem provides the displacement beneath the force at nodes A, B, C ... as:

$$\Delta_A = \frac{\partial U}{\partial F_A} = \sum_{i=1}^{n} \left[\left(F_A k' + F_B k'' + F_C k''' + \dots \right) \frac{k' L}{AE} \right]_i \tag{10.56a}$$

$$\Delta_B = \frac{\partial U}{\partial F_B} = \sum_{i=1}^{n} \left[\left(F_A k' + F_B k'' + F_C k''' + \dots \right) \frac{k'' L}{AE} \right]_i \tag{10.56b}$$

$$\Delta_C = \frac{\partial U}{\partial F_C} = \sum_{i=1}^{n} \left[\left(F_A k' + F_B k'' + F_C k''' + \dots \right) \frac{k''' L}{AE} \right]_i \tag{10.56c}$$

It should now be apparent how all nodal point displacements will follow from the theorem. For those nodes where no forces are applied the external forces are set to zero but the bar forces k', k'' ... etc, arising from applying unit loads to these nodes, are retained within the equations. For example, with no force applied at C, Eqs 10.56a–c become:

$$\Delta_A = \sum_{i=1}^{n} \left[\left(F_A k'^2 + F_B k' k'' + 0 + \ldots \right) \frac{L}{AE} \right]_i$$

$$\Delta_B = \sum_{i=1}^{n} \left[\left(F_A k' k'' + F_B k''^2 + 0 + \ldots \right) \frac{L}{AE} \right]_i$$

$$\Delta_C = \sum_{i=1}^{n} \left[\left(F_A k' k''' + F_B k'' k''' + 0 + \ldots \right) \frac{L}{AE} \right]_i$$

Equations 10.56a–c require a number of separate calculation of bar forces under isolated unit loads. The number of calculations may be reduced within the single calculation that supplies bar forces P with all external loads acting upon the structure. That is, P replaces $F_A k' + F_B k'' + F_C k''' + \ldots$ in Eqs 10.56a–c, giving the nodal displacements:

$$\Delta_A = \sum_{i=1}^{n} \left(\frac{P k' L}{AE} \right)_i \qquad (10.57a)$$

$$\Delta_B = \sum_{i=1}^{n} \left(\frac{P k'' L}{AE} \right)_i \qquad (10.57b)$$

$$\Delta_C = \sum_{i=1}^{n} \left(\frac{P k''' L}{AE} \right)_i \qquad (10.57c)$$

Equations 10.57a–c offer an alternative calculations of nodal displacements in which the calculations of bar forces k', k'' and k''' under unit loads applied separately to each load point form the P bar forces. The following example demonstrates these two approaches.

Example 10.18 Find the deflection at points B, D and E for the Warren structure in Fig. 2.11 (see p. 43), given that all bar areas are 50 mm^2 and $E = 207$ GPa.

As the bars are of similar material, length and area, Eqs 10.56a–c give the nodal displacements when subscripts B, D and E replace A, B and C respectively:

$$\Delta_B = \frac{L}{AE} \left(F_D \sum_{i=1}^{n} k' k'' + F_E \sum_{i=1}^{n} k' k''' \right) \qquad (i)$$

$$\Delta_D = \frac{L}{AE} \left(F_D \sum_{i=1}^{n} k''^2 + F_E \sum_{i=1}^{n} k'' k''' \right) \tag{ii}$$

$$\Delta_E = \frac{L}{AE} \left(F_D \sum_{i=1}^{n} k'' k''' + F_E \sum_{i=1}^{n} k'''^2 \right) \tag{iii}$$

To find k', let a unit load act vertically downwards at point B in Fig. 10.24.

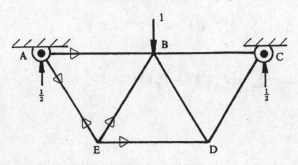

Figure 10.24 Bar forces in frame with unit load at B

From joint equilibrium:

At A:

$+\uparrow,$ $1/2 - k_{AE}' \sin 60° = 0,$ $\therefore k_{AE}' = 1/\sqrt{3} \ (= k_{CD}')$

$+\rightarrow,$ $k_{AB}' + k_{AE}' \cos 60° = 0,$ $\therefore k_{AB}' = -k_{AE}'/2 = -1/(2\sqrt{3}) \ (= k_{CB}')$

At E:

$+\uparrow,$ $k_{EA}' \sin 60° + k_{EB}' \sin 60° = 0,$ $\therefore k_{EB}' = -k_{EA}' = -1/\sqrt{3} \ (= k_{BD}')$

$+\rightarrow, -k_{EA}' \cos 60° + k_{EB}' \cos 60° + k_{ED}' = 0,$ $\therefore k_{ED}' = k_{AE}'/2 - k_{BE}'/2 = 1/\sqrt{3}$

To find k''', let a unit load act vertically downwards at point E in Fig. 10.25.

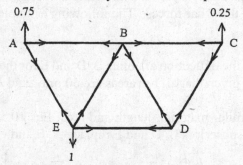

Figure 10.25 Bar forces in frame with unit load at E

The upward reactions at A and C follow from applying $\sum M_A = 0$ and then $\sum F_v = 0$. These give $R_C = 0.25$ and $R_A = 0.75$.

From joint equilibrium:

At A:

$\quad +\uparrow, \quad 0.75 - k_{AE}'''\sin 60° = 0, \qquad\qquad \therefore k_{AE}''' = \sqrt{3}/2$

$\quad +\rightarrow, \quad k_{AB}''' + k_{AE}'''\cos 60° = 0, \qquad\qquad \therefore k_{AB}''' = -\sqrt{3}/4$

At C:

$\quad +\uparrow, \quad 0.25 - k_{CD}'''\sin 60° = 0, \qquad\qquad \therefore k_{CD}''' = 1/(2\sqrt{3})$

$\quad +\rightarrow, \quad -k_{CB}''' - k_{CD}'''\cos 60° = 0, \qquad\qquad \therefore k_{CB}''' = -1/(4\sqrt{3})$

At E:

$\quad +\uparrow, \quad -1 + k_{EA}'''\sin 60° + k_{EB}'''\sin 60° = 0 \qquad \therefore k_{EB}''' = 1/(2\sqrt{3})$

$\quad +\rightarrow, \quad -k_{EA}'''\cos 60° + k_{EB}'''\cos 60° + k_{ED}''' = 0, \quad \therefore k_{ED}''' = 1/(2\sqrt{3})$

At D:

$\quad +\uparrow, \quad k_{DC}'''\sin 60° + k_{DB}'''\sin 60° = 0 \qquad\qquad \therefore k_{DB}''' = -1/(2\sqrt{3})$

(check)

$\quad +\rightarrow, \quad -k_{DB}'''\cos 60° + k_{DC}'''\cos 60° - k_{DE}''' = 1/(4\sqrt{3}) - 1/(2\sqrt{3}) + 1/(4\sqrt{3}) = 0 \checkmark$

Table 10.2 Bar forces under unit loading

BAR	k'	k''	k'''	k'''^2	k''^2	$k'k'''$	$k'k''$	$k'k'$
AB	$\dfrac{-1}{2\sqrt{3}}$	$\dfrac{-1}{4\sqrt{3}}$	$\dfrac{-\sqrt{3}}{4}$	$\dfrac{3}{16}$	$\dfrac{1}{48}$	$\dfrac{1}{16}$	$\dfrac{1}{8}$	$\dfrac{1}{24}$
BC	$\dfrac{-1}{2\sqrt{3}}$	$\dfrac{-\sqrt{3}}{4}$	$\dfrac{-1}{4\sqrt{3}}$	$\dfrac{1}{48}$	$\dfrac{3}{16}$	$\dfrac{1}{16}$	$\dfrac{1}{24}$	$\dfrac{1}{8}$
CD	$\dfrac{1}{\sqrt{3}}$	$\dfrac{\sqrt{3}}{2}$	$\dfrac{1}{2\sqrt{3}}$	$\dfrac{1}{12}$	$\dfrac{3}{4}$	$\dfrac{1}{4}$	$\dfrac{1}{6}$	$\dfrac{1}{2}$
DE	$\dfrac{1}{\sqrt{3}}$	$\dfrac{1}{2\sqrt{3}}$	$\dfrac{1}{2\sqrt{3}}$	$\dfrac{1}{12}$	$\dfrac{1}{12}$	$\dfrac{1}{12}$	$\dfrac{1}{6}$	$\dfrac{1}{6}$
BE	$\dfrac{-1}{\sqrt{3}}$	$\dfrac{-1}{2\sqrt{3}}$	$\dfrac{1}{2\sqrt{3}}$	$\dfrac{1}{12}$	$\dfrac{1}{12}$	$\dfrac{-1}{12}$	$\dfrac{-1}{6}$	$\dfrac{1}{6}$
BD	$\dfrac{-1}{\sqrt{3}}$	$\dfrac{1}{2\sqrt{3}}$	$\dfrac{-1}{2\sqrt{3}}$	$\dfrac{1}{12}$	$\dfrac{1}{12}$	$\dfrac{-1}{12}$	$\dfrac{1}{6}$	$\dfrac{-1}{6}$
AE	$\dfrac{1}{\sqrt{3}}$	$\dfrac{1}{2\sqrt{3}}$	$\dfrac{\sqrt{3}}{2}$	$\dfrac{3}{4}$	$\dfrac{1}{12}$	$\dfrac{1}{4}$	$\dfrac{1}{2}$	$\dfrac{1}{6}$

$$\sum = \frac{31}{24} \qquad \frac{31}{24} \qquad \frac{13}{24} \qquad 1 \qquad 1$$

The k'' bar forces refer to those when a unit force is applied at D. The symmetry of the frame allows these to be deduced from the k''' forces calculated above. The full set of bar forces, their products and summations, as required for Eqs i–iii, are listed in Table 10.2. Within the substitution of each summation into Eqs i–iii a common factor $L/(AE)$ appears. Taken with the conversion of kN to N, the multiplying factor required for consistent units of N and mm becomes:

$$\frac{(10 \times 10^3)}{50 \times (207 \times 10^3)} \times 10^3 = 0.966$$

and therefore

$$\Delta_B = [(4 \times 1) + (2 \times 1)]0.966 = 5.796 \text{ mm}$$

$$\Delta_D = [(4 \times 31/24) + (2 \times 13/24)]0.966 = 6.038 \text{ mm}$$

$$\Delta_E = [(2 \times 31/24) + (4 \times 13/24)]0.966 = 4.589 \text{ mm}$$

To confirm these deflections we may use Eqs 10.57a–c. Here the bar forces calculated previously in Example 2.7 correspond to P kN, whereas k', k'' and k''' appear in Table 10.2. The products required for Eqs 10.57a–c are tabulated and summed within Table 10.3.

Table 10.3 Bar forces under applied and unit loading

Bar	P	Pk'	Pk''	Pk'''
AB	$-\dfrac{5}{2\sqrt{3}}$	$\dfrac{5}{12}$	$\dfrac{5}{24}$	$\dfrac{5}{8}$
BC	$-\dfrac{7}{2\sqrt{3}}$	$\dfrac{7}{12}$	$\dfrac{7}{8}$	$\dfrac{7}{24}$
CD	$\dfrac{7}{\sqrt{3}}$	$\dfrac{7}{3}$	$\dfrac{7}{2}$	$\dfrac{7}{6}$
DE	$\sqrt{3}$	1	$\dfrac{1}{2}$	$\dfrac{1}{2}$
BE	$\dfrac{-1}{\sqrt{3}}$	$\dfrac{1}{3}$	$\dfrac{1}{6}$	$\dfrac{-1}{6}$
BD	$\dfrac{1}{\sqrt{3}}$	$\dfrac{-1}{3}$	$\dfrac{1}{6}$	$\dfrac{1}{6}$
AE	$\dfrac{5}{\sqrt{3}}$	$\dfrac{5}{3}$	$\dfrac{5}{6}$	$\dfrac{5}{2}$
Σ		6	$\dfrac{25}{4}$	$\dfrac{57}{12}$

The summations provide similar displacements as before

$$\Delta_B = \frac{(\sum Pk')L}{AE} = 6 \times 0.966 = 5.796 \text{ mm}$$

$$\Delta_D = \frac{(\sum Pk'')L}{AE} = \frac{25}{4} \times 0.966 = 6.038 \text{ mm}$$

$$\Delta_E = \frac{(\sum Pk''')L}{AE} = \frac{57}{12} \times 0.966 = 4.589 \text{ mm}$$

10.6 Stationary Potential Energy

Where forces displace those points at which they are applied a potential energy V refers to the force system in its undisplaced position. For the displaced system the strain energy conserved U becomes the potential energy of strain, giving the total potential energy for the force system as $U + V$. There are a number of variational principles which apply to the stationary value of the total potential energy that connects the displacements arising from a system of forces in equilibrium. One of these in the *Principle of Stationary Potential Energy* (SPE) which states that: *Of all the possible compatible displacement systems which satisfy the boundary conditions that which also satisfies the equilibrium condition gives a stationary value to the total potential energy*. This principle will be applied here to find the displacement of a beam and the critical buckling load for a strut. It will be seen that the principle is useful for finding very good approximations to the lateral displacement and buckling force within each structure.

10.6.1 Beam Displacement

Say we wish to find the displacement at any position z in the length of a beam with the offset concentrated load P shown in Fig. 10.26.

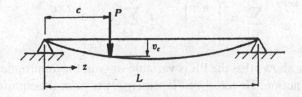

Figure 10.26 Simply supported beam with offset concentrated load

It can be seen that one compatible displacement function is the sine series:

$$v(z) = \sum_{n=1}^{N} a_n \sin \frac{n \pi z}{L} \qquad (10.58)$$

Referring to the principle stated above, compatibility is ensured because the beam's longitudinal and lateral strains depend upon a single transverse displacement. It is in the nature of uniaxial straining that this would remain the case for any alternative displacement function. More generally, in plane stressing where the three strain components ε_x, ε_x and γ_{xy} depend upon two displacements u and v, a condition of compatibility must exist between the three strains This can be seen in Chapter 13 on the triangular finite element adopted for plane stress (see §. 13.7, p.580). All that is required here is to satisfy the boundary conditions before the equilibrium is ensured under a stationary potential energy. Beam equilibrium was met earlier in Chapter 5 (see Eq. 5.2, p. 150), which shows how the stress is derived from the bending moment, which in turn depends upon the beam's transverse forces applied.

Clearly the boundary conditions, $v = 0$ for $z = 0$ and $v = 0$ for $z = L$, are met with the displacement function $v(z)$ assumed in Eq. 10.58. Next, for a uniform beam of length L the total potential energy for loading is written as:

$$U + V = \frac{1}{2EI} \int_0^L [M(z)]^2 dz - Pv \qquad (10.59a)$$

showing a loss in PE (V) from a displaced force and a gain in PE (U) from stored strain energy. For unloading the signs are interchanged as the PE of strain is released with the corresponding gain in PE for the load. From the theory of bending, Eq. 5.4 gives $M = EI/R$ where the radius of curvature R is given by Eq. 6.3. Hence Eq. 10.59b becomes:

$$U + V = \frac{EI}{2} \int_0^L \left(\frac{d^2 v}{dz^2} \right)^2 dz - Pv \qquad (10.59b)$$

where the beam's lateral displacement v applies to the offset load position at which $z = c$. Substituting from Eq. 10.58:

$$U + V = \frac{EI}{2} \int_0^L \left[\sum_{n=1}^N a_n \left(\frac{n\pi}{L} \right)^2 \sin \frac{n\pi z}{L} \right]^2 dz - P \sum_{n=1}^N a_n \sin \frac{n\pi c}{L}$$

$$= \frac{EIL}{4} \sum_{n=1}^N a_n^2 \left(\frac{n\pi}{L} \right)^4 - P \sum_{n=1}^N a_n \sin \frac{n\pi c}{L} \qquad (10.59c)$$

Equation 10.59c shows that the PE is variable only in the amplitude of the assumed displacement function. Hence, the SPE principle provides the equilibrium condition from Eq. 10.59c as:

$$\frac{d(U + V)}{da_n} = 0 \qquad (10.60a)$$

when from Eq. 10.59c

$$\frac{EIL}{2} \sum_{n=1}^N a_n \left(\frac{n\pi}{L} \right)^4 - P \sum_{n=1}^N \sin \frac{n\pi c}{L} = 0 \qquad (10.60b)$$

Equation 10.60b represents a decoupled system of N, linear simultaneously equations for which the solution is

$$a_n = \frac{2PL^3}{\pi^4 EI} \sum_{n=1}^{N} n^{-4} \sin \frac{n\pi c}{L} \qquad (10.61a)$$

Substituting Eq. 10.61a into Eq. 10.58 the required displacement function is correspondingly

$$v(z) = \frac{2PL^3}{\pi^4 EI} \sum_{n=1}^{N} n^{-4} \sin \frac{n\pi c}{L} \sin \frac{n\pi z}{L} \qquad (10.61b)$$

As a simple check upon the accuracy of Eq. 10.61b let the load act centrally with $c = L/2$. The central deflection then follows from the series expansion with $z = L/2$:

$$v(L/2) = \frac{2PL^3}{\pi^4 EI} \sum_{n=1}^{N} n^{-4} \sin^2 \frac{n\pi}{2}$$

$$v(L/2) = \frac{2PL^3}{\pi^4 EI} \left(1 + 3^{-4} + 5^{-4} + 7^{-4} + \ldots \right) = \frac{PL^3}{48.035\,EI}$$

This result is extremely close to the exact expression with 48 in the denominator.

10.6.2 Strut Buckling

Equation 10.58 is also a suitable displacement function for a pinned-end strut satisfying $v(0) = 0$ and $v(L) = 0$. However, Eq. 10.59a must be revised for the strut's axial compression under P and total in-line displacement λ (see Fig. 10.27).

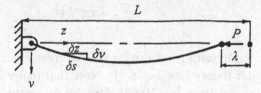

Figure 10.27 Simply supported strut showing axial and lateral displacements

An incremental length change $\delta\lambda$ may be approximated as follows

$$\delta\lambda = \delta s - \delta z = \sqrt{\delta z^2 + \delta v^2} - \delta z \doteq \delta z \left[1 + \left(\frac{\delta v}{\delta z} \right)^2 \right]^{\frac{1}{2}} - \delta z \approx \frac{1}{2} \left(\frac{\delta v}{\delta z} \right)^2 \delta z \qquad (10.62)$$

Equation 10.62 appears in the total PE's V term:

$$U + V = \frac{EI}{2} \int_0^L \left(\frac{d^2 v}{dz^2} \right)^2 dz - \frac{P}{2} \int_0^L \left(\frac{dv}{dz} \right)^2 dz \qquad (10.63a)$$

Substituting into Eq. 10.63a the first and second derivatives from Eq. 10.58:

$$U+V = \frac{EI}{2} \int_0^L \left[\sum_{n=1}^N a_n \left(\frac{n\pi}{L} \right)^2 \sin \frac{n\pi z}{L} \right]^2 dz - \frac{P}{2} \int_0^L \left[\sum_{n=1}^N a_n \left(\frac{n\pi}{L} \right) \cos \frac{n\pi z}{L} \right]^2 dz$$

$$= \frac{EIL}{4} \sum_{n=1}^N a_n^2 \left(\frac{n\pi}{L} \right)^4 - \frac{PL}{4} \sum_{n=1}^N a_n^2 \left(\frac{n\pi}{L} \right)^2 \qquad (10.63b)$$

Here a solution to Eq. 10.63b is required for the least value of the critical buckling load P. Taking P as a fixed quantity then the PE is variable within Eq. 10.63b only in the amplitude a_n of the assumed displacement function. Thus, the SPE principle ensures the equilibrium condition is met when:

$$\frac{d(U+V)}{da_n} = 0 \qquad (10.64a)$$

and therefore the buckling stress may be calculated from dividing P by the strut's section area. Applying Eq. 10.64a to Eq. 10.63b gives:

$$\frac{EIL}{2} \sum_{n=1}^N a_n \left(\frac{n\pi}{L} \right)^4 - \frac{PL}{2} \sum_{n=1}^N a_n \left(\frac{n\pi}{L} \right)^2 = 0 \qquad (10.64b)$$

Equation 10.64b provides the solution to P as:

$$P = EI \times \frac{\sum_{n=1}^N (n\pi/L)^4}{\sum_{n=1}^N (n\pi/L)^2} \qquad (10.64c)$$

The least P correspond to setting $n = 1$ in Eq 10.64c gives $P = \pi^2 EI/L^2$ which agrees exactly with the Euler theory (see Eq. 8.3). Note that many authors provide the same least buckling load from Eq. 10.63b directly by setting $U + V = 0$. Their interpretation of the total PE as a buckling criterion is that the complete loss in V is matched by the gain in U. Further useful applications of energy methods to finite element including the virtual work principle appear in §. 13.2, p. 544.

A simpler displacement function might be employed for a cantilever column in which one end is fixed and one end is free (see Table 8.1):

$$v(z) = \delta \left(1 - \cos \frac{\pi z}{2L} \right) \qquad (10.65a)$$

in which $v(L) = \delta$ is the free-end deflection. The derivatives required from Eq. 10.65a are:

$$\frac{dv}{dz} = \frac{\pi \delta}{2L} \sin \frac{\pi z}{2L}; \qquad \frac{d^2v}{dz^2} = \frac{\pi^2 \delta}{4L^2} \cos \frac{\pi z}{2L} \qquad (10.65b,c)$$

Substituting into Eq. 10.63a gives the total PE as:

$$U + V = \frac{EI}{2}\left(\frac{\pi^2\delta}{4L^2}\right)^2 \int_0^L \left(\cos\frac{\pi z}{2L}\right)^2 dz - \frac{P}{2}\left(\frac{\pi\delta}{2L}\right)^2 \int_0^L \left(\sin\frac{\pi z}{2L}\right)^2 dz \quad (10.66)$$

Once again, either one of the buckling criteria: $d(U+V)/d\delta = 0$ or $U+V = 0$, when applied to Eq. 10.66, agrees with this strut's Euler buckling load: $P = \pi^2 EI/4L^2$.

Exercises

Strain Energy

10.1 Calculate the stress and strain energy stored in a bronze buffer 250 mm long and 125 mm in diameter, when the compressive axial displacement is limited to 0.25 mm. Take $E = 207$ GPa. (Ans: 83.4 MPa, 121 Nm)

10.2 A load of 40 kN is applied slowly to a 25 mm diameter bar, 150 mm long. A second bar of similar material and initial dimensions is turned down to 22.5 mm for 50 mm of its length and a load is gradually applied. If the maximum stress in both bars is the same, find the strain energy in each bar. Take $E = 200$ GPa.

10.3 A solid steel shaft 50 mm in diameter is to be replaced by a solid phosphor bronze shaft of the same length that will transmit the same torque and store the same strain energy per unit volume. Find the diameter of the new shaft and the ratio between the maximum shear stress in each shaft. Take $G = 80$ GPa for steel and $G = 47$ GPa for phosphor bronze.

10.4 The stepped circular shaft in Fig. 10.28 is built in at the left-hand end and is to carry separately, a torque of 100 Nm and a moment of 100 Nm at its free end. Determine the respective free-end rotations from the energy stored. Under which mode of loading is the shaft the stiffer? Take $E = 2.6G = 100$ GPa.
(Ans: $\theta = 5.68°$, $\varphi = 4.36°$, bending)

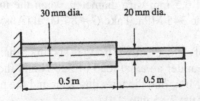

30 mm dia.   20 mm dia.

0.5 m   0.5 m

Figure 10.28

10.5 Use Eq. 10.2 to find the maximum deflection in each of the following beams:
(i) simply supported carrying a central concentrated force W. (Ans: $WL^3/48EI$)
(ii) simply supported carrying a uniformly distributed load w/unit length. (Ans: $5wL^4/384EI$)
(iii) cantilever carrying a concentrated force W at its free end. (Ans: $WL^3/3EI$)
(iv) cantilever carrying a uniformly distributed load w/unit length. (Ans: $wL^4/8EI$)

10.6 A solid steel shaft, diameter 63.5 mm and 0.75 m long, rotates at 200 rev/min. Calculate the power transmitted if the energy stored is 67.5 J. Take $G = 80$ GPa. (Ans: 100 kW)

428   MECHANICS OF ENGINEERING STRUCTURES

10.7 Calculate the deflection beneath the force for each of the cantilevered brackets in Figs 10.29a and b. Take $E = 200$ GPa, $G = 80$ GPa.
(Ans: 13 mm, 6.64 mm)

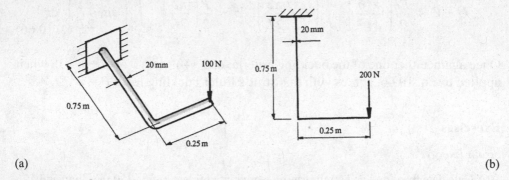

(a)   (b)

Figure 10.29

10.8 A steel beam, 5 m long, is simply supported at its ends and carries a concentrated force of 30 kN at 3.5 m from one end. Given that $I = 78 \times 10^4$ mm^4 and $E = 208$ GPa, employ an energy method to find the deflection beneath the force. (Ans: 3.3 mm)

10.9 Derive the maximum deflection due to shear for the following two beams, each having a rectangular cross section of area A:
   (i) simply supported carrying a central concentrated force, F. (Ans: $3FL/10AG$)
   (ii) cantilever carrying a uniformly distributed load, w. (Ans: $3wL^2/5AG$)

Springs

10.10 A close-coiled helical spring has a mean diameter of 100 mm with a 6.5 mm wire diameter. If the spring is to extend 25 mm under an axial force of 90 N, how many coils should the spring have? Take $G = 76$ GPa (Ans: 4.5 coils)

10.11 Determine the safe maximum load and the deflection of a spring with 10 close coils, 50 mm mean diameter and 6.5 mm wire diameter, when the maximum shear stress from torsional effects is limited to 345 MPa. Take $G = 83$ GPa. (Ans: 680 N, 53 mm)

10.12 The mean diameter of a close-coiled spring is six times that of the wire diameter. If the spring is to deflect 25 mm under an axial load of 220 N and the maximum torsional shear stress is not to exceed 413 MPa, find the wire diameter and the nearest number of complete coils. Take $G = 83$ GPa. (Ans: 2.85 mm, 16)

10.13 A close-coiled spring supports a load of 160 N. The mean coil diameter is 50 mm, the wire diameter is 6.5 mm. There are 15 coils and the spring stiffness is 8.4 N/mm. Find the rigidity modulus for the spring material and the maximum load the spring can support when (i) the torsional shear stress is limited to 275 MPa and (ii) the total shear stress is limited to 275 MPa. (Ans: 81 GPa, 545 N)

10.14 A close-coiled spring is subjected to a couple of 0.45 Nm. With 6 effective coils the angular rotation is to be 25° without the maximum bending stress in the wire exceeding 300 MPa. Find the wire and coil diameters given $E = 207$ GPa. (Ans: 2.47 mm, 19.7 mm)

10.15 A close-coiled spring, with 12 coils of 54 mm mean diameter, is made from 4.75 mm square section steel wire. The spring is subjected to an axial torque of 2.75 Nm. Calculate the maximum bending stress and the angular rotation between the ends. Take $E = 207$ GPa.

10.16 A close-coiled torsion spring with 8 coils is to twist by 45° under an axial couple of 550 Nm without the maximum bending stress exceeding 350 MPa. Determine the wire and coil diameters when the wire cross section is (a) circular and (b) square. Take $E = 210$ GPa.

10.17 An open-coiled helical spring is made from steel wire of 7.5 mm diameter with a mean diameter of 50 mm and a 30° helix angle. If the displacement under a maximum axial load is 15 mm whilst the stress due to bending is restricted to 103 MPa find the load and the number of coils required. Take $E = 207$ GPa and $G = 83$ GPa. (Ans: 354 N, 11)

10.18 The bending and shear stresses in a 10-coil open spring must not exceed 69 and 96.5 MPa, respectively, when it is subjected to an axial load of 225 N. If the mean coil diameter is ten times the wire diameter, find these dimensions together with the axial displacement and helix angle. Take $E = 207$ GPa, $G = 83$ GPa. (Ans: 8 mm, 23.8°)

10.19 An open-coiled spring has 12 coils each with a 35° helix angle, a mean diameter of 75 mm and a wire diameter of 10 mm. Calculate the axial load that will deflect the spring by 20 mm and find the corresponding stresses due to bending and twisting.

10.20 An open-coiled spring has 12 coils wound with a helix angle of 30°. The coil and wire diameters are 102 mm and 12.5 mm, respectively. Calculate the axial couple necessary to produce an angular twist of 2 radians. Take $E = 207$ GPa and $G = 83$ GPa. (Ans: 112.5 Nm)

10.21 A leaf spring is to span 0.75 m and carry a central concentrated load of 10 kN for a plate width of 75 mm. If the permissible stress due to bending is 465 MPa and the central deflection is not to exceed 50 mm, calculate the number and thickness of the plates required. Take $E = 207$ GPa. (Ans: 6.5 mm, 8)

10.22 A steel leaf spring spans 915 mm and lies flat under a central concentrated load of 3.3 kN. The width of the plates is 57 mm and they are each 4.75 mm thick. Find the number of plates required and the central deflection to give a maximum stress of 500 MPa.

10.23 Show that, by clamping a normal leaf spring at its centre, the end deflection and maximum stress, for a cantilever leaf spring construction, are given by $\Delta = 6FL^3/(nbt^3E)$ and $\sigma = 6FL/(nbt^2)$, respectively.

Impact Loading

10.24 A bar of 25 mm square section, 3 m long is fixed at the top and held vertically. A load of 450 N is dropped a distance of 150 mm onto a collar attached to the bottom of the bar (see Fig. 10.30). Determine: (i) the maximum tensile stress induced in the bar and (ii) the extension of the bar. Take $E = 207$ GPa.
(Ans: 121 MPa, 1.8 mm)

10.25 A load of 2.2 kN falls freely through a height of 19 mm on to a stop at the lower end of a vertical bar with section area 645 mm² and length 4.6 m as shown in Fig. 10.30.

Determine the stress in the bar at impact and the instantaneous tensile deflection. Take $E =$ 200 GPa. (Ans: 1.8 mm, 79.3 MPa)

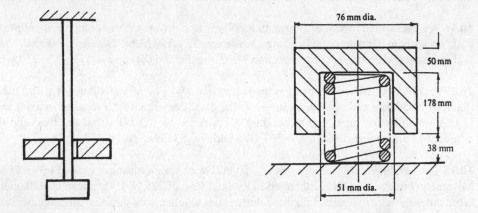

Figure 10.30   **Figure 10.31**

10.26 A 90 N weight falls through a height of 265 mm to strike the spring-loaded cylinder shown in Fig. 10.31. The impact causes the end of the cylinder to make contact with the base plate. If the spring stiffness is 35 N/mm, determine the maximum compressive stress induced in the cylinder due to impact. Take $E =$ 207 GPa. (Ans: 38.16 MPa)

10.27 The square, tubular-section, torsion bar in Fig. 10.32 is mounted as a cantilever with a 0.7 m rigid torsion arm as shown. Calculate the maximum shear stress in the tube when a force of 50 N falls 10 mm to impact with the arm end. Take $E = 2.5G = 200$ GPa. (Ans: 13.2 MPa)

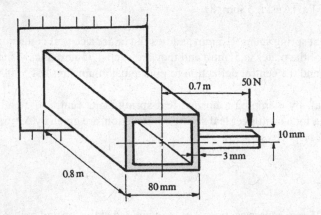

Figure 10.32

10.28 In Fig. 10.19 the weight $P = 150$ N falls by $h = 50$ mm to make elastic impact with the end of a 200 mm long arm, with 10 mm square section, attached to a torsion bar, 25 mm diameter × 500 mm long. The arm lies lies at right angles to the fall and to the bar axis. Accounting for shear, bending and torsional effects, determine the arm's end deflection, the bar's twist and maximum stress. Take $E = 207$ GPa and $G = 80$ GPa for both arm and bar.

10.29 Find the instantaneous deflection, the equivalent static load and the maximum stress reached when a weight of 135 N falls through a height of 50 mm to make elastic impact with each of the following. Where appropriate take $E = 207$ GPa and $G = 80$ GPa.

(i)   transverse loading the central 1 m span of a seven leaf spring with plates 65 mm wide × 6.5 mm thick. (Ans: 1109.4 N, 16.08 mm, 86.56 MPa)

(ii) axial loading of a close-coiled spring 3 m long with wire and coil diameters of 6.5 and 60 mm, respectively. (Ans: 432.2 N, 83.2 mm, 240.46 MPa)

(iii) transverse loading the end of a 75 mm long arm, attached at right angles to the axis of a close-coiled spring of length 3 m with wire and coil diameters 12.5 and 60 mm, respectively. (Ans: 600.5 N, 40.85 mm, 234.9 MPa)

Castigliano's Theorems

10.30 Derive the horizontal displacement at the free end of the davit in Fig. 10.23.

10.31 The circular steel ring in Fig. 10.33, with mean diameter 100 mm and 6 mm section diameter, has a narrow gap in the position shown. Determine the force required to open the gap by a further 0.5 mm. Take $E = 200$ GPa.

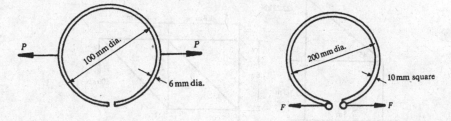

Figure 10.33   **Figure 10.34**

10.32 The split steel piston ring in Fig. 10.34 has a mean diameter of 200 mm with a 10 mm square cross section. Determine the equal opposing forces necessary to prise open the cut by 5 mm when they are applied tangentially at this point. Take $E = 210$ GPa.

10.33 The tubular body of the bow saw in Fig. 10.35 is 30 mm outer diameter and 2 mm thick. The blade section is 25 mm × 1.5 mm. How much shorter than the nominal length of 1 m should the blade be, so that when mounted in the frame the tensile stress does not exceed 20 MPa? Take $E = 200$ GPa for both materials.

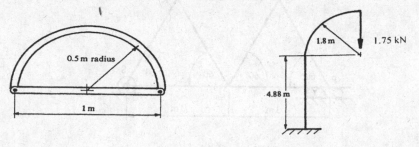

Figure 10.35   **Figure 10.36**

10.34 The lantern of the lamp post in Fig. 10.36 weighs 1.75 kN. Find the vertical and horizontal deflections due to this weight given $I = 210 \times 10^6$ mm^4 for the post's tubular cross section. Take $E = 13.8$ GPa. (Ans: 13 mm)

10.35 Determine the displacement in the direction of the load for each of the frames in Figs 10.37a–d given that all bar areas are 1280 mm^2 and $E = 207$ GPa.
[Ans: (b) 1.7 mm, (c) 2.96 mm, (d) 8.9 mm]

(a)   (b)

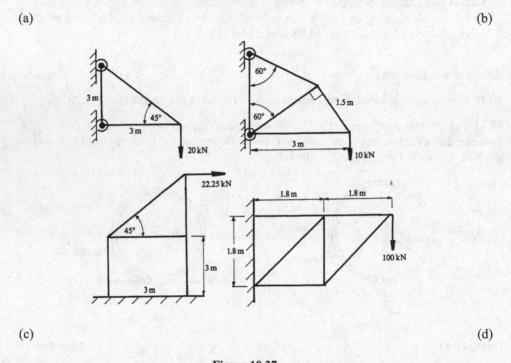

(c)   (d)

Figure 10.37

10.36 Find the vertical deflection at points Q and T for the 60° Warren bay girder in Fig. 10.38, when it is hinged at S and supported on rollers at P. The cross-sectional areas of each bar is 1280 mm^2. Take $E = 208$ GPa. (Ans: 1.36 mm, 1.51 mm)

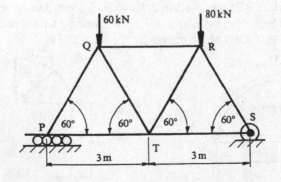

Figure 10.38

CHAPTER 11

PLANE STRESS AND STRAIN

Summary: In this chapter we consider methods for transforming stress and strain states from one orientation to another lying in the same plane. It will be seen that there are a number of methods available to make the required transformation. Among them are analytical, graphical and matrix methods all of which will be given here with illustrative examples. The choice between which of the alternative techniques to adopt for this type of analysis is left to the reader given that within themselves they serve to check one another. The topic becomes important to ensuring a safe engineering design based upon the greatest (principal) stress induced by any manner of combined loadings. The latter might include tension, bending, torsion and shear applied to components operating under service conditions. These analyses enable the principal stress to be factored safely to a lower level than the limiting stress imposed by the material under load.

11.1 Plane Stress Analyses

Say that we know the state of stress at a point for each of two perpendicular reference directions, x and y, consists of normal and shear stress (σ_x, σ_y and τ_{xy}) acting in combination. Here we determine the normal and shear stresses (σ_θ, τ_θ) along another direction inclined at θ to x. It will be shown how the three methods referred to above are applied to make this transformation. Simpler two-dimensional stress transformation equations arise where stress reductions apply to the reference directions, for example, where the shear stress is absent. In particular, the four stress states that follow may be identified in practice from the most common forms of externally applied loading. A knowledge of the latter enables the applied, or reference, stress components σ_x, σ_y and τ_{xy} to be found.

11.1.1 Simple Tension Compression

When a bar or plate is subjected to uniaxial tensile stress σ_x (see Fig. 11.1a), the transverse plane AB is subjected to σ_x as its normal stress component. The inclined plane AC has both normal and shear stress components, σ_θ and τ_θ, acting upon it as shown. Assume that the bar is of rectangular cross section with a unit thickness. This allows the forces produced by these stresses to be resolved in directions parallel and perpendicular to AC (see Fig. 11.1b).

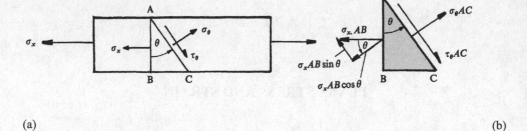

(a)   (b)

Figure 11.1 Wedge element under tension

The force components lying parallel and perpendicular to AC are in equilibrium with the resolved forces upon AB. Taking the equilibrium condition between forces perpendicular to AC and noting that $AB = AC \cos\theta$:

$$\sigma_\theta AC = \sigma_x AB \cos\theta$$

$$\sigma_\theta AC = \sigma_x AC \cos^2\theta$$

$$\sigma_\theta = \sigma_x \cos^2\theta \qquad (11.1a)$$

and with force equilibrium parallel to AC:

$$\tau_\theta AC = \sigma_x AB \sin\theta$$

$$\tau_\theta AC = \sigma_x AC \sin\theta \cos\theta$$

$$\tau_\theta = (\sigma_x/2) \sin 2\theta \qquad (11.1b)$$

Equation 11.1b shows that $\tau_\theta = \sigma_x/2$ is a maximum for the $\theta = 45°$ and 135° planes. Correspondingly, Eq. 11.1a shows that the normal stress on these planes of maximum shear is $\sigma_\theta = \sigma_x/2$. A similar result applies to compression when, conventionally, a minus sign accompanies the numerical value of σ_x. That is, planes of maximum shear lie at 45° and 135° to the axis of either an applied tension or compression. The magnitudes of both the normal and shear stress upon these plane of maximum shear are each one half the applied stress.

11.1.2 Pure Shear

A state of pure shear, i.e. where $\sigma_x = \sigma_y = 0$, is typical of a point on the surface of a shaft in torsion as in Fig. 11.2a. The stress analysis of this pure shear state has been dealt with earlier in § 4.2 (see p. 114). There it was shown that when τ_{xy} acts along the x-direction, then a *complementary* shear stress τ_{yx} of equal magnitude must act parallel to the y-direction, as is shown in Fig. 11.2b.

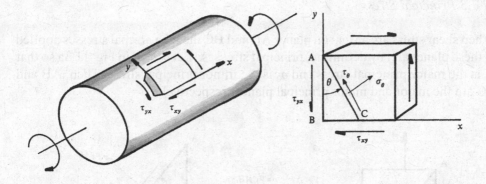

(a)   (b)

Figure 11.2 Element of a bar under torsion

Also, the force resolution provided within Figs 4.10a,b lead to the stress state for the inclined direction (see Eqs 4.20a,b). Thus the normal and shear stresses upon the plane AC in Fig. 11.2b are:

$$\sigma_\theta = \tau_{xy} \sin 2\theta \tag{11.2a}$$

$$\tau_\theta = \tau_{xy} \cos 2\theta \tag{11.2b}$$

Equations 11.2a,b show:

(i) when $\theta = 45°$; $\sigma_{45°} = \tau_{xy}$ and $\tau_{45°} = 0$,

(ii) when $\theta = 135°$; $\sigma_{135°} = - \tau_{xy}$ and $\tau_{135°} = 0$.

The planes aligned with the 45° and 135° directions, being free of shear stress, are called *principal planes*. Tensile and compressive stresses act normally to these planes, with magnitudes each equal to τ_{xy}. These stresses are called the *principal stresses*. The positive (tensile) principal stress is called the *major principal stress* σ_1 and this acts in a direction normal to the *major principal plane*. The negative (compressive) principal stress is called the *minor principal stress* σ_2 lying normal to the *minor principal plane*. The major and minor principal planes (and stresses) are perpendicular to one another. In summary, pure shear produces tension and compression on planes inclined at 45° to the shear planes. That is, an element with an equivalent principal stress state has a 45° inclination to the shear planes (see Fig. 4.9b). Note that pure shear arises in structures from two sources: transverse shear forces applied to plates and beams and axial torsion of bars and tubes. Each results in equal magnitudes of principal tension and compression, i.e.

$$\sigma_1 = \tau_{xy} \text{ (tension)}, \quad \sigma_2 = - \tau_{xy} \text{ (compression)}$$

For all other loading that combines tension, torsion and bending the principal stresses σ_1 and σ_2 that apply are not equal, with the major principal stress σ_1 taking the greater numerical value.

11.1.3 Principal Stress

When shear stress are absent on planes AB and BC then the normal stresses applied to these planes are, by definition, principal stresses. Let $\sigma_1 > \sigma_2$ in Fig. 11.3a so that σ_1 is the major principal stress and σ_2 is the minor principal stress. Then, AB and BC are the major and minor principal planes, respectively.

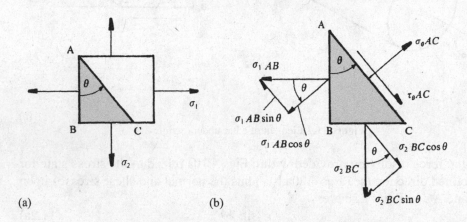

(a)   (b)

Figure 11.3 Element under two normal (principal) stresses

Figure 11.3b converts the wedge stresses into a force system assuming unit thickness. These forces are then resolved parallel and perpendicular to AC as shown. Noting that $AB = AC \cos\theta$ and $BC = AC \sin\theta$, the equilibrium equation that applies parallel to AC is:

$$\tau_\theta AC = \sigma_1 AB \sin\theta - \sigma_2 BC \cos\theta$$

$$= \sigma_1 AC \sin\theta \cos\theta - \sigma_2 AC \sin\theta \cos\theta$$

$$\tau_\theta = \tfrac{1}{2}(\sigma_1 - \sigma_2) \sin 2\theta \qquad (11.3a)$$

The equilibrium equation perpendicular to AC is:

$$\sigma_\theta AC = \sigma_1 AB \cos\theta + \sigma_2 BC \sin\theta$$

$$= \sigma_1 AC \cos^2\theta + \sigma_2 AC \sin^2\theta$$

$$\sigma_\theta = \sigma_1 \cos^2\theta + \sigma_2 \sin^2\theta \qquad (11.3b)$$

The following observations apply to Eqs 11.3a,b:

(i) Equation 11.3a shows that the shear stress τ_θ is a maximum for the planes $\theta = 45°$ and $135°$. On each maximum shear plane, inclined at $45°$ and $135°$ to the principal planes, the shear stress becomes one half the difference between the principal stresses, written as: $\tau_{max} = \tfrac{1}{2}(\sigma_1 - \sigma_2)$.

(ii) Equation 11.3b shows that the normal stress acting on the $\theta = 45°$ and $135°$ planes of maximum shear stress becomes one half the sum of the principal stresses, written as: $\sigma = \frac{1}{2}(\sigma_1 + \sigma_2)$.

(iii) The values of θ, giving maximum and minimum values of σ_θ, must correspond to $d\sigma_\theta/d\theta = 0$. Thus, from Eq. 11.3b:

$$-2\sigma_1 \cos\theta \sin\theta + 2\sigma_2 \sin\theta \cos\theta = 0$$

$$-(\sigma_1 - \sigma_2) \sin 2\theta = 0$$

Hence $\sin 2\theta = 0$ when $\theta = 0°$ or $90°$, giving maximum and minimum values: $\sigma_{0°} = \sigma_1$ and $\sigma_{90°} = \sigma_2$. This shows that when all inclined planes are considered, there is no normal stress greater than σ_1 nor less that σ_2. This coupling applies generally to principal stresses, associated with two perpendicular planes free of shear stress.

11.1.4 General Plane Stress

In general, normal stresses σ_x and σ_y ($\sigma_x > \sigma_y$) will act on the perpendicular planes AB and BC together with shear stresses τ_{xy} and τ_{yx}, as shown in Fig. 11.4a.

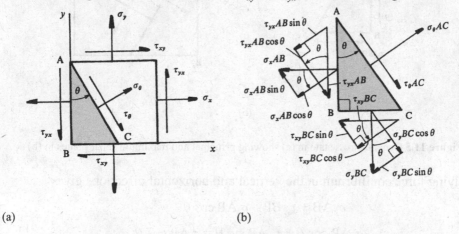

(a)   (b)

Figure 11.4 Element with combined normal and shear stress

Complementary shear ($\tau_{xy} = \tau_{yx}$) exists from a consideration of moment equilibrium about any point. Assuming a unit thickness, ABC is converted to the force system resolved parallel and perpendicular to AC, as shown in Fig. 11.4b. Using: AB = ACcosθ and BC = ACsinθ, the respective equilibrium equations become:

$$\sigma_\theta AC = \sigma_x AB \cos\theta + \tau_{yx} AB \sin\theta + \sigma_y BC \sin\theta + \tau_{xy} BC \cos\theta$$

$$\sigma_\theta AC = AC\sigma_x \cos^2\theta + AC\tau_{xy} \sin\theta \cos\theta + AC\sigma_y \sin^2\theta + AC\tau_{xy} \sin\theta \cos\theta$$

$$\sigma_\theta = \sigma_x \cos^2\theta + \sigma_y \sin^2\theta + \tau_{xy} \sin 2\theta \qquad (11.4a)$$

$$\tau_\theta AC = \sigma_x AB \sin\theta - \tau_{yx} AB \cos\theta - \sigma_y BC \cos\theta + \tau_{xy} BC \sin\theta$$

$$\tau_\theta AC = AC\sigma_x \sin\theta \cos\theta - AC\tau_{yx} \cos^2\theta - AC\sigma_y \sin\theta \cos\theta + AC \tau_{xy} \sin^2\theta$$

$$\tau_\theta = (\sigma_x - \sigma_y) \sin\theta \cos\theta - \tau_{xy}(\cos^2\theta - \sin^2\theta)$$

$$\tau_\theta = \tfrac{1}{2}(\sigma_x - \sigma_y) \sin 2\theta - \tau_{xy} \cos 2\theta \qquad\qquad (11.4b)$$

Equations 11.4a,b contain stress systems § 11.1.1–§ 11.1.3 as special cases when the absent stresses are set to zero. The inclinations of principal planes for this general state are found from setting $\tau_\theta = 0$ in Eq. 11.4b:

$$\tan 2\theta = 2\tau_{xy}/(\sigma_x - \sigma_y) \qquad\qquad (11.5)$$

The principal stresses σ_1 and σ_2 may then be found from eliminating θ between Eqs 11.4a and 11.5. Alternatively, simplified equilibrium equations apply when AP in Fig. 11.5a is a taken as principal plane.

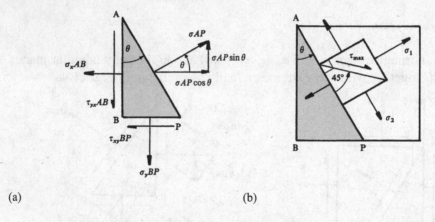

(a)   (b)

Figure 11.5 General stress state in (a) showing principal and maximum shear planes in (b)

Applying force equilibrium in the vertical and horizontal directions gives:

$$\sigma_x AB + \tau_{xy} BP = \sigma AP \cos \theta$$

$$\sigma_x AP \cos \theta + \tau_{xy} AP \sin \theta = \sigma AP \cos \theta$$

$$\sigma - \sigma_x = \tau_{xy} \tan \theta \qquad\qquad (11.6a)$$

$$\sigma AP \sin\theta = \tau_{yx} AB + \sigma_y BP$$

$$\sigma AP \sin\theta = \tau_{yx} AP \cos\theta + \sigma_y AP \sin\theta$$

$$\sigma - \sigma_y = \tau_{yx} \cot\theta \qquad\qquad (11.6b)$$

Multiplying Eqs 11.6a,b and setting $\tau_{xy} = \tau_{yx}$ leads to a quadratic in σ:

$$(\sigma - \sigma_x)(\sigma - \sigma_y) = \tau_{yx}^2$$

$$\sigma^2 - (\sigma_x + \sigma_y)\sigma + (\sigma_x\sigma_y - \tau_{yx}^2) = 0 \qquad\qquad (11.6c)$$

The roots to the quadratic Eq. 11.6c are the principal stresses magnitudes:

$$\sigma_{1,2} = \frac{1}{2}(\sigma_x + \sigma_y) \pm \frac{1}{2}\sqrt{(\sigma_x - \sigma_y)^2 + 4\tau_{xy}^2} \qquad (11.7)$$

in which the major stress σ_1 takes the positive discriminant and the minor stress σ_2 the negative discriminant. Figure 11.5b shows that σ_1 and σ_2 act on planes θ and $\theta + 90°$, as defined by Eq. 11.5. The previous analysis, given in § 11.1.3, will apply to the principal element shown inset. In particular, it follows from Eq. 11.7 that the maximum shear stress is given by:

$$\tau_{max} = \frac{1}{2}(\sigma_1 - \sigma_2) = \frac{1}{2}\sqrt{(\sigma_x - \sigma_y)^2 + 4\tau_{xy}^2} \qquad (11.8)$$

which acts along the 45° plane shown in Fig. 11.5b. Further points to note are:

i)   Equation 11.7 shows that the sum of the principal stresses equals the sum of the applied normal stresses: $\sigma_1 + \sigma_2 = \sigma_x + \sigma_y$.
ii) If σ_1 is a pre-determined design stress value then Eq. 11.6a gives the inclination of the major principal plane:

$$\tan\theta = \frac{\sigma_1 - \sigma_x}{\tau_{xy}}$$

iii) Zero or negative substitutions for normal stress σ_x or σ_y in Eqs 11.4–11.7 apply when absent or compressive respectively.
iv) If the directions of τ_{xy} and τ_{yx} in Fig. 11.4a are reversed they become negative in the foregoing equations. Alternatively, the wedge element ABC can be re-positioned to match the τ_{xy} and τ_{yx} directions given for the equations derived.

Example 11.1 A steel drive shaft of circular section is supported in bearings 2 m apart when rotating at $N = 500$ rev/min. The shaft is to transmits power $P = 900$ kW when carrying a central concentrated load $W = 20$ kN (see Fig. 11.6). Determine the shaft's design diameter d when based upon the following limiting stresses for the steel: (a) 75 MPa for the maximum shear stress and (b) 150 MPa for the principal tensile stress.

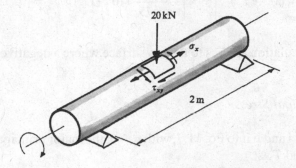

Figure 11.6 Rotating shaft carrying a central concentrated load

The applied stresses, σ_x and τ_{xy}, are due to bending and torsion, respectively. The bending stress σ_x attains its greatest values in tension and compression at the central bottom and top positions, respectively. Equation 5.4 provides σ_x as:

$$\sigma_x = \frac{My}{I} = \frac{M(d/2)}{(\pi d^4/64)} = \frac{32M}{\pi d^3} \tag{i}$$

where M is the maximum central bending moment upon the shaft. Under a simultaneous torque T the shear stress τ_{xy} attains its maximum everywhere on the shaft's outer surface. Equation 7.6 provides τ_{xy} as:

$$\tau_{xy} = \frac{Tr}{J} = \frac{T(d/2)}{(\pi d^4/32)} = \frac{16T}{\pi d^3} \tag{ii}$$

The central (sagging) bending moment M is

$$M = \frac{WL}{4} = \frac{20 \times 2}{4} = 10 \text{ kNm}$$

The torque T is found from the power P and speed N as:

$$P = \frac{2\pi NT}{60}, \quad \therefore T = \frac{60P}{2\pi N} = \frac{(60 \times 900)}{(2\pi \times 500)} = 17.19 \text{ kNm}$$

(a) Maximum Shear Stress

Substituting Eqs i and ii into Eq. 11.8, with $\sigma_y = 0$, the maximum shear stress appears as:

$$\tau_{max} = \frac{16}{\pi d^3} \left(M^2 + T^2 \right)^{1/2} \tag{iii}$$

Applying a multiplication factor of 10^6 to M and T (giving each a unit of of Nmm), the shaft diameter is found from Eq. iii:

$$d = \left[\frac{16}{\pi \tau_{max}} \left(M^2 + T^2 \right)^{\frac{1}{2}} \right]^{\frac{1}{3}} = \left[\frac{16 \times 10^6}{\pi \times 75} \left(10^2 + 17.19^2 \right)^{\frac{1}{2}} \right]^{\frac{1}{3}} = 110.53 \text{ mm}$$

An identical calculation applies to the top surface where a negative σ_x is squared in Eq. 11.8.

(b) Major Principal Stress

Substituting Eqs i and ii into Eq. 11.7, with $\sigma_y = 0$, the major principal stress appears in terms of M and T as:

$$\sigma_1 = \frac{16}{\pi d^3} \left(M + \sqrt{M^2 + T^2} \right) \tag{iv}$$

Where the bending stress is tensile, for the central bottom position, the shaft diameter is found from Eq. iv as:

$$d = \left[\frac{16}{\pi \sigma_1} \left(M + \sqrt{M^2 + T^2} \right) \right]^{\frac{1}{3}} = \left[\frac{16 \times 10^6}{\pi \times 150} \left(10 + \sqrt{10^2 + 17.19^2} \right) \right]^{\frac{1}{3}} = 100.49 \text{ mm}$$

Where the bending stress is compressive, for the central top position, the shaft diameter is found by taking M to be negative in Eq. iv:

$$d = \left[\frac{16}{\pi \sigma_1} \left(-M + \sqrt{M^2 + T^2} \right) \right]^{\frac{1}{3}} = \left[\frac{16 \times 10^6}{\pi \times 150} \left(-10 + \sqrt{10^2 + 17.19^2} \right) \right]^{\frac{1}{3}} = 69.48 \text{ mm}$$

The lesser diameter is found because the major principal tensile stress (from Eq. 11.7) differs between the top and bottom positions. However, as the difference between σ_1 and σ_2 is constant, the shear diameter calculation in (a) applies to both top and bottom stress states. Overall, shear provides the basis for selecting the greatest (safest) shaft diameter.

It can be seen that under combined bending and torsion of a solid circular shaft τ_{max} and σ_1 may be solved directly from the substitution of Eqs i and ii into Eqs 11.7 and 11.8. This gives:

$$\sigma_1 = \frac{32 M_E}{\pi d^3}, \qquad \tau_{max} = \frac{16 T_E}{\pi d^3}$$

where the *equivalent bending and twisting moments* are:

$$M_E = \frac{1}{2} \left(M + \sqrt{M^2 + T^2} \right), \qquad T_E = \sqrt{M^2 + T^2}$$

Example 11.2 Figure 11.7a shows the stress states (in MPa) upon two planes inclined at $\theta = 35°$. Determine τ_θ, τ_{max}, σ_1 and σ_2 and the planes on which they act.

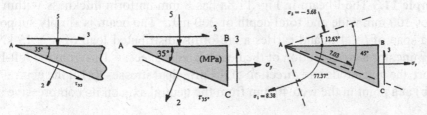

(a)   (b)   (c)

Figure 11.7 Stress states on two inclined planes AB and AC

Firstly, the equilibrium stress state for the wedge element ABC in Fig. 11.7b must be completed with the additions of a normal stress σ_y and a complementary shear stress of 3 MPa to plane BC as shown. Rotating the element in Fig. 11.7b to align with Fig. 11.4a, identifies the stress components as follows: $\sigma_x = -5$ MPa, $\tau_{xy} = \tau_{yx}$ = 3 MPa and $\sigma_{35°} = 2$ MPa. Substitution into Eqs 11.4a,b gives $\tau_{35°}$ and σ_y:

$$2 = -5\cos^2 35° + \sigma_y \sin^2 35° + 3\sin 70° = -0.329\sigma_y - 0.536,$$

$$\tau_{35°} = \tfrac{1}{2}(-5 - \sigma_y)\sin 70° - 3\cos 70° = \tfrac{1}{2}(-12.71)\sin 70° - 3\cos 70°$$

from which from which $\sigma_y = 7.71$ MPa (tensile) and $\tau_{35°} = -7.0$ MPa. The minus sign indicates that the shear direction is reversed. The principal stress directions are found from Eq. 11.5:

$$\tan 2\theta = \frac{2 \times 3}{(-5 - 7.71)} = -0.472$$

$$2\theta = -25.26° \text{ or } 154.73°$$

The positive value $\theta = 77.37°$ locates the major principal plane while the negative value, $\theta = -12.63°$ locates the minor principal plane, as shown in Fig. 11.7c. The principal stresses are found from Eq. 11.7:

giving
$$\sigma_{1,2} = \tfrac{1}{2}(-5 + 7.71) \pm \tfrac{1}{2}\sqrt{(-5 - 7.71)^2 + 4(3)^2}$$

$$\sigma_1 = 8.38 \text{ MPa} \quad \text{and} \quad \sigma_2 = -5.68 \text{ MPa}$$

acting upon their respective planes in Fig. 11.7c. Equation 11.8 provides the maximum shear stress:

$$\tau_{max} = \tfrac{1}{2}\sqrt{(-5 - 7.71)^2 + 4(3)^2} = 7.03 \text{ MPa}$$

which lies along those planes inclined at 45° and 135° to the major principal plane in Fig. 11.7c.

Example 11.3 The I-beam in Fig. 11.8a has 8 mm uniform thickness within two flanges 100 mm wide and total depth of 200 mm. The beam is simply supported over a span of $l = 4$ m and carries a uniformly distributed load of $w = 15$ kN/m. Determine, for a cross section of the beam at a distance $z = 1$ m from the left-hand support, the magnitude and direction of the principal stresses: (a) at the neutral axis and (b) at a point in the web 40 mm from the neutral axis on its compressive side.

The bending and shear stress in Eqs 5.4 and 5.26b apply to all beams where bending is produced by lateral loading:

$$\sigma_x = \frac{My}{I}, \qquad \tau_{xy} = \frac{F(A\bar{y})}{Ib} \qquad \text{(i, ii)}$$

in which F and M are extracted from Fig. 2.15b (see p. 49), where, for $z = 1$ m:

$$M = \frac{wz^2}{2} - \frac{wlz}{2} = \frac{wz}{2}(z - l) = \frac{15 \times 1}{2}(1 - 4) = -22.5 \text{ kNm (sagging)}$$

$$F = \frac{dM}{dz} = \frac{w}{2}(2z - l) = \frac{15}{2}(2 \times 1 - 4) = -15 \text{ kN}$$

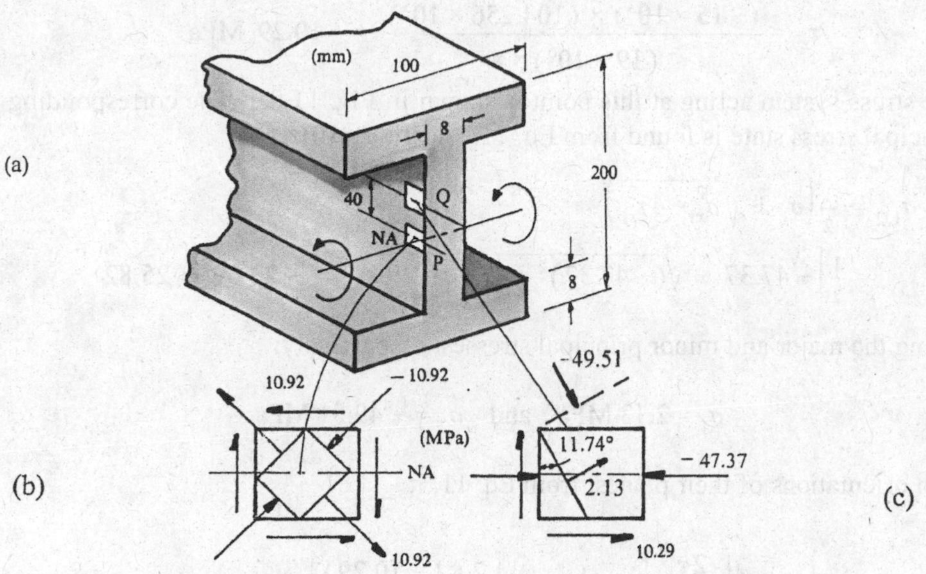

(a)

(b)

(c)

Figure 11.8 Stress states for points P and Q within an I-section

(a) Point P on the Neutral Axis (see Fig. 11.8a)

The bending stress is zero while the shear stress due to shear force F is a maximum. Within Eq. ii, for $b = 8$ mm:

$$I = \frac{100 \times 200^3}{12} - \frac{92 \times 184^3}{12} = 19 \times 10^6 \text{ mm}^4$$

$$A\bar{y} = (100 \times 8 \times 96) + (92 \times 8 \times 46) = 110.656 \times 10^3 \text{ mm}^3$$

$$\therefore \tau_{xy} = \frac{(-15 \times 10^3) \times (110.656 \times 10^3)}{(19 \times 10^6) \times 8} = -10.92 \text{ MPa}$$

At position P the stress system is pure shear where an equivalent principal stress system lies with 45° inclinations (see §. 4.2, p. 114). Hence tensile and compressive stresses, each of magnitude 10.92 MPa, act in 45° directions to the neutral axis, as shown in Fig. 11.8b.

(b) Position Q in the Web (see Fig. 11.8a)

At Q, where $y = 40$ mm, both bending and shear stresses exist. Equations i and ii provide their magnitudes:

$$\sigma_x = \frac{My}{I} = \frac{(-22.5 \times 10^6) \times 40}{(19 \times 10^6)} = -47.37 \text{ MPa}$$

$$A\bar{y} = (52 \times 8 \times 66) + (100 \times 8 \times 96) = 104.256 \times 10^3 \text{ mm}^3$$

$$\tau_{xy} = \frac{(-15 \times 10^3) \times (104.256 \times 10^3)}{(19 \times 10^6) \times 8} = -10.29 \text{ MPa}$$

The stress system acting at this point is shown in Fig. 11.8c. The corresponding principal stress state is found from Eq. 11.7 (with $\sigma_y = 0$):

$$\sigma_{1,2} = \frac{1}{2}\left(\sigma_x \pm \sqrt{\sigma_x^2 + 4\tau_{xy}^2}\right)$$

$$= \frac{1}{2}\left[-47.37 \pm \sqrt{(-47.37)^2 + 4(-10.29)^2}\right] = -23.69 \pm 25.82$$

giving the major and minor principal stresses, respectively:

$$\sigma_1 = 2.13 \text{ MPa} \quad \text{and} \quad \sigma_2 = -49.51 \text{ MPa}$$

with orientations of their planes, from Eq. 11.5:

$$\theta = \frac{1}{2}\tan^{-1}\left(\frac{2\tau_{xy}}{\sigma_x}\right) = \frac{1}{2}\tan^{-1}\left[\frac{2 \times (-10.29)}{(-47.37)}\right] = \frac{1}{2}\tan^{-1} 0.4345$$

giving $\theta = 11.74°$ and $(90 + 11.74)°$, as shown in Fig. 11.8c.

11.2 Mohr's Circle

Stress states $(\sigma_\theta, \tau_\theta)$ on inclined planes within the systems § 11.1.1–§ 11.1.4 given above may also be found graphically from a Mohr's circle construction. Squaring and adding Eqs 11.4a,b for the general stress system § 11.1.4, we have:

$$\left[\sigma_\theta - \frac{1}{2}(\sigma_x + \sigma_y)\right]^2 + \tau_\theta^2 = \frac{1}{4}(\sigma_x - \sigma_y)^2 + \tau_{xy}^2 \tag{11.9a}$$

Referring to Fig. 11.9a the known stress components for directions x and y are σ_x, σ_y and τ_{xy}. For any inclined direction θ the variables are σ_θ and τ_θ. Writing $X = \sigma_\theta$ and $Y = \tau_\theta$, it can be seen that Eq. 11.9a describes a circle of the form:

$$(X - A)^2 + Y^2 = C^2 \tag{11.9b}$$

with centre co-ordinates:

$$(A, 0) = [\tfrac{1}{2}(\sigma_x + \sigma_y), 0] \tag{11.9c}$$

and radius

$$C = \tau_{max} = \sqrt{\frac{1}{4}(\sigma_x - \sigma_y)^2 + \tau_{xy}^2} \tag{11.9d}$$

Equations 11.9a–d allow the circle in Fig. 11.9b to be constructed from the known stress state existing on perpendicular planes AB and BC according to the following convention:

Normal stresses σ_x and σ_y are plotted to the right (positive) when tensile and to the left when compressive (negative). Shear stress τ_{xy} is plotted upward (positive) when acting in a clockwise sense upon the material element and downward (negative) when acting in an anticlockwise sense.

Thus, for the material wedge element in Fig. 11.9a, the stress states upon planes AB and BC each appear as single points upon the circumference of the circle. Following the convention outlined above it is seen that these two points, shown as AB and BC in Fig. 11.9b, lie at opposite ends of the circle's diameter, enabling the circle to be drawn. The circle's *focus point* F is then located by projecting the normal to plane AB (i.e. direction x) through the corresponding stress point (AB) in the circle. Since the singular point F is the focus of all normal directions it may also be located from projecting the normal to plane BC (i.e. direction y) through the corresponding stress point (BC) on the circle. The stress state on plane AC is found from the converse construction by projecting the normal to AC (i.e. direction θ) through F to intersect with the circle. The circle shows positive co-ordinates for the intersection point AC ($\sigma_\theta, \tau_\theta$). These stress components act upon the physical plane AC (shaded within the circle's construction) in the directions shown, these having been returned to the physical element in Fig. 11.9a.

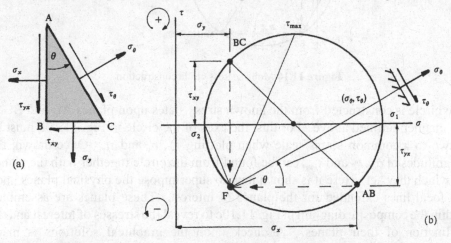

Figure 11.9 Mohr's circle for a general plane stress state

The normal to other planes of interest may be projected from the focus. Most important are the magnitudes and directions of the principal stresses σ_1 and σ_2 which lie at the ends of the horizontal diameter where the shear stress is zero. The normal to the major and minor principal planes join F to σ_1 and F to σ_2. The line which joins F to the top point (τ_{max}) is the normal to the maximum shear plane.

Example 11.4 Two perpendicular tensile stresses, with magnitudes of 30 and 120 MPa, act together with a shear stress of 75 MPa, as shown in Fig. 11.10a. Determine, graphically, the principal stresses, the maximum shear stress and the planes on which they act. Confirm the answers from the appropriate formulae.

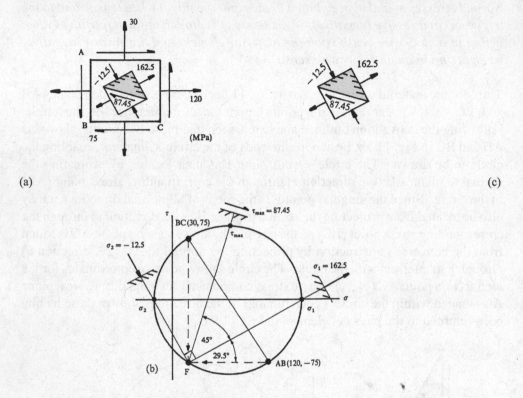

Figure 11.10 Mohr's stress circle construction

The circle is constructed from the known stress states upon planes AB and BC in the manner outlined above. For this, the axes of the circle in Fig. 11.10b must be drawn to a common stress scale when plotting σ_x, σ_y and τ_{xy}. Once drawn, the magnitudes of σ_1, σ_2 and τ_{max} can be found from the circle together with the planes on which they act. Here it is shown how to superimpose the physical planes upon the focal lines extended for the planes of interest. These planes are assembled within the composite diagram in Fig. 11.10c to reveal the stresses of interest and the inclination of their planes. A check upon the graphical solutions is made analytically from Eqs 11.5–11.8. Equation 11.7 provides the principal stresses:

$$\sigma_{1,2} = \tfrac{1}{2}\,(\sigma_x + \sigma_y) \pm \tfrac{1}{2}\sqrt{(\sigma_x - \sigma_y)^2 + 4\tau_{xy}^2}$$

$$= \tfrac{1}{2}\,(120 + 30) \pm \tfrac{1}{2}\sqrt{(120 - 30)^2 + 4(75)^2} = 75 \pm 87.45$$

$$\sigma_1 = 162.45 \quad \text{and} \quad \sigma_2 = -12.45 \text{ MPa}$$

Equation 11.5 provides the orientations of the principal planes

$$\tan 2\theta = \frac{2\tau_{xy}}{\sigma_x - \sigma_y} = \frac{2 \times 75}{120 - 30} = 1.667 \quad \therefore \quad \theta = 29.5° \text{ and } 119.5°$$

The maximum shear stress is found from Eq. 11.8

$$\tau_{max} = \frac{1}{2}\sqrt{(\sigma_x - \sigma_y)^2 + 4\tau_{xy}^2} = \frac{1}{2}(\sigma_1 - \sigma_2)$$

$$= \frac{1}{2}\left[162.45 - (-12.45)\right] = 87.45 \text{ MPa}$$

which lies at 45° to the principal planes (i.e. 74.5° to AB).

Example 11.5 A 60 mm-diameter aluminium alloy bar is subjected to three combined loadings consisting of an axial compressive load $W = 200$ kN, a bending moment $M = 600$ Nm and a torque $T = 1200$ Nm. Determine from Mohr's circle, for a point on the bar surface subjected to the greatest net compressive stress: (i) the state of stress on a plane inclined at 40° to the bar axis and (ii) the principal stresses and the orientation of their planes.

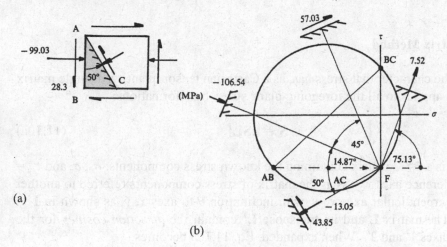

Figure 11.11 Mohr's circle construction for principal stresses

The shear stress due to torsion is found from Eq. 7.6:

$$\tau_{xy} = \frac{Tr}{J} = \frac{(1200 \times 10^3) \times 30}{\pi(60)^4/32} = 28.3 \text{ MPa}$$

The greatest compressive stress, arising from the combination of direct compression and bending, is found from Eq. 5.6d:

$$\sigma_x = -\frac{W}{A} - \frac{My}{I} = -\frac{(200 \times 10^3)}{\pi(60)^2/4} - \frac{(600 \times 10^3) \times 30}{\pi(60)^4/64}$$

$$= -70.736 - 28.294 = -99.03 \text{ MPa}$$

The stress state is shown in Fig. 11.11a, in which BC lies parallel to the shaft axis. The circle is constructed in Fig. 11.11b to give the major principal stress $\sigma_1 = 7.52$ MPa, the minor principal stress $\sigma_2 = -106.54$ MPa and the maximum shear stress $\tau_{max} = 57.03$ MPa. These act upon those physical planes lying normal to the focal lines, each plane having been separated in Fig. 11.11b.

The stress state for the 50° plane (i.e. 40° to the shaft axis) is found from projecting from the focus point F a line normal to the plane AC in Fig. 11.11a. At its intersection with the circle the normal and shear stress co-ordinates (–13.05, –43.85) apply to AC as shown. These may be checked from setting $\sigma_y = 0$ in Eqs 11.4a,b to give:

$$\sigma_\theta = \sigma_x \cos^2\theta + \sigma_y \sin^2\theta + \tau_{xy} \sin 2\theta$$

$$= -99.03 \cos^2 50° + 28.3 \sin 100° = -13.05 \text{ MPa}$$

$$\tau_\theta = \tfrac{1}{2}(\sigma_x - \sigma_y) \sin 2\theta - \tau_{xy} \cos 2\theta$$

$$= -(99.03/2)\sin 100° - 28.3 \cos 100° = -43.85 \text{ MPa}$$

11.3 Matrix Method

Within the character that stress has, as a Cartesian tensor quantity, a single matrix equation applies to all the foregoing plane stress transformations:

$$\mathbf{S'} = \mathbf{LSL}^T \qquad\qquad (11.10a)$$

where $\mathbf{S}$ is a 2×2 matrix containing the known stress components: σ_x, σ_y and $\tau_{xy} = \tau_{yx}$ for reference axes x, y. $\mathbf{S'}$ is the matrix of stress components referred to another pair of perpendicular axes x', y' with inclination θ to axes x, y, as shown in Fig. 11.12. The matrix $\mathbf{L}$ and its transpose $\mathbf{L}^T$ contain the *direction cosines* for the inclined axes 1' and 2'. When expanded, Eq. 11.10a becomes:

$$\begin{bmatrix} \sigma_x' & \tau_{xy}' \\ \tau_{yx}' & \sigma_y' \end{bmatrix} = \begin{bmatrix} l_{xx} & l_{xy} \\ l_{yx} & l_{yy} \end{bmatrix} \begin{bmatrix} \sigma_x & \tau_{xy} \\ \tau_{yx} & \sigma_y \end{bmatrix} \begin{bmatrix} l_{xx} & l_{yx} \\ l_{xy} & l_{yy} \end{bmatrix} \qquad (11.10b)$$

where, for each component of the 2×2 matrix $\mathbf{L}$ of direction cosines, the first subscript refers to the primed axis and the second to the unprimed axis. Hence, direction cosines apply to the rotation between these axes in Fig. 11.12 as follows:

$$l_{xx} = \cos \triangle(x'x) = \cos\theta, \quad l_{xy} = \cos \triangle(x'y) = \cos(90° - \theta) = \sin\theta$$

$$l_{yy} = \cos \triangle(y'y) = \cos\theta, \quad l_{yx} = \cos \triangle(y'x) = \cos(\theta + 90°) = -\sin\theta$$

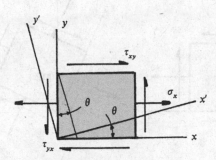

Figure 11.12 Rotation in stress axes

Written in full, Eq. 11.10b becomes:

$$
\begin{bmatrix} \sigma_x' & \tau_{xy}' \\ \tau_{yx}' & \sigma_y' \end{bmatrix} = \begin{bmatrix} \cos\theta & \sin\theta \\ -\sin\theta & \cos\theta \end{bmatrix} \begin{bmatrix} \sigma_x & \tau_{xy} \\ \tau_{yx} & \sigma_y \end{bmatrix} \begin{bmatrix} \cos\theta & -\sin\theta \\ \sin\theta & \cos\theta \end{bmatrix}
\qquad (11.10c)
$$

The matrix multiplication within Eq. 11.10c confirms that σ_θ by Eq. 11.4a provides both σ_x' and σ_y' for planes with orientations at θ and $\theta + 90°$, respectively. However, a sign change will be found between τ_θ from Eq. 11.4b and τ_{xy}' by the matrix Eq. 11.10c; this reflecting their differing conventions on positive shear.

The principal stress magnitudes identify with the roots to a quadratic equation that results from expanding the determinant to the **S** matrix:

$$
\det \begin{vmatrix} \sigma_x - \sigma & \tau_{xy} \\ \tau_{yx} & \sigma_y - \sigma \end{vmatrix} = (\sigma_x - \sigma)(\sigma_y - \sigma) - \tau_{xy}\tau_{yx} = 0
\qquad (11.11)
$$

and with $\tau_{xy} = \tau_{yx}$, Eq. 11.11 leads to a quadratic with roots $\sigma_{1,2}$ given by Eq. 11.7.

Example 11.6 Determine the full state of stress following a 30° anticlockwise rotation of the element with the stress state given in Fig. 11.13a. What are its principal stress values and directions?

The direction cosines for the rotation shown in Fig. 11.13b are:

$$l_{xx} = \cos \measuredangle (x'x) = \cos 30°$$

$$l_{xy} = \cos \measuredangle (x'y) = \cos (90° - 30°) = \sin 30°$$

$$l_{yy} = \cos \measuredangle (y'y) = \cos 30°$$

$$l_{yx} = \cos \measuredangle (y'x) = \cos (30° + 90°) = -\sin 30°$$

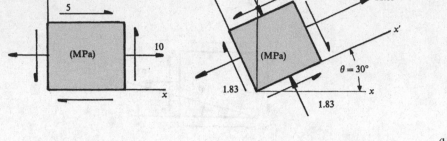

(a)   (b)

Figure 11.13 Plane stress transformation

Substituting into Eq. 11.10b, with $\sigma_x = 10$, $\tau_{xy} = \tau_{yx} = 5$ and $\sigma_y = 0$, gives:

$$\mathbf{S}' = \begin{bmatrix} \sqrt{3}/2 & 1/2 \\ -1/2 & \sqrt{3}/2 \end{bmatrix} \begin{bmatrix} 10 & 5 \\ 5 & 0 \end{bmatrix} \begin{bmatrix} \sqrt{3}/2 & -1/2 \\ 1/2 & \sqrt{3}/2 \end{bmatrix} = \begin{bmatrix} 11.83 & -1.83 \\ -1.83 & -1.83 \end{bmatrix}$$

The components of $\mathbf{S}'$ act as shown in Fig. 11.13b. Note that the matrix supplies the complete stress state for the rotated element, i.e. for the two inclined planes θ and $\theta + 90°$ in Eqs 11.4a,b. The principal stresses follow from Eq. 11.11 where:

$$\det \begin{vmatrix} 10-\sigma & 5 \\ 5 & 0-\sigma \end{vmatrix} = \sigma^2 - 10\sigma - 25 = 0$$

The roots are the principal stresses values: $\sigma_1 = 12.05$ and $\sigma_2 = -2.07$ MPa, which may be confirmed from Eq. 11.7. The directions of σ_1 and σ_2 follow from Eq. 11.5 as, respectively:

$$\theta = \tan^{-1}\left(\frac{2\tau_{xy}}{\sigma_x - \sigma_y}\right) = \tan^{-1}\left(\frac{2 \times 5}{10}\right) = 45° \text{ and } 135°$$

11.4 Plane Strain Analyses

We shall see that similar equations govern both two-dimensional stress and strain transformations. Hence, the strain state in any direction will follow from transforming strains from a reference state using similar analytical, graphical and matrix methods to those used for transforming stress. In practice, however, when finding the magnitude and directions of principal strain in a body the direct strains in three directions are measured within a rosette strain gauge whose elements are configured with orientations either at 0°, 45° and 90° or at 0°, 60° and 120°.

11.4.1 Geometrical Derivation

Though particular two-dimensional strain systems may receive separate analyses, as with stress, it is convenient to start with the most general plane strain system. This may then be reduced to simpler systems by setting one or more of its strain components to zero. Consider the resulting distortion of a rectilinear element ABDE as it is strained to A′B′D′E′, as shown in Figs 11.14a and b.

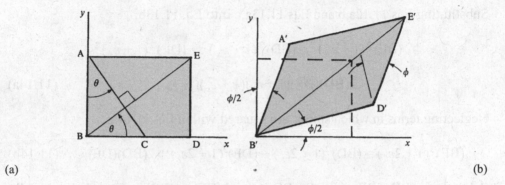

(a)   (b)

Figure 11.14 Normal and shear distortion of a rectangular element

(a) Reference Strains in Axes x and y

The distortion refers to the amount that the sides have changed in length and in their angular disposition. From the length changes to AB and BD, two *direct strain components* are derived.

$$\varepsilon_x = (B′D′ - BD)/BD = (A′E′ - AE)/AE$$

$$\varepsilon_y = (A′B′ - AB)/AB = (D′E′ - DE)/DE$$

From these we can write the element's strained, side lengths as:

$$B′D′ = BD(1 + \varepsilon_x), \quad D′E′ = DE(1 + \varepsilon_y) \qquad (11.12a,b)$$

The *shear strain component* is defined from the angular distortion φ:

$$\gamma_{xy} = \tan \varphi \approx \varphi^c \qquad (11.12c)$$

where φ is the net angular change (in radians) to the right-angle ABC.

(b) Normal Strain: Plane AC

Associated with the oblique plane AC, there will be a strain ε_θ along its normal direction BE when it extends to:

$$B′E′ = BE(1 + \varepsilon_\theta) \qquad (11.13a)$$

The strained length B′E′ is found from applying the cosine rule to triangle B′D′E′:

$$(B'E')^2 = (B'D')^2 + (D'E')^2 - 2(B'D')(D'E') \cos(\pi/2 + \theta) \qquad (11.13b)$$

where, in using Eq. 11.12c:

$$\cos(\pi/2 + \varphi) = -\sin\varphi \simeq -\varphi \ (\text{rad}) = -\gamma_{xy} \qquad (11.13c)$$

Substituting Eqs 11.12a,b and Eqs 11.13a,c into Eq. 11.13b:

$$(BE)^2(1 + \varepsilon_\theta)^2 = (BD)^2(1 + \varepsilon_x)^2 + (DE)^2(1 + \varepsilon_y)^2$$

$$- 2(BD)(DE)(1 + \varepsilon_x)(1 + \varepsilon_y)(-\gamma_{xy}) \qquad (11.14a)$$

Neglecting terms in which strains are squared within Eq. 11.14a:

$$(BE)^2(1 + 2\varepsilon_\theta) = (BD)^2(1 + 2\varepsilon_x) + (DE)^2(1 + 2\varepsilon_y) + 2(BD)(DE)\gamma_{xy} \qquad (11.14b)$$

Substituting: $BD = BE\cos\theta$, $DE = BE\sin\theta$ and $(BE)^2 = (BD)^2 + (DE)^2$ in Eq. 11.14b:

$$\varepsilon_\theta = \varepsilon_x \cos^2\theta + \varepsilon_y \sin^2\theta + \gamma_{xy} \sin\theta \cos\theta$$

$$\varepsilon_\theta = \varepsilon_x \cos^2\theta + \varepsilon_y \sin^2\theta + \tfrac{1}{2}\gamma_{xy} \sin 2\theta \qquad (11.14c)$$

(c) Shear Strain: Planes AC and BE

The relative rotation (in radians) that occurs between directions AC and BE in Fig. 11.14a defines the shear strain γ_θ. This is the change in the right angle, formed by the intersection between AC and BE. The rotation in BE is the sum of contributions, respectively, from ε_x, ε_y and γ_{xy}, shown separated in Figs 11.15a, b and c.

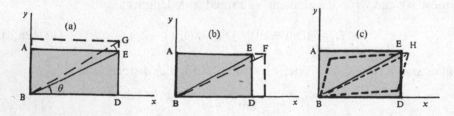

Figure 11.15 Components of the shear strain

Then net clockwise (positive) rotation in BE arising from the three distortions is

$$+\varphi_{BE} = -\angle GBE + \angle EBF + \angle EBH$$

$$\simeq -\frac{EG\cos\theta}{BE} + \frac{EF\sin\theta}{BE} + \frac{EH\sin\theta}{BE} \ (\text{rad})$$

Substituting for the three strain components $\varepsilon_x = EF/BD$, $\varepsilon_y = EG/ED$ and $\gamma_{xy} = EH/ED$, in which $ED = BE\sin\theta$ and $BD = BE\cos\theta$:

$$+\varphi_{BE} \approx -\varepsilon_y \sin\theta\cos\theta + \varepsilon_x \sin\theta\cos\theta + \gamma_{xy}\sin^2\theta \qquad (11.15a)$$

Correspondingly, the rotation in AC is found by replacing θ in Eq. 11.15a with $\theta + \pi/2$. This gives:

$$+\varphi_{AC} \approx +\varepsilon_y \sin\theta\cos\theta - \varepsilon_x \sin\theta\cos\theta + \gamma_{xy}\cos^2\theta \qquad (11.15b)$$

Subtracting Eq. 11.15b from Eq. 11.15a gives the *engineering* shear strain:

$$\gamma_\theta = \varphi_{BE} - \varphi_{AC} = 2(\varepsilon_x - \varepsilon_y)\sin\theta\cos\theta - \gamma_{xy}(\cos^2\theta - \sin^2\theta)$$

$$\gamma_\theta = (\varepsilon_x - \varepsilon_y)\sin 2\theta - \gamma_{xy}\cos 2\theta \qquad (11.16a)$$

Dividing Eq. 11.16a by 2 gives the semi-shear strain:

$$\tfrac{1}{2}\gamma_\theta = \tfrac{1}{2}(\varepsilon_x - \varepsilon_y)\sin 2\theta - \tfrac{1}{2}\gamma_{xy}\cos 2\theta \qquad (11.16b)$$

Table 11.1 Plane Stress and Strain Transformation Equations

System	Stress Transformation	Strain Transformation
(1) Fig. 11.1	$\sigma_\theta = \sigma_x \cos^2\theta$ $\tau_\theta = \tfrac{1}{2}\sigma_x \sin 2\theta$ $\tau_{max} = \tfrac{1}{2}\sigma_x$ $\sigma_{45°} = \tfrac{1}{2}\sigma_x$	$\varepsilon_\theta = \varepsilon_x \cos^2\theta$ $\tfrac{1}{2}\gamma_\theta = \tfrac{1}{2}\varepsilon_x \sin 2\theta$ $\tfrac{1}{2}\gamma_{max} = \tfrac{1}{2}\varepsilon_x$ $\varepsilon_{45°} = \tfrac{1}{2}\varepsilon_x$
(2) Fig. 11.2	$\sigma_\theta = \tau_{xy}\sin 2\theta$ $\tau_\theta = \tau_{xy}\cos 2\theta$ $\sigma_1 = \tau_{xy}$ $\sigma_2 = -\tau_{xy}$	$\varepsilon_\theta = \tfrac{1}{2}\gamma_{xy}\sin 2\theta$ $\tfrac{1}{2}\gamma_\theta = \tfrac{1}{2}\gamma_{xy}\cos 2\theta$ $\varepsilon_1 = \tfrac{1}{2}\gamma_{xy}$ $\varepsilon_2 = -\tfrac{1}{2}\gamma_{xy}$
(3) Fig. 11.3	$\sigma_\theta = \sigma_1 \cos^2\theta + \sigma_2 \sin^2\theta$ $\tau_\theta = \tfrac{1}{2}(\sigma_1 - \sigma_2)\sin 2\theta$ $\tau_{max} = \tfrac{1}{2}(\sigma_1 - \sigma_2)$ $\sigma_{45°} = \tfrac{1}{2}(\sigma_1 + \sigma_2)$	$\varepsilon_\theta = \varepsilon_1 \cos^2\theta + \varepsilon_2 \sin^2\theta$ $\tfrac{1}{2}\gamma_\theta = \tfrac{1}{2}(\sigma_1 - \sigma_2)\sin 2\theta$ $\tfrac{1}{2}\gamma_{max} = \tfrac{1}{2}(\varepsilon_1 - \varepsilon_2)$ $\varepsilon_{45°} = \tfrac{1}{2}(\varepsilon_1 + \varepsilon_2)$
(4) Fig. 11.4 Fig. 11.5	$\sigma_\theta = \sigma_y \sin^2\theta + \sigma_x \cos^2\theta + \tau_{xy}\sin 2\theta$ $\quad = \tfrac{1}{2}(\sigma_x + \sigma_y) + \tfrac{1}{2}(\sigma_x - \sigma_y)\cos 2\theta + \tau_{xy}\sin 2\theta$ $\tau_\theta = \tfrac{1}{2}(\sigma_x - \sigma_y)\sin 2\theta - \tau_{xy}\cos 2\theta$ $\sigma_{1,2} = \tfrac{1}{2}(\sigma_x + \sigma_y) \pm \tfrac{1}{2}\sqrt{[(\sigma_x - \sigma_y)^2 + 4\tau_{xy}^2]}$ $\tan 2\theta = 2\tau_{xy}/(\sigma_x - \sigma_y)$ $\tau_{max} = \tfrac{1}{2}(\sigma_1 - \sigma_2) = \tfrac{1}{2}\sqrt{[(\sigma_x - \sigma_y)^2 + 4\tau_{xy}^2]}$	$\varepsilon_\theta = \varepsilon_y \sin^2\theta + \varepsilon_x \cos^2\theta + \tfrac{1}{2}\gamma_{xy}\sin 2\theta$ $\quad = \tfrac{1}{2}(\varepsilon_x + \varepsilon_y) + \tfrac{1}{2}(\varepsilon_x - \varepsilon_y)\cos 2\theta + \tfrac{1}{2}\gamma_{xy}\sin 2\theta$ $\tfrac{1}{2}\gamma_\theta = \tfrac{1}{2}(\varepsilon_x - \varepsilon_y)\sin 2\theta - \tfrac{1}{2}\gamma_{xy}\cos 2\theta$ $\varepsilon_{1,2} = \tfrac{1}{2}(\varepsilon_x + \varepsilon_y) \pm \tfrac{1}{2}\sqrt{[(\varepsilon_x - \varepsilon_y)^2 + \gamma_{xy}^2]}$ $\tan 2\theta = \gamma_{xy}/(\varepsilon_x - \varepsilon_y)$ $\tfrac{1}{2}\gamma_{max} = \tfrac{1}{2}(\varepsilon_1 - \varepsilon_2) = \tfrac{1}{2}\sqrt{[(\varepsilon_x - \varepsilon_y)^2 + \gamma_{xy}^2]}$

Derived geometrically in this manner, it is seen that strain Eqs 11.14c and 11.16b are similar to stress Eqs 11.4a,b. That is, stress and strain have identical transformation properties when shear stress τ corresponds to the semi-shear strain $\frac{1}{2}\gamma$. The engineering definition, $\gamma = \varphi$ in Fig. 11.14b, identifies with the net change in the right angle. Only by identifying shear strain with $\frac{1}{2}\varphi$, which imposes the symmetry shown in Fig. 11.14b, is there an exact correspondence between the stress and strain transformation equations. This is also apparent from Table 11.1 where Eqs 11.14 and 11.16a,b have been reduced to the four basic strain systems corresponding to those derived in § 11.1.1–§ 11.1.4 for plane stress. In system 4, corresponding to Fig. 11.5, principal strains ε_1 and ε_2 lie along perpendicular directions for which shear strain is absent. Hence principal co-ordinates 1 and 2 have replaced x and y.

11.4.2 Co-ordinate Rotation

The separate geometrical derivations of the normal and shear strains given above marry when the strains accompanying a rotation in the x, y co-ordinate frame are required. This means that these strains become the transformed strains in rotated axes x' and y'. In Fig. 11.14a, let BE refer to a rotation θ in the co-ordinate axis from x to x'. It follows that Eq. 11.14c provides the strain in the direction x' as

$$\varepsilon_x' = \varepsilon_x \cos^2\theta + \varepsilon_y \sin^2\theta + \frac{1}{2}\gamma_{xy}\sin 2\theta \qquad (11.17\text{a})$$

Correspondingly, the second co-ordinate axis y rotates to y'. The strain in the direction y' follows from replacing θ in Eq. 11.14c with $\theta + 90°$. This gives:

$$\varepsilon_y' = \varepsilon_x \sin^2\theta + \varepsilon_y \cos^2\theta - \frac{1}{2}\gamma_{xy}\sin 2\theta \qquad (11.17\text{b})$$

Perversely, another way to derive the shear strain in axes x', y' is with an inverse transformation from x', y' to x, y. This requires that θ in Eq. 11.17a be replaced with $-\theta$, accompanied by an interchange in the co-ordinates:

$$\varepsilon_x = \varepsilon_x' \cos^2\theta + \varepsilon_y' \sin^2\theta - \frac{1}{2}\gamma_{xy}'\sin 2\theta \qquad (11.18\text{a})$$

Rearranging Eq. 11.18a for γ_{xy}' and using trig identities:

$$\frac{1}{2}\gamma_{xy}'\sin 2\theta = -\varepsilon_x + \frac{1}{2}(\varepsilon_x' + \varepsilon_y') + \frac{1}{2}(\varepsilon_x' - \varepsilon_y')\cos 2\theta \qquad (11.18\text{b})$$

Substituting for ε_x' and ε_y' from Eqs 11.17a,b into Eq. 11.18b:

$$\frac{1}{2}\gamma_{xy}'\sin 2\theta = -\varepsilon_x + \frac{1}{2}(\varepsilon_x + \varepsilon_y) + \frac{1}{2}(\varepsilon_x - \varepsilon_y)\cos^2 2\theta + \frac{1}{2}\gamma_{xy}\sin 2\theta\cos 2\theta$$

gives the transformed *mathematical shear strain*

$$\frac{1}{2}\gamma_{xy}' = -\frac{1}{2}(\varepsilon_x - \varepsilon_y)\sin 2\theta + \frac{1}{2}\gamma_{xy}\cos 2\theta \qquad (11.18\text{c})$$

Here we see a sign change when Eq. 11.18c is compared with the semi-shear strain in Eq. 11.16b by the engineering definition. Only the latter ensure a correspondence between equations of stress and strain transformation for systems 1-4. Thus, across the two columns in Table 11.1, $\sigma_x \rightarrow \varepsilon_x$, $\sigma_y \rightarrow \varepsilon_y$, $\tau_{xy} \rightarrow \frac{1}{2}\gamma_{xy}$; $\sigma_1 \rightarrow \varepsilon_1$ $\sigma_2 \rightarrow \varepsilon_2$.

11.4.3 Graphical Method

The correspondence recognised above ensures that Mohr's strain circle can be constructed in a similar manner to his stress circle. This amounts to taking an ordinate for the strain circle that is one-half the engineering shear strain $\frac{1}{2}\gamma_{xy}$, i.e. its 'semi-shear strain' axis. The circle's normal strain axis accommodates direct engineering strains ε_x and ε_y. Conventionally, these three strain components are taken to be positive when they accompany positive stress (i.e. tension and clockwise shear). The construction adopted here requires that a focus F be located within the normal and semi-shear strain axes of Fig. 11.16.

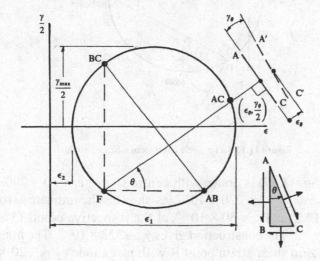

Figure 11.16 Mohr's strain circle with focus point F

Referring to stressed element ABC (shown inset) ε_x is positive, γ_{xy} is negative for plane AB while both ε_y and γ_{xy} are positive for the plane BC. These locate two diametrically opposite points (AB) and (BC) on the circle, which enable its construction, given that the centre will always lie along the normal strain axis. The focus F is then found from projecting the normal to the plane AB (or BC) through the corresponding point on the strain circle. The strain state $(\varepsilon_\theta, \frac{1}{2}\gamma_\theta)$ for any inclined plane AC is found from projecting the normal to AC through F as shown. The intersection point (AC) shows tensile strain (ε_θ) in a direction normal to AC accompanied by γ_θ, the angular change between AC and its normal. The physical interpretation of the two strains (inset) shows the shift and rotation of the plane AC to its strained position A'C'. The circle also provides principal strains ε_1 and ε_2, the maximum shear strain γ_{max} and their respective planes relative to F. Since the normals to planes AB, BC and AC are the directions in which ε_x, ε_y and ε_θ lie, the normal strain directions are applied to F directly in a Mohr's strain circle.

Example 11.7 At a point on the surface of a strained body the strain state, referred to axes x, y, consists of normal strains are: $\varepsilon_x = 350 \times 10^{-6}$ and $\varepsilon_y = 50 \times 10^{-6}$, together with an unknown shear strain γ_{xy}. The major principal strain is known to be $\varepsilon_1 = 420 \times 10^{-6}$ but the minor principal strain ε_2 is unknown. Find, both graphically and analytically, γ_{xy} and ε_2 and their directions relative to x. What are the principal stresses? Take $E = 210$ GPa and $v = 0.3$.

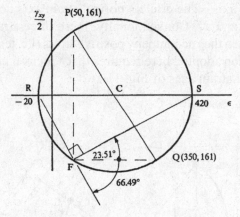

Figure 11.17 Strain circle construction (micro-strains)

The circle (see Fig. 11.17) is drawn with centre C = $\frac{1}{2}(\varepsilon_x + \varepsilon_y) = 200 \times 10^{-6}$ and radius CS = $\varepsilon_1 - \frac{1}{2}(\varepsilon_x + \varepsilon_y) = 220 \times 10^{-6}$, as shown. The ordinate associated with both $\varepsilon_x = 350 \times 10^{-6}$ and $\varepsilon_y = 50 \times 10^{-6}$, at the respective points Q and P on the circle, becomes $\gamma_{xy}/2$. The construction gives $\gamma_{xy} = 322 \times 10^{-6}$. The minor principal strain lies at the zero shear strain point R with magnitude $\varepsilon_2 = -20 \times 10^{-6}$. The focus is found from projecting either the direction of ε_y through P or the direction of ε_x through Q. The major and minor principal strain directions then lie parallel to FS and FR, respectively. Their orientations are shown relative to FQ in the circle, which aligns with the x-direction in the body.

In the analytical solution a direct substitution into the major principal strain expression (System 4, Table 11.1) yields both γ_{xy} and ε_2 directly:

$$\varepsilon_1 = \frac{1}{2}(\varepsilon_x + \varepsilon_y) + \frac{1}{2}\sqrt{(\varepsilon_x - \varepsilon_y)^2 + \gamma_{xy}^2}$$

$$420 = 200 + \frac{1}{2}\sqrt{[(300)^2 + \gamma_{xy}^2]}$$

$$\therefore \; \gamma_{xy} = \pm 321.9 \times 10^{-6}$$

The minor principal strain then follows:

$$\varepsilon_1 = \tfrac{1}{2}(\varepsilon_x + \varepsilon_y) - \tfrac{1}{2}\sqrt{(\varepsilon_x - \varepsilon_y)^2 + \gamma_{xy}^2}$$

$$= 200 - \tfrac{1}{2}\sqrt{[(300)^2 + (321.9)^2]} = -20 \times 10^{-6}$$

The directions of the major and minor principal planes of strain are coincident with those of stress. From Table 11.1:

$$\tan 2\theta = \frac{\gamma_{xy}}{\varepsilon_x - \varepsilon_y}$$

$$\theta = \frac{1}{2}\tan^{-1}\left(\frac{\pm 321.9}{50 - 350}\right) = 23.51° \text{ or } -66.49°$$

Of the two orientations, $\theta = +23.51°$ locates the minor principal plane and $\theta = -66.49°$ the major principal plane, relative to the y-direction. The principal strain directions lie normal to their planes, i.e. $23.51°$ and $-66.49°$, relative to the x-direction, consistent with Fig. 11.17. Equations 4.34a,b (see p. 121) are used to convert the principal strains to principal stresses:

$$\sigma_1 = \frac{E}{(1-v^2)}(\varepsilon_1 + v\varepsilon_2)$$

$$= \frac{(210 \times 10^3)}{(1-0.3^2)}[420 - (0.3 \times 20)] \times 10^{-6} = 95.54 \text{ MPa}$$

$$\sigma_2 = \frac{E}{(1-v^2)}(\varepsilon_2 + v\varepsilon_1)$$

$$= \frac{210 \times 10^3}{(1-0.3^2)}[-20 + (0.3 \times 420)] \times 10^{-6} = 24.46 \text{ MPa}$$

Example 11.8 The three elements a, b and c within a strain gauge rosette record three direct strains: 600 µε, -200 µε and 250 µε (1 µε = 1×10^{-6}) in a body for orientations lying at $0°$, $60°$ and $120°$ anticlockwise to the x-axis. Find the magnitude and direction of the principal strains both graphically and analytically.

In constructing Mohr's circle, the three gauges are rotated so that any one (here gauge c) lies in the vertical position, as shown inset in Fig. 11.18a. The recorded direct strains are scaled horizontally in Fig. 11.18b from the vertical semi-shear strain axis. Parallel, vertical lines are drawn through these points as shown. Any focus point F is chosen along the line representing the vertically aligned gauge c. From F, lines are drawn parallel to the re-positioned directions of the remaining two gauges a and b. These intersect their corresponding vertical lines in P and Q.

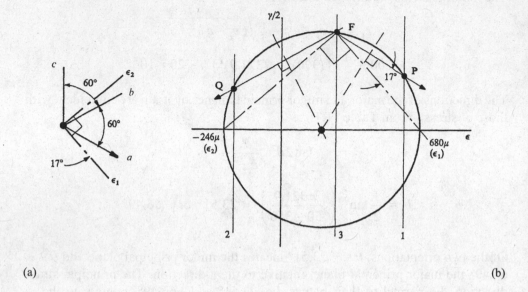

Figure 11.18 Mohr's circle for a strain gauge rosette

The intersection between the perpendicular bisectors of chords FP and FQ locates the circle's centre, lying on the normal strain ε-axis. At the ends of the circle's horizontal diameter along the ε-axis are the principal strains. The chosen scale provides these as: $\varepsilon_1 = 680$ µε and $\varepsilon_2 = -246$ µε. The directions of the principal strains within Figs 11.18a,b lie parallel to the chain lines $F\varepsilon_1$ and $F\varepsilon_2$. The true principal directions are found from rotating Fig. 11.18a back into the original configuration in which gauge a aligns with x. This, giving ε_2 at 73° and ε_1 at 163° anticlockwise to x.

In the analytical solutions, we substitute for θ and ε_θ in Eq. 11.14c to give three simultaneous equations as follows:

$$\varepsilon_x = 600 \text{ µε} \tag{i}$$

$$-200 \text{ µε} = \varepsilon_y \sin 260° + \varepsilon_x \cos 260° + (\gamma_{xy}/2) \sin 120°$$

$$0.75\varepsilon_y + 0.433\gamma_{xy} = -350 \text{ µε} \tag{ii}$$

$$250 \text{ µε} = \varepsilon_y \sin^2 120° + \varepsilon_x \cos^2 120° + (\gamma_{xy}/2) \sin 240°$$

$$0.75\varepsilon_y - 0.433\gamma_{xy} = 100 \text{ µε} \tag{iii}$$

The solution to Eqs i, ii and iii gives the normal strains in the x and y directions as: $\varepsilon_x = 600$ µε, $\varepsilon_y = -166.67$ µε and $\gamma_{xy} = -519.60$ µε. Substituting into the principal strain expression (see Table 11.1) gives their magnitudes:

$$\varepsilon_{1,2} = \frac{1}{2}(\varepsilon_x + \varepsilon_y) \pm \frac{1}{2}\sqrt{(\varepsilon_x - \varepsilon_y)^2 + (\gamma_{xy})^2} \qquad \text{(iv)}$$

$$= \frac{1}{2}(600 - 166.67) \pm \frac{1}{2}\sqrt{(600 + 166.67)^2 + (519.63)^2}$$

$$\varepsilon_1 = 679.75 \ \mu\varepsilon \quad \text{and} \quad \varepsilon_2 = -246.42 \ \mu\varepsilon$$

and their directions, also from Table 11.1, are:

$$\theta = \frac{1}{2}\tan^{-1}\frac{\gamma_{xy}}{\varepsilon_x - \varepsilon_y} = \frac{1}{2}\tan^{-1}\left[\frac{-519.6}{600 - (-166.67)}\right]$$

giving:

$$\theta = -17.06° \quad \text{and} \quad 72.94°$$

These inclinations define the directions of the principal planes relative to y, or the directions of the principal strains relative to x.

11.4.4 Rosette Analyses

Example 11.8 adopted a two-step approach transforming strain between directions a, b and c from the rosette to reference axes x and y and then to principal directions 1 and 2. The principal strains are found directly from a single equation combining Eqs i–iv above. In a three-element, 60° delta rosette the strains within each gauge are denoted by ε_a, ε_b and ε_c when their inclinations relative to the major principal strain direction-1 are as shown in Fig. 11.19a.

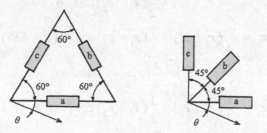

(a)   (b)

Figure 11.19 Orientations of strain gauge rosette elements

When the major and minor principal strains are ε_1 and ε_2 we may write each rosette's strain from Table 11.1 (System 3) as:

$$\varepsilon_\theta = \varepsilon_1 \cos^2\theta + \varepsilon_2 \sin^2\theta \qquad (11.19a)$$

$$= \frac{1}{2}(\varepsilon_1 + \varepsilon_2) + \frac{1}{2}(\varepsilon_1 - \varepsilon_2)\cos 2\theta \qquad (11.19b)$$

The alternative Eq. 11.19b gives the three rosette strains more conveniently when substituting θ, $\theta + 120°$ and $\theta + 240°$ for gauges a, b and c, respectively:

$$\varepsilon_a = \frac{1}{2}(\varepsilon_1 + \varepsilon_2) + \frac{1}{2}(\varepsilon_1 - \varepsilon_2)\cos 2\theta \tag{11.20a}$$

$$\varepsilon_b = \frac{1}{2}(\varepsilon_1 + \varepsilon_2) + \frac{1}{2}(\varepsilon_1 - \varepsilon_2)\cos[2(\theta + 120°)]$$

$$= \frac{1}{2}(\varepsilon_1 + \varepsilon_2) - \frac{1}{2}(\varepsilon_1 - \varepsilon_2)\cos(2\theta + 60°) \tag{11.20b}$$

$$\varepsilon_c = \frac{1}{2}(\varepsilon_1 + \varepsilon_2) + \frac{1}{2}(\varepsilon_1 - \varepsilon_2)\cos[2(\theta + 240°)]$$

$$= \frac{1}{2}(\varepsilon_1 + \varepsilon_2) - \frac{1}{2}(\varepsilon_1 - \varepsilon_2)\cos(2\theta - 60°) \tag{11.20c}$$

Equations 11.20a–c may be solved simultaneously for the magnitudes of the principal strains ε_1, ε_1:

$$\varepsilon_{1,2} = \frac{1}{3}(\varepsilon_a + \varepsilon_b + \varepsilon_c) \pm \frac{2}{3}\sqrt{(\varepsilon_a - \varepsilon_b)^2 + (\varepsilon_a - \varepsilon_c)^2 - (\varepsilon_a - \varepsilon_b)(\varepsilon_a - \varepsilon_c)} \tag{11.21a,b}$$

and their orientations θ

$$\theta = \frac{1}{2}\tan^{-1}\left[\frac{\sqrt{3}(\varepsilon_b - \varepsilon_c)}{2\varepsilon_a - \varepsilon_b - \varepsilon_c}\right] \tag{11.21c}$$

In a three-element 45° rosette (see Fig. 11.19b) the inclinations of the three gauges a, b and c, relative to the major principal strain direction-1, are: θ, $\theta + 45°$ and $\theta + 90°$, respectively, as shown. The gauge strains follow from Eq. 11.19b as:

$$\varepsilon_a = \frac{1}{2}(\varepsilon_1 + \varepsilon_2) + \frac{1}{2}(\varepsilon_1 - \varepsilon_2)\cos 2\theta \tag{11.22a}$$

$$\varepsilon_b = \frac{1}{2}(\varepsilon_1 + \varepsilon_2) + \frac{1}{2}(\varepsilon_1 - \varepsilon_2)\cos[2(\theta + 45°)]$$

$$= \frac{1}{2}(\varepsilon_1 + \varepsilon_2) - \frac{1}{2}(\varepsilon_1 - \varepsilon_2)\sin 2\theta \tag{11.22b}$$

$$\varepsilon_c = \frac{1}{2}(\varepsilon_1 + \varepsilon_2) + \frac{1}{2}(\varepsilon_1 - \varepsilon_2)\cos[2(\theta + 90°)]$$

$$= \frac{1}{2}(\varepsilon_1 + \varepsilon_2) - \frac{1}{2}(\varepsilon_1 - \varepsilon_2)\cos 2\theta \tag{11.22c}$$

Equation 11.22a–c may be solved for ε_1, ε_1 and θ:

$$\varepsilon_{1,2} = \frac{1}{2}(\varepsilon_a + \varepsilon_c) \pm \frac{1}{\sqrt{2}}\sqrt{(\varepsilon_a - \varepsilon_b)^2 + (\varepsilon_b - \varepsilon_c)^2} \tag{11.23a}$$

$$\theta = \frac{1}{2}\tan^{-1}\left(\frac{\varepsilon_a + \varepsilon_c - 2\varepsilon_b}{\varepsilon_a - \varepsilon_c}\right) \tag{11.23b}$$

In fact, the analyses of each rosette that lead to the required principal strain magnitudes and their orientations (Eqs 11.20a–11.23b) are slightly inaccurate. This is because they ignore the gauge's *cross sensitivity* and the low insulation resistance of its polymer backing (see § 11.5.5). The early wire-wound gauges (pre-1960) were susceptible to such inaccuracies. A modern gauge grid etched from thin foil has minimal material in the transverse direction which serves to reduce the cross-sensitivity error. Therefore, the equations above are sufficiently accurate for practical purposes. Here the strains ε_a, ε_b and ε_c would be measured directly within a Wheatstone bridge circuit as described in § 11.5. These circuits in their ¼, ½ and full-bridge configurations, have been adopted widely for strain measurement within commercial multi-channel meters and data logging hardware.

11.4.5 Matrix Method

The matrix method for two-dimensional strain transformation is similar to that given in § 11.3 for two-dimensional stress transformation. The resulting stress and strain component expressions accompanying a co-ordinate rotation can be made to correspond when a mathematical definition of shear strain applies. The latter requires that the angular change φ (in radians) to a right angle be equally disposed to axes x and y, as shown in Fig. 11.14b. The required symmetry amounts to the removal of any rigid body rotation that does not contribute to the strain. Thus, in the x-y reference plane shown, each *mathematical shear strain* $\varepsilon_{xy} = \varepsilon_{yx} = \frac{1}{2}\varphi$ corresponds to a single *engineering shear strain* $\gamma_{xy} = \varphi$, giving the relationship between the three shear strains: $\varepsilon_{xy} = \varepsilon_{yx} = \frac{1}{2}\gamma_{xy}$. The components of the symmetrical 2×2 strain matrix $\mathbf{E}$, referred to axes x and y, consist of the two shear strains ε_{xy} and ε_{yx} and two direct strains ε_x and ε_y. When referred to rotated axes x' and y' the four strain components transform according to the matrix transformation law:

$$\mathbf{E}' = \mathbf{LEL}^\mathsf{T} \qquad (11.24a)$$

in which $\mathbf{E}'$ provides the transformed strain components and $\mathbf{L}$ describes the rotation with its direction cosines. The correspondence with Eq. 11.10a is evident when transforming stress components, as we might expect, to give a 2×2 stress matrix with a similar symmetry. In full, Eq. 11.24a becomes

$$\begin{bmatrix} \varepsilon_x' & \varepsilon_{xy}' \\ \varepsilon_{yx}' & \varepsilon_y' \end{bmatrix} = \begin{bmatrix} l_{xx} & l_{xy} \\ l_{yx} & l_{yy} \end{bmatrix} \begin{bmatrix} \varepsilon_x & \varepsilon_{xy} \\ \varepsilon_{yx} & \varepsilon_y \end{bmatrix} \begin{bmatrix} l_{xx} & l_{yx} \\ l_{xy} & l_{yy} \end{bmatrix} \qquad (11.24b)$$

The direction cosines that apply between axes x, y and x', y' were described previously in Fig. 11.12. Thus, with a rotation θ from x to x' Eq. 11.24b becomes:

$$\begin{bmatrix} \varepsilon_x' & \varepsilon_{xy}' \\ \varepsilon_{yx}' & \varepsilon_y' \end{bmatrix} = \begin{bmatrix} \cos\theta & \sin\theta \\ -\sin\theta & \cos\theta \end{bmatrix} \begin{bmatrix} \varepsilon_x & \varepsilon_{xy} \\ \varepsilon_{yx} & \varepsilon_y \end{bmatrix} \begin{bmatrix} \cos\theta & -\sin\theta \\ \sin\theta & \cos\theta \end{bmatrix} \qquad (11.24c)$$

Expanding Eq. 11.24c and setting $\varepsilon_{xy}' = \varepsilon_{yx}' = \frac{1}{2}\gamma_{xy}'$ gives:

$$\varepsilon_x' = \varepsilon_x \cos^2\theta + \varepsilon_y \sin^2\theta + \frac{1}{2}\gamma_{xy}\sin 2\theta \qquad (11.25a)$$

$$\varepsilon_y' = \varepsilon_x \cos^2\theta + \varepsilon_y \sin^2\theta - \frac{1}{2}\gamma_{xy}\sin 2\theta \qquad (11.25b)$$

$$\frac{1}{2}\gamma_{xy}' = -\frac{1}{2}(\varepsilon_x - \varepsilon_y)\sin 2\theta + \frac{1}{2}\gamma_{xy}\cos 2\theta \qquad (11.25c)$$

Equation 11.25a agrees with ε_θ in Eq. 11.14c. Equation 11.25b follows from setting $\theta = \theta + 90°$ in Eq. 11.14c. Note, again the sign change between the shear strain expression, Eq. 11.25c and $\frac{1}{2}\gamma_\theta$ in Eq. 11.16b, is consistent with the difference between the engineering and mathematical (matrix) derivations. This means that positive shear by one method is negative shear by the other. If the mathematical definition of γ is adopted for Mohr's strain circle this would require plotting an ordinate of $-\frac{1}{2}\gamma$ versus ε to achieve an exact correspondence with the plot of τ versus σ in the stress circle. This was not used for constructing the strain circle here, however, where a positive semi-shear strain was taken to accompany positive (clockwise) shear stress (see § 11.4.3).

Finally, the reader should also note the similarity between the 2D transformation matrices for stress (Eq. 11.10a), strain (Eq. 11.24a) and second moments of area in Chapter 1 (Eq. 1.15a, p. 25). This reveals that these quantities transform in an identical manner, so identifying their commonality as *symmetric, second-order tensors* having their components assembled within 2×2 matrices.

11.4.6 Strain-Displacement Relations

An elegant derivation of the general 2D strain transformation equations follows from the mathematical definition of the three Cartesian strain components ε_x, ε_x and γ_{xy}. Firstly, referring to Fig. 11.20, let a point $P(x, y)$ displace to P' in which the components of that displacement u and v align with x and y.

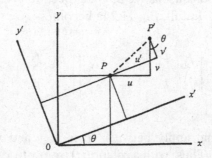

Figure 11.20 Displacement of a point with co-ordinate rotation

Now let the x, y co-ordinates rotate by θ into the new co-ordinates x' and y'. The displacements of P to P', in alignment with x' and y', become then u' and v', which are to be expressed as follows:

$$x = x(x', y') = x' \cos\theta - y' \sin\theta \qquad (11.26a)$$

$$y = y(x', y') = x' \sin\theta + y' \cos\theta \qquad (11.26b)$$

$$u' = u'(u, v) = u\cos\theta + v\sin\theta \qquad (11.26c)$$

$$v' = v'(u, v) = -u\sin\theta + v\cos\theta \qquad (11.26d)$$

Given the displacement functions $u = u(x, y)$ and $v = v(x, y)$, the three, two-dimensional Cartesian strain components: ε_x, ε_y and γ_{xy} refer the displacements u and v to co-ordinates (x, y) in their mathematical definitions as strain-displacement derivatives

$$\varepsilon_x = \frac{\partial u}{\partial x} \qquad (11.27a)$$

$$\varepsilon_y = \frac{\partial v}{\partial y} \qquad (11.27b)$$

$$\gamma_{xy} = \frac{\partial v}{\partial x} + \frac{\partial u}{\partial y} \qquad (11.27c)$$

Transforming ε_x to x', y' with the functions $x = x(x', y')$ and $y = y(x', y')$ connects the two co-ordinate systems:

$$\varepsilon_x' = \frac{\partial u'}{\partial x'} = \frac{\partial u'}{\partial x} \times \frac{\partial x}{\partial x'} + \frac{\partial u'}{\partial y} \times \frac{\partial y}{\partial x'} \qquad (11.28a)$$

Similarly, the u and u' displacements of P within the two co-ordinate frames bear a geometrical relationship in functional form $u' = u'(u, v)$. Eq. 11.28a becomes

$$\varepsilon_x' = \left(\frac{\partial u}{\partial x}\frac{\partial u'}{\partial u} + \frac{\partial v}{\partial x}\frac{\partial u'}{\partial v} \right) \frac{\partial x}{\partial x'} + \left(\frac{\partial u}{\partial y}\frac{\partial u'}{\partial u} + \frac{\partial v}{\partial y}\frac{\partial u'}{\partial v} \right) \frac{\partial y}{\partial x'} \qquad (11.28b)$$

Substituting into Eq. 11.28b, the partial derivatives from Eqs 11.26a–d:

$$\varepsilon_x' = \left(\frac{\partial u}{\partial x}\cos\theta + \frac{\partial v}{\partial x}\sin\theta \right) \cos\theta + \left(\frac{\partial u}{\partial y}\cos\theta + \frac{\partial v}{\partial y}\sin\theta \right) \sin\theta \qquad (11.28c)$$

and then substituting from Eqs 11.27a–c into Eq. 11.28c:

$$\varepsilon_x' = \varepsilon_x \cos^2\theta + \varepsilon_y \sin^2\theta + \tfrac{1}{2}\gamma_{xy}\sin 2\theta \qquad (11.28d)$$

Functions $x = x(x', y')$ and $y = y(x', y')$ transform ε_y to x', y':

$$\varepsilon_y' = \frac{\partial v'}{\partial y'} = \frac{\partial v'}{\partial y} \times \frac{\partial y}{\partial y'} + \frac{\partial v'}{\partial x} \times \frac{\partial x}{\partial y'} \qquad (11.29a)$$

and with $v' = v'(u, v)$, Eq. 11.29a becomes

$$\varepsilon_y' = \left(\frac{\partial v}{\partial y} \frac{\partial v'}{\partial v} + \frac{\partial u}{\partial y} \frac{\partial v'}{\partial u} \right) \frac{\partial y}{\partial y'} + \left(\frac{\partial v}{\partial x} \frac{\partial v'}{\partial v} + \frac{\partial u}{\partial x} \frac{\partial v'}{\partial u} \right) \frac{\partial x}{\partial y'} \qquad (11.29b)$$

Substituting partial derivatives from Eqs 11.26a–d into Eq. 11.29b:

$$\varepsilon_y' = \left(\frac{\partial v}{\partial y} \cos\theta - \frac{\partial u}{\partial y} \sin\theta \right) \cos\theta - \left(\frac{\partial v}{\partial x} \cos\theta - \frac{\partial u}{\partial x} \sin\theta \right) \sin\theta \qquad (11.29c)$$

Substituting Eqs 11.27a–c into Eq. 11.29c, the transformed strain becomes

$$\varepsilon_y' = \varepsilon_x \sin^2\theta + \varepsilon_y \cos^2\theta - \tfrac{1}{2} \gamma_{xy} \sin 2\theta \qquad (11.29d)$$

The shear strain in axes x', y' employs the transformation functions $x = x(x', y')$ and $y = y(x', y')$:

$$\gamma_{xy}' = \frac{\partial v'}{\partial x'} + \frac{\partial u'}{\partial y'} = \left(\frac{\partial v'}{\partial x} \frac{\partial x}{\partial x'} + \frac{\partial v'}{\partial y} \frac{\partial y}{\partial x'} \right) + \left(\frac{\partial u'}{\partial x} \frac{\partial x}{\partial y'} + \frac{\partial u'}{\partial y} \frac{\partial y}{\partial y'} \right) \qquad (11.30a)$$

Applying both displacement relationships: $v' = v'(u, v)$ and $u' = u'(u, v)$, in turn to Eq. 11.30a:

$$\gamma_{xy}' = \left(\frac{\partial v}{\partial x} \frac{\partial v'}{\partial v} + \frac{\partial u}{\partial x} \frac{\partial v'}{\partial u} \right) \frac{\partial x}{\partial x'} + \left(\frac{\partial v}{\partial y} \frac{\partial v'}{\partial v} + \frac{\partial u}{\partial y} \frac{\partial v'}{\partial u} \right) \frac{\partial y}{\partial x'}$$

$$+ \left(\frac{\partial u}{\partial x} \frac{\partial u'}{\partial u} + \frac{\partial v}{\partial x} \frac{\partial u'}{\partial v} \right) \frac{\partial x}{\partial y'} + \left(\frac{\partial u}{\partial y} \frac{\partial u'}{\partial u} + \frac{\partial v}{\partial y} \frac{\partial u'}{\partial v} \right) \frac{\partial y}{\partial y'} \qquad (11.30b)$$

Substituting from Eqs 11.26a–d into Eq. 11.30b:

$$\gamma_{xy}' = \left[\frac{\partial v}{\partial x} \cos\theta + \frac{\partial u}{\partial x} (-\sin\theta) \right] \cos\theta + \left[\frac{\partial v}{\partial y} \cos\theta + \frac{\partial u}{\partial y} (-\sin\theta) \right] \sin\theta$$

$$+ \left[\frac{\partial u}{\partial x} \cos\theta + \frac{\partial v}{\partial x} \sin\theta \right] (-\sin\theta) + \left[\frac{\partial u}{\partial y} \cos\theta + \frac{\partial v}{\partial y} \sin\theta \right] \cos\theta$$

Simplifying:

$$\gamma_{xy}' = \left(\frac{\partial v}{\partial x} + \frac{\partial u}{\partial y} \right)(\cos^2\theta - \sin^2\theta) - 2\left(\frac{\partial u}{\partial x} - \frac{\partial v}{\partial y} \right)\sin\theta\cos\theta \qquad (11.30c)$$

and finally, Eq. 11.30c is written from Eqs 11.27a–c as

$$\frac{1}{2}\gamma_{x'y'} = -\frac{1}{2}(\varepsilon_x - \varepsilon_y)\sin 2\theta + \frac{1}{2}\gamma_{xy}\cos 2\theta \qquad (11.30d)$$

Satisfyingly, in this mathematical derivation of two-dimensional strain transformation all three strain components (Eqs 11.28d, 11.29c and 11.30d) agree with those found earlier from matrix multiplication (see Eq. 11.25a–c).

11.5 Wheatstone Bridge Theory

The bridge circuit attributed to Sir Charles Wheatstone (1802–1875) is commonly used to measure an unknown resistance, but more usefully here the bridge serves to record strains universally in commercial strain meters. The bridge has four arms into which resistors R_1, R_2, R_3 and R_4 are connected, a supply voltage V and a detector voltage e, as shown in Fig. 11.21.

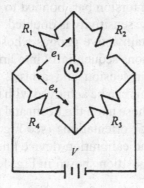

Figure 11.21 Wheatsone's bridge circuit

The bridge is said to be balanced when zero volts are detected with there being no difference between the voltages across its arms 1 and 4. That is:

$$e = e_1 - e_4 = 0 \qquad (11.31a)$$

where e_1 and e_2 divide V in the ratio of the bridge's series resistance connections across arms 1-2 and 3-4. Equation 11.31a gives

$$\left(\frac{R_1}{R_1 + R_2} - \frac{R_4}{R_3 + R_4} \right)V = 0 \qquad (11.31b)$$

from which

$$R_1 R_3 = R_2 R_4 \qquad \text{or} \qquad \frac{R_1}{R_2} = \frac{R_4}{R_3} \qquad (11.32a,b)$$

11.5.1 Strain Gauge Bridges

This basic Eqs 11.32a,b for a balanced bridge must apply when all four resistance are made the same in its application to strains gauges within ¼, ½ and full bridge arrangements. These refer respectively to where one, two and four active strain gauges are connected from their installations into arm 1, arms 1 and 2 and all four arms. Bridge completion resistors, equal to the gauge resistance (usually 120 or d Ω), are required for the ¼ and ½ bridges. The active arms change their resistance in proportion to the strain in the gauge. Therefore, as the bridge can measure resistance change to one or more of its arms so the strain in the gauge can be found. With many strain gauges bonded to a structure at different locations, each may be switched into a ¼ bridge to record its strain. A multi-channel data logger is capable of scanning and recording strains very rapidly in this manner. The ½ bridge would normally be adopted to connect strain gauges into arms 1 and 2 that are known to be of equal magnitude but of opposite sense: as for strain gauge orientations: (i) aligned with the length of a rectangular beam, bonded to its top and bottom surfaces and (ii) at ± 45° to the axis of a torsion bar, bonded to its outer diameter. In this way it will be seen that the bridge sensitivity is doubled. The sensitivity is doubled further when all four arms contain active gauges whose equal-magnitude strains alternate in sense with connections sequentially into arms 1–4. Typically, arms 1 and 3 would contain gauges in tension and arms 2 and 4 contain gauges in compression (or vice versa). This can be achieved with the attachment of a second pair of gauges lying along side those upon the beam and anywhere along the length of the torsion bar, again at ± 45° orientations (see Fig. 11.22a). To achieve the equal magnitude strains in a load calibration device known as a *proving ring* the four gauges are attached at the position shown in Fig. 11.22b

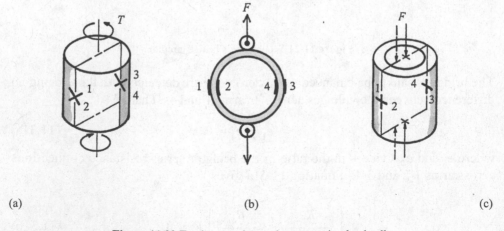

(a)   (b)   (c)

Figure 11.22 Torsion, tension and compression load cells

The enhanced sensitivity that a four-arm bridge offers is the basis for load cell design. We could see, for example, how the beam and torsion bar could be used as load or torque measuring devices given a respective bridge output voltage proportional to say a load applied to the end of cantilever weighing scale or to the tightening torque applied to a wrench. A far greater range of load cells is accommodated by a four-arm bridge in which the positive strains in gauges 1 and 3 are identical and the compressive strains in gauges 2 and 4 are also identical but not restricted to equal those in gauges 1 and 3. Thus, in Fig. 11.22c, a short tubular compression transducer would have its two axial gauges within arms 1 and 3 and its two lateral gauges in arms 2 and 4, (or vice versa). Here the four gauges are placed with equal spacing around the circumference to minimise bending error arising from off-central thrusts (see Example 11.9).

11.5.2 Voltage from an Unbalance Bridge

For an unbalanced bridge in Fig. 11.21 it follows from Eqs 11.31a,b that the detector will record a voltage that is found from:

$$e = e_1 - e_4 = \left(\frac{R_1}{R_1 + R_2} - \frac{R_4}{R_3 + R_4} \right) V \qquad (11.33a,b)$$

In a strain gauge bridge all four resistance elements are made the same so that an output occurs only when the resistance of one or more of its arms is altered by strain. A tensile strain within an active gauge causes the resistance to increase from R to $R + \delta R$ whilst a compressive strain causes the resistance to decrease from R to $R - \delta R$. Where such changes occur within the bridge arrangements given in Figs 11.23a–d, the following voltage outputs are to be found from Eq. 11.33b.

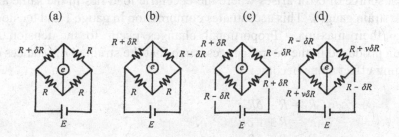

Figure 11.23 Wheatstone bridges used with strain gauges

(*a*) ¼ *Bridge* - Active Gauge 1 Under Tension (Figure 11.23a)

$$e = \left(\frac{R + \delta R}{2R + \delta R} - \frac{R}{2R} \right) V = \left[\frac{2R(R + \delta R) - R(2R + \delta R)}{2R(2R + \delta R)} \right] V \approx \frac{V}{4} \frac{\delta R}{R} \qquad (11.34a)$$

(b) ½ Bridge - Gauge 1 Under Tension; 2 Under Compression (Fig. 11.23b)

$$e = \left[\frac{(R + \delta R)}{(R + \delta R) + (R - \delta R)} - \frac{R}{2R} \right] V = \left[\frac{(R + \delta R)}{2R} - \frac{R}{2R} \right] V . \approx \frac{V}{2} \frac{\delta R}{R} \quad (11.34b)$$

(c) Full Bridge - Gauges 1 and 3 in Tension; 2 and 4 in Compression (Fig. 11.23c)

$$e = \left[\frac{(R + \delta R)}{(R + \delta R) + (R - \delta R)} - \frac{(R - \delta R)}{(R - \delta R) + (R + \delta R)} \right] V = \left[\frac{(R + \delta R)}{2R} - \frac{(R - \delta R)}{2R} \right] V$$

$$e = V \frac{\delta R}{R} \quad (11.34c)$$

(d) Full Bridge - Compression Load Cell with Gauges 1 and 3 Under Compression; Gauges 2 and 4 Under Lateral Tension (Fig. 11.23d)

$$e = \left[\frac{(R - \delta R)}{(R - \delta R) + (R + v\delta R)} - \frac{(R + v\delta R)}{(R + v\delta R) + (R - \delta R)} \right] V = \left[\frac{- \delta R(1 + v)}{2R - \delta R(1 - v)} \right] V$$

$$e \approx \frac{- V(1 + v)}{2} \frac{\delta R}{R} \quad (11.34d)$$

Example 11.9 Show how the compressive load cell design given in Fig. 11.22c serves to minimise the error arising from the eccentric compressive load F.

The greatest source of error arises where the eccentric load lies in the same axial plane as the strain gauges. This accentuates compression in gauge 1 and tension in gauge 3 to their maxima. Proportional changes occur to the tension and compression produced in the lateral gauges, such that the strained resistances due to the bending effect alone become:

$$R_1 = R - \delta R$$
$$R_2 = R + v \, \delta R$$
$$R_3 = R + \delta R$$
$$R_4 = R - v \, \delta R$$

The bridge output follows from Eq. 11.33a as:

$$e = \left[\frac{(R - \delta R)}{(R - \delta R) + (R + v\delta R)} - \frac{(R - v\delta R)}{(R - v\delta R) + (R + \delta R)} \right] V = \left[\frac{- (\delta R)^2 (1 - v^2)}{4R^2 - (\delta R)^2 (1 - v)^2} \right] V$$

$$e \approx - \frac{V(1 - v)^2}{4} \left(\frac{\delta R}{R} \right)^2 \quad (i)$$

Equation i shows that the output is therefore very small, which has been confirmed experimentally. Similarly, a bridge connected to gauges attached axially and laterally to its top and bottom surfaces would show little response when placed under four-point bending.

Should it be required to account for both compression and bending outputs in the eccentrically loaded cell, Eqs 11.34d and Eq. i are summed to give:

$$e = \left[\frac{-\delta R(1+v)}{2R - \delta R(1-v)} - \frac{(\delta R)^2(1-v^2)}{4R^2 - (\delta R)^2(1-v)^2} \right] V$$

$$= - \frac{2V(1+v)[R\delta R + (1-v)(\delta R)^2]}{4R^2 - (\delta R)^2(1-v)^2}$$

which returns to Eq. 11.34d when terms in $(\delta R)^2$ are neglected.

Example 11.10 Show how the four active arms within a torsion cell in Fig. 11.22a eliminate error in the presence of (a) tension and (b) bending.

(a) When the cell is subjected to an axial tensile strain ε the strains within the $\pm 45°$ gauges are found from substituting $\theta = 45°$ and $135°$ into $\varepsilon_\theta = \varepsilon \cos^2\theta$ (system (1), Table 11.1). This shows that all four gauges strain by the same amount $\frac{1}{2}\varepsilon$ and therefore the resistance of each gauge increases by a similar amount δR. Thus with $R_1 = R_2 = R_3 = R_4 = R + \delta R$ in Eq. 11.33b, the bridge output is zero.

(b) Let the neutral plane of bending lie at $90°$ to the plane of the gauges. When the bending moment produces tension in gauges 1 and 2, the analysis given in (a) again provides their strains as $\frac{1}{2}\varepsilon$ where ε is the axial strain. With the compression in gauges 3 and 4, their strains are of equal magnitude and opposite sense, i.e. $-\frac{1}{2}\varepsilon$. Correspondingly, the resistance changes to the four gauges become:

$$R_1 = R + \delta R$$
$$R_2 = R + \delta R$$
$$R_3 = R - \delta R$$
$$R_4 = R - \delta R$$

A similar result applies for all orientations of the gauge plane in Fig. 11.22a to the neutral plane. This is a consequence of the linear strain distribution across any diameter perpendicular to the neutral plane between maximum tension and maximum compression. Substituting R_1 to R_4 into Eq. 11.33b gives the bridge output as

$$e = \left[\frac{(R + \delta R)}{2(R + \delta R)} - \frac{(R - \delta R)}{2(R - \delta R)} \right] V = 0$$

Dutifully, this cell will detect only torsion in the presence of unavoidable tension and bending, with the bridge output being given by Eq. 11.34b.

11.5.3 Strain Conversion

The resistance R of a wire is proportional to its density ρ and length L and inversely proportional to its cross sectional area A, as follows

$$R = \frac{\rho L}{A} \tag{11.35a}$$

Hence, increasing L in a grid with a large number of closely packed parallel wires or etched strips, gives a strain gauge a high resistance. Differentiating Eq. 11.35a

$$\frac{\delta R}{R} = \frac{\delta \rho}{\rho} + \frac{\delta L}{L} - \frac{\delta A}{A} \tag{11.35b}$$

The area and length changes of a wire under tension are connected within a relationship for compressible elasticity (see Exercise 11.29, p. 485):

$$\frac{\delta A}{A} + 2v\frac{\delta L}{L} = 0 \tag{11.35c}$$

Combining Eqs 11.35b,c

$$\frac{\delta R}{R} = \frac{\delta \rho}{\rho} + (1 + 2v)\frac{\delta L}{L} \tag{11.36a}$$

A strain sensitivity factor (gauge factor) is defined from Eq. 11.36a as

$$S = \frac{\delta R/R}{\delta L/L} = (1 + 2v) + \frac{\delta \rho/\rho}{\delta L/L} \tag{11.36b}$$

For metals the final term is constant along with Poisson's ratio v, giving a constant sensitivity factor S. Most strain gauges are manufactured from a range of nickel, copper and chrome alloys, having a gauge factor $S \approx 2.1$. Consequently, the gauge's resistance change is converted to strain from the simple linear relation

$$\frac{\delta R}{R} = S\varepsilon \tag{11.37}$$

Thus, in converting the voltage output from a ¼-bridge (Fig. 11.23a) to strain Eqs 11.34a and 11.37 are combined to give

$$e = \frac{V}{4}\left(\frac{\delta R}{R}\right) = \frac{V}{4}S\varepsilon, \qquad \therefore \ \varepsilon = \frac{4e}{VS} \tag{11.38a}$$

Similarly, the strain conversions for the ½- and full-bridges (Figs 11.23b and c) follow from Eqs 11.34b,c and 11.37 as

$$\varepsilon = \frac{2e}{VS} \quad \text{and} \quad \varepsilon = \frac{e}{VS} \tag{11.38b,c}$$

11.5.4 Cross Sensitivity

The manufacturer's gauge factor S is calibrated for the uniaxial stress field found at the surface of a beam in four-point loading or at the surface of a bar under tension. The former (see Fig. 4.4, p. 106) is most common method of calibration in which one gauge from a batch is bonded axially to the beam surface at mid-position between its supports. Equation 11.34a is used to calculate the resistance change from the voltage detected by a quarter-bridge as it becomes unbalanced from a known strain in the gauge. Here the strain is calculated independently from the beam's central displacement δ which is given by Eq. 4.7b:

$$\varepsilon = \frac{4\delta t}{L^2} \qquad (11.39a)$$

where t is the beam thickness and L is its support length. If a tensile bar is used for the calibration it is advisable to measure its extension x over a gauge length l of at least 50 mm when calculating strain as $\varepsilon = x/l$. Alternatively, using the known load W, the testpiece's area A and tensile modulus E allow the strain to be found indirectly, i.e. $\varepsilon = W/AE$. It follows from Eq. 11.37 that the sensitivity factor becomes the gradient of the loading plot between $\delta R/R$ versus ε. For the beam, S can be found from the gradient of the plot between the measured parameters, e and δ, by dividing Eqs 11.34a and 11.39a as follows:

$$S = \frac{\delta R/R}{\varepsilon} = \frac{L^2}{Vt}\left(\frac{e}{\delta}\right) \qquad (11.39b)$$

The gauge factor S, as supplied by the gauge manufacturer, is calibrated in a similar manner. The resistance change to the gauge under a given uniaxial stress σ follows from Eq. 11.37 as:

$$\frac{\delta R}{R} = S\frac{\sigma}{E} \qquad (11.40a)$$

Now, because of the manner in which the gauge grid is constructed, its resistance change is due to both the axial and lateral strains which occur under a uniaxial stress σ calibration. We may refer strains ε_1 and ε_2, respectively, along and across the gauge, to their own axial and cross transverse sensitivity factors S_A and S_C. The net resistance is then the sum of two contributions:

$$\frac{\delta R}{R} = S_A \varepsilon_1 + S_C \varepsilon_2 \qquad (11.40b)$$

The two strains are in proportion to the stress according to the definition of Poisson's ratio v and Young's modulus E, as follows:

$$\varepsilon_2 = - v\varepsilon_1 = -\frac{v\sigma}{E} \qquad (11.40c)$$

Consequently, it may be assumed that the ratio between the two sensitivities is constant:

$$\lambda = \frac{S_C}{S_A} \qquad (11.41a)$$

Combining Eqs 11.40a–c and 11.41a provides a relationship between the three sensitivities:

$$S = S_A(1 - \lambda v) \qquad (11.41b)$$

Fortunately, the relative cross-sensitivity factor λ is not greater than 0.05 and when taken with v in the range 0.23–0.33, Eq. 11.41b shows $S \approx 0.98 S_A$. It can be seen from Eq. 11.41b that the true principal strains under a biaxial stress can only be measured with two gauges aligned with the principal directions. The resistance changes in each gauge is due to both of the principal strains, when from Eqs 11.40b and 11.41b

$$\left(\frac{\delta R}{R} \right)_1 = S_A(\varepsilon_1 + \lambda \varepsilon_2) \qquad (11.42a)$$

$$\left(\frac{\delta R}{R} \right)_2 = S_A(\varepsilon_2 + \lambda \varepsilon_1) \qquad (11.42b)$$

from which ε_1 and ε_2 may be solved. Strain meters convert each resistance change to strain using the manufacturer's sensitivity factor:

$$\varepsilon_1 = \frac{1}{S} \left(\frac{\delta R}{R} \right)_1 \quad \text{and} \quad \varepsilon_2 = \frac{1}{S} \left(\frac{\delta R}{R} \right)_2 \qquad (11.42c,d)$$

In general, Eqs 11.42a,b and 11.42c,d will not give identical strains in a biaxial strain field. The calibration for S ensures that Eqs 11.42a and 11.43a do provide identical *axial strain* arising from a uniaxial tension (or compression). Thus, setting $\varepsilon_2 = -v\varepsilon_1$ for a gauge under axial tension confirms an equality between Eqs 11.42a and c:

$$\left(\frac{\delta R}{R} \right)_1 = S_A[\varepsilon_1 + \lambda(-v\varepsilon_1)] = S_A(1 - \lambda v)\varepsilon_1 = S\varepsilon_1 \qquad (11.43a)$$

However, the *lateral strain* under uniaxial tension, as calculated from S and the change in resistance of a lateral gauge will be in error. The apparent lateral strain ε_2^* is calculated from:

$$\left(\frac{\delta R}{R} \right)_2 = S\varepsilon_2^* = S_A(1 - \lambda v)\varepsilon_2^* \qquad (11.43b)$$

Substitute $\varepsilon_1 = -\varepsilon_2/v$ into Eq. 11.42b:

$$\left(\frac{\delta R}{R} \right)_2 = S_A\left[\varepsilon_2 + \lambda\left(\frac{-\varepsilon_2}{v} \right) \right] = S_A\left(1 - \frac{\lambda}{v} \right)\varepsilon_2 \qquad (11.43c)$$

As the resistance change within Eqs 11.43b,c must remain the same they provide the following ratio between apparent strain and true lateral strains:

$$\frac{\varepsilon_2^*}{\varepsilon_2} = \frac{1 - \lambda/v}{1 - \lambda v} \qquad (11.43d)$$

It can be seen that agreement between ε_2^* and ε_2 applies only when $\lambda = 0$. Equation 11.43d gives a lateral strain ratio of 0.85 when $\lambda = 0.5$ and $v = 0.3$. This amounts to an error in lateral strain measurement of 15%, which is the largest that error can arise from ignoring cross sensitivity.

Example 11.11 Establish the % error associated with application of the manufacturer's gauge factor to calculate the apparent strain ε^* from a single gauge in a biaxial stress field.

The resistance change may be written in terms of the three sensitivities as:

$$\frac{\delta R}{R} = S\varepsilon^* = S_A(1 - \lambda v)\varepsilon^* \qquad (i)$$

Let the gauge lie in any arbitrary 1-direction. The cross sensitivity requires that the proportion of the resistance change arising from transverse strain in the gauge is accounted for:

$$\frac{\delta R}{R} = S_A(\varepsilon_1 + \lambda\varepsilon_2) \qquad (ii)$$

Equating i and ii gives the apparent strain in terms of the true strains ε_1 and ε_2:

$$\varepsilon^* = \frac{\varepsilon_1 + \lambda\varepsilon_2}{1 - \lambda v} \qquad (iii)$$

Equation iii is in error by the % measure of the ratio

$$E_1 = \frac{\varepsilon^* - \varepsilon_1}{\varepsilon_1} \qquad (iv)$$

Substituting Eq. iii into Eq. iv

$$E_1 = \frac{\lambda(v + \varepsilon_2/\varepsilon_1)}{1 - \lambda v} \qquad (v)$$

in which a second, transverse gauge is required to measure ε_2. Equation v confirms that $E_1 = 0$ for uniaxial tension where $\varepsilon_2/\varepsilon_1 = -v$. Taking $\lambda = 0.05$, when $\varepsilon_2/\varepsilon_1 = 1$ for equi-biaxial straining in steel and aluminium, shows error for each strain as 6–7 %, with v in the range $\frac{1}{4}$–$\frac{1}{3}$.

Finally, we should note that if the precision in strain measurement, as provided by the two sensitivities S_A and λ is required, then one of them must be found. Applying Eqs 11.42a,b to a *plane strain field* where $\varepsilon_2 = 0$ gives:

$$\left(\frac{\delta R}{R}\right)_1 = S_A\varepsilon_1, \qquad \left(\frac{\delta R}{R}\right)_2 = S_A\lambda\varepsilon_1 \qquad (11.44a,b)$$

Dividing Eqs 11.44a,b provides λ from the resistance changes within two gauges aligned with the 1- and 2-directions:

$$\lambda = \frac{(\delta R/R)_2}{(\delta R/R)_1} \qquad (11.44c)$$

Knowing λ, Eq.11.41b allows S_A to be calculated from the manufacturer's gauge factor S. To achieve the necessary elastic plane strain condition the two gauges may be attached circumferentially (direction 1) and axially (direction 2) to the outer surface of a long, thick-walled cylinder, internally pressurised within rigid end closures (see Fig. 11.24a and § 4.6.3, p. 134).

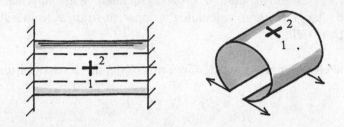

(a)   (b)

Figure 11.24 Plane strain calibration tests for cross-sensitivity factor

Alternatively, the wide curved beam in Fig. 11.24b ensures no strain across its width (direction 2) when flexed as the gap is opened and closed. With gauge attachments in alignment with the circumferential and width directions, 1 and 2 respectively, Eq. 11.44c provides λ from the resistance changes arising in each gauge. The following example shows there is one method of removing the error arising from ignoring cross sensitivity without the need for the gauge sensitivity's two factors, S_A and λ.

Example 11.12 When a uniaxial elastic stress $\sigma = 200$ MPa is applied to a plate, its axial and transverse strain gauges recorded resistance changes: $\delta R_1/R_1 = 2088 \times 10^{-6}$ and $\delta R_2/R_2 = -567 \times 10^{-6}$, respectively. When the same plate was subjected to a more complex, biaxial elastic loading, the gauge resistance changes (non-principal) were $\delta R_x/R_x = 880 \times 10^{-6}$ and $\delta R_y/R_y = -610 \times 10^{-6}$, respectively. Show that by taking the sum and difference of each set of gauge readings it is possible to calculate the biaxial stress state (σ_x, σ_y) aligned with the gauge directions, neither knowing the gauge sensitivity factors, S_A and λ nor the plate material's elastic constants, E and v.

The two resistance changes arising from uniaxial tension are written from Eqs 11.42a,b and Eq. 11.40c as:

$$\frac{\delta R_1}{R_1} = S_A(\varepsilon_1 + \lambda \varepsilon_2) = \frac{S_A \sigma}{E}(1 - \lambda v) \tag{i}$$

$$\frac{\delta R_2}{R_2} = S_A(\varepsilon_2 + \lambda \varepsilon_1) = \frac{S_A \sigma}{E}(\lambda - v) \tag{ii}$$

Adding and subtracting Eqs i and ii:

$$\frac{\delta R_1}{R_1} + \frac{\delta R_2}{R_2} = \frac{S_A \sigma}{E}(1 - v)(1 + \lambda) \tag{iii}$$

$$\frac{\delta R_1}{R_1} - \frac{\delta R_2}{R_2} = \frac{S_A \sigma}{E}(1 + v)(1 - \lambda) \tag{iv}$$

The two resistance changes arising from biaxial stressing are written from Eqs 11.42a,b:

$$\frac{\delta R_x}{R_x} = S_A(\varepsilon_x + \lambda \varepsilon_y), \qquad \frac{\delta R_y}{R_y} = S_A(\varepsilon_y + \lambda \varepsilon_x) \tag{v, vi}$$

where the following biaxial stress–strain relations apply (see § 4.4, p.120):

$$\varepsilon_x = (\sigma_x - v \sigma_y)/E, \qquad \varepsilon_y = (\sigma_y - v \sigma_x)/E \tag{vii, viii}$$

Combining Eqs v–viii and then taking their sum and difference leads to:

$$\frac{\delta R_x}{R_x} + \frac{\delta R_y}{R_y} = \frac{S_A(\sigma_x + \sigma_y)}{E}(1 - v)(1 + \lambda) \tag{ix}$$

$$\frac{\delta R_x}{R_x} - \frac{\delta R_y}{R_y} = \frac{S_A(\sigma_x - \sigma_y)}{E}(1 + v)(1 - \lambda) \tag{x}$$

Eliminating their common factors, Eq. iii is combined with Eq. ix and Eq. iv combined with Eq. x:

$$\sigma\left(\frac{\delta R_x}{R_x} + \frac{\delta R_y}{R_y}\right) = (\sigma_x + \sigma_y)\left(\frac{\delta R_1}{R_1} + \frac{\delta R_2}{R_2}\right) \tag{xi}$$

$$\sigma\left(\frac{\delta R_x}{R_x} - \frac{\delta R_y}{R_y}\right) = (\sigma_x - \sigma_y)\left(\frac{\delta R_1}{R_1} - \frac{\delta R_2}{R_2}\right) \tag{xii}$$

It can be seen that the resulting simultaneous Eqs xi and xii provide a solution to σ_x and σ_x given the gauge resistance changes arising from their combination and the resistance changes under a known uniaxial stress. Applying the numerical values given in the example to Eqs xi and xii:

$$200\,[880 + (-610)] = (\sigma_x + \sigma_y)\,[2088 + (-567)]$$

$$200\,[880 - (-610)] = (\sigma_x - \sigma_y)\,[2088 - (-567)]$$

from which

$$\sigma_x = 73.87 \text{ MPa}, \quad \sigma_y = -38.37 \text{ MPa}$$

Note that as σ_x and σ_y are stresses aligned with the gauge directions they become principal stresses when the gauge directions x and y are aligned with the principal directions (if known). Otherwise, x and y define the direction of planes upon which normal stresses σ_x and σ_y act together with an unknown shear stress τ_{xy}. The full picture is revealed from the resistance changes measured within a three-element strain gauge rosette, as the following analyses show.

11.5.5 Cross Sensitivity in Strain Gauge Rosettes

The earlier analyses (see § 11.4.4) of principal strain magnitudes and their directions, provided by three-element 45° and 60° rosettes, ignored the cross-sensitivity within each gauge. Accounting for cross sensitivity begins with the principal strain transformation Eq. 11.19b

$$\varepsilon_\theta = \frac{1}{2}(\varepsilon_1 + \varepsilon_2) + \frac{1}{2}(\varepsilon_1 - \varepsilon_2)\cos 2\theta \qquad (11.45a)$$

whereupon it can be seen that on setting:

$$m = \frac{1}{2}(\varepsilon_1 + \varepsilon_2) \text{ and } n = \frac{1}{2}(\varepsilon_1 - \varepsilon_2)$$

these simultaneous equations provide magnitudes of the principal strains

$$\varepsilon_1 = m + n, \qquad \varepsilon_2 = m - n \qquad (11.45b,c)$$

(a) Three-Element, Delta Rosette

Equation 11.45a provides the strains in gauges a, b and c inclined at θ, $\theta + 120°$ and $\theta + 240°$ respectively to the major principal strain (direction-1 in Fig. 11.19a):

$$\varepsilon_a = m + n\cos 2\theta \qquad (11.46a)$$

$$\varepsilon_b = m + n\cos 2(\theta + 120°) = m - n\cos(2\theta + 60°) \qquad (11.46b)$$

$$\varepsilon_c = m + n\cos 2(\theta + 240°) = m - n\cos(2\theta - 60°) \qquad (11.46c)$$

The cross strain (primed) in each gauge align with their perpendicular directions: $\theta + 90°$, $\theta + 210°$ and $\theta + 330°$:

$$\varepsilon_a' = m + n\cos 2(\theta + 90°) = m - n\cos 2\theta \qquad (11.47a)$$

$$\varepsilon_b' = m + n\cos 2(\theta + 210°) = m + n\cos(2\theta + 60°) \qquad (11.47b)$$

$$\varepsilon_c' = m + n\cos 2(\theta + 330°) = m + n\cos(2\theta - 60°) \qquad (11.47c)$$

Writing, from Eqs 11.42a,b, $r_i = (\delta R/R)_i/S_A$, where $i = a, b, c$, accounts for resistance change in each gauge due to an in-line and cross-straining. Thus, the net strains measured by the rosette's gauges follow from Eqs 11.46a–c and 11.47a–c:

$$r_a = \varepsilon_a + \lambda\varepsilon_a' = m(1 + \lambda) - n(1 - \lambda)\cos 2\theta \tag{11.48a}$$

$$r_b = \varepsilon_b + \lambda\varepsilon_b' = m(1 + \lambda) - n(1 - \lambda)\cos(2\theta + 60°) \tag{11.48b}$$

$$r_c = \varepsilon_c + \lambda\varepsilon_c' = m(1 + \lambda) - n(1 - \lambda)\cos(2\theta - 60°) \tag{11.48c}$$

The following relationships apply to Eqs 11.48a–c

$$r_a + r_b + r_c = 3m(1 + \lambda) \tag{11.49a}$$

$$r_a - r_c = \frac{n}{2}(1 - \lambda)(3\cos 2\theta + \sqrt{3}\sin 2\theta) \tag{11.49b}$$

$$r_a - r_b = \frac{n}{2}(1 - \lambda)(3\cos 2\theta - \sqrt{3}\sin 2\theta) \tag{11.49c}$$

Adding and subtracting Eqs 11.49b,c:

$$3n(1 - \lambda)\cos 2\theta = (r_a - r_c) + (r_a - r_b) = 2r_a - r_b - r_c \tag{11.50a}$$

$$\sqrt{3}n(1 - \lambda)\sin 2\theta = (r_a - r_c) - (r_a - r_b) = r_b - r_c \tag{11.50b}$$

Now m follows directly from Eq. 11.49a and n is found from eliminating θ in Eqs 11.50a,b by squaring and adding. Substituting m and n into Eqs 11.45b,c gives the principal strain magnitudes:

$$\varepsilon_{1,2} = \frac{r_a + r_b + r_c}{3(1 + \lambda)} \pm \frac{2\sqrt{(r_a - r_b)^2 + (r_a - r_c)^2 - (r_a - r_b)(r_a - r_c)}}{3(1 - \lambda)} \tag{11.51a,b}$$

which return to Eqs 11.21a,b for $\lambda = 0$. Dividing Eq. 11.50b by Eq. 11.50a for the principal strain directions:

$$\tan 2\theta = \frac{\sqrt{3}(r_b - r_c)}{2r_a - r_b - r_c} \tag{11.51c}$$

(b) Three-Element, 45° Rosette

Equation 11.45a provides the strains in gauges a, b and c inclined at θ, $\theta + 45°$ and $\theta + 90°$, respectively, to the major principal strain (direction-1 in Fig. 11.19b):

$$\varepsilon_a = m + n\cos 2\theta \tag{11.52a}$$

$$\varepsilon_b = m + n\cos 2(\theta + 45°) = m - n\sin 2\theta \tag{11.52b}$$

$$\varepsilon_c = m + n\cos 2(\theta + 90°) = m - n\cos 2\theta \tag{11.52c}$$

The cross strains (primed below) apply to directions $\theta + 90°$, $\theta + 135°$ and $\theta + 180°$:

$$\varepsilon_a' = m + n\cos2(\theta + 90°) = m - n\cos2\theta \qquad (11.53a)$$

$$\varepsilon_b' = m + n\cos2(\theta + 135°) = m + n\sin2\theta \qquad (11.53b)$$

$$\varepsilon_c' = m + n\cos2(\theta + 180°) = m + n\cos2\theta \qquad (11.53c)$$

Equations 11.52a–c and 11.53a–c contribute to the net strains for the rosette's elements as

$$r_a = \varepsilon_a + \lambda\varepsilon_a' = m(1+\lambda) + n(1-\lambda)\cos2\theta \qquad (11.54a)$$

$$r_b = \varepsilon_b + \lambda\varepsilon_b' = m(1+\lambda) - n(1-\lambda)\sin2\theta \qquad (11.54b)$$

$$r_c = \varepsilon_c + \lambda\varepsilon_c' = m(1+\lambda) - n(1-\lambda)\cos2\theta \qquad (11.54c)$$

Solving Eqs 11.54a–c for m and n, then substituting into Eq. 11.45b,c the principal magnitudes are found

$$\varepsilon_{1,2} = \frac{r_a + r_c}{2(1+\lambda)} \pm \frac{\sqrt{(r_a - r_b)^2 + (r_b - r_c)^2}}{\sqrt{2}(1-\lambda)} \qquad (11.55a,b)$$

Equations 11.54a–c also provide the strain directions

$$\tan2\theta = \frac{r_a + r_c - 2r_b}{r_a - r_c} \qquad (11.55c)$$

It can be seen that when cross-sensitivity is ignored from setting $\lambda = 0$, then $S_A = S$ and Eqs 11.55a–c simplify to the principal strain Eqs 11.23a-c.

Example 11.13 The following micro-voltages were detected in the 0°, 45° and 90° arms of a strain gauge rosette as each gauge a, b and c (see Fig. 11.19b) was switched into a quarter bridge supplied with 4 volts:

$$e_a = 1290 \ \mu V, \ e_b = 430 \ \mu V, \ e_c = 350 \ \mu V$$

The rosette is bonded to a steel structure in which the direction of gauge a provides a convenient reference direction with these readings taken at full load. Determine the magnitudes and directions of the principal stresses: (a) when based upon the manufacturer's sensitivity factor 2.1 and (b) in accounting for the effect of cross sensitivity in each gauge, given a relative cross-sensitivity factor of 0.05. Take, for steel: $E = 207$ GPa and $v = 0.3$

(a) Using only the manufacturer's gauge factor for the rosette analysis ignores the cross sensitivity of each gauge. The strains follow from Eq. 11.38a as:

$$\varepsilon_i = \frac{4e_i}{VS} = \frac{4e_i}{4 \times 2.1} = \frac{e_i}{2.1}$$

giving

$$\varepsilon_a = 614.3 \times 10^{-6}, \quad \varepsilon_b = 204.8 \times 10^{-6}, \quad \varepsilon_c = 166.7 \times 10^{-6}$$

Here Eq. 11.23a,b provide the principal strains from a 45° rosette as follows

$$\varepsilon_{1,2} = \frac{1}{2}(\varepsilon_a + \varepsilon_c) \pm \frac{1}{\sqrt{2}}\sqrt{(\varepsilon_a - \varepsilon_b)^2 + (\varepsilon_b - \varepsilon_c)^2}$$

$$= \left[\frac{1}{2}(614.3 + 166.7) \pm \frac{1}{\sqrt{2}}\sqrt{(614.3 - 204.8)^2 + (204.8 - 166.7)^2}\right] \times 10^{-6}$$

$$\varepsilon_1 = 681.3 \times 10^{-6}, \quad \varepsilon_2 = 99.7 \times 10^{-6}$$

Equation 11.23c gives the inclination of the major and minor principal strains to the gauge *a* direction, respectively:

$$\theta = \frac{1}{2}\tan^{-1}\left(\frac{\varepsilon_a + \varepsilon_c - 2\varepsilon_b}{\varepsilon_a - \varepsilon_c}\right)$$

$$= \frac{1}{2}\tan^{-1}\left(\frac{614.3 + 166.7 - 2 \times 204.8}{614.3 - 166.7}\right) = 19.84°, \ 109.84°$$

The principal stresses follow from Eqs 4.34a,b (p. 121):

$$\sigma_1 = \frac{E}{1 - v^2}(\varepsilon_1 + v\varepsilon_2)$$

$$= \frac{(210 \times 10^3)}{(1 - 0.3^2)}[681.3 + (0.3 \times 99.7)]10^{-6} = 164.13 \text{ MPa}$$

$$\sigma_2 = \frac{E}{1 - v^2}(\varepsilon_2 + v\varepsilon_1)$$

$$= \frac{(210 \times 10^3)}{(1 - 0.3^2)}[99.7 + (0.3 \times 681.3)]10^{-6} = 70.18 \text{ MPa}$$

(b) The cross-sensitivity factor λ must be used with the axial sensitivity factor S_A, which is found from Eq. 11.41b as:

$$S_A = \frac{S}{1 - \lambda v} = \frac{2.1}{1 - (0.05 \times 0.3)} = 2.132$$

In this case Eqs 11.55a–c provide the principal strains, where again from Eqs 11.42a,b: $r_i = (\delta R/R)_i/S_A$, $(i = a, b, c)$. Substituting from Eq. 11.34a for the resistance change in a quarter-bridge gives:

$$r_i = \frac{4e_i}{VS_A} = \frac{4e_i}{4 \times 2.1} = \frac{e_i}{2.132}$$

giving

$$r_a = 605.07 \times 10^{-6}, \; r_b = 201.69 \times 10^{-6}, \; r_c = 164.17 \times 10^{-6}$$

Here Eq. 11.55a,b provide the principal strains from a 45° rosette as follows

$$\varepsilon_{1,2} = \frac{r_a + r_c}{2(1 + \lambda)} \pm \frac{\sqrt{(r_a - r_b)^2 + (r_b - r_c)^2}}{\sqrt{2}(1 - \lambda)}$$

$$= \left[\frac{(605.07 + 164.17)}{2(1 + 0.05)} \pm \frac{\sqrt{(605.07 - 201.69)^2 + (201.69 - 164.17)^2}}{\sqrt{2}(1 - 0.05)} \right] \times 10^{-6}$$

$$\therefore \; \varepsilon_1 = 667.9 \times 10^{-6}, \quad \varepsilon_2 = 64.8 \times 10^{-6}$$

Equation 11.55c provides the principal strain directions:

$$\theta = \frac{1}{2} \tan^{-1} \left(\frac{r_a + r_c - 2r_b}{r_a - r_c} \right)$$

$$= \frac{1}{2} \tan^{-1} \left(\frac{605.07 + 164.17 - 2 \times 201.69}{605.07 - 164.17} \right) = 19.84°, \; 109.84°$$

Equations 4.23a,b provide the major and minor principal stresses as, respectively:

$$\sigma_1 = \frac{(210 \times 10^3)}{(1 - 0.3^2)} \left[667.9 + (0.3 \times 64.8) \right] \times 10^{-6} = 158.6 \, \text{MPa}$$

$$\sigma_2 = \frac{(210 \times 10^3)}{(1 - 0.3^2)} \left[64.8 + (0.3 \times 667.9) \right] \times 10^{-6} = 61.2 \, \text{MPa}$$

A comparison between the principal strain provided by the rosette analyses given in (a) and (b) shows that cross sensitivity influences the magnitudes but not the directions of the principal strains. The difference found between the strain magnitudes warrants the account of cross sensitivity given. However, the influence upon principal stress may not be of a similar concern when scaled down with generous safety factor for a design that places a limit on the working stress level.

EXERCISES

Two-Dimensional Stress Analyses

11.1 A 30 mm diameter bar is subjected to an axial force of 50 kN. Calculate the normal and shear stresses (σ_θ, τ_θ) on planes making angles θ of 8°, 38° and 68° with the bar axis.
(Ans: σ_θ, τ_θ (in MPa): 1.37, 9.8; 26.8, 34.7; 61, 24.5)

11.2 Plane stresses σ_x = 62 MPa, σ_y = − 46 MPa and τ_{xy} = 39 MPa act at a point. Determine the magnitude and direction of the maximum shear stress and the major and minor principal stresses.
(Ans: 66.61 MPa, 74.61 MPa, −58.61 MPa, 18°, 63°)

11.3 Given the applied principal stresses σ_1 = 40 MPa and σ_2 = − 20 MPa, find σ_θ and θ for a plane on which τ_θ = 15 MPa.

11.4 Given that the major principal tensile stress is 4 MPa and that a normal compressive stress of 1 MPa acts on a plane inclined at 60° to the major principal plane, determine both graphically and analytically, the shear stress on the inclined plane and the minor principal stress. (Ans: 2.9 MPa, −2.7 MPa)

11.5 The principal stresses at a point are 30 and 50 kPa in tension. Determine, for a plane inclined at 40° to the major principal plane: (a) the normal and shear stresses, and (b) the magnitude and direction of the resultant stress.
(Ans: 4.18 kPa, 0.985 kPa, 4.3 kPa, 53.5°)

11.6 Normal stresses, σ_x = 80 kPa in tension and σ_y = 60 kPa in compression, are applied in the x- and y-directions. If the major principal stress is limited to 100 kPa what magnitude of shear stress is aligned with x and y. What is the magnitude of the maximum shear stress and the minor principal stress?
(Ans: 28.28 kPa, 30 kPa, 40 kPa)

11.7 A glass-fibre sheet is loaded along a plane lying at 25° to the fibre axis with combined direct and shear stresses of 15 MPa tensile and 5 MPa, respectively. If the allowable working stresses are 60 MPa tension for a plane normal to the fibre axis and 10 MPa in parallel shear, determine the safety factors used in the design.
(Ans: 1.12)

11.8 A block of brittle material with rectangular section, 50 mm wide × 100 mm deep, is loaded axially under a compressive force of 71.2 kN. What are the normal and shear stresses on an oblique plane making 60° with the vertical depth? If the ultimate compressive and shear strengths are 50 and 20 MPa respectively, determine the compressive fracture force and the orientation of the failure planes.
(Ans: 10.68 MPa, 6.17 MPa, 200 kN, 45°, 135°)

11.9 A thin-walled steel tube 150 mm i.d. with thickness 2 mm contains gas at a pressure of 11 bar. Find the tensile and shear stresses on a helical seam inclined at 25° to the cross section. Note that 1 MPa = 10 bar.
(Ans: 24.4 MPa, 7.92 MPa)

11.10 The prism in Fig. 11.25 has a resultant stress of 77 MPa on face AB and a principal stress of 123.5 MPa on face AC. Determine θ, the state of stress on plane BC and the other principal stress.

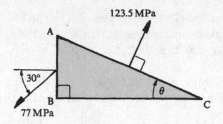

Figure 11.25

11.11 Draw the Mohr's circles for the stress systems (MPa) in Fig. 11.26a–c. Locate the focus and find the normal and shear stress on the 30° plane. Determine the principal stresses in (b) and (c) and the planes on which they act. Check your answers numerically.
[Ans in MPa: (a) 12.5, 13; (b) 13, 7.5, ± 15, ± 45°; (c) 25.5, 5.5, 26.2, 22.5°, – 16.2, 112.5°]

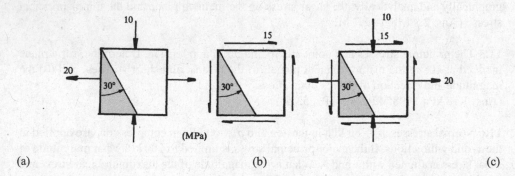

(a)   (b)   (c)

Figure 11.26

11.12 Different pulley tensions rotate the shaft in Fig. 11.27 within its encastré-bearing support. If the major principal stress is not to exceed 80 MPa, determine the shaft diameter based upon bending and torsional effects.
(Ans: 14.2 mm)

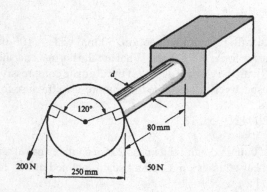

Figure 11.27

11.13 A shaft 100 mm in diameter, transmits 375 kW when rotating at 300 rev/min whilst supporting simultaneously an axial compressive force of 400 kN. Find graphically the principal stresses induced.
(Ans: 38.1, −87.7 MPa)

11.14 A 250 mm propeller shaft transmits a torque of 162.8 kNm. In addition, stresses of 9.65 MPa in compression and ± 5.5 MPa in bending, arise due to the respective actions of an axial thrust and self-weight. Given that the allowable direct and shear stress are 165.4 and 103.35 MPa respectively, determine the safety factor used in the shaft design.
(Ans: 10.9, 13.6)

11.15 A 20 mm diameter shaft is subjected to a bending moment of 25 Nm and an axial torque of 15 Nm. Determine, graphically for the o.d., the principal stresses and the angles between the maximum shear planes and the shaft axis. Check your answers analytically.

11.16 A solid steel propeller shaft 152.5 mm in diameter absorbs 447.5 kW at 180 rev/min in developing a thrust of 1.5 MN. Determine the principal stresses, the maximum shear stress in the shaft and the planes on which they act.
(Ans: 12.35 MPa, −94.2 MPa, 53.3 MPa, 19.92°, 64.92°)

11.17 A shaft is required to transmit simultaneously a torque of 36.4 kNm and a bending moment of 15.2 kNm. Given that the major principal stress must not exceed 92.5 MPa, find the minimum external shaft diameter when the shaft section is: (a) solid and (b) hollow, with a diameter ratio of 0.6. (Ans: 143.8 mm, 90.3 mm)

11.18 The bracket in Fig. 11.28 carries a load of 4 kN acting on a pin passing through the hole. Neglecting the effect of the hole, find that cross section which is subjected to the greatest bending stress. What is the maximum induced shear stress at this section due to the combined effects of bending and shear?
(Ans: 10 mm from wall, 7.3 MPa)

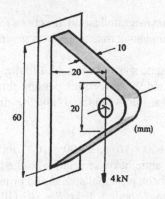

Figure 11.28

11.19 An I-section beam is 255 mm deep with $I = 50.8 \times 10^6$ mm^4. At a section where the bending moment is 36.4 kNm, the shear force produces shear stresses of 23.3 MPa at the neutral axis and 19.3 MPa on the tensile side, 100 mm from the neutral axis. Find the principal stresses and planes at each point.
(Ans: 77.8 MPa, 4.83 MPa, 14.03°, 23.3 MPa, ± 45°)

484   MECHANICS OF ENGINEERING STRUCTURES

11.20 A 1.5 m-long steel cantilever with an I-section has flanges 150 mm wide and 25 mm thick. The overall depth is 305 mm and the web is 25 mm thick. If the cantilever carries a uniformly distributed load of 202.5 kN/m determine, for the fixed end, the principal stresses at the top and bottom of the web and at the neutral axis of bending. Indicate with sketches the planes on which the principal stresses act.

11.21 A simply supported, I-section beam, 4 m long, carries a uniformly distributed load of 65 kN/m. Calculate, for a section of the beam 0.5 m from the left-hand support, the principal stresses at the neutral axis and at the top of the web on the compressive side. The flanges are 100 mm wide, the total depth of section is 200 mm and the web and flange thicknesses are each 7.5 mm.

11.22 A simply supported beam carries a central concentrated load of 300 kN within its span of 3.05 m. The dimensions of its I-section are: flanges 152.5 × 25 mm and web 255 × 12.5 mm. Find the principal stresses at the top of the tension side of the web at a section 1.22 m from a support.
(Ans: 146.8 MPa, –9.8 MPa)

Two-Dimensional Strain Analyses

11.23 A point in a material is under the strains: $\varepsilon_x = 400$ με, $\varepsilon_y = 200$ με and $\gamma_{xy} = 350$ με (where 1 με $= 10^{-6}$). Determine analytically and graphically the magnitude and orientation of the principal strains. Find also the state of strain for an axis inclined at 45° to the x-axis.
(Ans: 500με, 30° to x, 100με, 120° to x, 470με, 212με)

11.24 If the strain state at a point is given by $\varepsilon_x = -250$ με, $\varepsilon_y = 375$ με and $\gamma_{xy} = 400$ με, calculate the magnitude and direction of the principal strains. Determine the state of strain for perpendicular directions aligned with a clockwise rotation of 50° in the x, y co-ordinates.
(Ans: 433.5 με, –308.5 με, 73.7°, –80 με, 205 με, –685 με)

11.25 Confirm the principal strain magnitudes and directions found in Example 11.8 by using the appropriate, alternative strain gauge rosette analysis given in § 11.4.4, (see p. 459).

11.26 The 45° strain gauge rosette shown in Fig.11.19b measures direct strains of 250με, 200με and 350με for 0°, 45° and 90° directions passing through a point. What is the % change in the right angle in radians between the 0° and 90° directions?
(Ans: 200με)

11.27 Rosette element strains: 600×10^{-6}, -200×10^{-6} and 250×10^{-6} are measured in directions of 0°, 60° and 120° anticlockwise relative to the horizontal direction. Find the magnitude and direction of both the principal strains and principal stresses at the surface of a steel structure to which rosette is bonded. Take $E = 207$ GPa and $v = 0.3$.
(Ans: 680×10^{-6} at –17°, -246×10^{-6} at 73°, 137.7 MPa, –9.73 MPa)

11.28 A strain gauge rosette, with 60° inclinations between its three arms, recorded the direct strains 500 με, 200 με and –300 με upon the surface of a loaded structure. Calculate the principal stresses given $E = 100$ GPa and $v = 0.3$ for the structure material.
(Ans: 57 MPa, –16.6 MPa)

11.29 Confirm Eq. 11.35c, using Eqs 4.16a and 4.30, based upon the volume change under hydrostatic component of stress in simple tension.

11.30 A 60° strain gauge rosette measured direct strains of $-50\mu\varepsilon$, $250\mu\varepsilon$ and $500\mu\varepsilon$ in directions inclined at 0°, 60° and 120°, respectively, to the horizontal x-direction. Determine the complete state of strain for the x and y-directions.
(Ans: $\varepsilon_x = -500\mu\varepsilon$, $\varepsilon_y = 515\mu\varepsilon$, $\gamma_{xy} = -289\mu\varepsilon$)

11.31 A rosette of three strain gauge elements spaced 120° apart is bonded to the web of a loaded I-beam. The gauge elements record tensile strains of 3.22×10^{-4}, 2.18×10^{-4} and 0.21×10^{-4}. Determine the magnitude of the principal strains and stresses and the direction of the maximum principal stress relative to the direction of the maximum strain gauge reading. Take $E = 207$ GPa and $v = 0.28$.

11.32 The three-element, 60° strain gauge rosette, shown in Fig. 11.29, measures its strains as $\varepsilon_a = 4 \times 10^{-4}$, $\varepsilon_b = 3 \times 10^{-4}$ and $\varepsilon_c = 2 \times 10^{-4}$. Determine the shear strain between the x and y-directions and the normal strain in the y-direction.
(Ans: $\gamma_{xy} = -1.16 \times 10^{-4}$, $\varepsilon_y = 2 \times 10^{-4}$)

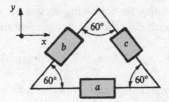

Figure 11.29

11.33 A strain gauge rosette consists of three gauges a, b and c arranged in an equilateral triangle as shown in Fig. 11.29, and is attached to an isotropic sheet. Find the principal strains and their inclination to gauge a, given that the sheet is under a state of plane stress for which the strains are: $\varepsilon_a = \varepsilon$, $\varepsilon_b = 2\varepsilon$ and $\varepsilon_c = 3\varepsilon$.

11.34 A certain steel is linear elastic, having a Young's modulus of 200 GPa and a Poisson's ratio of 0.3. A thin sheet of this material is marked in the unstressed condition with three crosses as shown in Fig. 11.30, each at a different orientation relative to the x-axis as shown. The legs of each cross are perpendicular and of unit length. Determine the lengths of the legs of each cross and the angle between them when a uniaxial stress of 200 MPa is applied in in the x-direction within the plane of the sheet.

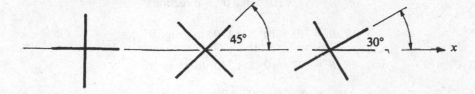

Figure 11.30

11.35 In a laboratory test a short steel cantilever with a thin-walled, tubular square section was subjected to combined bending, shear force and torsion due to the application of an offset vertical load applied to its free end as shown in Fig. 11.31.

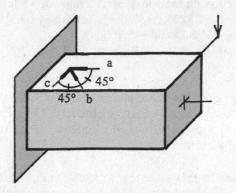

Figure 11.31

Under an elastic, incremental loading the following strains per unit load (units $\mu\varepsilon$/kg) were found from a 45° rosette positioned near the end fixing upon the central top surface:

$$\varepsilon_a = 9.8, \quad \varepsilon_b = 3.3, \quad \varepsilon_c = 2.7 \text{ (units: } \mu\varepsilon/\text{kg)}$$

The axes of gauges a, b and c lay at 0°, 45° and 90° to the tube's longitudinal axis as shown. Given $E = 210$ GPa and $v = 0.27$ for steel and the manufacture's gauge factor $S = 2.1$, find the magnitude and direction of both the principal strains and stresses when the following two conditions are applied to the strain gauges:

(i) cross sensitivity is neglected

(ii) cross sensitivity is accounted for within a factor $\lambda = 0.05$.

11.36 It is required to find the direct and shear strain components on orthogonal axes x, y from the strains recorded by a three-element, 120° rosette in which x is parallel to strain gauge element a. Derive expressions for the component x, y strains taking the effect of cross sensitivity into account. Use these expressions to find the strains and hence the stresses on planes parallel and perpendicular to the axis of gauge a given the following resistance changes for the rosette's elements a, b and c:

$$\delta R_a/R_a = 57 \times 10^{-6} \text{ (the 0°, x-direction)}$$

$$\delta R_b/R_b = 703 \times 10^{-6} \text{ at 120° to gauge } a$$

$$\delta R_c/R_c = 877 \times 10^{-6} \text{ at 240° to gauge } a$$

Each gauge has an axial sensitivity factor of $S_A = 2.22$ and a relative cross-sensitivity factor $\lambda = 0.02$. Take $E = 160$ GPa and $v = 0.34$ for the material to which the rosette is bonded.

CHAPTER 12

YIELD, STRENGTH AND FAILURE CRITERIA

Summary: Despite the advances in developing new composite materials for load-bearing applications, ductile metals and alloys have retained their usefulness for a similar role unceasingly. This is largely due to the capacity of metals to survive overloads without resulting in failure. The same cannot be said of man-made composites that would either fail or become unserviceable with the permanent damage caused through overloading. While simple stress states can be factored for safety, there remains the risk of overloading a material when it has to bear a combination of loads, arising from, say, tension, bending, shear and torsion in action simultaneously. Across these two classes of material, metals and composites, a safe design would avoid yielding and fracture. Hence this chapter's title refers to yield criteria for metals and strength criteria for brittle non-metals. They are placed together here because the respective criteria for their limiting elasticity often take similar forms. For example, the von Mises criterion may refer to both the yielding and fracture of an isotropic solid. However, real materials are often anisotropic through their processing, as with the rolling of metals, or by design, as with the enhanced strength aligned with the fibres in a composite. Certain one- and two-dimensional anisotropic criteria are material specific but general three-dimensional criteria limit both directional yielding and strength in a common formulation.

12.1 Yielding of Ductile Metals

A metallic material will begin to deform plastically under a tensile or a compressive stress, i.e. a uniaxial stress, after the yield stress has been reached. It has been seen that simple tensile or compressive yielding can be avoided when a safety factor has imposed a safe working elastic stress for a material to attain. In practice, however, stress states are more often biaxial or triaxial. For example, a thin plate with forces distributed over its edges, presents a two-dimensional (plane) stress state while a point in the wall of a thick-walled cylinder under internal pressure, is under three-dimensional stress. The question arises as to what magnitudes of each stress condition will produce yielding in the plate and cylinder materials? The answer requires that a suitable criterion be found, based upon stress, strain or strain energy, that connects yielding under two- and three-dimensional stress conditions to uniaxial yielding. It will be shown that a yield criterion can be related to the uniaxial yield stress (symbol Y), which is measured most conveniently from a tension test. Alternatively, these criteria may be related to the shear yield stress (symbol k) found from a simple torsion test. In the following section a summary is given of the various *yield criteria* that have been proposed over the past two centuries. Of these, it is now accepted that the *von Mises* and *Tresca* criteria are most representative of initial yielding in metals. Experimental evidence is presented in support of this.

Firstly, all criteria will be given in their principal, triaxial stress forms. These forms refer to a stress system σ_1, σ_2 and σ_3 in which $\sigma_1 > \sigma_2 > \sigma_3$ (see Fig. 12.1). It will then be shown how to reduce each yield criterion to any two-dimensional combination of direct and shear stresses using an appropriate stress transformation taken from Chapter 11.

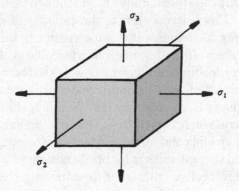

Figure 12.1 Principal triaxial stress system

12.1.1 Maximum Principal Stress Theory

The simplest criterion of yielding (William J. M. Rankine, 1820–1872) places a limit on the major principal stress σ_1. Yielding of the element in Fig. 12.1 commences when σ_1 attains the value of the tensile yield stress Y. When the minor principal stress σ_3 is compressive and of greater numerical magnitude than σ_1, Rankine's criterion places a similar limit on σ_3 such that yielding in compression commences when σ_3 attains the value $-Y$. That is:

$$\sigma_1 = Y \quad \text{or} \quad \sigma_3 = -Y \qquad (12.1\text{a,b})$$

The Rankine theory assumes, along with other *isotropic* yield criteria, that the tensile and compressive yield stresses are equal. Equations 12.1a,b ignore the influence of the intermediate principal stress σ_2. For yielding to occur under a plane biaxial stress system, where σ_x, σ_y and τ_{xy} exist in combination, the principal stresses σ_1 and σ_3 are taken from Table 11.1 (System 4, p. 453) as follows:

$$\sigma_1 = \tfrac{1}{2}(\sigma_x + \sigma_y) + \tfrac{1}{2}\sqrt{(\sigma_x - \sigma_y)^2 + 4\tau_{xy}^2} \qquad (12.2\text{a})$$

$$\sigma_3 = \tfrac{1}{2}(\sigma_x + \sigma_y) - \tfrac{1}{2}\sqrt{(\sigma_x - \sigma_y)^2 + 4\tau_{xy}^2} \qquad (12.2\text{b})$$

The minor principal stress in Eq. 12.2b is compressive and therefore identifies with σ_3 in Eq. 12.1b. Taken with $\sigma_2 = 0$, the convention: $\sigma_1 > \sigma_2 > \sigma_3$ is obeyed. Equations 12.2a,b must be reduced to predict the onset of yielding under simpler biaxial stress states. For example, by setting $\sigma_y = 0$ in Eq. 12.2a, Rankine predicts that yielding under combined tension-torsion occurs when:

$$\tfrac{1}{2}\sigma_x + \tfrac{1}{2}\sqrt{\sigma_x^2 + 4\tau_{xy}^2} = Y \tag{12.2c}$$

Equation 12.2c reduces to $\sigma_x = Y$ correctly for simple tensile yielding. However, its prediction, $\tau_{xy} = k = Y$ for yielding under simple torsion, is unrealistic. Despite this and its obvious failure to account for the influences of the minor and intermediate principal stresses upon triaxial yielding, Rankine's criterion is appropriate for strength failures that are principal-stress directed. Such behaviour has been observed in copper and nickel base alloys where cavities form and grow under creep conditions eventually linking into cracks aligned with the major principal plane. 'Creep' in these materials refers to their accumulation of strain with the passage of time as they endure constant loading at moderately high temperatures.

12.1.2 Maximum Principal Strain Theory

Barre de St. Venant (1797–1886) postulated that a material yields under multi-axial straining when the principal strain of greatest magnitude equals the uniaxial strain at the elastic limit. This criterion employs the larger numerical value between the major or the minor principal strains, so recognising the possibility that $|\varepsilon_3| > |\varepsilon_1|$ whilst satisfying $\varepsilon_1 > \varepsilon_2 > \varepsilon_3$. Thus, the limiting tensile strain (Y/E) is made equal to either the major or the minor principal strain. For yielding to occur under a principal triaxial stress state (see Fig. 12.1), the yield strain under simple tension and compression sets the limit, where from Eqs 4.32a,b, respectively:

$$\varepsilon_1 = \frac{Y}{E} = \frac{1}{E}\left[\sigma_1 - v(\sigma_2 + \sigma_3)\right] \tag{12.3a}$$

$$\varepsilon_3 = -\frac{Y}{E} = \frac{1}{E}\left[\sigma_3 - v(\sigma_1 + \sigma_2)\right] \tag{12.3b}$$

The question as to which yield strain Y/E is reached first depends upon the relative magnitudes of the principal stresses in Eqs 12.3a,b. It follows that the St. Venant yield criterion is based upon one of the following:

$$\sigma_1 - v(\sigma_2 + \sigma_3) = Y \tag{12.4a}$$

$$\sigma_3 - v(\sigma_1 + \sigma_2) = -Y \tag{12.4b}$$

Of the two Eqs 12.4a,b that which has the greater magnitude of left-hand side will predict the multi-axial yield condition. In one reduction of Eqs 12.4a,b to a general biaxial stress state where $\sigma_2 = 0$, then σ_1 and σ_3 are given by Eqs 12.2a,b. For a particular combination of a tensile stress and a shear stress (σ_y, τ_{xy}), where $\sigma_y = 0$, Eqs 12.4a,b become

$$(1 - v)\sigma_x \pm (1 + v)\sqrt{\sigma_x^2 + 4\tau_{xy}^2} = \pm 2Y \qquad (12.4c)$$

which reduces correctly to the condition required $(\sigma_x = Y)$ for simple tension. For simple shear and pure torsion, Eq. 12.4c predicts that yielding occurs for a shear stress magnitude: $k = Y/(1 + v)$. This value is known to overestimate the shear yield stress which experiment shows lies within the range $Y/2$ and $Y/\sqrt{3}$.

12.1.3 Maximum Shear Stress Theory

This more useful theory, attributed variously to Coloumb (1773), Tresca (1868) and Guest (1900), takes yielding to begin when the maximum shear stress reaches a critical value. Correspondingly, the latter is taken as the maximum shear stress k, at the yield point in simple tension (or compression). Now, from Table 11.1 (System 1, p. 453), it follows that $k = Y/2$, which acts along planes at 45° to the uniaxial stress axis (see Fig. 12.2a). When for Fig. 12.1 $\sigma_1 > \sigma_2 > \sigma_3$ then the greatest shear stress is $\tau_{max} = \frac{1}{2}(\sigma_1 - \sigma_3)$ and this acts along a plane inclined at 45° to the 1 and 3 directions (see Fig. 12.2b).

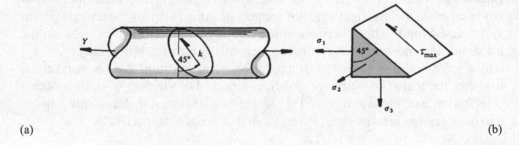

(a)   (b)

Figure 12.2 Planes of maximum shear stress

Equating τ_{max} to k, leads to a yield criterion that normally bears Tresca's name:

$$\sigma_1 - \sigma_3 = Y \qquad (12.5a)$$

The numerical values of σ_1 and σ_3 must be substituted into the left-hand side of Eq. 12.5a with the signs appropriate to tension and compression, e.g. if $\sigma_1 = 200$ MPa, $\sigma_2 = -100$ MPa and $\sigma_3 = -150$ MPa, then the left-hand side of Eq. 12.5a becomes: $200 - (-150`) = 350$ MPa. Since the intermediate stress value is always absent, Eq. 12.5a is remembered simply as:

Greatest Principal Stress - Least Principal Stress = Tensile Yield Stress   (12.5b)

For yielding under a general biaxial stress condition, Eqs 12.2a,b are substituted into Eq. 12.5b to give:

$$(\sigma_x - \sigma_y)^2 + 4\tau_{xy}^2 = Y^2 \qquad (12.6a)$$

Where $\sigma_y = 0$ in Eq. 12.6a but σ_x and τ_{xy} exist, the Tresca criterion becomes

$$\sigma_x^2 + 4\tau_{xy}^2 = Y^2 \qquad (12.6b)$$

It will be shown in § 12.3 that Eq. 12.6b has many applications where tension and torsion are combined. Eq. 12.6b predicts the shear yield stress in pure torsion as:

$$k = Y/2 \qquad (12.6c)$$

It will be seen that Eq. 12.6c, along with Tresca's other predictions to yielding under combined stresses, usually provide the most conservative (safest) estimate compared to other criteria considered here.

12.1.4 Total Strain Energy Theory

The theory of Beltrami (1885) and Haigh (1920) assumes that yielding of the element in Fig. 12.1 commences when the total strain energy stored attains the value of the strain energy for uniaxial yielding. For the principal stress system the strain energy density (see Eq. 10.4b) becomes:

$$u = \frac{U}{V} = \int_\varepsilon \sigma_i \, d\varepsilon_i$$

where the repeated subscript implies a summation for $i = 1, 2$ and 3:

$$u = \int_{\varepsilon_1} \sigma_1 \, d\varepsilon_1 + \int_{\varepsilon_2} \sigma_2 \, d\varepsilon_2 + \int_{\varepsilon_3} \sigma_3 \, d\varepsilon_3 \qquad (12.7a)$$

Equation 12.7a represents the sum of the areas beneath each principal stress–strain plot in Figs 12.3a–c.

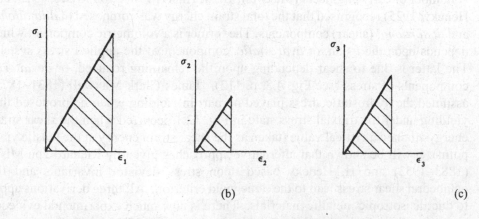

(a)   (b)   (c)

Figure 12.3 Principal, triaxial components of the strain energy density

For linear elasticity in each plot this sum becomes

$$u = \tfrac{1}{2}(\sigma_1\varepsilon_1 + \sigma_2\varepsilon_2 + \sigma_3\varepsilon_3) \qquad (12.7b)$$

Substituting Eqs 4.32a,b,c into Eq. 12.7b leads to an equivalent u expression in terms of principal stresses:

$$u = \frac{1}{2E}(\sigma_1^2 + \sigma_2^2 + \sigma_3^2) - \frac{v}{E}(\sigma_1\sigma_2 + \sigma_2\sigma_3 + \sigma_1\sigma_3) \qquad (12.7c)$$

Setting $\sigma_1 = Y$ and $\sigma_2 = \sigma_3 = 0$, Eq. 12.7c gives $u = Y^2/2E$ corresponding to the yield point for simple tension. Equating u for tension to Eq. 12.7c leads to Haigh's total strain energy criterion:

$$(\sigma_1^2 + \sigma_2^2 + \sigma_3^2) - 2v(\sigma_1\sigma_2 + \sigma_2\sigma_3 + \sigma_1\sigma_3) = Y^2 \qquad (12.8a)$$

Note here that all three principal stresses contribute to yielding. For principal biaxial stresses σ_1 and σ_3, $(\sigma_2 = 0)$, Eq. 12.8a becomes:

$$\sigma_1^2 - 2v\sigma_1\sigma_3 + \sigma_3^2 = Y^2 \qquad (12.8b)$$

Substituting σ_1 and σ_3 from Eq. 12.2a,b into Eq. 12.8b leads to a general plane stress yield criterion. In particular, when $\sigma_y = 0$, yielding arising from combining axial loading with shear commences when:

$$\sigma_x^2 + 2(1 + v)\tau_{xy}^2 = Y^2 \qquad (12.8c)$$

Equation 12.8c predicts the shear yield stress $k = Y/\sqrt{2}(1 + v)$.

12.1.5 Distortion Energy Theory

A number of early engineers (Maxwell, 1856; Huber, 1904; von Mises, 1913; and Hencky, 1925) recognised that the total strain energy was composed of *dilatational* and *distortional* (shear) components. The former is a volumetric component which depends upon the mean, or *hydrostatic*, component of the applied stress system. The latter is due to shear depending upon the remaining reduced, or *deviatoric*, components of stress (see Fig. 4.8, p. 112). James Clark Maxwell (1831–1879) assumed the hydrostatic stress played no part in yielding when he proposed that yielding under the triaxial stress state in Fig. 12.1 occurred when the shear strain energy attained a critical value (taken as the shear strain energy at the tensile yield point). It will be shown that alternative approaches given by Richard von Mises (1883–1953) and H. Hencky based upon stress deviator invariants and the octahedral shear stress lead to the same yield criterion. All three derivations apply to ductile isotropic metallic materials. There is now much experimental evidence to support the popular *von Mises yield criterion* (see §. 12.2 and 12.3).

(a) Shear Strain Energy

Referring to Fig. 4.8b, the mean (or hydrostatic) stress σ_m results in a volume change for which the strain energy density is given by:

$$u_v = \int_\varepsilon \sigma_m \, d\varepsilon + \int_\varepsilon \sigma_m \, d\varepsilon + \int_\varepsilon \sigma_m \, d\varepsilon$$

$$= \frac{1}{2}\left(\sigma_m \varepsilon + \sigma_m \varepsilon + \sigma_m \varepsilon\right) \tag{12.9a}$$

where, from Eq. 4.16c, $\varepsilon = \sigma_m/3K$ for the strain in each principal direction. Then using Eq. 4.30, Eq. 12.9a becomes:

$$u_v = \frac{\sigma_m^2}{2K} = \frac{3(1-2v)\sigma_m^2}{2E} \tag{12.9b}$$

Substituting Eq. 4.17a into Eq. 12.9b:

$$u_v = \frac{(1-2v)}{6E}(\sigma_1 + \sigma_2 + \sigma_3)^2 \tag{12.9c}$$

Subtracting Eq. 12.9c from Eq. 12.7c leads to the shear strain energy:

$$u_s = u - u_v$$

$$u_s = \frac{1}{2E}(\sigma_1^2 + \sigma_2^2 + \sigma_3^2) - \frac{v}{E}(\sigma_1\sigma_2 + \sigma_2\sigma_3 + \sigma_1\sigma_3) - \frac{(1-2v)}{6E}(\sigma_1 + \sigma_2 + \sigma_3)^2$$

$$= \frac{(1+v)}{6E}\left[(\sigma_1 - \sigma_2)^2 + (\sigma_2 - \sigma_3)^2 + (\sigma_1 - \sigma_3)^2\right] \tag{12.10a}$$

The value of u_s at the tensile yield point is found from substituting $\sigma_1 = Y$ and $\sigma_2 = \sigma_3 = 0$ in Eq. 12.10a:

$$u_s = \frac{(1+v)}{3E}Y^2 \tag{12.10b}$$

Equating 12.10a and 12.10b provides the usual form of the *von Mises* yield criterion in terms of the three principal stresses:

$$(\sigma_1 - \sigma_2)^2 + (\sigma_2 - \sigma_3)^2 + (\sigma_1 - \sigma_3)^2 = 2Y^2 \tag{12.11a}$$

If the intermediate principal stress $\sigma_2 = 0$, is zero Eq. 12.11a reduces to its principal biaxial form:

$$\sigma_1^2 - \sigma_1\sigma_3 + \sigma_3^2 = Y^2 \tag{12.11b}$$

Substituting Eqs 12.2a,b, for σ_1 (tensile) and σ_3 (compressive), Eq. 12.11b, will provide a general biaxial yield criterion in terms of σ_x, σ_y and τ_{xy}:

$$\frac{1}{4}(\sigma_x + \sigma_y)^2 + \frac{3}{4}(\sigma_x - \sigma_y)^2 + 3\tau_{xy}^2 = Y^2 \tag{12.12a}$$

Setting $\sigma_y = 0$ in Eq. 12.12a, leads to a yield criterion for an axial stress combined with a shear stress:

$$\sigma_x^2 + 3\tau_{xy}^2 = Y^2 \qquad (12.12b)$$

Subscripts x and y in Eq. 12.12b are often omitted to account for yielding under all combinations of direct stress and shear. A direct stress will arises from tension, compression and bending; a shear stress from shear force and torsion. Under simple torsion, where $\sigma_x = 0$, Eq. 12.12b predicts the shear yield stress

$$k = \frac{Y}{\sqrt{3}} \qquad (12.12c)$$

Equations 12.6c and 12.12c are very useful when shear yield stresses are to be calculated from the tensile yield stress. Taking an average k between $Y/2$ and $Y/\sqrt{3}$ provides a good estimate when, as is often the case, shear properties for a material are not listed among published data.

(b) Octahedral Shear Stress

Clark Maxwell proposed that yielding under the principal stress system in Fig. 12.1 began when the *octahedral shear stress* τ_o reached its critical value at the tensile yield point. Here τ_o acts upon all planes that are equally inclined to the principal planes to form the regular octahedron shown in Fig. 12.4a.

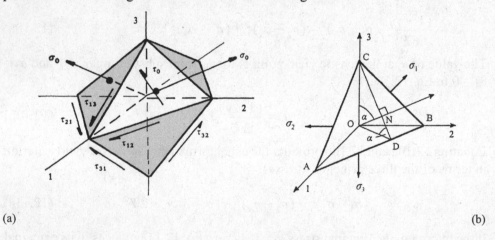

(a)   (b)

Figure 12.4 Octahedral planes

Taking one such plane, a triangle ABC with surface area A in Fig. 12.4b, its normal ON lies with equal inclination α to the principal directions 1, 2 and 3. Constructing the perpendicular CD to AB as shown, the surface areas become:

$$\text{ABC} = \frac{1}{2}\,\text{AB} \times \text{CD}, \quad \text{OAB} = \frac{1}{2}\,\text{AB} \times \text{OD} \qquad (12.13a,b)$$

Dividing Eqs 12.13a,b shows that:

$$\frac{OAB}{ABC} = \frac{OD}{CD} = \cos \alpha \qquad (12.13c)$$

and therefore the area of OAB (and the areas of OAC and OBC) are each equal to $ABC \times \cos\alpha = A\cos\alpha$. When the length of the normal ON is given as:

$$ON = \sqrt{OA^2 + OB^2 + OC^2} \qquad (12.14a)$$

its *direction cosines* are:

$$\cos\alpha = \frac{OA}{ON} = \frac{OB}{ON} = \frac{OC}{ON} \qquad (12.14b)$$

Combining Eqs 12.14a,b provides direction of normal ON as:

$$\cos\alpha = \frac{1}{\sqrt{3}}, \qquad \therefore \ \alpha = 54.74° \qquad (12.14c)$$

The tetrahedron is placed in equilibrium by three equal orthogonal forces acting upon ABC and aligned with 1, 2 and 3 such that

$$S_1 \times A = \sigma_1 \times A\cos\alpha \qquad (12.15a)$$

$$S_2 \times A = \sigma_2 \times A\cos\alpha \qquad (12.15b)$$

$$S_3 \times A = \sigma_3 \times A\cos\alpha \qquad (12.15c)$$

Expressed per unit area these forces become:

$$S_1 = \frac{1}{\sqrt{3}}\sigma_1, \quad S_1 = \frac{1}{\sqrt{3}}\sigma_2, \quad S_3 = \frac{1}{\sqrt{3}}\sigma_3 \qquad (12.16a,b,c)$$

for which their resultant is:

$$S = \sqrt{S_1^2 + S_2^2 + S_3^2} = \sqrt{\frac{1}{3}(\sigma_1^2 + \sigma_2^2 + \sigma_3^2)} \qquad (12.16d)$$

Resolving each component force into the normal direction gives the normal stress upon the octahedral plane:

$$\sigma_o = S_1\cos\alpha + S_2\cos\alpha + S_3\cos\alpha \qquad (12.17a)$$

Substituting Eqs 12.14c and 12.16a–c into Eq. 12.17a gives:

$$\sigma_o = \frac{1}{3}(\sigma_1 + \sigma_2 + \sigma_3) \qquad (12.17b)$$

which is the mean or hydrostatic stress that was recognised earlier in §. 12.1.5 as having no influence upon yielding as it produces elastic dilatation only. It follows that the remaining shear component of the resultant will control yielding. This is the octahedral shear stress τ_o which is found from S and σ_o as follows:

$$\tau_o^2 = S^2 - \sigma_o^2 \tag{12.18a}$$

in which τ_o lies tangential to the octahedral plane perpendicular to σ_o, as shown in Fig. 12.4b. Substituting Eqs 12.16d and 12.17b into 12.18a:

$$\tau_o^2 = \frac{1}{3}(\sigma_1^2 + \sigma_2^2 + \sigma_3^2) - \left[\frac{1}{3}(\sigma_1 + \sigma_2 + \sigma_3)\right]^2$$

$$= \frac{1}{9}\left[(\sigma_1 - \sigma_2)^2 + (\sigma_2 - \sigma_3)^2 + (\sigma_1 - \sigma_3)^2\right] \tag{12.18b}$$

Taking the square root of Eq. 12.18b:

$$\tau_o = \pm\frac{1}{3}\left[(\sigma_1 - \sigma_2)^2 + (\sigma_2 - \sigma_3)^2 + (\sigma_1 - \sigma_3)^2\right]^{\frac{1}{2}} \tag{12.18c}$$

in which the signs indicate the opposing sense of complementary shear on each side of the octahedral plane. The distortion which occurs at yield is due solely to τ_o. The inclinations $(\beta_1, \beta_2, \beta_3)$ of τ_o, with respect to the co-ordinates 1, 2 and 3, follow from the equilibrium equations aligned with each co-ordinate direction:

$$S_1 = \sigma_1\cos\alpha = \sigma_o\cos\alpha + \tau_o\cos\beta_1 \tag{12.19a}$$

$$S_2 = \sigma_2\cos\alpha = \sigma_o\cos\alpha + \tau_o\cos\beta_2 \tag{12.19b}$$

$$S_3 = \sigma_3\cos\alpha = \sigma_o\cos\alpha + \tau_o\cos\beta_3 \tag{12.19c}$$

Equations 12.19a–c give three direction cosines for τ_o

$$\cos\beta_1 = \frac{\sigma_1 - \sigma_o}{\sqrt{3}\,\tau_o}, \quad \cos\beta_2 = \frac{\sigma_2 - \sigma_o}{\sqrt{3}\,\tau_o}, \quad \cos\beta_3 = \frac{\sigma_3 - \sigma_o}{\sqrt{3}\,\tau_o} \tag{12.20a-c}$$

Equations 12.20a–c conform to the relationship between these directions:

$$\cos^2\beta_1 + \cos^2\beta_2 + \cos^2\beta_3 = 1$$

At yield in simple tension, $\sigma_1 = Y$ and $\sigma_2 = \sigma_3 = 0$ in Eq. 12.18c, to give $\tau_o = \sqrt{2}Y/3$. For yielding under multi-axial stress τ_o in Eq. 12.18c is limited by this critical value for tension, thereby forming the yield criterion:

$$\frac{1}{3}\left[(\sigma_1 - \sigma_2)^2 + (\sigma_2 - \sigma_3)^2 + (\sigma_1 - \sigma_3)^2\right]^{\frac{1}{2}} = \frac{\sqrt{2}Y}{3}$$

giving

$$(\sigma_1 - \sigma_2)^2 + (\sigma_2 - \sigma_3)^2 + (\sigma_1 - \sigma_3)^2 = 2Y^2$$

which is identical to Eq. 12.11a. Clearly, energy considerations are unnecessary when formulating the von Mises criterion on this basis. Here the removal of the mean stress from the principal stress state as required to formulate the von Mises yield criterion applies to the *octahedral planes* (eight in Fig. 12.4a). The following provides an alternative method by which mean stress may be removed for the derivation of this yield criterion.

(c) Deviatoric Stress Invariants

Because the mean or hydrostatic component of the stress state plays no part in yielding the latter depends on a reduced or *deviatoric*, state of stress (see Fig. 4.8c, p. 112). In its general formulation, the yield criterion is taken to be a function of the *invariants* of the deviatoric stress. These invariants are independent of the co-ordinates used to define the stress state. This must be so if yielding is to be a property of the material and not the chosen co-ordinates. Recall from Example 11.6 (p. 449) that the two roots which the determinant in Eq. 11.11 provides are the principal stresses for a general plane stress state referred to co-ordinates x and y. Expanding this determinant leads to the quadratic equation

$$\sigma^2 - (\sigma_x + \sigma_y)\sigma + (\sigma_x \sigma_y - \tau_{xy}^2) = 0$$

The two roots, σ_1 and σ_2, of this equation must be independent of the reference axes x and y. It follows that the two bracketed coefficients are the invariants. Moreover, when the principal directions become the reference axes the invariants apply to the principal stresses as $\sigma_1 + \sigma_1$ and $\sigma_1 \sigma_2$, which we may confirm numerically using the stress magnitudes in Example 11.4 say (p. 446). That is: $\sigma_1 + \sigma_1 = \sigma_x + \sigma_y$ and $\sigma_1 \sigma_2 = \sigma_x \sigma_y - \tau_{xy}^2$. Similarly, the invariants for any triaxial stress state are the coefficients of a cubic equation whose roots supply the principal stress magnitudes. For simplicity, the co-ordinates are aligned with the principal stress directions. Then, for the reduced (deviatoric) stress state in Fig. 4.8c, the principal stress deviators: σ_1', σ_2' and σ_3' must conform to the roots of a cubic equation found from expanding the determinant:

$$\det \begin{vmatrix} (\sigma_1' - \sigma') & 0 & 0 \\ 0 & (\sigma_2' - \sigma') & 0 \\ 0 & 0 & (\sigma_3' - \sigma') \end{vmatrix} = 0 \qquad (12.21a)$$

Equation 12.21a gives the *deviatoric stress cubic*:

$$\sigma'^3 - (\sigma_1' + \sigma_2' + \sigma_3')\sigma'^2 + (\sigma_1'\sigma_2' + \sigma_1'\sigma_3' + \sigma_2'\sigma_3')\sigma' - \sigma_1'\sigma_2'\sigma_3' = 0 \qquad (12.21b)$$

Equation 12.21b is written as follows:

$$\sigma'^3 - J_1'\sigma'^2 - J_2'\sigma' - J_3' = 0 \qquad (12.21c)$$

in which the *deviatoric stress invariants*, J_1', J_2' and J_3' are the coefficients:

$$J_1' = \sigma_1' + \sigma_2' + \sigma_3' \tag{12.22a}$$

$$J_2' = -\left(\sigma_1'\sigma_2' + \sigma_1'\sigma_3' + \sigma_2'\sigma_3'\right) \tag{12.22b}$$

$$J_3' = \sigma_1'\sigma_2'\sigma_3' \tag{12.22c}$$

Substituting $\sigma_i' = \sigma_i - \sigma_m$ for i = 1, 2, 3 and $\sigma_m = \frac{1}{3}(\sigma_1 + \sigma_2 + \sigma_3)$ into Eqs 12.22a–c:

$$J_1' = (\sigma_1 - \sigma_m) + (\sigma_2 - \sigma_m) + (\sigma_3 - \sigma_m) = (\sigma_1 + \sigma_2 + \sigma_3) - 3\sigma_m = 0 \tag{12.23a}$$

$$J_2' = -(\sigma_1 - \sigma_m)(\sigma_2 - \sigma_m) - (\sigma_1 - \sigma_m)(\sigma_3 - \sigma_m) - (\sigma_2 - \sigma_m)(\sigma_3 - \sigma_m)$$

$$= -(\sigma_1\sigma_2 + \sigma_1\sigma_3 + \sigma_2\sigma_3) + 3\sigma_m^2$$

$$= -\frac{1}{3}\left[(\sigma_1\sigma_2 + \sigma_1\sigma_3 + \sigma_2\sigma_3) - (\sigma_1^2 + \sigma_2^2 + \sigma_3^2)\right]$$

$$= \frac{1}{6}\left[(\sigma_1 - \sigma_2)^2 + (\sigma_2 - \sigma_3)^2 + (\sigma_1 - \sigma_3)^2\right] \tag{12.23b}$$

$$J_3' = (\sigma_1 - \sigma_m)(\sigma_2 - \sigma_m)(\sigma_3 - \sigma_m)$$

$$= \sigma_1\sigma_2\sigma_3 - \sigma_m(\sigma_1\sigma_2 + \sigma_2\sigma_3 + \sigma_1\sigma_3) + 2\sigma_m^3$$

$$= \sigma_1\sigma_2\sigma_3 - \frac{1}{3}(\sigma_1 + \sigma_2 + \sigma_3)(\sigma_1\sigma_2 + \sigma_2\sigma_3 + \sigma_1\sigma_3) + \frac{2}{27}(\sigma_1 + \sigma_2 + \sigma_3)^3$$

$$= \left(\frac{1}{3}\right)^4 \left[(2\sigma_1 - \sigma_2 - \sigma_3)^3 + (2\sigma_2 - \sigma_1 - \sigma_3)^3 + (2\sigma_3 - \sigma_1 - \sigma_2)^3\right] \tag{12.23c}$$

Since $J_1' = 0$, the cubic Eq. 12.21c becomes:

$$\sigma'^3 - J_2'\sigma' - J_3' = 0 \tag{12.24a}$$

Yielding begins when a function f of the two, non-zero deviatoric invariants in Eq. 12.24a attains a critical constant value, say C:

$$f\left(J_2', J_3'\right) = C \tag{12.24b}$$

where C is normally defined from the reduction of the chosen function to yielding under simple tension or torsion. For tensile yielding under stress Y, Eqs 12.23b,c reduce to: $J_2' = Y^2/3$, $J_3' = 2Y^3/27$. For torsional yielding under shear stress k then $J_2' = k^2$, $J_3' = 0$ (where $\sigma_1 = k$, $\sigma_2 = 0$ and $\sigma_3 = -k$ in Eqs 12.23b,c). There are many specific forms for the yield function in Eq. 12.24b. Three functions, homogenous in stress σ', have been given with the limiting shear yield stress k as follows:

$$J_2'^3 - a J_3'^2 = k^6 \qquad (12.25a)$$

$$J_2'^{\frac{3}{2}} - b J_3' = k^3 \qquad (12.25b)$$

$$J_2' - c\left(J_3'/J_2'\right) = k^2 \qquad (12.25c)$$

where a, b and c are constants. Equations 12.25a-c have been used to predict the onset of yielding in many metallic materials when placed under any manner of combined loading. When J_3' is omitted and f becomes J_2', Eq. 12.24b appears in a simplified form:

$$J_2' = k^2 \qquad (12.26a)$$

where $k^2 = Y^2/3$ is found from putting $\sigma_1 = Y$ and $\sigma_2 = \sigma_3 = 0$ in Eq. 12.23b. From Eq. 12.26a and 12.12b this yield criterion is equivalent to:

$$\frac{1}{6}\left[(\sigma_1 - \sigma_2)^2 + (\sigma_2 - \sigma_3)^2 + (\sigma_1 - \sigma_3)^2\right] = k^2 = \frac{Y^2}{3} \qquad (12.26b)$$

which again has the von Mises form (Eq. 12.11a). In this interpretation of multi-axial yielding we say that the second invariant of deviatoric stress reaches a critical value equal to its corresponding value at the yield point in simple tension.

Example 12.1 A mild steel tube has a 100 mm mean diameter and 3 mm wall thickness. Determine the torque that can be transmitted by the tube according to the five yield criteria described in § 12.1.1–§ 12.1.5 above using a safety factor $S = 2.25$. Take the tensile yield stress $Y = 230$ MPa and Poisson's ratio $v = 0.3$.

Taking k to be the shear yield stress, the working shear stress becomes: $\tau_w = k/S$. The Batho theory (Eq. 7.29b) gives the torque as:

$$T = 2At\tau_w = 2(\pi \times 50^2) \times 3 \times \tau_w = \frac{(15 \times 10^3)\pi k}{S \times 10^3} \text{ Nm} \qquad (i)$$

When applying multi-axial criteria to predict yielding under pure torsion the principal stresses will conform to $\sigma_1 > \sigma_2 > \sigma_3$ when:

$$\sigma_1 = k, \quad \sigma_2 = 0 \quad \text{and} \quad \sigma_3 = -k \qquad (ii)$$

The five yield criteria are now applied in the order in which they were explained.

(a) Maximum Principal Stress Theory

Equations 12.1a and b give: $k = Y$ and $-k = -Y$. Hence, from Eq. i:

$$T = \frac{(15 \times 10^3)(\pi \times 230)}{2.25 \times 10^3} = 4820 \text{ Nm}$$

(b) *Maximum Principal Strain Theory*.

Equations 12.4a,b reduce to:

$$\sigma_1 - v\sigma_3 = k(1 + v) = Y$$

$$\sigma_3 - v\sigma_1 = -k(1 + v) = -Y$$

Substituting from Eq. ii leads to a common result:

$$k = \frac{Y}{1 + v}$$

when, from Eq. i, the torque is:

$$T = \frac{(15 \times 10^3)(\pi \times 230)}{(1 + 0.3) \times 2.25 \times 10^3} = 3706 \text{ Nm}$$

(c) *Maximum Shear Stress Theory*

Tresca's Eq. 12.5a becomes:

$$\sigma_1 - \sigma_3 = 2k = Y$$

from which

$$k = \frac{Y}{2}$$

Substituting k into Eq. i leads to:

$$T = \frac{(15 \times 10^3)(\pi \times 230)}{2 \times 2.25 \times 10^3} = 2408 \text{ Nm}$$

(d) *Maximum Strain Energy Theory*

Haigh's Eq. 12.8b applies to torsion:

$$\sigma_1^2 - 2v\sigma_1\sigma_3 + \sigma_3^2 = Y^2$$

giving

$$k^2 - 2vk(-k) + (-k)^2 = Y^2$$

from which

$$k = \frac{Y}{\sqrt{2(1 + v)}}$$

Substituting into Eq. i:

$$T = \frac{(15 \times 10^3)(\pi \times 230)}{\sqrt{2(1 + 0.3)} \times 2.25 \times 10^3} = 2987.8 \text{ Nm}$$

(e) Maximum Shear Strain Energy Theory

The von Mises Eq. 12.11a becomes:

$$(k - 0)^2 + (0 + k)^2 + (k + k)^2 = 2Y^2$$

giving

$$k = Y/\sqrt{3}$$

From Eq. i:

$$T = \frac{(15 \times 10^3)(\pi \times 230)}{\sqrt{3} \times 2.25 \times 10^3} = 2778 \text{ Nm}$$

which is also the prediction that would be given by each of the homogenous forms in Eqs 12.25a–c. Note that Tresca provides the safest prediction. That this will often be the case can be seen from the graphical presentations of yield loci that follow. Tresca's yield locus is inscribed within most of the other yield loci and therefore will be one of the first to predict yield under combined stress states.

12.2 Principal Biaxial Stress

When the third principal stress σ_3 is zero each yield criterion given above describes a closed boundary, called the yield locus, in principal stress axes σ_1 and σ_2. Five yield loci are compared graphically within normalised axes in Fig. 12.5.

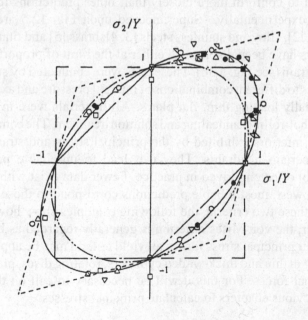

Figure 12.5 Comparison between five strength criteria

Keys :(Theory : – – – MP Stress; - - - MP Strain; —– - — Total SE; ——— Maxwell-Mises; —– · —— Tresca)
(Data: ×, ■ carbon steels [1, 2]; ∆, □ stainless steels [3]; ○, ▽, • alloy steels [3 ,4]; ¬ brass [5]; ∟ Ti-alloy [6])

Setting $\sigma_3 = 0$ in the energy-based criteria, Eqs 12.8a and 12.11a, both describe ellipses with a 45° inclination in their major axes

$$\left(\frac{\sigma_1}{Y}\right)^2 - 2v\left(\frac{\sigma_1}{Y}\right)\left(\frac{\sigma_2}{Y}\right) + \left(\frac{\sigma_2}{Y}\right)^2 = 1 \quad \text{and} \quad \left(\frac{\sigma_1}{Y}\right)^2 - \left(\frac{\sigma_1}{Y}\right)\left(\frac{\sigma_2}{Y}\right) + \left(\frac{\sigma_2}{Y}\right)^2 = 1$$

Setting $\sigma_3 = 0$ in Eqs 12.1a,b, 12.4a,b and 12.5a,b results in the square, parallelogram and hexagon loci respectively, as shown. To construct these, each equation must be applied separately to the stress state existing within each quadrant. For example, the vertical and horizontal sides of Tresca's hexagon that lie in quadrant 1 are respectively $\sigma_1/Y = 1$ and $\sigma_2/Y = 1$. This is because in Eq. 12.5b the *minor principal stress* is $\sigma_3 = 0$ when both σ_1 and σ_2 are tensile. Tresca and Rankine loci coincide in quadrants 1 and 3 but not in quadrants 2 and 4 where it is left to the reader to derive the equations for the sides, The sloping sides (dotted) of St. Venant's maximum principal strain locus that lie in quadrant 1, are expressed from Eq. 12.4a,b as:

$$\frac{\sigma_1}{Y} - \frac{v\,\sigma_2}{Y} = 1 \quad \text{and} \quad \frac{\sigma_2}{Y} - \frac{v\,\sigma_1}{Y} = 1$$

The Tresca, Rankine and St. Venant loci are discontinuous, consisting of linear segments showing corners. In contrast the Haigh and von Mises elliptical loci are smooth and continuous. Elastic conditions prevail within the interior region. Plastic behaviour occurs to the exterior of each locus. The inscribed Tresca locus provides the most conservative predictions and is therefore the safest. The von Mises ellipse has been found to conform more closely than other predictions to biaxial yield stresses found experimentally. Superimposed upon Fig. 12.5 are test data for carbon steels [1, 2], alloy and stainless steels [3, 4], brass [5] and titanium alloy [6]. The yield points have been determined either at the limit of proportionality or for a given offset strain (see Fig. 3.3b). These tests were conducted by subjecting thin-walled cylinders to different combinations of internal pressure and axial loading and also from biaxially loading thin, flat plates. All materials were in a heat-treated condition from hot-rolling, annealing and solution treatment. The comparison shows that the elastic interiors exhibited by the principal stress and strain theories are unrealistic in certain quadrants. They can lead to unreliable predictions and therefore are not often employed in practice. Fewer data exist within quadrants 2 and 4 but the lowest, most reliable predictions correspond to those of Tresca and von Mises. Of these two criteria, the following examples show how Tresca is the safer. However, the von Mises criterion is generally regarded as being the more realistic. In their principal stress forms, all yield criteria may be applied directly to predict yielding in thin and thick-walled cylinders, rotating discs, plates and blocks under orthogonal forces. For this it will be necessary to call on the appropriate theory from previous chapters to calculate principal stresses.

Example 12.2 A thin-walled cylinder 750 mm inside diameter and 15 mm wall thickness is subjected to an internal pressures of 20 bar ($\equiv$ 2 MPa). Calculate the additional safe elastic axial load that may be applied from Tresca and von Mises based upon safety factor of $S = 5$ and a tensile yield stress $Y = 300$ MPa.

In a thin-walled cylinder the radial stress is negligible compared to the circumferential and axial stresses, which are, from Eqs 4.38a,b:

$$\sigma_\theta = \frac{pd}{2t} = \frac{2 \times 750}{2 \times 15} = 50 \, \text{MPa} \tag{i}$$

$$\sigma_z = \frac{pd}{4t} + \frac{W}{\pi dt} = \frac{2 \times 750}{4 \times 15} + \frac{W}{\pi \times 750 \times 15} = 25 + \frac{W}{(35.34 \times 10^3)} \, \text{(MPa)} \tag{ii}$$

Taking $\sigma_z > \sigma_\theta > \sigma_r$ (with $\sigma_r \approx 0$), Tresca's equation, Eq. 12.5a becomes, in polar coordinates and with a safety factor S:

$$\sigma_z - \sigma_r = \frac{Y}{S}$$

when from Eq. ii

from which

$$\left[25 + \frac{W}{(35.34 \times 10^3)} \right] - 0 = \frac{300}{5}$$

$$W = 1.237 \, \text{MN}$$

The von Mises criterion (Eq. 12.11b) is written here as:

$$\sigma_\theta^2 - \sigma_\theta \sigma_z + \sigma_z^2 = \left(\frac{Y}{S} \right)^2$$

Substituting from Eqs i and ii leads to a quadratic with its positive root:

$$50^2 - 50 \left[25 + \frac{W}{(35.34 \times 10^3)} \right] + \left[25 + \frac{W}{(35.34 \times 10^3)} \right]^2 = 60^2$$

$$25 + \frac{W}{(35.34 \times 10^3)} = 66.53$$

giving

$$W = 1.468 \, \text{MN}$$

Example 12.3 Derive expressions for the pressure p required to initiate yielding in the bore of a thick-walled, closed-end cylinder with an outside/inside diameter ratio K, according to the Tresca and Mises yield criteria. Determine p for a closed cylinder with 30 mm inner diameter and 50 mm outer diameter given $Y = 250$ MPa. What is the effect of 'open ends' upon yielding when the end force is sustained by bore pistons?

Closed Ends. If we substitute separately $\sigma_r = 0$ for $r = r_o$ and $\sigma_r = -p$ for $r = r_i$ in Eq. 4.46a the two simultaneous equations can be solved for a and b. Eqs 4.48a,b, together with Eq. 4.49a provide the solution to the principal, triaxial stress state in the wall of a closed-end cylinder. This state refers to the hoop, radial and axial stress expressed as:

$$\sigma_\theta = \frac{pr_i^2}{(r_o^2 - r_i^2)}\left(1 + \frac{r_o^2}{r^2}\right) \qquad \text{(i)}$$

$$\sigma_r = \frac{pr_i^2}{(r_o^2 - r_i^2)}\left(1 - \frac{r_o^2}{r^2}\right) \qquad \text{(ii)}$$

$$\sigma_z = \frac{pr_i^2}{r_o^2 - r_i^2} \qquad \text{(iii)}$$

Since $\sigma_\sigma > \sigma_z > \sigma_r$, the Tresca criterion, Eq. 12.5b becomes:

$$\sigma_\theta - \sigma_r = Y \qquad \text{(iv)}$$

Substituting Eqs i and ii into Eq. iv:

$$\left[\frac{pr_i^2}{r_o^2 - r_i^2}\right]\left[\left(1 + \frac{r_o^2}{r^2}\right) - \left(1 - \frac{r_o^2}{r^2}\right)\right] = Y$$

$$\frac{2pr_i^2 r_o^2}{r^2(r_o^2 - r_i^2)} = Y$$

from which

$$p = \frac{Yr^2(r_o^2 - r_i^2)}{2r_i^2 r_o^2}$$

Clearly, the lowest yield pressure occurs in the cylinder bore, where for $r = r_i$:

$$p = \frac{Y}{2}\left(1 - \frac{r_i^2}{r_o^2}\right) = \frac{Y(R^2 - 1)}{2R^2} \qquad \text{(v)}$$

and when $R = r_o/r_i = 50/30 = 1.667$, Eq. v gives $p = 800$ bar.

The von Mises criterion, Eq. 12.11a is written in polar co-ordinates as:

$$(\sigma_\theta - \sigma_z)^2 + (\sigma_\theta - \sigma_r)^2 + (\sigma_r - \sigma_z)^2 = 2Y^2 \qquad \text{(vi)}$$

Substituting Eqs i–iii into Eq. vi, with $r = r_i$ leads to a condition for bore yielding

$$6p^2\left(\frac{r_o^2}{r_o^2 - r_i^2}\right)^2 = 2Y^2$$

from which

$$p = \frac{Y}{\sqrt{3}}\left(1 - \frac{r_i^2}{r_o^2}\right) = \frac{Y}{\sqrt{3}}\left(\frac{R^2 - 1}{R^2}\right) \qquad \text{(vii)}$$

and for $R = 1.667$, Eq. vii gives $p = 923.8$ bar.

Open Ends. When bore pistons seal and react internal pressure there will be no axial stress in the cylinder. It can be seen from Eq. iv that an end condition for which $\sigma_z = 0$ will not alter the Tresca prediction since σ_z will remain intermediate to σ_r and σ_θ. The von Mises criterion, Eq. vi, does depend upon the end condition by becoming:

$$\sigma_\theta^2 - \sigma_\theta \sigma_r + \sigma_r^2 = Y^2$$

Substituting from Eqs i and ii, with $r = r_i$, leads to the bore yield pressure

$$p^2 \left[\frac{(r_o^2 + r_i^2)}{(r_o^2 - r_i^2)} \right]^2 + p^2 \frac{(r_o^2 + r_i^2)}{(r_o^2 - r_i^2)} + (-p)^2 = Y^2$$

and with $R = r_o/r_i$

$$\frac{p^2}{(R^2 - 1)^2} \left[(R^2 + 1)^2 + (R^2 - 1)(R^2 + 1) + (R^2 - 1)^2 \right] = Y^2$$

giving

$$p^2 \frac{(1 + 3R^4)}{(R^2 - 1)^2} = Y^2$$

from which

$$p = \frac{Y(R^2 - 1)}{\sqrt{(3R^4 + 1)}}$$

which predicts, for $R = 1.667$, a bore yield pressure $p = 904.6$ bar. This shows that Tresca's safest common prediction will subsume both von Mises predictions in which the open cylinder appears as the weaker design.

12.3 Combined Axial and Shear Stresses

The principal stress forms of the various yield criteria given in the previous section are more convenient to apply when reduced to particular plane stress states. For example, the loading applied to many structures is defined by combined axial and shear stress states. Here it is usual to omit the subscripts x and y on σ_x and τ_{xy} to account for a similar combination within both Cartesian and polar co-ordinates. Figure 12.6 compares our five yield loci (see § 12.1.1 – § 12.1.5) in this stress space using normalised stress axes σ/Y and τ/Y. The three loci, Tresca, Haigh and von Mises, which describe ellipses (Eqs 12.6b, 12.8c and 12.12b) are circumscribed by the non-elliptical envelopes of Rankine and St Venant (Eqs 12.2c and 12.4c). The shear yield point, i.e. the intersection with the τ/Y axis, differs between each theory with Tresca predicting the smallest ratio $k/Y = 1/2$. Once again, the Tresca prediction to yielding under combined σ, τ is the most conservative of the five criteria. Yield stresses found experimentally under various combination of σ, τ are superimposed upon Fig. 12.6.

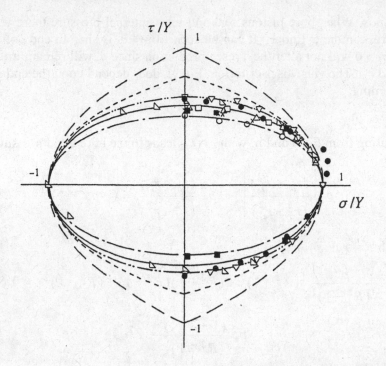

Figure 12.6 Comparison between the five strength criteria in combined σ, τ space

Keys: (Theory: – – – MP Stress; - - - - MP Strain; — - - — Total SE; —— von Mises; — - — Tresca)

(Data: ×, • carbon steels [1, 7]; ▽ Al-alloy [8]; ∟ brass [9]; ○ Ti-alloy [10]; △ Cu [11]; ■ Al [11, 12])

Data lying in quadrants 1 and 4 were obtained from subjecting thin-walled, annealed and stress-relieved tubes of ductile metallic materials to combined tension and torsion. Fewer data exist for quadrants 2 and 3 by combining compression with torsion. The materials tested include carbon steels [1, 7], aluminium alloy [8], brass [9], titanium alloy [10], copper and aluminium [11, 12]. Where no yield point was revealed a small offset strain was used to determine these yield points (as shown in Fig. 3.3b). We see that within all four quadrants most yield points lie closest to the von Mises ellipse. In practice, the σ, τ stress state arises from a number of combined loading from among: axial tension/compression with torsion or shear and bending with torsion or shear. Shear and bending arising from the transverse loading of a beam give a combination of direct stress and shear stress at points away from its neutral axis over the whole length. The following examples illustrate how yield criteria are applied to a number of these loadings.

Example 12.4 Bolt loading, shown in Fig. 12.7, consists of an axial tensile force $W = 10$ kN combined with a transverse shear force of $F = 5$ kN. Estimate, from the Tresca and von Mises yield criteria, a safe bolt diameter d when the bearing pressure between the bolt shank and the 10 mm thick contacting plate is (a) ignored and (b) accounted for. Base the design upon a safety factor of 3 given that the tensile yield stress of the bolt material is 270 MPa.

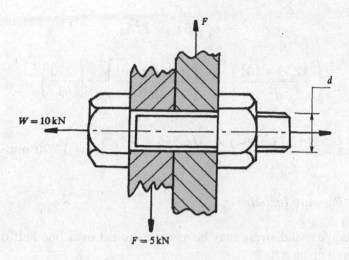

Figure 12.7 Bolt loaded under tension and shear

Within the sectional view shown let x and y define the axial and transverse directions, respectively. The axial stress is given by:

$$\sigma_x = \frac{W}{A} = \frac{(10 \times 10^3)}{(\pi d^2/4)} = \frac{(12.73 \times 10^3)}{d^2} \text{ MPa} \qquad \text{(i)}$$

and for a bolt in *single shear*, the average shear stress is:

$$\tau_{xy} = \frac{F}{A} = \frac{(5 \times 10^3)}{(\pi d^2/4)} = \frac{(6.366 \times 10^3)}{d^2} \text{ MPa} \qquad \text{(ii)}$$

(a) Bearing Pressure Ignored

Substituting Eqs i and ii into the Tresca yield criterion, Eq.12.6b for this case

$$\sigma_x^2 + 4\tau_{xy}^2 = Y^2$$

$$\left(\frac{12.73 \times 10^3}{d^2}\right)^2 + 4\left(\frac{6.366 \times 10^3}{d^2}\right)^2 = \left(\frac{270}{3}\right)^2$$

from which

$$d = \left[\frac{(12.73 \times 10^3)^2 + 4(6.366 \times 10^3)^2}{90^2}\right]^{\frac{1}{4}} = 14.15 \text{ mm}$$

Applying the von Mises criterion in Eq. 12.12b to Eqs i and ii:

$$\sigma_x^2 + 3\tau_{xy}^2 = Y^2$$

$$\left(\frac{12.73 \times 10^3}{d^2}\right)^2 + 3\left(\frac{6.366 \times 10^3}{d^2}\right)^2 = \left(\frac{270}{3}\right)^2$$

from which

$$d = \left[\frac{(12.73 \times 10^3)^2 + 3(6.366 \times 10^3)^2}{90^2}\right]^{\frac{1}{4}} = 13.70 \text{ mm}$$

(b) Bearing Pressure Included

Here an average radial stress may be assumed to act over one-half of each bolt diameter being given by:

$$\sigma_y = \frac{F}{\pi d t/2} = \frac{5 \times 10^3}{\pi d \times 10/2} = \frac{318.31}{d} \text{ MPa} \tag{iii}$$

The general plane stress form of the Tresca yield criterion in Eq. 12.6a is now applied with σ_x and τ_{xy} and σ_y taken from Eqs i and ii and iii respectively:

$$\left[\frac{(12.73 \times 10^3)}{d^2} - \frac{318.31}{d}\right]^2 + 4\left(\frac{6.366 \times 10^3}{d^2}\right)^2 = \left(\frac{270}{3}\right)^2$$

for which a trial solution gives $d = 13.1$ mm.

Making similar substitutions in the von Mises plane stress Eq. 12.12a:

$$\frac{1}{4}\left[\frac{(12.73 \times 10^3)}{d^2} + \frac{318.31}{d}\right]^2 + \frac{3}{4}\left[\frac{(12.73 \times 10^3)}{d^2} - \frac{318.31}{d}\right]^2 + 3\left(\frac{6.366 \times 10^3}{d^2}\right)^2 = \left(\frac{270}{3}\right)^2$$

for which another trial solution gives $d = 13.2$ mm, showing there is very little difference between the yield criteria. The reduced diameter provided by each analysis shows that the bearing pressure raises the hydrostatic stress component which does not influence yielding. An effect of greater significance would be the axial force upon the bolt arising from an initial clamping pressure when taken in combination with the external loading shown.

Example 12.5 A solid steel drive shaft 32 mm in diameter transmits 100 kW at 3000 rev/min whilst sustaining a bending moment M simultaneously. Calculate the allowable value of M for a safety factor of 3 according to the Tresca and von Mises yield criteria. Take $Y = 400$ MPa.

The second moments of area are required about the shaft's axis and its diameter:

$$J = \frac{\pi d^4}{32} = \frac{\pi \times 32^4}{32} = 102.943 \times 10^3 \text{ mm}^4$$

$$I = \frac{\pi \times d^4}{64} = \frac{\pi \times 32^4}{64} = 51.472 \times 10^3 \text{ mm}^4$$

Equation 7.7b allows the torque to be found from the power and the speed:

$$T = \frac{60P}{2\pi N} = \frac{60 \times (100 \times 10^3)}{2\pi \times 3000} = 318.31 \text{ Nm}$$

Equation 7.6 provides the maximum shear stress at the outer radius of the shaft under this torque

$$\tau = \frac{Tr}{J} = \frac{(318.31 \times 10^3) \times 16}{(102.943 \times 10^3)} = 49.47 \text{ MPa} \tag{i}$$

The bending stress is expressed in terms of M (Nm) from the bending Eq. 5.4:

$$\sigma = \frac{My}{I} = \frac{(M \times 10^3) \times 16}{(51.472 \times 10^3)} = \frac{M}{3.217} \text{ MPa} \tag{ii}$$

Substituting Eqs i and ii into Eq. 12.6b, Tresca gives:

$$\sigma^2 + 4\tau^2 = Y^2$$

$$\left(\frac{M}{3.217}\right)^2 + 4(49.47)^2 = \left(\frac{400}{3}\right)^2$$

from which Tresca's moment prediction is $M = 287.53$ Nm.

Substituting Eqs (i) and (ii) into Eq. 12.12b, von Mises gives:

$$\sigma^2 + 3\tau^2 = Y^2$$

$$\left(\frac{M}{3.217}\right)^2 + 3(49.47)^2 = \left(\frac{400}{3}\right)^2$$

from which $M = 328.64$ Nm, showing that Tresca's value is more conservative.

Example 12.6 Calculate the maximum uniformly distributed elastic loading a 2 m-long steel cantilever can bear with the T cross section shown in Fig. 12.8. To do so apply separately a von Mises yield criterion to: (a) the flange top, (b) the neutral axis and (c) the web top each with a tensile yield stress of 310 MPa.

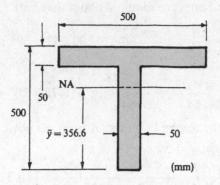

Figure 12.8 T-section showing the neutral axis

A consideration of yielding occurring at each position requires a calculation of both the bending and shear stresses. These stresses arise from the bending moment M and a shear force F that exist in beams under transverse loading:

$$\sigma = \frac{My}{I}, \quad \tau = \frac{F(A\bar{y})}{Ib} \qquad \text{(i, ii)}$$

The design requires that M and F are taken as the maximum values for this beam. Also, Eq. i refers y to the neutral axis (NA) which passes through the centroid of the section. The T-section's centroid position $\bar{y}$ from the base follows from

$$A\bar{y} = \Sigma A_i y_i$$

$$[(450 \times 50) + (500 \times 50)]\bar{y} = (450 \times 50 \times 225) + (500 \times 50 \times 475)$$

$$\bar{y} = 356.58 \text{ mm}$$

The T-section's second moment of area I is also required in both Eqs i and ii. This refers to the neutral axis (NA) as follows:

$$I = \Sigma \frac{bd^3}{3} = \frac{50 \times 356.58^3}{3} + \frac{500 \times 143.42^3}{3} - \frac{450 \times 93.42^3}{3}$$

$$I = 1125.03 \times 10^6 \text{ mm}^4$$

In Eq. ii, b is the breadth of the section at a position where the shear stress is required and $A\bar{y}$ is the first moment of area above that position. Here the first moment $A\bar{y}$ applies to areas above the neutral axis and the web top, respectively

$$A\bar{y} = (500 \times 50)118.42 + (93.42 \times 50)46.71 = 3.179 \times 10^6 \text{ mm}^3$$

$$A\bar{y} = 500 \times 50 \times 118.42 = 2.96 \times 10^6 \text{ mm}^3$$

(a) At the *flange top* (where, $\tau = 0$) $y = 143.42$ mm from the NA. Figure 2.15d shows that the maximum moment is $M = wL^2/2$ and thus the maximum bending stress is found from Eq. i:

$$\sigma = \frac{My}{I} = \frac{wL^2y}{2I} = \frac{w \times (2 \times 10^3)^2 \times 143.42}{2 \times (1125.03 \times 10^6)} = 0.255w \text{ MPa}$$

Hence the flange top begin to yield when: $0.255w = Y$. That is

$$0.255w = 310 \quad \therefore \quad w = 1216 \text{ N/mm}$$

(b) At the *neutral axis* (NA, where $\sigma = 0$) the maximum shear stress is found from Eq. ii with $b = 50$ mm and $|F| = wL$ (see Fig. 2.15d):

$$\bar{Ay} = 3.179 \times 10^6 \text{ mm}^3$$

$$\tau = \frac{wL(A\bar{y})}{Ib} = \frac{w(2 \times 10^3) \times (3.179 \times 10^6)}{(1125.03 \times 10^6) \times 50} = 0.113w \text{ MPa}$$

Taking the shear yield stress from Eq. 12.12c as $k = Y/\sqrt{3}$, the distributed load w follows as:

$$0.113w = \frac{310}{\sqrt{3}} \quad \therefore \quad w = 1584 \text{ N/mm}$$

(c) At the *web top*, where $y = 93.42$ mm and $b = 50$ mm both bending and shear stresses appear. They are found from Eqs i and ii as follows:

$$\sigma = \frac{My}{I} = \frac{wL^2y}{2I} = \frac{w \times (2 \times 10^3)^2 \times 93.42}{2 \times (1125.03 \times 10^6)} = 0.166w \text{ MPa}$$

$$\bar{Ay} = 2.96 \times 10^6 \text{ mm}^3$$

$$\tau = \frac{wL(A\bar{y})}{Ib} = \frac{w(2 \times 10^3) \times (2.96 \times 10^6)}{(1125.03 \times 10^6) \times 50} = 0.105w \text{ MPa}$$

Yielding occurs under these combined stresses when, according to the von Mises criterion in Eq. 12.12b

$$(0.166w)^2 + 3(0.105w)^2 = 310^2 \quad \therefore \quad w = 1259 \text{ N/mm}$$

Yielding starts at the flange top under the least value of the distributed loading, i.e. $w = 1216$ N/mm (kN/m). Despite the web top being near to yield, bending controls the design of a 2 m long beam. In a shorter-length cantilever, similar analyses would show how shear influences its loading capacity (see Exercise 12.38).

Example 12.7 A tubular alloy beam has a closed, thin-walled, equilateral triangular section: 30 mm side and 1 mm thick (see Fig. 12.9a). It is mounted as a short cantilever 20 mm long. Calculate the magnitude of the end load W that may be applied according to the von Mises and Tresca yield criteria when the load is applied (a) along the vertical side and (b) vertically at the shear centre. Use a safety factor of 2 with a tensile yield stress value of 180 MPa.

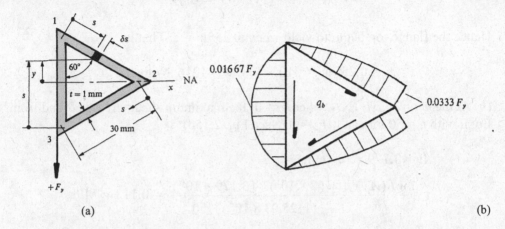

(a) (b)

Figure 12.9 Shear flow in a triangular tube

As with the previous example this design should consider both the bending moment and the shear force in the beam. Figure 2.15c shows that their maximum magnitudes are $M = WL$ and $|F| = W$, respectively. Referring to Example 1.5 (see p. 15), the second moment of area for this tube about the neutral axis x is:

$$I_x = \frac{a^3 t}{4} = \frac{30^3 \times 1}{4} = 6750 \text{ mm}^4$$

The maximum bending stress occurs in the cross section at points 1 and 3, where $y = \pm 15$ mm and is given by:

$$\sigma = \frac{My}{I_x} = \frac{WLy}{I_x} = \frac{20 W \times (\pm 15)}{6750} = \pm 0.0444 W$$

The signs refer points 1 and 3, respectively, to the equal magnitude, axial tensile and compressive bending stresses which occur at the cantilever's fixed end section under the cantilever's hogging moment. Since shear stress is absent at points 1 and 3, the allowable design stress $\sigma = 180/2 = 90$ MPa must limit each bending stress. This gives, by both the von Mises and Tresca yield criteria

$$0.0444 W = 90, \quad \therefore \quad W = 2025 \text{ N}$$

To find the magnitude of the maximum shear stress τ due to shear force in a closed, thin-walled tube a shear flow parameter q is adopted so that $\tau = qt$, where t is the tube thickness (see § 5.5.2, p. 182). The shear flow parameter also applies to the

Batho theory of torsion for this thin-walled, closed section (see § 7.3.1, p. 264), when the shear force is not applied at its shear centre. Once established, the maximum shear stress will appear within the net q distribution for the thin tube. The net shear flow is found from applying Eqs 5.39a,b with x as a principal symmetry axis for the triangular section within which a vertical shear force F_y is transmitted in a positive downward direction (see Fig. 12.9a):

$$q = q_b + q_o = \frac{F_y}{I_x} D_{xb} + q_o \tag{i}$$

The first q_b term in Eq. i refers to the basic shear flow in the tube:

$$q_b = \frac{F_y}{I_x} D_{xb} = \frac{F_y}{I_x} \int_s yt\,ds \tag{ii}$$

where the integral $\int yt\,ds$ is the first moment of a wall element (shaded in Fig. 12.9a) about the x-axis. The second q_o term is the shear flow in the wall at the origin chosen for the wall's perimeter dimension s (here point 1 in Fig. 12.9a). The magnitude of q_o will depend upon where in the cross section the shear force F_y is taken to be applied. In this question, the two load positions, given in (a) and (b), are to be examined. Mathematically, q_o provides the required constant when integrating Eq. ii for q_b without limits in the following manner.

Basic Shear Flow - q_b distribution

Working clockwise from 1, the following integration refers the first moments D_{xb} to the axis x for each side of the tube:

$1 \rightarrow 2$, $y = 15 - s/2$, $t = 1$ mm

$$\therefore D_{xb} = \int_s (15 - s/2)\,ds = 15s - s^2/4 \tag{iii}$$

   At 1, $s = 0$, $D_{xb1} = 0$

   At 2, $s = 30$ mm, $D_{xb2} = 225$ mm^3

$2 \rightarrow 3$, $y = -s/2$, $t = 1$ mm

$$\therefore D_{xb} = -\int_s (-s/2)\,ds + 225 = 225 - s^2/4 \tag{iv}$$

   At 2, $s = 0$, $D_{xb2} = 225$ mm^3

   At 3, $s = 30$ mm, $D_{xb3} = 0$

$3 \rightarrow 1$, $y = s - 15$, $t = 1$ mm

$$\therefore D_{xb} = \int_s (s - 15)\,ds = s^2/2 - 15s \tag{v}$$

At 3, $D_{xb3} = 0$

At NA, $s = 15$ mm, $D_{xb} = -112.5$ mm^3

At 1, $s = 30$ mm, $D_{xb1} = 0$

Then from multiplying these D_{xb} values by $F_y/6750$ the q_b distribution becomes that shown in Fig. 12.9b.

(a) F_y Applied Along the Vertical Side

Here F_y both bends and twists the section. To find q_o in Eq. i it is assumed to follow 1-2-3 in a clockwise direction. Taking moments about point 2 eliminates moments arising from shear flow in the sloping sides. The resultant moment, $(30 \sin 60°)F_y$ becomes the sum of the moments due to the q_b and q_o shear flows in the web. Here the q_o identifies with the Batho torque:

$$(30 \sin 60°)F_y = (30 \sin 60°)(2/3 \times 30 \times 0.01667 F_y) - 2Aq_o$$

from which

$$q_o = -0.02222 F_y$$

This shows that q_o is a constant, anticlockwise shear flow: reinforcing the basic shear flow in the web whilst reducing that in the sloping sides. Adding q_o to Fig. 12.9b in this manner, the maximum shear flow is found at the NA, midway between points 1 and 3, where $\sigma = 0$. That is:

$$q_{max} = 0.01667 F_y + 0.02222 F_y = 0.03889 F_y$$

when

$$\tau_{max} = \frac{q_{max}}{t} = 0.03889 F_y \ \text{MPa}$$

Introducing the safety factor into the Mises criterion (Eq. 12.12c):

$$\sqrt{3} \, \tau_{max} = \frac{Y}{2}$$

$$\sqrt{3} \times 0.03889 F_y = \frac{180}{2}$$

giving

$$F_y = 1335.8 \ \text{N}$$

The Tresca criterion (Eq. 12.6c) becomes:

$$2\tau_{max} = \frac{Y}{2}$$

giving

$$F_y = 1156.8 \ \text{N}$$

(b) F_y Applied at the Shear Centre

Here the beam bends without twisting. The particular q_o that applies to the shear centre appears within Eq. 5.39a as q_E. The latter is found from applying the two path integrals in Eq. 5.40a as follows:

$$\left[\int_s q_{12} \frac{ds}{t} + \int_s q_{23} \frac{ds}{t} + \int_s q_{31} \frac{ds}{t} \right] + q_E \oint_s \frac{ds}{t} = 0$$

Substituting into Eq. vi the q_b shear flows q_{12}, q_{23} and q_{31}, that follow from the corresponding D_{xb} in Eqs iii–v, gives:

$$\frac{F_y}{I_x t} \left[\int_0^{30} \left(15s - \frac{s^2}{4} \right) ds + \int_0^{30} \left(225 - \frac{s^2}{4} \right) ds + \int_0^{30} \left(\frac{s^2}{2} - 15s \right) ds \right] + (3 \times 30) \frac{q_E}{t} = 0$$

$$\frac{F_y}{I_x t} \left[\left| \frac{15s^2}{2} - \frac{s^3}{12} \right|_0^{30} + \left| 225s - \frac{s^3}{12} \right|_0^{30} + \left| \frac{s^3}{6} - \frac{15s^2}{2} \right|_0^{30} \right] + \frac{90 q_E}{t} = 0$$

All but the linear term in s cancel. Substituting its limits together with I_x gives q_E:

$$q_E = - \frac{225 \times 30\, F_y}{6750 \times 90} = - 0.0111 F_y$$

The negative shear flow opposes the clockwise direction 1-2-3 in which q_b was established. With the actual directions for q_b that this revealed in Fig. 12.9b, q_E subtracts from q_b in the sloping sides but adds to q_b in the vertical side. This is similar to the effect that q_o had upon modifying q_b in Part (a). Here again the maximum, net shear flow is found midway between points 1 and 3 but of a lesser magnitude:

$$q_{max} = 0.01667 F_y + 0.01111 F_y = 0.02778 F_y$$

when

$$\tau_{max} = \frac{q_{max}}{t} = 0.02778 F_y \text{ MPa}$$

and this permits an increased end load by each yield criterion. From von Mises:

$$\sqrt{3} \times 0.02778 F_y = \frac{180}{2}$$

giving

$$F_y = 1870.5 \text{ N}$$

and from Tresca:

$$2 \times 0.02778 F_y = \frac{180}{2}$$

giving

$$F_y = 1619.9 \text{ N}$$

Note that the analysis in (b) does not require the position of the shear centre along the x-axis but it is of interest to know where F_y should be applied for this optimal loading condition. The torque due to F_y, applied at distance e from point 2, is counterbalanced by the moment produced by the net shear flow distribution. In referring this torque-moment balance to point 2 the effect of shear flow in the sloping sides is eliminated. Consequently only the q_b and q_E shear flows in the web are required as follows:

$$F_y \times e = \frac{\sqrt{3}}{2} \times 30 \int_0^{30} (q_{31} + q_E)\, ds$$

and with the integral taken as the sum of the areas beneath each distribution:

$$F_y\, e = \frac{\sqrt{3}}{2} \times 30 \left[\left(\frac{2}{3} \times 0.01667 F_y \times 30 \right) + (0.0111 F_y \times 30) \right]$$

from which

$$e = 17.32 \text{ mm}$$

which is usually identified as $30/\sqrt{3}$. Note, finally, by equating any of the F_y values found in (a) and (b) above to the end load W, all appear less than the limiting value of 2025 N which bending provides. Here then, in contrast to the previous example, we see that shear and not bending governs the choice of an allowable load value.

12.4 Failure Criteria for Brittle Solids

Consider the intercepts that the yield loci shown in Figs 12.5 and 12.6 make with their axes. They all show equal magnitudes of tensile and compressive yield stresses ($\pm Y$) and forward and reversed shear yield stresses ($\pm k$). This is a reasonable assumption to make upon those initial yield strengths found in ductile metallic materials, particularly in their annealed condition. However, a similar assumption cannot be made for many brittle solids that are inherently *anisotropic*, e.g, where their strengths differ between tension and compression. It is worth noting here that the tensile and compressive strengths of a plastically pre-strained ductile metal can differ due to a phenomenon known as the *Bauschinger effect*. This refers to the reduction in reversed yield stress that occurs following a forward deformation into the plastic range. Such deformation induced anisotropy should not be confused with the natural initial anisotropy present in brittle solids, rock, ceramic, concrete, cast iron and dry soil. Differences between the tensile and compressive strengths in each of these materials can be accounted for with the internal friction (*Coloumb–Mohr*) and *modified Mohr* criteria. Each criterion modifies *Rankine's* major principal stress theory (see § 12.1.1, p. 488) to allow for differences between the tensile and compressive yield strengths. Figure 12.10 shows how these modifications apply to a principal biaxial stress space. The three fracture criteria coincide in quadrants 1 and 3 but in quadrants 2 and 4 it is seen that the modified Mohr prediction is less conservative than Coloumb–Mohr but more conservative than Rankine.

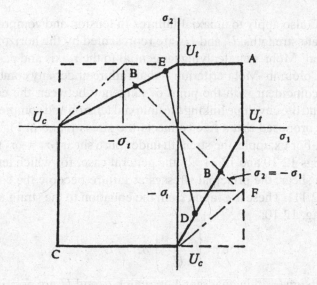

Figure 12.10 Strength criteria for brittle solids in principal biaxial stress space

Key: Coloumb–Mohr ——— ; Rankine – – – – – ; Modified Mohr —— · ——

12.4.1 Stress States by Coloumb–Mohr

Combined stress points, A, B, C, D and E, upon the Coloumb-Mohr failure locus in Fig. 12.10 take their corresponding positions upon the Mohr's circles in Figure 12.11. The latter shows that major principal stress failures under either σ_1 or σ_2 in biaxial tension and compression, appear as the extreme points A and C on their respective circles.

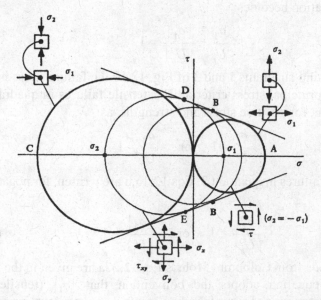

Figure 12.11 Uniaxial and biaxial strength criteria by Coloumb–Mohr

In fact, A and C also apply to uniaxial failures in tension and compression. Their different ultimate strengths, U_t and U_c, are represented by the horizontal diameter of each 'uniaxial' Mohr's circle, lying tangential to the τ-axis and passing through A and C. The Coloumb–Mohr criterion is that failure under any combined, biaxial stress state is coincident with the point of tangency between the corresponding Mohr's circle and the envelope linking the uniaxial tension and compression circles. Duguet (1885) proposed that a line connects the stress plane in Fig. 12.10 to the Mohr's circles. For example, the strength under pure shear ($\sigma_1 = -\sigma_2$) corresponds to point B in Figs 12.10 and 12.11. In the general case, for which tension (say) is combined with shear, the principal stresses at failure become the co-ordinates of point D (Fig. 12.11). These are found from the equation to the straight side passing through D in Fig. 12.10:

$$\sigma_2 = \left(\frac{|U_c|}{U_t} \right) \sigma_1 - U_c$$

Accounting for signs within quadrant 4, in which σ_1 and U_t are positive and σ_2 and U_c are both negative, this strength equation becomes:

$$\frac{\sigma_1}{U_t} + \frac{|\sigma_2|}{|U_c|} = 1 \tag{12.27a}$$

At point E in each of Figs 12.10 and 12.11 compression is combined with shear. The equation to the straight side passing through E in Fig. 12.10 becomes:

$$\sigma_2 = \left(\frac{U_t}{|U_c|} \right) \sigma_1 + U_t$$

Within quadrant 2, where both σ_2 and U_t are positive and σ_1 and U_c are both negative, this equation becomes:

$$\frac{\sigma_2}{U_t} + \frac{|\sigma_1|}{|U_c|} = 1 \tag{12.27b}$$

Within the remaining quadrants 1 and 3 of Fig. 12.10 the failure stress is given by Rankine's major principal stress criteria. For tensile failures in quadrant 1, Eqs 12.1a,b are written, for positive stress and strengths, as:

$$\frac{\sigma_1}{U_t} = 1 \quad \text{or} \quad \frac{\sigma_2}{U_c} = 1 \tag{12.27c,d}$$

For compressive failures in quadrant 3, Eqs 12.1a,b are written, for negative stress and strengths as:

$$\frac{|\sigma_1|}{|U_c|} = 1 \quad \text{or} \quad \frac{|\sigma_2|}{|U_t|} = 1 \tag{12.27e,f}$$

Fracture predictions from Coloumb–Mohr's Eq. 12.27a are given in the following examples. This equation adopts the convention that $|\sigma_1|$ (tensile) $> |\sigma_2|$ (compressive), irrespective of signs for when $\sigma_3 = 0$. Less conservative predictions are available from the two alternative theories referred to as 'Rankine' and

'Modified Mohr' in Fig. 12.10. The latter predicts failure more accurately than Rankine and Coloumb-Mohr under combined stress states within quadrants two and four in Fig. 12.10, where they are compared graphically.

Example 12.8 A brittle material is to be subjected to a direct stress of magnitude $\sigma = 75$ MPa and a shear stress τ. What value of τ would cause failure when the direct stress is (a) tensile and (b) compressive? The tensile and compressive yield strengths for the material are 100 and 200 MPa, respectively.

(a) When σ is tensile the principal stresses are given by Eqs 12.2a,b as:

$$\sigma_1 = \frac{\sigma}{2} + \frac{1}{2}\sqrt{\sigma^2 + 4\tau^2} = 37.5 + \sqrt{37.5^2 + \tau^2} \quad \text{(tensile)}$$

$$\sigma_2 = \frac{\sigma}{2} - \frac{1}{2}\sqrt{\sigma^2 + 4\tau^2} = 37.5 - \sqrt{37.5^2 + \tau^2} \quad \text{(compressive)}$$

Knowing that σ_2 is negative its substitution into Eq. 12.27a requires that it be made positive with a sign change:

$$\frac{37.5 + \sqrt{37.5^2 + \tau^2}}{100} - \frac{37.5 + \sqrt{37.5^2 + \tau^2}}{200} = 1$$

$$3\sqrt{37.5^2 + \tau^2} = 200 + 37.5 - 75$$

from which

$$\tau = 39.09 \text{ MPa}$$

when

$$\sigma_1 = 91.67 \text{ MPa}, \quad \sigma_2 = -16.67 \text{ MPa}$$

This principal stress combination corresponds to the intersection point P shown in Fig. 12.12.

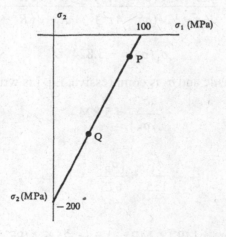

Figure 12.12 Quadrant four fracture line by Coloumb–Mohr

(b) When σ is compressive the principal stresses are given by Eqs 12.2a,b as:

$$\sigma_1 = \frac{\sigma}{2} + \frac{1}{2}\sqrt{\sigma^2 + 4\tau^2} = -37.5 + \sqrt{37.5^2 + \tau^2} \quad \text{(tensile)}$$

$$\sigma_2 = \frac{\sigma}{2} - \frac{1}{2}\sqrt{\sigma^2 + 4\tau^2} = -37.5 - \sqrt{37.5^2 + \tau^2} \quad \text{(compressive)}$$

The substitution into Eq. 12.27a again requires a sign change for σ_2:

$$\frac{-37.5 + \sqrt{37.5^2 + \tau^2}}{100} + \frac{37.5 + \sqrt{37.5^2 + \tau^2}}{200} = 1$$

$$3\sqrt{37.5^2 + \tau^2} = 200 - 37.5 + 75$$

from which

$$\tau = 69.72 \text{ MPa}$$

when

$$\sigma_1 = 41.67 \text{ MPa}, \quad \sigma_2 = -116.67 \text{ MPa}$$

This combination corresponds to the intersection point Q in Fig. 12.12. Combining a direct stress with shear usually results in a compressive minor principal stress, an assumption that is true in each of (a) and (b) above. If, however, the analysis reveals that both principal stresses are either tensile or compressive then the failure will lie in their respective quadrants 1 and 3, for which the failure criteria in Eqs 12.27c–f apply.

Example 12.9 A brittle material is loaded under a constant stress ratio $R = \sigma/\tau = 2$. Determine the stress combination at the point of failure given ultimate strengths in tension and compression: $U_t = 155$ MPa and $U_c = 675$ MPa.

The corresponding principal stress ratio is from Eqs 12.2a,b:

$$\frac{\sigma_1}{\sigma_2} = \frac{\frac{1}{2}\sigma + \frac{1}{2}\sqrt{(\sigma^2 + 4\tau^2)}}{\frac{1}{2}\sigma - \frac{1}{2}\sqrt{(\sigma^2 + 4\tau^2)}} = \frac{R + \sqrt{(R^2 + 4)}}{R - \sqrt{(R^2 + 4)}}$$

and for $R = 2$:

$$\sigma_1/\sigma_2 = -5.824 \tag{i}$$

Assuming that σ_1 is tensile and σ_2 is compressive, Eq. i is written as:

$$\frac{\sigma_1}{|\sigma_2|} = 5.824 \tag{ii}$$

and Eq. 12.27a becomes:

$$\frac{\sigma_1}{155} + \frac{|\sigma_2|}{675} = 1 \tag{iii}$$

Solving Eqs ii and iii:

$$\sigma_1 = 149.12 \text{ MPa}, \quad |\sigma_2| = 25.6 \text{ MPa}$$

and from Eq. 12.2a

$$\frac{\sigma_1}{\tau} = \frac{R}{2} + \frac{1}{2}\sqrt{R^2 + 4} = 2.4142$$

giving the required stress combination

$$\tau = \frac{\sigma_1}{2.4142} = \frac{149.12}{2.4142} = 61.77 \text{ MPa}$$

$$\sigma = R\tau = 2 \times 61.77 = 123.54 \text{ MPa}$$

Assuming that σ_1 is compressive and σ_2 is tensile, Eq. i becomes:

$$\frac{|\sigma_1|}{\sigma_2} = 5.824 \tag{iv}$$

And Eq. 12.27b is written as:

$$\frac{\sigma_2}{155} + \frac{|\sigma_1|}{675} = 1 \tag{v}$$

Solving Eqs iv and v gives:

$$\sigma_2 = 66.31 \text{ MPa}, \quad |\sigma_1| = 386.21 \text{ MPa}$$

from which

$$\tau = \frac{\sigma_1}{2.4142} = \frac{-386.21}{2.4142} = -159.98 \text{ MPa}$$

$$\sigma = R\tau = 2 \times (-159.98) = -319.96 \text{ MPa}$$

This is also the solution for $R = -2$, i.e. with a reversal in the direction of τ.

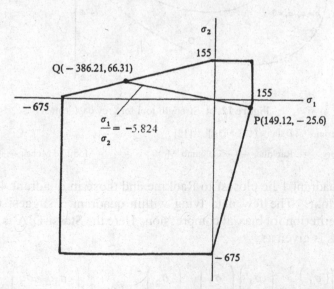

Figure 12.13 Coulomb–Mohr locus showing failure points P and Q under a combined stress path

The solutions show that the material is stronger when shear is combined with a direct compressive stress. Figure 12.13 shows that principal stress values from the two solutions. They appear at the respective intersection points (P and Q) between the stress path: $\sigma_1/\sigma_2 = -5.824$ and the strength locus in quadrants four and two.

12.4.2 Comparisons with Experiment

Figure 12.14 shows experimental results for fracture of grey cast iron under various combinations of principal, biaxial stress [13–15]. These results appear to suggest that different fracture criteria apply within the quadrants the data occupies.

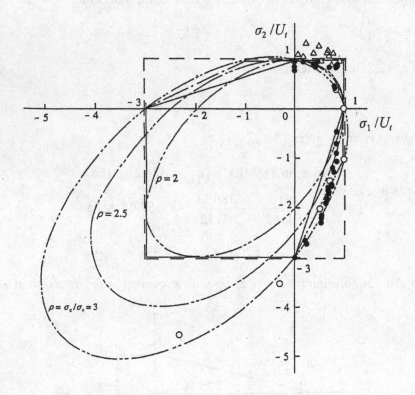

Figure 12.14 Strength loci for grey cast iron

Keys: Experimental Data: ○ [13], ● [14], △ [15]

Theory: – – – Rankine; ——— Coloumb–Mohr ; ——·—— Modified Mohr ; ——··—— Stassi

Points in quadrant 1 lie closest to Rankine and those in quadrant 4 lie closest to modified Mohr. The few data lying within quadrant 3 suggest the use of an alternative criterion for biaxial compression. Here the Stassi-D'Alia [16] elliptical failure locus is given as:

$$\left(\frac{\sigma_1}{U_t}\right)^2 - \left(\frac{\sigma_1}{U_t}\right)\left(\frac{\sigma_2}{U_t}\right) + \left(\frac{\sigma_2}{U_t}\right)^2 + (\rho - 1)\left(\frac{\sigma_1}{U_t} + \frac{\sigma_2}{U_t}\right) = \rho \quad (12.28a)$$

where $\rho = |U_c|/U_t$, as a positive ratio between tensile and compressive strengths, has been taken as 3. Equation 12.28a appears as the principal stress form of a general invariant function for brittle failure (see Example 12.10):

$$J_2' + \alpha J_1 = C \tag{12.28b}$$

where α and C are constants and J_2' is given in Eq. 12.23b. The inclusion of the first invariant in Eq. 12.28b admits a fracture dependent upon the hydrostatic component of the applied stress (the mean stress σ_m in Fig. 4.8b). This is because

$$J_1 = \sigma_1 + \sigma_2 + \sigma_3 \tag{12.29a}$$

when, the mean stress follows

$$\sigma_m = \frac{1}{3}(\sigma_1 + \sigma_2 + \sigma_3) = \frac{J_1}{3} \tag{12.29b}$$

Equation 12.29a applies to absolute stress and not deviatoric stress components, since Eq. 12.23a shows that $J_1' = 0$.

Example 12.10 Reduce the invariant form of the fracture function in Eq. 12.28b to (a) the Stassi-D'Alia criterion for failure under biaxial principal stress and (b) a criterion for brittle failure under combined axial loading and shear.

(a) Substituting Eqs 12.23b and 12.29a into Eq. 12.28b leads to a fracture criterion for cast iron under principal triaxial stresses:

$$\frac{1}{6}\left[(\sigma_1 - \sigma_2)^2 + (\sigma_2 - \sigma_3)^2 + (\sigma_1 - \sigma_3)^2\right] + \alpha(\sigma_1 + \sigma_2 + \sigma_3) = C \tag{i}$$

where the constant C is found from failure under a uniaxial condition, i.e. setting $\sigma_2 = \sigma_3 = 0$ in Eq. i:

$$\frac{1}{6}(2\sigma_1^2) + \alpha\sigma_1 = C \tag{ii}$$

The roots of Eq. ii must become the tensile and compressive strengths, i.e. $\sigma_1 = U_t$ and $\sigma_1 = -U_c$, respectively. This means:

$$(\sigma_1 - U_t)(\sigma_1 + U_c) = \sigma_1^2 + 3\alpha\sigma_1 - 3C \tag{iii}$$

Equating the coefficients in Eq. iii (see Exercise 12.49) gives:

$$\alpha = \frac{1}{3}(U_c - U_t) \qquad C = \frac{1}{3}U_t U_c$$

Substituting into Eq. i with $\sigma_3 = 0$, leads to the Stassi-D'Alia criterion Eq. 12.28a.

(b) Substituting Eqs 12.2a,b into Eq.12.28a leads to a failure criterion for when axial stress is combined with shear (here σ_3 replaces σ_2 and $\sigma_y = 0$):

$$\sigma^2 + 3\tau^2 + \sigma U_t(\rho - 1) = \rho U_t^2 \tag{iv}$$

When $\rho = U_c/U_t = 1$, Eq. iv recovers the von Mises locus (Eq. 12.12b). This amounts to an omission of J_1 in Eq. 12.28b when it is known that hydrostatic stress does not influence fracture.

12.5 Failure Criteria for Fibrous Materials

Looking again at Fig. 12.14, whilst all failure criteria for brittle solids admit differences between tensile and compressive strengths they do not account for the strength differences between perpendicular directions 1 and 2. Many directional materials occurring in nature and those that are man-made, including wood and reinforced composites, require a description of their inherent anisotropy. An obvious example is where the tensile strength aligned with the grain (fibres) is many times greater than the transverse strength, these being the material directions 1 and 2 respectively. Coupled to this is the likelihood that tensile and compressive strengths aligned with the grain (fibres) will differ between the fibre breakage and buckling stress that defines their respective failures. The literature is abundant with uniaxial failure criteria that account for directional strength variations in anisotropic materials. The review that follows is restricted to plane 2D stress criteria both isotropic and anisotropic. Isotropic criteria for brittle fracture are identical to the yield criteria for metals considered above. Plane anisotropic failure criteria also admit directional strength variations but are few in number. Their applications to a woven glass mat and a uni-directional carbon-fibre reinforced composite follow.

12.5.1 Plane Principal Stress Criteria

Normal stresses aligned with the material directions 1 and 2 in the absence of shear stress are, by definition, the principal stresses σ_1 and σ_2. Experimentally, the corresponding 2D failure locus, σ_1 versus σ_2, is established from radial or stepped loadings using combinations of internal pressure, pure shear and biaxial tension applied to tube and plate. Similar test procedures were adopted in the determination of the yield loci for metals where von Mises and Tresca apply (see § 12.2, p. 501).

12.5.2 Plane Symmetrical Functions

Consider, firstly, a restricted class of failure criteria in which the tensile and compressive strengths are the same for the 1-direction and the 2-direction but dissimilar between 1 and 2. Symbolically, this condition is expressed as: $\sigma_{1c} = -\sigma_{1t}$ and $\sigma_{2c} = -\sigma_{2t}$ but with $\sigma_{1c} \neq \sigma_{2c}$ and $\sigma_{1t} \neq \sigma_{2t}$. Simplifying, σ_{1f} denotes both tensile and compressive strengths for direction 1 and σ_{2f} the same for direction 2. Non-dimensional variables for the applied stresses: $x = \sigma_1/\sigma_{1f}$ and $y = \sigma_2/\sigma_{2f}$, are positive for tension, negative for compression, within each prediction given in Table 12.1. Here the stress-based functions 1–4 employ the strengths σ_{1f} and σ_{2f} aligned with the principal stress directions. The elastic constants E and v, which appear within the strain-based functions 5 and 6, conform to a relation $v_2/v_1 = E_2/E_1$.

Table 12.1 Symmetrical Failure Functions

ID	Function f	Reference
1.	$x^2 + y^2 - (\sigma_{2f}/\sigma_{1f})xy = 1$	Hill [17]
2.	$x^2 + y^2 + c(\sigma_{2f}/\sigma_{1f})xy = 1$	Griffith and Baldwin [18]
3.	$x^2 + y^2 = 1$	Norris A [19]
4.	$x^2 - xy + y^2 = 1$	Norris B [20]
5.	$y = (\sigma_{1f}/\sigma_{2f})(x - 1)/v_1$	Waddoups [21]
6.	$y = 1 + v_1(E_2/E_1)(\sigma_{1f}/\sigma_{2f})x$	Waddoups [21]

Consider a woven glass-fibre, reinforced-epoxy matrix composite, with equal tensile and compressive strengths where: $\sigma_{1f} = 307$ MPa and $\sigma_{2f} = 264$ MPa. Elastic constants aligned with the fibres are: $E_1 = 18$ GPa, $E_2 = 19.2$ GPa and $v_1 = 0.143$. Using these material constants, a comparison between the failure criteria in Table 12.1 is made in Fig. 12.15.

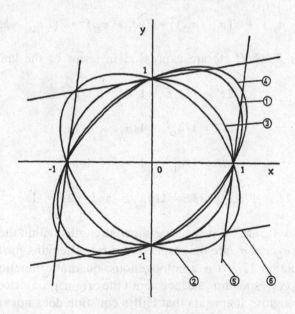

Figure 12.15 Symmetrical failure loci (**Key**: see Table 12.1)

This shows that the elliptical loci 1 and 2 [17, 18] have contrasting orientations. However, the inclination of Griffin and Baldwin's ellipse 2 can be altered with an additional material property, as required to determine the constant c. Hill's description of transverse isotropy (locus 1) lies between the Norris's circle 3 [19] and ellipse 4 [20] with a similar inclination. The circular locus 3 [19] appears relatively safe across all four quadrants compared to other stress-based predictions.

Waddoups, [21] strain-based failure envelope is formed from the two straight lines 5 and 6. The expression for each line, given in Table 12.1, requires both strength and elasticity properties. All loci 1–6 appear in Fig. 12.15 with unit intercepts along the normalised axes x and y. Taking the shear strength as 250 MPa gives $c = 0.84$, which allows lower strengths than other predictions in quadrants 1 and 3. The moduli employed in the derivation of this theory (locus 2) imply that linear anisotropic elasticity prevails to the point of failure. The assumption that a sudden fracture interrupts a linear stress versus strain response will not apply universally.

12.5.3 Three-Dimensional Strength Criteria

Further plane fracture criteria involve an appropriate reduction to more general forms of three-dimensional criteria but only a few are available. In the first of these, Hill's 1948 criterion [17] accounts for yielding and fracture under any triaxial stress state. Though originally proposed as the yield criterion for anisotropic metals, Hill's versatile function has been adopted by many as a failure criterion for brittle anisotropic solids. The six, independent stress components appear in the *Hill failure function f* as:

$$2f(\sigma_{ij}) = F(\sigma_{22} - \sigma_{33})^2 + G(\sigma_{11} - \sigma_{33})^2 + H(\sigma_{11} - \sigma_{22})^2 + 2(L\sigma_{23}^2 + M\sigma_{13}^2 + N\sigma_{12}^2) = 1$$
$$(12.30a)$$

where constants F, G, H etc are expressed in terms of the known orthotropic strengths:

$$2F = 1/\sigma_{2f}^2 + 1/\sigma_{3f}^2 - 1/\sigma_{1f}^2$$

$$2G = 1/\sigma_{3f}^2 + 1/\sigma_{1f}^2 - 1/\sigma_{2f}^2 \qquad (12.30b\text{-}d)$$

$$2H = 1/\sigma_{1f}^2 + 1/\sigma_{2f}^2 - 1/\sigma_{3f}^2$$

$$2L = 1/\sigma_{23f}^2, \quad 2M = 1/\sigma_{13f}^2 \quad \text{and} \quad 2N = 1/\sigma_{12f}^2 \qquad (12.30e\text{-}g)$$

Here σ_{12f}, σ_{23f} and σ_{13f} are the respective shear strengths within the planes 1-2, 2-3 and 1-3 and σ_{1f}, σ_{2f} and σ_{3f} are the respective direct strengths for the 1-, 2- and 3-directions. Equation 12.30a is a homogenous, quadratic function, in which the symmetry of the corresponding surface about the origin, provides equal tensile and compressive strengths. It appears that Hill's equation does not stipulate whether uniaxial strengths σ_{1f}, σ_{2f} and σ_{3f} in Eqs 12.30b–d are tensile or compressive. We have seen, from applying other criteria in § 12.4 to brittle solids, that where the two strengths differ, σ_{1f} has been replaced by σ_{1t} or $-\sigma_{1c}$ and σ_{2f} by σ_{2t} or $-\sigma_{2c}$, according to the sign of σ_{11} and σ_{22}, respectively. We may interpret Eq. 12.30a in this manner only when it is applied separately to individual quadrants of the failure envelope. Thus Eq. 12.30a retains its capability to predict fracture across all four quadrants though it is less convenient to apply compared to the following alternatives that provide a single, continuous failure criterion.

Two such alternatives [22, 23] have been proposed in which linear stress terms are added to the quadratic terms in Eq. 12.30a. In the first of these Hoffmann [22] proposed the criterion in six-dimensional stress space:

$$C_1(\sigma_{22} - \sigma_{33})^2 + C_2(\sigma_{33} - \sigma_{11})^2 + C_3(\sigma_{11} - \sigma_{22})^2$$
$$+ C_4\sigma_{11} + C_5\sigma_{22} + C_6\sigma_{33} + C_7\sigma_{23}^2 + C_8\sigma_{13}^2 + C_9\sigma_{12}^2 = 1 \qquad (12.31a)$$

The stress dimension is reduced to three with principal stresses as co-ordinates. Thus, the principal, triaxial stress failure criterion follows from setting the shear stress terms to zero in Eq. 12.31a:

$$C_1(\sigma_2 - \sigma_3)^2 + C_2(\sigma_3 - \sigma_1)^2 + C_3(\sigma_1 - \sigma_2)^2 + C_4\sigma_1 + C_5\sigma_2 + C_6\sigma_3 = 1 \qquad (12.31b)$$

In Eq. 12.31b, the constants C_i ($i = 1, 2 \ldots 6$) are:

$$2C_1 = 1/(\sigma_{2c}\sigma_{2t}) + 1/(\sigma_{3c}\sigma_{3t}) - 1/(\sigma_{1c}\sigma_{1t})$$
$$2C_2 = 1/(\sigma_{3c}\sigma_{3t}) + 1/(\sigma_{1c}\sigma_{1t}) - 1/(\sigma_{2c}\sigma_{2t}) \qquad (12.32a\text{-}c)$$
$$2C_3 = 1/(\sigma_{1c}\sigma_{1t}) + 1/(\sigma_{2c}\sigma_{2t}) - 1/(\sigma_{3c}\sigma_{3t})$$
$$C_4 = 1/\sigma_{1t} - 1/\sigma_{1c}, C_5 = 1/\sigma_{2t} - 1/\sigma_{2c} \text{ and } C_6 = 1/\sigma_{3t} - 1/\sigma_{3c} \qquad (12.32d\text{-}f)$$

in which σ_{1t}, σ_{1c} are the uniaxial tensile and compressive strengths in the 1-direction etc. Remaining constants C_7, C_8 and C_9, in Eq. 12.31a relate to the shear strengths σ_{12f}, σ_{23f} and σ_{13f} within the material's orthogonal planes as follows:

$$C_7 = 1/\sigma_{23f}^2, \; C_8 = 1/\sigma_{13f}^2 \text{ and } C_9 = 1/\sigma_{12f}^2 \qquad (12.32g\text{-}i)$$

Hoffman's general and principal failure criteria are useful when strength differences arising from tension and compression aligned with each material direction are to be represented in a single continuous expression.

Marin [23] proposed another continuous failure criterion in principal stress form, again one that combines linear and quadratic stress terms, but with fewer empirical constants than Eq. 12.31b:

$$(\sigma_1 - a)^2 + (\sigma_2 - b)^2 + (\sigma_3 - c)^2$$
$$+ q[(\sigma_1 - a)(\sigma_2 - b) + (\sigma_2 - b)(\sigma_3 - c) + (\sigma_3 - c)(\sigma_1 - a)] = \sigma^2 \qquad (12.33)$$

where a, b, c and σ are determined from the uniaxial tensile and compressive strengths in orthogonal directions. The remaining floating constant q must be evaluated under a complex stress condition, which enables Eq. 12.33 to fit data better than were it to rely upon uniaxial data alone. This is shown for plane stress reductions in the following section. In its absence of linear terms it will also be shown that when Hill's equation, Eq. 12.30a, is adapted to each quadrant discontinuously it presents a distorted fracture surface. Potentially, this is useful should distortion be found to characterise the effect of anisotropy upon the failure surface of fibrous composites. This irregular shape feature does not appear in the ellipsoidal surfaces that the Hoffman and Marin criteria describe.

Note that distortion has often been observed in the subsequent yield surface for plastically pre-strained metals under laboratory conditions. Distortion is one manifestation of strain induced anisotropy [24] that has also appeared within the initial yield surface for rolled and extruded material with residual strain.

12.5.4 Plane Stress Reductions

These reductions are required when a uniaxial stress $\sigma_{1'}$ is not aligned with the material's fibre directions 1 and 2. For example, consider an off-axis tensile test, with an orientation $0 < \theta < 90°$ of $\sigma_{1'}$ to the fibre 1-direction (see Fig. 12.16a).

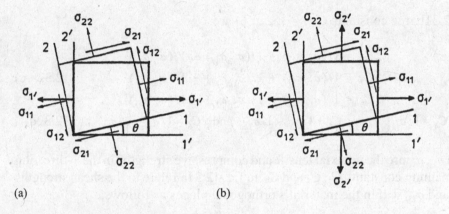

(a)   (b)

Figure 12.16 Off-axes tests under (a) uniaxial tension and (b) plane stress

Table 11.1 (stress system 1; p. 453) shows that each stress component aligned with 1 and 2 transforms from an off-axis tensile stress ital $\sigma_{1'}$ as

$$\sigma_{11} = \sigma_{1'} \cos^2\theta; \quad \sigma_{22} = \sigma_{1'} \sin^2\theta; \quad \sigma_{12} = \tfrac{1}{2}\sigma_{1'} \sin 2\theta \qquad (12.34\text{a-c})$$

Now consider an off-axis, plane principal stress state $(\sigma_{1'}, \sigma_{2'})$ in Fig. 12.16b with similar orientation to fibre 1. The corresponding stress transformations are taken from Table 11.1 (see stress system 3):

$$\sigma_{11} = \sigma_{1'} \cos^2\theta + \sigma_{2'} \sin^2\theta$$

$$\sigma_{11} = \sigma_{2'} \cos^2\theta + \sigma_{1'} \sin^2\theta \qquad (12.35\text{a-c})$$

$$\sigma_{12} = \tfrac{1}{2}(\sigma_{1'} - \sigma_{2'}) \sin 2\theta$$

Hill's equation, Eq. 12.30a, is sufficiently general to allow for the two-dimensional reduction required. For example, a plane stress reduction for a transversely isotropic material, with $\sigma_{2f} = \sigma_{3f}$, was considered by Azzi and Tsai [25]. When all stress components, except σ_{11}, σ_{22} and σ_{12}, are set to zero Eq. 12.30a gives:

$$\frac{\sigma_{11}^2 - \sigma_{11}\sigma_{22}}{\sigma_{1f}^2} + \frac{\sigma_{22}^2}{\sigma_{2f}^2} + \frac{\sigma_{12}^2}{\sigma_{12f}^2} = 1 \qquad (12.36a)$$

Equation 12.36a provides the failure envelope for each off-axis test when it is combined with Eqs 12.34a–c and 12.35a–c. The authors showed from Eq. 12.36a that the direction of the failure path was not aligned with the major principal stress plane. The latter, which lies perpendicular to $\sigma_{1'}$ in Fig. 12.16b, is given as $90° - \theta$ to the 1-direction where, from Eq. 11.5:

$$\theta = \frac{1}{2} \tan^{-1}\left(\frac{2\sigma_{12}}{\sigma_{11} - \sigma_{22}} \right) \qquad (12.36b)$$

in which σ_{11}, σ_{22} and σ_{12} are the plane stress components aligned with the material directions, The fracture path is taken to align with the major principal strain plane, having its inclination to the fibre 1-direction as the complementary angle $(90° - \beta)$, where β is given as

$$\beta = \frac{1}{2} \tan^{-1}\left(\frac{\gamma_{12}}{\varepsilon_{11} - \varepsilon_{22}} \right) \qquad (12.36c)$$

Equations 12.36b,c show no influence of shear strain upon the fracture path when $\theta = 0°$ or $90°$. Here we should expect the fracture path to lie normal to the applied stress (i.e. $\theta = \beta$), this being consistent with the application of fracture criteria composed only of normal stresses when $\theta = 0°$ or $90°$. However, for intermediate orientations $(0° < \theta < 90°)$ of the stress axes in Fig. 12.16b, shear strain γ_{12} exists and the principal axes of stress and strain will not coincide. That is, generally, $\theta \neq \beta$ for orthotropic and transversely isotropic materials. This implies that the absence of a shear stress term in Eq. 12.33 restricts the application of Marin's function to where principal stresses σ_1, σ_2 and σ_3, are aligned with the material's orthogonal axes. If shear stress is assumed to have no effect on the failure condition then σ_1, σ_2 and σ_3 can be replaced with the normal stress components σ_{11}, σ_{22} and σ_{33}, that lie in the material co-ordinates. Correspondingly, the plane stress reduction to Eq. 12.33 is:

$$\sigma_{11}^2 + K_1\sigma_{11}\sigma_{22} + \sigma_{22}^2 + K_2\sigma_{11} + K_3\sigma_{22} = K_4 \qquad (12.37a)$$

Substituting known uniaxial strengths into Eq. 12.37a provides three of the constants K_i in Eq. 12.37a (see Exercise 12.49):

$$K_2 = \sigma_{1c} - \sigma_{1t}, \quad K_3 = \sigma_{2c} - \sigma_{2t} \text{ and } K_4 = \sigma_{1c}\sigma_{1t} = \sigma_{2c}\sigma_{2t} \qquad (12.37b\text{-}d)$$

where subscripts t and c denote the tensile and compressive strengths respectively. The 'floating constant' K_1 $(= a)$ may be found from knowing the failure stress σ_b under equi-biaxial tension (i.e. a pressure bulge test) where $\sigma_1 = \sigma_b$ and $\sigma_2 = \sigma_b$:

$$K_1 = \frac{\sigma_{1c}\sigma_{1t} + [(\sigma_{1t} + \sigma_{2t}) + (\sigma_{1c} + \sigma_{2c})]\sigma_b}{\sigma_b^2} - 2 \qquad (12.37e)$$

Alternatively, pure shear gives $\sigma_1 = \sigma_{12f}$, $\sigma_2 = -\sigma_{12f}$, for which K_1 becomes

$$K_1 = 2 - \sigma_{1c}\sigma_{1t}/\sigma_{12f}^2 + (\sigma_{1c} - \sigma_{1t} - \sigma_{2c} + \sigma_{2t})/\sigma_{12f} \qquad (12.37f)$$

When a shear stress is known to influence fracture, Marin [23] modified the plane stress criterion in Eq. 12.37a to become:

$$A_1\sigma_{11}^2 + A_2\sigma_{11}\sigma_{22} + A_3\sigma_{22}^2 + A_4\sigma_{11} + A_5\sigma_{22} + A_6\sigma_{12}^2 = 1 \qquad (12.38a)$$

The constants A_i in Eq. 12.38a are determined from known strengths in a similar manner to K_i above. The additional constant A_2 in Eq. 12.38a is 'floating' and A_6 applies to an in-plane shear failure. In full, the constants A_i allow Eq. 12.38a to appear in the form:

$$\frac{(\sigma_{11} - A_2\sigma_{22})\sigma_{11}}{\sigma_{1c}\sigma_{1t}} + \frac{\sigma_{22}^2}{\sigma_{2c}\sigma_{2t}} + \frac{(\sigma_{1c} - \sigma_{1t})\sigma_{11}}{\sigma_{1c}\sigma_{1t}} + \frac{(\sigma_{2c} - \sigma_{2t})\sigma_{22}}{\sigma_{2c}\sigma_{2t}} + \frac{\sigma_{12}^2}{\sigma_{12f}^2} = 1$$

$$(12.38b)$$

where A_2 identifies with K_1 in Eqs 12.37e,f. Reducing Eq. 12.38b to a plane principal stress state in which A_2 may be different from unity:

$$\frac{\sigma_1^2 - A_2\sigma_1\sigma_2}{\sigma_{1c}\sigma_{1t}} + \frac{\sigma_2^2}{\sigma_{2c}\sigma_{2t}} + \frac{\sigma_1(\sigma_{1c} - \sigma_{1t})}{\sigma_{1c}\sigma_{1t}} + \frac{\sigma_2(\sigma_{2c} - \sigma_{2t})}{\sigma_{2c}\sigma_{2t}} = 1 \qquad (12.38c)$$

Corresponding plane stress forms of Hoffman's equations, Eqs 12.31a,b, become:

$$\frac{(\sigma_{11} - \sigma_{22})\sigma_{11}}{\sigma_{1c}\sigma_{1t}} + \frac{(\sigma_{22} - \sigma_{11})\sigma_{22}}{\sigma_{2c}\sigma_{2t}} + \frac{(\sigma_{1c} - \sigma_{1t})\sigma_{11}}{\sigma_{1c}\sigma_{1t}} + \frac{(\sigma_{2c} - \sigma_{2t})\sigma_{22}}{\sigma_{2c}\sigma_{2t}} + \frac{\sigma_{12}^2}{\sigma_{12f}^2} = 1$$

$$(12.39a)$$

$$\frac{(\sigma_1 - \sigma_2)\sigma_1}{\sigma_{1c}\sigma_{1t}} + \frac{(\sigma_2 - \sigma_1)\sigma_2}{\sigma_{2c}\sigma_{2t}} + \frac{(\sigma_{1c} - \sigma_{1t})\sigma_1}{\sigma_{1c}\sigma_{1t}} + \frac{(\sigma_{2c} - \sigma_{2t})\sigma_2}{\sigma_{2c}\sigma_{2t}} = 1$$

$$(12.39b)$$

Equation 12.39a was used to describe the fracture surface for a unidirectional glass-fibre composite and a syntatic foam of micro-spheres embedded in a resin matrix [22]. Equations 12.39a,b are not quite identical to Eqs 12.38b,c even when $A_2 = 1$. The following example shows how Eqs 12.38c and 12.39b can account for differences between tensile and compressive strengths that exist within a composite.

Example 12.11 Apply the plane principal fracture criteria of Hoffman and Marin to a unidirectional glass-fibre, reinforced composite. The in-plane tensile and compressive strengths for directions parallel and perpendicular to the fibres (directions 1 and 2, respectively) were found to be: $\sigma_{1c} = 476.8$ MPa, $\sigma_{1t} = 529$ MPa, $\sigma_{2c} = 136.3$ MPa and $\sigma_{2t} = 49.2$ MPa. An additional compression test gave the through-thickness strength of this composite as $\sigma_{3c} = 167.7$ MPa.

The compressive strengths given for two directions perpendicular to the fibre suggest an approximate condition of transverse isotropy (i.e. $\sigma_{2c} \approx \sigma_{3c}$). Marin's assumption: $\sigma_{1c}\,\sigma_{1t} = \sigma_{2c}\,\sigma_{2t}$, within the fourth constant K_4 in Eq. 12.37d does not apply here. However, with his 'floating constant' modification, Eq. 12.38c provides a suitable plane principal stress failure criterion. Substituting the strengths (in MPa) gives the elliptical function:

$$(\sigma_1^2 - A_2\sigma_1\sigma_2) + 37.61\,\sigma_2^2 - 52.2\,\sigma_1 + 3275.83\,\sigma_2 = 25.22 \times 10^4 \qquad \text{(i)}$$

which is composed of linear and quadratic terms. The reader should check the uniaxial strengths given by setting σ_1 and σ_2 within Eq. i to zero in turn. The usefulness of a floating constant, $A_2 \neq 1$, in Eq. i now becomes apparent. It enables a fit to a further test, e.g. pure shear or biaxial tension. This facility is lost from setting $A_2 = 1$ in Eq. 12.38b and therefore does will not appear within a plane principal stress reduction to Hoffmann's Eq. 12.39b:

$$\sigma_1^2 - 38.61\,\sigma_1\sigma_2 + 37.61\,\sigma_2^2 - 52.2\,\sigma_1 + 3275.83\,\sigma_2 = 25.22 \times 10^4 \qquad \text{(ii)}$$

Care should be taken that the A_2-value chosen for Eq. i results in a closed locus. A comparison is made in Fig. 12.17 of failure loci corresponding to $A_2 = -1$, 1 and 10.

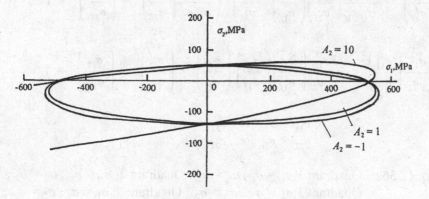

Figure 12.17 Unsymmetrical failure loci

It is seen that the change in sign of A_2 from +1 to -1 alters marginally the orientation of the safe load-bearing region enclosed within the closed loci. In contrast, loci with large A_2-values accentuate orientation but may become invalid where they remain open under biaxial compression in quadrant four. Closed loci apply here where: $-12.27 < A_2 < 12.27$ [26]. Hence, the three A_2 values used with Eq. i lie within this range but $A_2 = 38.61$, taken from Eq. ii does not (see Ex. 12.50).

12.5.5 Unsymmetrical Functions

Azzi and Tsai [25] constructed the failure envelope for transversely isotropic composites by applying Eq. 12.36a discontinuously to each of its quadrants. For this they allowed tensile and compressive strengths (σ_c or σ_t) to differ within each quadrant, as shown in Table 12.2. Kaminski and Lantz [27] fitted symmetrical functions to unsymmetrical test data within individual quadrants in a similar manner. Franklin [28] showed that these methods distort the failure locus when the stress axes σ_1 and σ_2 are normalised with directional strengths; σ_{1t}, σ_{1c}, σ_{2t} and σ_{2c} appropriate to their sense in each quadrant. This approach differs from adopting a continuous function for all quadrants as in Marin's equation, Eq. 12.38b. Normalising with a common reference stress the summation of linear and quadratic terms in the Marin and Hoffman plane principal stress equations, Eqs 12.38c and 12.39b, retains their elliptical forms. There are a number of further plane stress failure criteria that apply to each quadrant where differences in tensile and compressive strengths occur [29–31]. They are given in their general plane stress forms in Table 12.2.

Table 12.2 Unsymmetrical, plane stress failure functions

No. Ref.	Failure Function

1. [29] $$\left(\frac{\sigma_{11}}{\sigma_{1f}}\right)^2 - \frac{3\sigma_{12f}^2}{\sigma_{1f}\sigma_{2f}}\left(\frac{\sigma_{1f}}{\sigma_{2f}}\right)\left(\frac{\sigma_{11}}{\sigma_{1f}}\right)\left(\frac{\sigma_{22}}{\sigma_{2f}}\right) + \frac{3\sigma_{12f}^2}{\sigma_{1f}\sigma_{2f}}\left(\frac{\sigma_{22}}{\sigma_{2f}}\right)^2 + \left(\frac{\sigma_{12}}{\sigma_{12f}}\right)^2 = 1$$

2. [29] $$\frac{3\sigma_{12f}^2}{\sigma_{1f}\sigma_{2f}}\left(\frac{\sigma_{11}}{\sigma_{1f}}\right)^2 - \frac{3\sigma_{12f}^2}{\sigma_{1f}\sigma_{2f}}\left(\frac{\sigma_{2f}}{\sigma_{1f}}\right)\left(\frac{\sigma_{11}}{\sigma_{1f}}\right)\left(\frac{\sigma_{22}}{\sigma_{2f}}\right) + \left(\frac{\sigma_{22}}{\sigma_{2f}}\right)^2 + \left(\frac{\sigma_{12}}{\sigma_{12f}}\right)^2 = 1$$

3. [30] $$\sigma_{11} = \sigma_{1t} \quad \text{or} \quad \sigma_{11} = -\sigma_{1c}$$
4. [30] $$\sigma_{22} = \sigma_{2t} \quad \text{or} \quad \sigma_{22} = -\sigma_{2c}$$

5. [25] Eq. 12.36a: Quadrant 1, $\sigma_{1f} = \sigma_{1t}$, $\sigma_{2f} = \sigma_{2t}$; Quadrant 2, $\sigma_{1f} = \sigma_{1c}$, $\sigma_{2f} = \sigma_{2t}$; Quadrant 3, $\sigma_{1f} = \sigma_{1c}$, $\sigma_{2f} = \sigma_{2c}$; Quadrant 4, $\sigma_{1f} = \sigma_{1t}$, $\sigma_{2f} = \sigma_{2c}$;

6. [31] $$F_{1111}\sigma_{11}^2 + F_{2222}\sigma_{22}^2 + 2F_{1122}\sigma_{11}\sigma_{22} + 4F_{1212}\sigma_{12}^2 + F_{11}\sigma_{11} + F_{22}\sigma_{22} = 1$$
for: $F_{1111} = 1/(\sigma_{1c}\sigma_{1t})$, $F_{2222} = 1/(\sigma_{2c}\sigma_{2t})$, $F_{1212} = 1/\sigma_{12f}^2$
$$F_{11} = 1/\sigma_{1t} - 1/\sigma_{1c}, \quad F_{22} = 1/\sigma_{2t} - 1/\sigma_{2c}$$

Also, in 6, F_{1122} must be found from a biaxial test, preferably equi-tension, though a torsion test will also suffice. Each test gives respectively:

$$2F_{1122} = \left[1/\sigma_b^2 - (F_{1111} + F_{2222}) - (F_{11} + F_{22})/\sigma_b\right]$$
$$2F_{1122} = \left[(F_{1111} + F_{2222}) + (F_{11} + F_{22})/\sigma_{12f} - 1/\sigma_{12f}^2\right]$$

where σ_b and σ_{12f} are the strengths in equi-biaxial tension and torsion.

The shear stress terms appearing with plane stress criteria 1–4 in Table 12.2 are dropped when applied to failure under principal biaxial stress (see Exercise 12.51). Thus, Fig. 12.18 overlays six criteria in Table 12.2 for a composite material whose four uniaxial strengths depend upon their sense and their orientation as follows: $\sigma_{1t} = 200$, $\sigma_{2t} = 94$, $\sigma_{1c} = 160$ and $\sigma_{2c} = 140$ (in MPa).

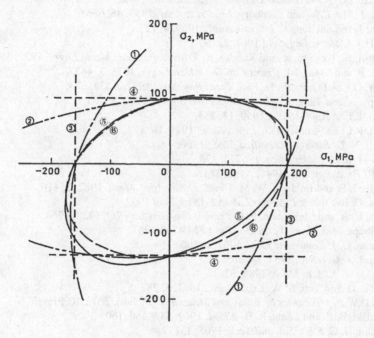

Figure 12.18 Comparison of quadrant-based failure loci in σ_1, σ_2 space (See Table 12.2)
(**Key**: 1, 2 - Puppo–Evensen [29]; 3, 4 - Puck [30]; 5 - Azzi and Tsai [25]; 6 - Tsai–Wu [31])

The two Puppo–Evensen criteria 1 and 2 [29] apply only to the envelope contained within the intersection between its limbs. Puck's criteria 3 and 4 [30] appear in principal space as the anisotropic equivalent to the Rankine theory. Figure 12.18 confirms that a quadrant-based strength fit from an elliptical function (Eq. 12.36a) produces the distorted locus 5 [25], after having set: $\sigma_{1f} = \sigma_{1t}$ and σ_{1c}; $\sigma_{2f} = \sigma_{2t}$ and σ_{2c} This contrasts with Figs. 12.15 and 12.17, where the differing strengths are matched within the orientation and shift of a closed, continuous, elliptical equation. There are advantages in adopting a continuous function provided it has the required generality. Here the Tsai–Wu [31] multi-axial failure criterion is a reduction to a general tensor function involving linear and quadratic stress terms:

$$F_{ijkl}\,\sigma_{ij}\,\sigma_{kl} + F_{ij}\,\sigma_{ij} = 1 \tag{12.40}$$

This versatile criterion has universal appeal through its representation of all reduced stress states. For example, in Table 12.2, Eq. 12.40 has allowed for the plane, principal stress reduction given in 6 whilst matching the strength variations.

References

[1]   Shahabi, S.N and Shelton, A. J. *Mech. Eng. Sci.*, 1975, **17**, 82.
[2]   Johnson, K.R. and Sidebottom, O. M. *Expl Mech.*, 1972, **12**, 264.
[3]   Moreton, D.N., Moffat, D.G. and Parkinson, D.B. *J. Strain Anal.*, 1981, **16**, 127.
[4]   Lessels, J.M. and MacGregor, C.W. *J. Franklin Inst.*, 1940, **230**, 163.
[5]   Miastkowski, J. and Szczepinski, W. *Int. J. Solid and Struc*, 1965, **1**, 189.
[6]   Marin, J., Hu, L.W. and Hamburg, J.F. *Proc. ASM*, 1953, **45**, 686.
[7]   Rogan, J. and Shelton, A. *J. Strain Anal.*, 1969, **4**, 127.
[8]   Ivey, H.J. *J. Mech. Eng. Sci.*, 1961, **3**, 15.
[9]   Shiratori, E., Ikegami, K. and Kaneko, K. *Trans Japan Soc. Mech. Engrs*, 1973, **39**, 458.
[10]   Ellyin, F. and Grass, J-P, *Trans Can Soc. Mech Engrs*, 1975, **3**, 463.
[11]   Taylor, G.I. and Quinney, H. *Phil. Trans Roy. Soc.*, 1931, A**230**, 323.
[12]   Phillips, A. and Tang, J-L. *Int. J. Solids and Struct.*, 1972, **8**, 463.
[13]   Coffin, L.F. *J. Appl. Mech.*, 1950, **17**, 233.
[14]   Grassi, R.C. and Cornet, I. *J. Appl. Mech.*, 1949, **16**, 178.
[15]   Alberti, N. *La Ricerca Scientifica*, Dec., 1960.
[16]   Stassi-D'Alia, F. *Meccanica*, 1967, **2**, 178.
[17]   Hill, R. *Proc. Roy. Soc*, 1948, A**193**, 281.
[18]   Griffith, J. E. and Baldwin, W. M. *Devel. Theor. Appl. Mech.* 1962, **1**, 410.
[19]   Norris, C. B. *Forrest Products Lab Rpt*, 1816, May 1962.
[20]   Norris, C. B. and McKinnon, P. F. *Forrest Products Lab Rpt*, 1328, 1964.
[21]   Waddoups, M. E. *Fort Worth Div. Rpt* FZM 4763,1967.
[22]   Hoffman, O. *J. Composite Mats*, 1967, **1**, 200.
[23]   Marin, J. *J. Aero Sci.*, 1957, **24**, 265.
[24]   Rees, D. W. A. *Acta Mech*, 1984, **52**, 16.
[25]   Azzi, V. D. and Tsai, S. W. *Expl. Mech.*, 1965, **5**, 283.
[26]   Rees, D.W.A. *Mechanics of Solids and Structures* (2nd Ed.), 2015, IC Press.
[27]   Kaminski, B. E. and Lantz, R. B. *ASTM*, 1969, STP 460, 160.
[28]   Franklin, H. G. *Fibre Sci. and Tech.* 1968, **1**, 137.
[29]   Puppo, A. H. and Evensen, H. A. *AIAA Jl*, 1972, **10**(4).
[30]   Puck, A. *Kunststoffe*, 1969, Bd **59**, 780.
[31]   Tsai, S. W. and Wu, E. M. *Jl. Composite Mats*, 1971, **5**, 58.

Exercises

Yield Criteria in Principal Stress Form

Hint: Refer to Ref. [26] for Exercises 12.13–12.17.

12.1 At a point in a material the major principal stress is 194 MPa. If the tensile yield stress is 253 MPa, find the minor principal stress that would cause yielding according to the total strain energy (for $v = 0.3$), the Tresca and von Mises yield criteria.
(Ans: 113.8, 58.67; 91.87 MPa)

12.2 Given the principal stresses $\sigma_1 = 185$ MPa and $\sigma_3 = 0$, determine the magnitude of the compressive stress σ_2 that would cause yielding according to the Tresca and von Mises yield criteria. Take $Y = 260$ MPa.

12.3 The tensile yield stress of a material is 232 MPa. If the principal biaxial stresses set up in service are 123.5 MPa in tension and 38.5 MPa in compression, calculate the operating safety factors from the Tresca and von Mises yield criteria. (Ans: 1.43, 1.58)

12.4 A steel cube, subjected to a constant uniaxial tensile stress, is lowered to increasing depth in the sea. If the steel's tensile yield stress is 350 MPa at atmospheric pressure and the density of sea water is 1020 kg/m^3, find the tensile stress at which yielding occurs for 1 km, 10 km and 100 km depths of immersion. (Ans: 350 MPa)

12.5 A thin-walled cylinder inner diameter d and thickness t, is subjected to an internal pressure p. Derive expressions for the pressure required to initiate yielding according to the von Mises and Tresca yield criteria when (i) the radial stress is neglected and (ii) when the magnitude of the radial stress is $-p$. The tensile yield stress for the cylinder material is Y. Compare the yield pressures when $d/t = 25$ and $Y = 300$ MPa. Ans: (i) $\sigma_r = 0$, $4Yt/\sqrt{3}d$, $2Yt/d$; (ii) $\sigma_r = -p$, $\sqrt{[2Y^2/(3d^2/8t^2 + 3d/2t + 2)]}$, $Y/(1 + d/2t)$; 27.7, 24, 25.63 and 22.2].

12.6 A thick-walled cylinder, 25 mm inner and 50 mm outer diameter, is subjected to a constant axial tensile load of 20 kN together with an internal pressure p. If the ends of the cylinder are closed and the external pressure is atmospheric, calculate the value of p that would initiate yielding within the bore according to the Tresca and von Mises yield criteria. Take $Y = 240$ MPa. (Ans: 103.8 MPa, 90 MPa)

12.7 A thick-walled cylinder, 100 mm inner diameter, withstands an internal pressure of 100 bar and a uniform bending moment of 3.5 kNm. Using a factor of safety of 3 for the most highly stressed point in the section, determine the cylinder outer diameter based upon the Mises yield criterion. Take $Y = 300$ MPa

12.8 A thick-walled cylinder, 100 mm inner diameter and 200 mm outer diameter, is subjected to an internal/external pressure ratio 5:1. Calculate the pressure that would initiate yielding in the bore according to Tresca's yield criterion for a tensile yield stress of 250 MPa. (Ans: 117.2 MPa)

12.9 The 2 m internal diameter of a thick cylinder is maintained at atmospheric pressure while the exterior surface is subjected to a fluid pressure of 45 bar. Given that the tensile yield stress for the cylinder material is 250 MPa, determine the wall thickness according to the Tresca yield criterion incorporating a safety factor of 3.

12.10 A high-pressure steel pipeline weighing 300 N/m is simply supported at 5 m intervals. If the inner and outer diameters are 30 and 50 mm respectively, estimate the internal pressure and point in bore where yielding begins. Assume a von Mises yield criterion and a cylinder with open ends over the support length. Take $Y = 250$ MPa. (Ans: 84.68 MPa, top of i.d.)

12.11 A uniformly thick brass cylinder has an outer radius of 125 mm and an internal radius of 75 mm. It has rigid end closures and is subjected to an internal pressure only. Using the Tresca and Maxwell–von Mises yield criteria, a tensile yield stress of 450 MPa, determine the pressure at which yielding occurs. Assume that the axial stress is uniform. (CEI)

12.12 A long thick cylinder of outside diameter 2a and inside diameter 2b, is subjected to an internal pressure p. Determine the maximum pressure permitted by the Tresca criterion if the cylinder is not to deform plastically. Calculate the upper and lower bound values for the longitudinal stress for one closed end containing a pressurized fluid that causes a projectile to be ejected from the open end. Under what conditions will the longitudinal stress affect the onset of yielding? (CEI)

12.13 A steel rotor disc of uniform thickness 50 mm has an outer rim diameter of 750 mm and a central hole of diameter 150 mm. There are 200 blades each of weight 2.25 N at an effective radius of 425 mm pitched evenly around the periphery. At what rotational speed does yielding first occur according to the Tresca yield criterion given that the yield stress in simple tension is 690 MPa? Take $E = 207$ GPa, $\rho = 7840$ kg/m^3 and $\nu = 0.29$. (CEI)

12.14 A steel tube has an internal diameter of 25 mm and an outside diameter of 50 mm. When a second steel tube is shrunk over the outer diameter of the first tube, a condition of yield exists at the inner diameter of each tube. Determine the interference required before shrinkage and the outer diameter of the second tube assuming a Tresca yield criterion. Ignore axial stress. Take $Y = 413.5$ MPa and $E = 207$ GPa. (Ans: 0.13 mm, 100 mm)

12.15 A collar having 200 mm inner diameter slides freely on a hollow column 160 mm i.d. and 200 mm o.d. When an internal pressure p is introduced into the column to create an interface pressure of 9 MPa the collar locks and is able to support an axial load of 200 kN. Determine p and check from the Tresca yield criterion that the stress states at each inner diameter are elastic within a safety factor of 3 on a tensile yield stress of 300 MPa.

12.16 Show that the von Mises criterion may be written in principal deviatoric stress form:

$$\sigma_1'^2 + \sigma_2'^2 + \sigma_3'^2 = \frac{2Y^2}{3}$$

12.17 Show that the two non-zero (positive) deviatoric stress invariants can be expressed in the matrix forms:

$$J_2' = \tfrac{1}{2}\mathrm{tr}\,(S')^2 \quad \text{and} \quad J_3' = \tfrac{1}{3}\,\mathrm{tr}\,(S')^3$$

where S' is a matrix of principal, deviatoric stress components and tr(S'), the trace of matrix S', is the sum of its leading diagonal components.

Yield Criteria in General Biaxial Stress Form

12.18 A specimen of steel has a yield stress of 330 MPa. A 50 mm diameter shaft made from this material has a axial torque of 2.5 kNm applied to it. Assuming that the material yields according to (i) the maximum shear strain energy criterion and (ii) the maximum shear stress criterion, calculate the value of the additional axial tensile load that would cause the shaft material to pass the yield point. (Ans: 547 kN, 508 kN)

12.19 A screwdriver has a steel shank with a tensile yield stress of 300 MPa. Using a safety factor of 10, calculate the minimum shank diameter for a maximum torque of 3 Nm, based upon the Tresca and von Mises criteria of elastic strength. (Ans: 10 mm, 9.6 mm)

12.20 Find an expression for the stress at the yield point of a solid circular shaft carrying a uniform bending moment M. Find expressions for the same shaft under the action of a torque T according to the five theories of yielding for ductile materials. What are the numerical ratios of M/T in each case? (Ans: 0.5, 0.65, 0.806, 1 and 0.865)

12.21 A composite tube, 100 mm external diameter, 50 mm internal diameter and 250 mm long, consists of a brass tube which just fits inside a steel tube. The tubes are bonded to each other at an interface diameter of 75 mm. Samples of brass and steel tested separately are found to yield in uniaxial tension at 150 MPa and 200 MPa, respectively. If yielding is to be avoided in both materials, determine: (i) the maximum axial tensile load and (ii) the maximum torque that may be applied to the composite tube. Assume, where necessary, a Tresca yield criterion and take the moduli E for brass and steel as 100 GPa and 200 GPa, respectively. (CEI)

12.22 A solid shaft 75 mm in diameter, is subjected to an axial force of 80 kN and a torque of 3.8 kNm. If the yield stress for the shaft material is 278 MPa, calculate the safety factors that have been used in the design according to the Tresca and von Mises yield criteria. (Ans: 3.13, 3.59)

12.23 A 30 mm diameter rod is subjected to a bearing pressure of 120 MPa around its outer surface. Examine the additional loading from torsion and from bending that would yield the shaft. Take $Y = 310$ MPa.

12.24 A steel shaft carries a bending moment of 2 kNm and a torque of 1.5 kNm. If the tensile yield stress for the shaft material is 280 MPa, determine a suitable shaft diameter based upon a safety factor of 2 and each of the following criteria of elastic strength: (i) maximum shear stress (ii) maximum total strain energy (take $v = 0.25$) and (iii) maximum shear strain energy.

12.25 A mild steel tube has a mean diameter of 100 mm and a thickness of 3 mm. Using a safety factor of 2 and a tensile yield stress of 230 MPa, calculate the torque which can be transmitted by this tube if the criterion of failure is: (i) the maximum shear stress and (ii) the maximum shear strain energy. (Ans: 2.5 kNm, 2.9 kNm)

12.26 A thin-walled closed tube with internal diameter d and thickness t reaches its yield point under an internal pressure p_y. Working in terms of p_y, determine the torque that would yield the cylinder when (i) p_y is removed and (ii) $p_y/2$ remains.
(Ans: $\pi p_y d^3/8$, $\sqrt{3}\pi p_y d^3/16$)

12.27 A 20 mm diameter steel shaft is loaded with an axial torque of 60 Nm and an axial tensile force of 15 kN. Given that the tensile yield stress is 300 MPa, determine the safety factors employed with the Tresca and von Mises yield criteria. (Ans: 3.33, 2.61)

12.28 A 32 mm diameter shaft is required to transmit 100 kW at 3000 rev/min and to carry simultaneously a bending moment of 300 Nm. Determine which of the five theories of yielding predicts that the shaft has exceeded its yield point when the tensile yield stress is 365 MPa. (Ans: Tresca and both energy theories)

12.29 A thin circular pipe withstands an axial torque of 1 kNm and an internal pressure of 25 bar simultaneously. Calculate the thickness according to the Tresca and von Mises yield criteria using a safety factor of 2.5. The tensile yield stress of the pipe material is 200 MPa. (Ans: 2.7, 2 mm)

12.30 A stirrer rotates at 600 rev/min under a thrust of 4 kN inside a chemical vessel where the internal pressure is 120 MPa. Given that the power absorbed by the stirrer is 4 kW, calculate its diameter using a safety factor of 3 and the von Mises yield criterion. The yield stress of the stirrer material is 180 MPa.
(Ans: 19 mm)

12.31 A thin-walled 120 mm outer square tube with 5 mm wall thickness withstands simultaneously a torque of 4.5 kNm and an axial force of 40 kN. Given that the tensile yield stress of the tube material is 310 MPa, find the safety factor that has been employed in the design according to the von Mises yield criterion.

12.32 A box beam section is 150 mm wide, 200 mm deep, with respective wall thicknesses of 20 mm and 15 mm, is riveted lengthwise along one vertical side 60 mm above the neutral axis. Given that the maximum shear force and bending moment in the beam are 200 kN and 65 kNm respectively, find the design safety factors according to the application of the maximum principal strain and total strain energy theories. Take $Y = 300$ MPa, $v = 0.25$ for the beam material.
(Ans: 3.34, 3.37)

12.33 State the Tresca and von Mises yield criteria. A particular steel is found to yield at 300 MPa in tension. Determine, for 25 mm diameter samples of the same steel: (i) the torque required to induce yielding according to the Tresca criterion (ii) the maximum bending moment to initiate yield by both criteria and (iii) the maximum bending moment required to induce yielding by the von Mises yield criterion, when the cylindrical surface only is subjected to a pressure of 100 MPa. (CEI)

12.34 A 5 mm diameter rod, which exhibits elastic–perfectly plastic behaviour, is found to yield in pure tension under a force of 3.5 kN. Determine from the Tresca and von Mises yield criteria the axial torque required to: (i) yield the outer fibres and (ii) produce a fully plastic bar. (CEI)

12.35 A solid circular section shaft 125 mm in diameter rotates at 30 rad/s. Due to the configuration of the bearings and the gearing, it is subjected to a maximum bending moment of 9 kNm. If the elastic limit stress in simple tension is 300 MPa, calculate the power which the shaft may transmit using the Tresca yield criterion. When the shaft is transmitting this power, check the safety of the system using the von Mises criterion. (CEI)

12.36 An element in a loaded structure is subjected to the stress state shown in Fig. 12.19. Determine the complete state of stress on planes AC and BC and find the safety factor that has been used against yielding according to von Mises yield criterion. Take $Y = 300$ MPa.

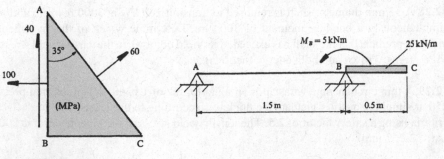

Figure 12.19   **Fig ure 12.20**

12.37 Calculate the safety factors that have been used against yielding at the web-flange interface for the beam in Fig. 12.20 with I-section outer dimensions: 50 mm wide × 100 mm deep and 5 mm thickness throughout. Take $Y = 250$ MPa.

12.38 Examine how shear influences the allowable loading that can be distributed uniformly upon the cantilever beam with T-section given in Example 12.7 (see p. 512), when the beam length is (a) halved and (b) quartered.

12.39 The shaft in Fig. 12.21 rotates in a plain bearing at its fixed end and carries a pulley belt drive with differing tensions at its free end. Determine a suitable pulley diameter d using a safety factor of 8 with a tensile yield stress of 300 MPa based upon the Tresca and von Mises criteria of elastic strength.

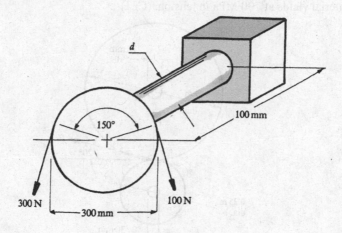

Figure 12.21

12.40 Given that the depth D of the cantilever in Fig. 12.22 varies with length according to $D = 10 + kz^2$, determine the shear and bending stresses at the hole A in terms of the width B. What should be the safe value of B according to Rankine's maximum principal stress yield criterion and Haigh's total strain energy yield criterion when a safety factor of 5 is used with a yield stress of 100 MPa and $v = 0.3$?
(Ans: $\sigma = 4.72/B$, $\tau = 9.45/B$, 0.6 mm, 0.8 mm)

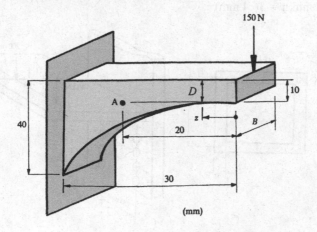

Figure 12.22

12.41 A gear wheel 1 m in diameter is mounted on a short stub axle 75 mm in diameter as shown in Fig. 12.23. It is driven by a small 0.25 m-diameter pinion wheel through an axle of 30 mm diameter. The gears mesh on a vertical axis and the line of force F between the two gears makes an angle of 20° to a horizontal axis in the plane x, y through the contact point. The 75 mm and 30 mm diameter axles are constructed from the same material with overhanging cantilever lengths of 200 mm and 100 mm, respectively. Determine the maximum power that the unit can transmit according to the Tresca and von Mises criteria when the larger gear wheel rotates at 300 rev/min without yielding of the shaft. A sample of the axle material yields at 700 MPa in tension. (CEI)

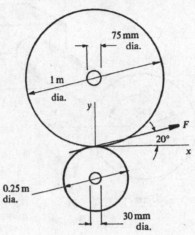

Figure 12.23

12.42 What are the sectional dimensions of the cheapest tube in Fig.12.24 to carry a torque of 50 Nm given that its cost c (pence/m length) is $100c = A(60 + y/t)$ where A is the section area in mm², t is the thickness and y is the mean side length? The design is to be based upon the total strain energy criterion with $v = 0.3$ and a tensile yield stress of 150 MPa. (Ans: $t = 1.21$ mm, $y = 36.4$ mm)

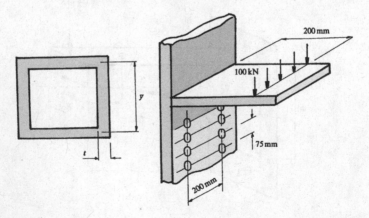

Figure 12.24   **Fig ure 12.25**

12.43 Are 22 mm diameter bolts suitable for the bracket assembly in Fig. 12.25, given their allowable stresses values of 120 and 80 MPa in tension and shear?

12.44 The belt drive system in Fig. 12.26 transmits power from 300 mm diameter pulleys A to B through a 100 mm diameter shaft. Belts from A are arranged horizontally while those from B are vertical each with tension forces 2P and P on their tight and slack sides. The shaft is simply supported in spherical bearings at C and D. Determine the position and magnitude of the maximum bending moment on the shaft, and also the position and magnitude of the maximum principal stress in the shaft, when transmitting 75 kW at a shaft speed of 500 rev/min. If the yield stress of the shaft material in uniaxial tension is 200 MPa, determine the maximum power that the pulley system can transmit at 500 rev/min without the shaft yielding, assuming a Tresca yield criterion. (CEI)

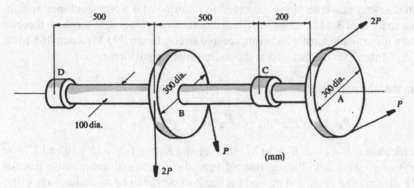

Figure 12.26

12.45 The cross section of the aluminium alloy beam in Fig. 12.27 is fabricated by riveting together three similar plates with four identical equal angle sections with the following local properties of their areas:

Plates: 125×5 mm, $\bar{x} = 62.5$ mm, $\bar{y} = 2.5$ mm, $I_x = 1302$ mm^4, $I_y = 813000$ mm^4.

Angles: $50 \times 50 \times 5$ mm, $\bar{x} = \bar{y} = 14.3$ mm, $I_x = I_y = 113000$ mm^4.

If the 3 m long, simply supported beam is to carry a central concentrated load W, determine the maximum permissible load W when the rivet holes give rise to an elastic stress concentration (i.e. a magnification factor) of 3 and the maximum vertical deflection is not to exceed 2 mm. The Tresca yield criterion is applicable for a maximum uniaxial working stress of 70 MPa. Take $E = 80$ GPa. (CEI)

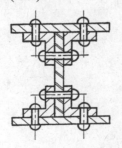

Figure 12.27

Failure of Brittle and Fibrous Materials

12.46 If the maximum principal stress is to be limited to one third of the 370 MPa tensile strength of cast iron, what is the minor principal stress by Coloumb–Mohr given that its compressive strength is 1390 MPa?
(Ans: -926 MPa)

12.47 If the maximum shear stress in a cast-iron component is not to exceed 385 MPa, find from the Coloumb–Mohr strength criterion, the magnitudes of the major and minor principal stresses given that the ultimate strengths are 1390 MPa and 370 MPa in compression and tension respectively.
(Ans: 224.3 MPa, -545.7 MPa)

12.48 A 6 mm diameter cast-iron pin is subjected to a torque of 9.8 Nm combined with an axial compressive force of 3.5 kN. Compare the safety factors according to the three theories of brittle rupture for ultimate tensile and compressive strengths are 293 MPa and 965 MPa.
(Ans: Coulomb–Mohr: 1.05; Mohr: 1.35 and maximum principal stress: 1.6)

12.49 Confirm that the plane principal stress reduction to Eq. 12.33 is given by:

$$\sigma_1^2 + K_1 \sigma_1 \sigma_2 + \sigma_2^2 + K_2 \sigma_1 + K_3 \sigma_2 = K_4$$

where the coefficients are: $K_1 = q$, $K_2 = -(2a + b + c)$ and $K_3 = -(aq + 2b + c)$ and $K_4 = \sigma^2 - (a^2 + b^2 + c^2 + abq + bc + ac)$. Taking, respectively, the tensile and compressive fracture stress for the 1-direction to be σ_{1t} and σ_{1c} and for the 2-direction to be σ_{2t} and σ_{2c}, show that Eqs 12.37b-d apply to the four fracture stresses.
Hint: Let the appropriate fracture stresses be the roots to the quadratic equation that the plane fracture criterion above provides for each direction.

12.50 Starting with Marin and Hoffman's principal, plane stress fracture criteria, Eqs 12.38c and 12.39b, show: (i) how each fits the uniaxial strengths that are the co-ordinate intercepts in Fig. 12.17 and (ii) that the Hoffman fit produces an open locus in quadrant 3.

12.51 Apply the six unsymmetrical failure functions given in Table 12.2 to the uniaxial strengths for the glass-fibre composite given in Fig. 12.18. Use the co-ordinate intercepts given for the quadrant-based criteria 1–5: $\sigma_{1t} = 200$, $\sigma_{2t} = 94$, $\sigma_{1c} = 160$ and $\sigma_{2c} = 140$ (each in MPa). In addition, take the shear strength in torsion as 250 MPa for an application of the Tsai-Wu criterion (No. 6 in Table 12.2).

CHAPTER 13

FINITE ELEMENTS

Summary: This introduction to finite elements will show how to derive the appropriate stiffness matrix, connecting 'force' to 'displacement', for: (i) a bar under uniaxial stressing, (ii) a beam in bending, (iii) a shaft under torsion and (iv) plates under plane stress and plane strain, using the theory of previous chapters. The approach used in the finite element method (FEM) refers to part of the uniform bar, beam, shaft or plate as an element and to its ends/corners as the nodes. When using the *stiffness or displacement method* of analysis the displacements at the element's nodes are the unknowns. These *nodal-point displacements* can be found from their relation to the *nodal forces*. The *stiffness matrix*, which connects nodal forces to displacements, has a unique form depending upon the number of degrees of freedom for the element in question. Once the displacements are known, the strains follow from the strain–displacement relations. Finally, knowing the material's elastic constants the stresses corresponding to these strains are found from applying the appropriate form of Hooke's law.

It will be shown how to develop the stiffness matrices for simple bar, beam and shaft elements for (i)–(iii) above using a routine procedure. The FEM requires the sub-division of a body into many smaller elements interconnected throughout at the nodal points. Examples will show how to assemble an overall stiffness matrix for two or more bar or beam elements connected in series. The initial unstrained shape of the body is thus discretised within an assemblage of elements. When the assembly is elastically strained, the behaviour of the entire body may be computed from the known elastic behaviour of its elements. We shall see that the stiffness method assumes a displacement function for the loaded body. This is to provide for equilibrium and compatibility of its internal stress and strain distributions and to match external boundary conditions for force and displacement. Having assembled the overall stiffness matrix, the displacements at the nodes will follow from finding the inverse of the stiffness matrix.

13.1 Stiffness Matrix

The approach adopted here lies in the derivation of an element stiffness matrix, $\mathbf{K}^e$ where e refers to the element number. For each element $e = $ I, II, III, etc., the matrix $\mathbf{K}^e$ connects its nodal point force and displacement vectors, $\mathbf{f}^e$ and $\boldsymbol{\delta}^e$ respectively:

$$\mathbf{f}^e = \mathbf{K}^e \boldsymbol{\delta}^e \tag{13.1a}$$

It will be seen that the element stiffness matrix is always square and symmetrical about its leading diagonal. The dimension of the square matrix $\mathbf{K}^e$ depends upon the type of element used to define a structure. Simple bar elements apply to beams in bending and shafts under torsion. Plane triangular elements are suited to thin plates and shells. The process of discretisation is that of interconnecting many such elements to match the structure's geometry. Bar elements may need only be connected in series but plane triangular elements often require a change in their size to fit an irregular boundary. The 'vectors' $\mathbf{f}^e$ and $\boldsymbol{\delta}^e$ in Eq. 13.1a are to be taken as column matrices of an element's nodal point forces and displacements as follows:

$$\mathbf{f}^e = \begin{bmatrix} f_1 \\ f_2 \\ f_3 \\ \cdot \\ \cdot \\ \cdot \end{bmatrix} \quad \text{and} \quad \boldsymbol{\delta}^e = \begin{bmatrix} \delta_1 \\ \delta_2 \\ \delta_3 \\ \cdot \\ \cdot \\ \cdot \end{bmatrix} \qquad (13.1\text{b,c})$$

Here the transpose may be used as $\mathbf{f} = [\,f_1 \ f_2 \ f_3 \ ...\,]^{\mathrm{T}}$ and $\boldsymbol{\delta} = [\,\delta_1 \ \delta_2 \ \delta_3 \ ...\,]^{\mathrm{T}}$ to save space where these components should otherwise be stacked vertically within the brackets as in Eqs 13.1b,c. Using selected examples it will be shown how the structure's overall stiffness matrix $\mathbf{K}$ is assembled from individual stiffness matrices $\mathbf{K}^e$ of elements that sub-divide the structure. This gives the full description of the displaced structure under the forces it bears within a single matrix equation

$$\mathbf{f} = \mathbf{K}\,\boldsymbol{\delta} \qquad (13.1\text{d})$$

In the analysis that follows, bold upper case Roman letters, $\mathbf{A}$, $\mathbf{B}$, $\mathbf{C}$ and $\mathbf{K}$, denote matrices and bold lowercase Roman and Greek symbols, $\mathbf{f}$, $\boldsymbol{\alpha}$, $\boldsymbol{\varepsilon}$, $\boldsymbol{\delta}$ and $\boldsymbol{\sigma}$, denote column matrices or column vectors. The vector has physical meaning as with force $\mathbf{f}$ and displacement $\boldsymbol{\delta}$ but otherwise the column matrix is a grouping of components or coefficients of a scalar quantity to which the laws of matrix algebra apply.

13.2 Energy Methods

Equations 13.1a-c show that the stiffness method reduces to finding the elemental stiffness matrices $\mathbf{K}^e$. Employing a suitable energy method provides the basis for determining $\mathbf{K}^e$. The two methods most often employed are the *principle of virtual work* (PVW) and *stationary potential energy* (SPE). Each method leads to a similar result where $\mathbf{K}^e$ is expressed in terms of the product of separate matrices expressing the nodal point co-ordinates and the elastic constants for the material. The choice between PVW and SPE will depend upon which is the more convenient to apply.

13.2.1 Principle of Virtual Work

Consider a system of real forces, f_k ($k = 1, 2, 3...$), in equilibrium. Let these forces experience virtual in-line displacements $\delta_k{}^v$ from an independent 'external agency' that does not change the magnitude of f_k. The external agency is taken here as another *virtual force*, i.e. outside those *real forces* that constitute the equilibrium system. The PVW states that the net virtual work of the system is zero. Thus, by imposing *virtual displacements* at each element node $\mathbf{K}^e$ can be found from a *stiffness method* in which real forces do no virtual work. In an alternative *flexibility method* for finding $\mathbf{K}^e$, virtual forces and real displacements are employed though this method is less often used. In a deformable body the net work is composed of external and internal components. The external work done is written as:

$$W_E = f_k \delta_k = f_1 \delta_1 + f_2 \delta_2 + f_3 \delta_3 + \dots \qquad (13.2a)$$

or, in matrix notation, as

$$W_E = \mathbf{f}^T \boldsymbol{\delta} = \boldsymbol{\delta}^T \mathbf{f} \qquad (13.2b)$$

where $\mathbf{f} = [f_1 \ f_2 \ f_3 \dots]^T$ and $\boldsymbol{\delta} = [\delta_1 \ \delta_2 \ \delta_3 \dots]^T$ denote column vectors. The work of the internal stresses is the negative of the internal store of strain energy $W_I = -U$. Now U follows from integrating the strain energy density, $u = \delta U/\delta V$ with respect to volume, as in Eqs 10.4b and 10.7b. To account for any combination of direct and shear stresses the internal work integral becomes:

$$W_I = - \int \int \sigma_{ij}\, d\varepsilon_{ij}\, dV \qquad (13.3a)$$

In Eq. 13.3a, Einstein's summation convention is implied over the subscripts i and j for coordinates 1, 2 and 3. This means that Eq. 13.3a provides the internal work generally where three independent normal stresses, σ_{11}, σ_{22} and σ_{33}, and three independent shear stresses, σ_{12}, σ_{23} and σ_{31}, act upon a material element. The corresponding incremental strains are three normal components $d\varepsilon_{11}$, $d\varepsilon_{22}$ and $d\varepsilon_{33}$ and three (tensor) shear components $d\varepsilon_{12}$ $d\varepsilon_{23}$ and $d\varepsilon_{31}$. Hence the integrand within Eq. 13.3a becomes:

$$\sigma_{ij}\, d\varepsilon_{ij} = \sigma_{11}\, d\varepsilon_{11} + \sigma_{22}\, d\varepsilon_{22} + \sigma_{33}\, d\varepsilon_{33} + 2(\sigma_{12}\, d\varepsilon_{12} + \sigma_{23}\, d\varepsilon_{23} + \sigma_{31}\, d\varepsilon_{31})$$

$$= \sigma_{11}\, d\varepsilon_{11} + \sigma_{22}\, d\varepsilon_{22} + \sigma_{33}\, d\varepsilon_{33} + \sigma_{12}\, d\gamma_{12} + \sigma_{23}\, d\gamma_{23} + \sigma_{31}\, d\gamma_{31} \qquad (13.3b)$$

where the engineering shear strains are $d\gamma_{12} = 2 \times d\varepsilon_{12}$ etc. The scalar work measure, equivalent to Eq. 13.3b, appears in the matrix form of Eq. 13.3a:

$$W_I = - \int_v \int_\varepsilon \boldsymbol{\sigma}^T d\boldsymbol{\varepsilon}\, dV \qquad (13.3c)$$

in which $\boldsymbol{\sigma}$ and $\boldsymbol{\varepsilon}$ represent column matrices. Again, to save space $\boldsymbol{\sigma}$ and ε may appear as a transpose of each row of their elements: $\boldsymbol{\sigma} = [\sigma_{11} \ \sigma_{22} \ \sigma_{33} \ \sigma_{12} \ \sigma_{23} \ \sigma_{31}]^T$

and $\varepsilon = [\,d\varepsilon_{11} \;\; d\varepsilon_{22} \;\; d\varepsilon_{33} \;\; d\gamma_{12} \;\; d\gamma_{23} \;\; d\gamma_{31}\,]^{T}$. Either of the alternative forms for W_{E} and W_{I}, given in Eqs 13.2a,b and 13.3a,c, may be used when connecting work and energy as follows:

$$W_{E} + W_{I} = 0 \tag{13.3d}$$

There are two ways of expressing the virtual work principle in the index notation. For an element with virtual nodal displacements the *principle of virtual displacements* (PVD) is written as

$$\delta_{k}^{ev} f_{k}^{e} - \int_{V} \varepsilon_{ij}^{v} \sigma_{ij} \, dV = 0 \tag{13.4a}$$

in which superscript e refers to the element number, containing the element's nodal forces and displacements. Superscript v refers to the virtual displacements and their accompanying strains. For an element with virtual nodal forces and stresses the *principle of virtual forces* (PVF) becomes:

$$\delta_{k}^{e} f_{k}^{ev} - \int_{V} \varepsilon_{ij} \, \sigma_{ij}^{v} \, dV = 0 \tag{13.4b}$$

In the matrix notation, PVD is written with its virtual displacements and strains in a form equivalent to Eq. 13.4a:

$$\left(\boldsymbol{\delta}^{ev}\right)^{T} \mathbf{f}^{e} - \int_{V} \left(\boldsymbol{\varepsilon}^{v}\right)^{T} \boldsymbol{\sigma} \, dV = 0 \tag{13.5a}$$

and PVF, with virtual forces and stresses, equivalent to Eq. 13.4b:

$$\left(\boldsymbol{\delta}^{e}\right)^{T} \mathbf{f}^{ev} - \int_{V} \boldsymbol{\varepsilon}^{T} \boldsymbol{\sigma}^{v} \, dV = 0 \tag{13.5b}$$

where superscript T is the transpose of the displacement vector and the strain matrix in each work term. The reversal in the order of matrix multiplication within the internal work integral governs only the choice of which matrix to transpose. It does not alter the magnitude of the resulting scalar products since and $\boldsymbol{\varepsilon}^{T}\boldsymbol{\sigma} = \boldsymbol{\sigma}^{T}\boldsymbol{\varepsilon}$. Similarly, a reversal to the product defining external work applies for: $\mathbf{f}^{T}\boldsymbol{\delta} = \boldsymbol{\delta}^{T}\mathbf{f}$. Each integral in Eq. 13.4a,b (or 13.5a,b) is applied over a volume V, through which the stress and strain will generally vary. In Eq. 13.4a (or 13.5a) the nodal force vector (f_{k}^{e} or $\mathbf{f}^{e}$) and internal stress (σ_{ij} or $\boldsymbol{\sigma}$) are real and in equilibrium. Thus the force vector contains the applied forces acting at the element nodes. The virtual nodal displacements (δ_{k}^{ev} or $\boldsymbol{\delta}^{ev}$) and the virtual internal strains (ε_{ij}^{v} or $\boldsymbol{\varepsilon}^{v}$) must be compatible but independent of the real force/stress system. Thus, in arriving at Eq. 13.4a (or Eq. 13.5a) strain has been integrated independently of stress. The alternative PVF equation, Eq. 13.4b (or Eq. 13.5b), is interpreted in a similar manner when force/stress are virtual and displacement/strain are real.

Specifically, Eq. 13.4a (or Eq. 13.5a) applies to the stiffness method derivation of $\mathbf{K}^{e}$ that follows. The matrix notation Eq. 13.5a will be adopted to derive $\mathbf{K}^{e}$ most effectively. In this equation $\left(\boldsymbol{\delta}^{ev}\right)^{T}$ means the transpose of the element's virtual nodal displacement vector whose components are generally the virtual displacements, rotations and twists at the nodal points. Correspondingly, the 'force' vector $\mathbf{f}^e$ will contain the real nodal forces, moments and torques applied at these nodal points. In the alternative derivation of $\mathbf{K}^e$ by the flexibility method, Eq. 13.4b (or Eq. 13.5b) applies.

13.2.2 Stationary Potential Energy (SPE)

The SPE principal states that of all the compatible displacements which satisfy the boundary conditions, that which also satisfies equilibrium, gives a stationary value to the total potential energy P (see § 10.6, p. 423). It follows that the stationary value P for compatible strains must apply to an equilibrium condition. For a deformable solid we write P as a sum of the strain energy stored U and the negative of the work of external forces, i.e. $T_E = -W_E$. This is a re-statement of Eq. 10.2 in which external work done and strain energy stored were equated in leading to the external/internal work balance in Eq. 13.3d. Here the total potential energy becomes $P = U + T_E$ so the condition for it to attain a stationary value becomes:

$$\Delta P = \Delta(U + T_E) = 0 \tag{13.6}$$

The strain energy stored within a single element of volume V is:

$$U = \int_V \int_\varepsilon \sigma_{ij}\, d\varepsilon_{ij}\, dV = \int_V \int_\varepsilon \boldsymbol{\sigma}^{\mathrm{T}} d\boldsymbol{\varepsilon}\, dV \tag{13.7a}$$

The superscript e (element number) must appear within the negative work done upon each element. With T_E the negative of W_E, Eq. 13.2a becomes:

$$T_E = -f_k^e\, \delta_k^e = -\left(\mathbf{f}^e\right)^{\mathrm{T}} \boldsymbol{\delta}^e \tag{13.7b}$$

Substituting Eqs 13.7a,b into 13.6 gives two alternative forms of SPE that adopt summations over subscripts i, j and k or matrix multiplications respectively

$$\Delta\left(\int_V \int_{\varepsilon_{ij}} \sigma_{ij}\, d\varepsilon_{ij}\, dV - f_k^e\, \delta_k^e \right) = 0 \tag{13.8a}$$

$$\Delta\left(\int_V \int_\varepsilon \boldsymbol{\sigma}^{\mathrm{T}} d\boldsymbol{\varepsilon}\, dV - \left(\mathbf{f}^e\right)^{\mathrm{T}} \boldsymbol{\delta}^e \right) = 0 \tag{13.8b}$$

For a Hookean material there is proportionality between stress σ_{ij} and strain ε_{ij} as both are real in SPE. Hence the strain integration in Eqs 13.8a,b gives:

$$\Delta\left(\frac{1}{2} \int_V \sigma_{ij}\, \varepsilon_{ij}\, dV - f_k^e\, \delta_k^e \right) = 0 \tag{13.9a}$$

$$\Delta\left(\frac{1}{2} \int_V \boldsymbol{\sigma}^{\mathrm{T}} \boldsymbol{\varepsilon}\, dV - \left(\mathbf{f}^e\right)^{\mathrm{T}} \boldsymbol{\delta}^e \right) = 0 \tag{13.9b}$$

In the *stiffness method* of FEM the change in SPE is applied as its partial derivative with respect to the displacement: $\Delta = \partial/\partial\delta_k^e$, when Eq. 13.9a becomes

$$\frac{1}{2}\int_V\left[\sigma_{ij}\left(\frac{\partial\varepsilon_{ij}}{\partial\delta_k^e}\right) + \varepsilon_{ij}\left(\frac{\partial\sigma_{ij}}{\partial\delta_k^e}\right)\right]\mathrm{d}V - f_k^e = 0 \qquad (13.10a)$$

Alternatively, writing $\Delta = \partial/\partial\delta^e$, Eq. 13.9b gives this stationary condition:

$$\frac{1}{2}\int_V\left[\sigma^{\mathrm{T}}\left(\frac{\partial\varepsilon}{\partial\delta^e}\right) + \varepsilon^{\mathrm{T}}\left(\frac{\partial\sigma}{\partial\delta^e}\right)\right]\mathrm{d}V - \left(\mathbf{f}^e\right)^{\mathrm{T}} = 0 \qquad (13.10b)$$

in which the energy density conforms to the scalar relationship: $\boldsymbol{\sigma}^{\mathrm{T}}\boldsymbol{\varepsilon} = \boldsymbol{\varepsilon}^{\mathrm{T}}\boldsymbol{\sigma}$.

In the *flexibility method* of FEM the change in SPE is applied to its partial derivative with respect to the force: $\Delta = \partial/\partial f_k^e$ when Eq. 13.9a becomes

$$\frac{1}{2}\int_V\left[\sigma_{ij}\left(\frac{\partial\varepsilon_{ij}}{\partial f_k^e}\right) + \varepsilon_{ij}\left(\frac{\partial\sigma_{ij}}{\partial f_k^e}\right)\right]\mathrm{d}V - \delta_k^e = 0 \qquad (13.11a)$$

Alternatively, writing $\Delta = \partial/\partial\mathbf{f}^e$, Eq. 13.9b gives this stationary condition:

$$\frac{1}{2}\int_V\left[\sigma^{\mathrm{T}}\left(\frac{\partial\varepsilon}{\partial\mathbf{f}^e}\right) + \varepsilon^{\mathrm{T}}\left(\frac{\partial\sigma}{\partial\mathbf{f}^e}\right)\right]\mathrm{d}V - \left(\boldsymbol{\delta}^e\right)^{\mathrm{T}} = 0 \qquad (13.11b)$$

in which the following scalar relationships have been applied to the scalar measure of work done:

$$\left(\mathbf{f}^e\right)^{\mathrm{T}}\boldsymbol{\delta}^e = \left(\boldsymbol{\delta}^e\right)^{\mathrm{T}}\mathbf{f}^e$$

13.3 Bar Element Under Axial Stress

This simplest loading of a uniform bar element arises from an axial tensile or a compressive force. This element is particularly useful when finding the forces and displacements in the ties and struts of framed structures. Consider a bar element of length L with uniform cross-sectional area A connecting nodal points 1 and 2, in Fig. 13.1

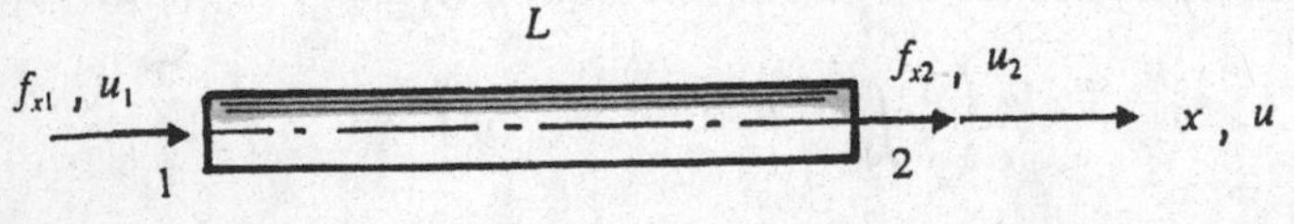

Figure 13.1 Bar element under tension

Let the axis of the bar be aligned with the x - direction. Nodal forces, f_{x1} at node 1 and f_{x2} at node 2, act in the positive x - direction. These produce positive displacements u_1 and u_2. In this case $\mathbf{K}^e$ is obvious from inspection since $f = (AE/L)u$. This reveals that the form of Eq. 13.1a is simply:

$$\begin{bmatrix} f_{x1} \\ f_{x2} \end{bmatrix} = \frac{AE}{L} \begin{bmatrix} 1 & -1 \\ -1 & 1 \end{bmatrix} \begin{bmatrix} u_1 \\ u_2 \end{bmatrix} \tag{13.12}$$

where $\mathbf{f}^e = [\, f_{x1} \ f_{x2} \,]^{\mathrm{T}}$ and $\boldsymbol{\delta}^e = [u_1 \ u_2]^{\mathrm{T}}$. Now let us confirm Eq. 13.12 with the formulation of the element stiffness matrix $\mathbf{K}^e$ using PVD. The latter employs the relationships between column vectors of the bar displacement $\boldsymbol{\delta}$, its strain ε and stress σ.

13.3.1 Displacements

The displacement u in the direction of x is assumed to be linear:

$$u = \alpha_1 + \alpha_2 x \tag{13.13a}$$

The corresponding, full and contracted matrix forms of Eq. 13.13a are:

$$u = [1 \ \ x] \begin{bmatrix} \alpha_1 \\ \alpha_2 \end{bmatrix} \qquad\qquad u = \mathbf{A}\boldsymbol{\alpha} \tag{13.13b}$$

where $\boldsymbol{\alpha} = [\alpha_1 \ \alpha_2]^{\mathrm{T}}$ are constants. Substituting $x = 0$ and $x = L$ into Eq. 13.13a gives:

$$u_1 = \alpha_1 \tag{13.13c}$$

$$u_2 = \alpha_1 + \alpha_2 L \tag{13.13d}$$

Equations 13.13c,d appear within the column matrix of the element's nodal displacements, $\boldsymbol{\delta}^e = [u_1 \ u_2]^{\mathrm{T}}$, as the matrix equations:

$$\begin{bmatrix} u_1 \\ u_2 \end{bmatrix} = \begin{bmatrix} 1 & 0 \\ 1 & L \end{bmatrix} \begin{bmatrix} \alpha_1 \\ \alpha_2 \end{bmatrix} \qquad\qquad \boldsymbol{\delta}^e = \mathbf{A}^e \boldsymbol{\alpha} \tag{13.14a}$$

Now from Eqs 13.13c,d, $\alpha_1 = u_1$ and $\alpha_2 = (u_2 - u_1)/L$. These define the components of the inverse of matrix $\mathbf{A}^e$ as:

$$\begin{bmatrix} \alpha_1 \\ \alpha_2 \end{bmatrix} = \begin{bmatrix} 1 & 0 \\ -1/L & 1/L \end{bmatrix} \begin{bmatrix} u_1 \\ u_2 \end{bmatrix} \qquad\qquad \boldsymbol{\alpha} = (\mathbf{A}^e)^{-1} \boldsymbol{\delta}^e \qquad (13.14b)$$

Combining Eqs 13.13b and 13.14b gives the displacement u in terms of $\boldsymbol{\delta}^e$:

$$u = \mathbf{A}(\mathbf{A}^e)^{-1} \boldsymbol{\delta}^e \qquad (13.15)$$

13.3.2 Strains

The axial strain is $\varepsilon = \partial u/\partial x = \alpha_2$ is found from Eq. 13.13a. Expressing this in general matrix form:

$$\varepsilon = [0 \quad 1] \begin{bmatrix} \alpha_1 \\ \alpha_2 \end{bmatrix} \qquad\qquad \varepsilon = \mathbf{C}\,\boldsymbol{\alpha} \qquad (13.16a,b)$$

Substituting from Eq. 13.14b gives ε in terms of $\boldsymbol{\delta}^e$:

$$\varepsilon = [0 \quad 1] \begin{bmatrix} 1 & 0 \\ -1/L & 1/L \end{bmatrix} \begin{bmatrix} u_1 \\ u_2 \end{bmatrix} \qquad\qquad \varepsilon = \mathbf{C}(\mathbf{A}^e)^{-1} \boldsymbol{\delta}^e \qquad (13.17a,b)$$

$$= [-1/L \quad 1/L] \begin{bmatrix} u_1 \\ u_2 \end{bmatrix} \qquad\qquad \varepsilon = \mathbf{B}\boldsymbol{\delta}^e \qquad (13.18a)$$

where $\mathbf{B} = \mathbf{C}(\mathbf{A}^e)^{-1}$. Since $\mathbf{B}$ does not contain x, it follows from Eq. 13.18a that the nodal strains ε_1 and ε_2 are equal:

$$\begin{bmatrix} \varepsilon_1 \\ \varepsilon_2 \end{bmatrix} = \begin{bmatrix} -1/L & 1/L \\ -1/L & 1/L \end{bmatrix} \begin{bmatrix} u_1 \\ u_2 \end{bmatrix} \qquad\qquad \boldsymbol{\varepsilon}^e = \mathbf{B}^e \boldsymbol{\delta}^e \qquad (13.18b)$$

13.3.3 Stresses

For a uniaxial stress field, stress and strain are related linearly through Hooke's law: $\sigma = E\varepsilon$. Therefore, multiplying Eqs 13.18a,b by E gives both the general and nodal axial stress expressions in their matrix forms, respectively:

$$\sigma = E\left(\mathbf{B}\boldsymbol{\delta}^e\right), \quad \boldsymbol{\sigma}^e = E\left(\mathbf{B}^e\boldsymbol{\delta}^e\right) \qquad (13.19\text{a,b})$$

Equation 13.19a is required for the formulation of the element stiffness matrix. Equations 13.18b and 13.19b enables the nodal strain $\boldsymbol{\varepsilon}^e = [\varepsilon_1 \quad \varepsilon_2]^{\mathrm{T}}$ and then the nodal stress $\boldsymbol{\sigma}^e = E\,\boldsymbol{\varepsilon}^e$ to be found from nodal displacements $\boldsymbol{\delta}^e = [u_1 \quad u_2]^{\mathrm{T}}$.

13.3.4 Stiffness Matrix

Applying PVD to the bar element, Eq. 13.5a is written as:

$$\left(\boldsymbol{\delta}^{ev}\right)^{\mathrm{T}} \mathbf{f}^e = \int_V \varepsilon^v \sigma \, \mathrm{d}V \qquad (13.20\text{a})$$

Substituting the axial strain and stress, Eqs 13.18a and 13.19a, into Eq. 13.20a:

$$\left(\boldsymbol{\delta}^{ev}\right)^{\mathrm{T}} \mathbf{f}^e = \int_V \left(\mathbf{B}\boldsymbol{\delta}^{ev}\right)^{\mathrm{T}} E\left(\mathbf{B}\boldsymbol{\delta}^e\right) \mathrm{d}V = \left(\boldsymbol{\delta}^{ev}\right)^{\mathrm{T}} \left(\int_V \mathbf{B}^{\mathrm{T}} E \mathbf{B} \, \mathrm{d}V\right) \times \boldsymbol{\delta}^e \qquad (13.20\text{b})$$

Since both the real and virtual displacements are not dependent upon the volume of the element they have been removed from the integral. Cancelling $\boldsymbol{\delta}^{ev}$ and putting $\mathbf{f}^e = \mathbf{K}^e \boldsymbol{\delta}^e$, the bar element stiffness matrix $\mathbf{K}^e$ is defined by the integral Eq. 13.20b:

$$\mathbf{K}^e = \int_V \mathbf{B}^{\mathrm{T}} E \mathbf{B} \, \mathrm{d}V \qquad (13.21\text{a})$$

For this bar element E, $\mathbf{B}$ and $\mathbf{B}^{\mathrm{T}}$ are constants in Eq. 13.21a so they may be removed from the integration. This is not the case for, say, a plane triangular element where E is replaced with an elasticity matrix $\mathbf{D}$ that must be retained for matrix multiplication $\mathbf{B}^{\mathrm{T}}\mathbf{D}\mathbf{B}$ within the integral (see § 13.7, p. 580). Substituting $\mathbf{B}$ from Eq. 13.18a with $\mathrm{d}V = A \times \mathrm{d}x$, for an element of uniform cross section, Eq. 13.21a gives:

$$\mathbf{K}^e = AE \begin{bmatrix} -1/L \\ 1/L \end{bmatrix} [-1/L \quad 1/L] \int_0^L \mathrm{d}x = \frac{AE}{L} \begin{bmatrix} 1 & -1 \\ -1 & 1 \end{bmatrix} \qquad (13.21\text{b})$$

This confirms the stiffness matrix given earlier in Eq. 13.12. The matrix dimension, 2×2, describes two nodes each with a single degree of freedom.

13.3.5 Elastic Spring

A bar of length L and uniform section area A, which deflects by the amount δ under a tensile force W, may be regarded as a spring with stiffness $k = W/\Delta = AE/L$. Hence when an elastic spring of stiffness, k, connects nodes 1 and 2 as in Fig. 13.2, a stiffness matrix, similar to Eq. 13.21b, will apply.

Figure 13.2 Single elastic spring with in-line forces

For the spring k replaces AE/L in Eq. 13.21b so that:

$$\begin{bmatrix} f_1 \\ f_2 \end{bmatrix} = k \begin{bmatrix} 1 & -1 \\ -1 & 1 \end{bmatrix} \begin{bmatrix} u_1 \\ u_2 \end{bmatrix} \qquad (13.22)$$

When node 1 is fixed, $u_1 = 0$ and Eq. 13.22 gives: $f_1 = -ku_2 = -f_2$. When node 2 is fixed, $u_2 = 0$ and $f_1 = k u_1 = -f_2$. Next, consider two springs a and b connected in series (see Fig. 13.3).

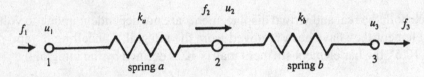

Figure 13.3 Two elastic springs in a series assembly

To assemble the overall stiffness matrix $\mathbf{K}$ in Eq. 13.1d, $\mathbf{f}^e = \mathbf{K}^e \boldsymbol{\delta}^e$ is applied to each spring element. Equation 13.22 gives each individual stiffness matrix:

$$\begin{bmatrix} f_1 \\ f_2 \end{bmatrix} = \begin{bmatrix} k_a & -k_a \\ -k_a & k_a \end{bmatrix} \begin{bmatrix} u_1 \\ u_2 \end{bmatrix}, \qquad \begin{bmatrix} f_2 \\ f_3 \end{bmatrix} = \begin{bmatrix} k_b & -k_b \\ -k_b & k_b \end{bmatrix} \begin{bmatrix} u_2 \\ u_3 \end{bmatrix} \qquad (13.23a,b)$$

Equations 13.23a,b are assembled within $\mathbf{f} = \mathbf{K}\boldsymbol{\delta}$ as follows:

$$\begin{bmatrix} f_1 \\ f_2 \\ f_3 \end{bmatrix} = \begin{bmatrix} k_a & -k_a & 0 \\ -k_a & k_a + k_b & -k_b \\ 0 & -k_b & k_b \end{bmatrix} \begin{bmatrix} u_1 \\ u_2 \\ u_3 \end{bmatrix} \qquad (13.23c)$$

Note that the overall stiffness matrix in Eq. 13.23c is symmetrical with a zero sum for the elements within each row and column. Also, the positive terms forming the leading diagonal are composed of the sum of the stiffness between adjacent elements. Equation 13.23c is solved for the displacements by an inversion of the overall stiffness matrix (i.e. $\boldsymbol{\delta} = \mathbf{K}^{-1} \mathbf{f}$). The solution must match prescribed displacements (the boundary conditions). For example, let this spring assembly be anchored at its left end. Then the nodal displacement $u_1 = 0$ and the spring forces p_a and p_b follow from the relative displacement between their nodes:

$$p_a = k_a(u_2 - u_1) \qquad\qquad (13.24a)$$

$$p_b = k_b(u_3 - u_2) \qquad\qquad (13.24b)$$

A similar procedure applies to the assembly of the overall stiffness matrix for interconnected finite elements.

13.4 Plane Frame

In Chapter 2 the bar forces in a plane, pin-jointed frame were found from the equations of force and moment equilibrium (see § 2.4, p. 43). To illustrate the application of Eq. 13.21b when finding bar forces consider three bars I, II and III in a frame with hinges at 1 and 3, connected with a pin at joint 2 (see Fig. 13.4).

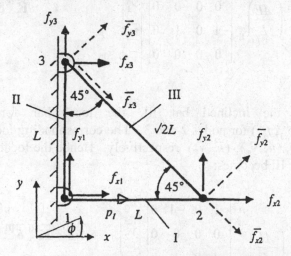

Figure 13.4 Three-bar plane frame

If an inclined load is applied at node 2 this resolves into forces f_{x2} and f_{y2} as shown. These are reacted by the component forces f_{x1}, f_{y1} and f_{x3}, f_{y3} within the hinge connections at nodes 1 and 3 respectively. With bars I being aligned with the frame's global co-ordinates (x, y), its stiffness matrix may be formed directly from Eq. 13.21b. Thus, for bar I, connecting nodes 1 and 2:

$$\begin{bmatrix} f_{x1} \\ f_{y1} \\ f_{x2} \\ f_{y2} \end{bmatrix} = \left(\frac{AE}{L}\right)_I \begin{bmatrix} 1 & 0 & -1 & 0 \\ 0 & 0 & 0 & 0 \\ -1 & 0 & 1 & 0 \\ 0 & 0 & 0 & 0 \end{bmatrix} \begin{bmatrix} u_1 \\ v_1 \\ u_2 \\ v_2 \end{bmatrix} \qquad \mathbf{f}^I = \mathbf{K}^I \boldsymbol{\delta}^I \qquad (13.25a,b)$$

where u_1 and v_1 are components of the displacement of node 1 aligned with *Cartesian co-ordinate* directions x and y, respectively. Similarly, u_2 and v_2 are the Cartesian components of the displacement for node 2.

The stiffness of both 'inclined' bars II and III must first be expressed from Eq. 13.21b in terms of their *local co-ordinates* $(\bar{x}, \bar{y})$, with $\bar{x}$ lying parallel to and $\bar{y}$ lying perpendicular to each bar's direction. For bar II the local bar force components become: $(\bar{f}_{x1}, \bar{f}_{x1})$, $(\bar{f}_{x3}, \bar{f}_{x3})$ and the local displacement components are: $(\bar{u}_1, \bar{v}_1)$, $(\bar{u}_3, \bar{v}_3)$ for nodes 1 and 3, respectively. The local stiffness matrix of bar II connects force and displacement locally between nodes 1 and 3 as:

$$\begin{bmatrix} \bar{f}_{x1} \\ \bar{f}_{y1} \\ \bar{f}_{x3} \\ \bar{f}_{y3} \end{bmatrix} = \left(\frac{AE}{L}\right)_{II} \begin{bmatrix} 1 & 0 & -1 & 0 \\ 0 & 0 & 0 & 0 \\ -1 & 0 & 1 & 0 \\ 0 & 0 & 0 & 0 \end{bmatrix} \begin{bmatrix} \bar{u}_1 \\ \bar{v}_1 \\ \bar{u}_3 \\ \bar{v}_3 \end{bmatrix} \qquad \bar{\mathbf{f}}^{II} = \bar{\mathbf{K}}^{II} \bar{\boldsymbol{\delta}}^{II} \qquad (13.26a,b)$$

Similarly, for the inclined bar III, the local bar force components are, $(\bar{f}_{x2}, \bar{f}_{x2})$, $(\bar{f}_{x3}, \bar{f}_{x3})$ for nodes 2 and 3. The corresponding local displacement components are, $(\bar{u}_2, \bar{v}_2)$, $(\bar{u}_3, \bar{v}_3)$ respectively. Hence the local stiffness matrix equation for bar III becomes:

$$\begin{bmatrix} \bar{f}_{x2} \\ \bar{f}_{y2} \\ \bar{f}_{x3} \\ \bar{f}_{y3} \end{bmatrix} = \left(\frac{AE}{L}\right)_{III} \begin{bmatrix} 1 & 0 & -1 & 0 \\ 0 & 0 & 0 & 0 \\ -1 & 0 & 1 & 0 \\ 0 & 0 & 0 & 0 \end{bmatrix} \begin{bmatrix} \bar{u}_2 \\ \bar{v}_2 \\ \bar{u}_3 \\ \bar{v}_3 \end{bmatrix} \qquad \bar{\mathbf{f}}^{III} = \bar{\mathbf{K}}^{III} \bar{\boldsymbol{\delta}}^{III} \qquad (13.27a,b)$$

13.4.1 Local to Global Co-ordinates

Before the overall stiffness matrix **K** for the frame can be assembled, the local co-ordinate forces and displacement vectors, **f** and $\boldsymbol{\delta}$, must be converted to their global equivalents **f** and **δ**. The transformation diagram between local and global force components is given in Fig. 13.5 for an inclination φ between their local and global co-ordinates.

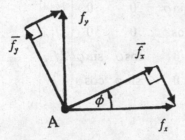

Figure 13.5 Local and global co-ordinate forces for a single node

The diagram shows that for any given node A (say):

$$\bar{f}_x = f_x \cos\varphi + f_y \sin\varphi \qquad (13.28a)$$

$$\bar{f}_y = f_y \cos\varphi - f_x \sin\varphi \qquad (13.28b)$$

Expressing Eqs 13.28a,b in matrix form provides the relation between the local and global force vectors $\bar{\mathbf{f}}$ and $\mathbf{f}$ is:

$$\begin{bmatrix} \bar{f}_x \\ \bar{f}_y \end{bmatrix} = \begin{bmatrix} \cos\varphi & \sin\varphi \\ -\sin\varphi & \cos\varphi \end{bmatrix} \begin{bmatrix} f_x \\ f_y \end{bmatrix} \qquad \bar{\mathbf{f}} = \mathbf{M}\mathbf{f} \qquad (13.29a,b)$$

where **M** is a skew-symmetric matrix of *direction cosines* m_{ij}. The specific components m_{ij} appearing in Eq. 13.29a follow from the cosine of the angle between the local and global co-ordinates as follows

$$m_{11} = \cos(\bar{x}x) = \cos\varphi, \; m_{12} = \cos(\bar{x}y) = \cos(90° - \varphi) = \sin\varphi,$$

$$m_{21} = \cos(\bar{y}x) = \cos(90° + \varphi) = -\sin\varphi \text{ and } m_{22} = \cos(\bar{y}y) = \cos\varphi$$

These components show that **M** is *orthogonal*, which means $\mathbf{M}^T = \mathbf{M}^{-1}$. This matrix also transforms displacements between co-ordinates such that the relation between the local and global displacement vectors $\bar{\mathbf{δ}}$ and $\mathbf{δ}$ is:

$$\begin{bmatrix} \bar{u} \\ \bar{v} \end{bmatrix} = \begin{bmatrix} \cos\varphi & \sin\varphi \\ -\sin\varphi & \cos\varphi \end{bmatrix} \begin{bmatrix} u \\ v \end{bmatrix} \qquad \bar{\mathbf{δ}} = \mathbf{M}\mathbf{δ} \qquad (13.30a,b)$$

Applying Eq. 13.29a to the two nodes A and B (say) for an inclined bar e, gives:

$$
\begin{bmatrix} \bar{f}_{xA} \\ \bar{f}_{yA} \\ \bar{f}_{xB} \\ \bar{f}_{yB} \end{bmatrix} = \begin{bmatrix} \cos\varphi & \sin\varphi & 0 & 0 \\ -\sin\varphi & \cos\varphi & 0 & 0 \\ 0 & 0 & \cos\varphi & \sin\varphi \\ 0 & 0 & -\sin\varphi & \cos\varphi \end{bmatrix} \begin{bmatrix} f_{xA} \\ f_{yA} \\ f_{xB} \\ f_{yB} \end{bmatrix} \qquad \bar{\mathbf{f}}^e = \mathbf{T}\mathbf{f}^e \qquad (13.31a,b)
$$

Equation 13.31a shows how $\mathbf{T}$ is formed from $\mathbf{M}$ so retaining both skew-symmetry and orthogonality: $\mathbf{T}^{\mathrm{T}} = \mathbf{T}^{-1}$. Similarly, from Eq. 13.30a, the nodal displacements for this bar are:

$$
\begin{bmatrix} \bar{u}_A \\ \bar{v}_A \\ \bar{u}_B \\ \bar{v}_B \end{bmatrix} = \begin{bmatrix} \cos\varphi & \sin\varphi & 0 & 0 \\ -\sin\varphi & \cos\varphi & 0 & 0 \\ 0 & 0 & \cos\varphi & \sin\varphi \\ 0 & 0 & -\sin\varphi & \cos\varphi \end{bmatrix} \begin{bmatrix} u_A \\ v_A \\ u_B \\ v_B \end{bmatrix} \qquad \bar{\boldsymbol{\delta}}^e = \mathbf{T}\boldsymbol{\delta}^e \qquad (13.32a,b)
$$

Substituting Eqs 13.31b and 13.32b into $\bar{\mathbf{f}}^e = \bar{\mathbf{K}}^e \bar{\boldsymbol{\delta}}^e$ gives:

$$
\mathbf{T}\mathbf{f}^e = \mathbf{K}^e\left(\mathbf{T}\boldsymbol{\delta}^e\right)
$$

$$
\mathbf{f}^e = \mathbf{T}^{-1}\mathbf{K}^e\left(\mathbf{T}\boldsymbol{\delta}^e\right) = \mathbf{T}^{\mathrm{T}}\mathbf{K}^e\left(\mathbf{T}\boldsymbol{\delta}^e\right)
$$

and since $\mathbf{f}^e = \mathbf{K}^e\boldsymbol{\delta}^e$, a local to global transformation of the stiffness matrix follows:

$$
\mathbf{K}^e = \mathbf{T}^e \bar{\mathbf{K}}^e \mathbf{T} \qquad (13.33a)
$$

The transformation: $\bar{\mathbf{K}}^e \to \mathbf{K}^e$ for the inclined bars $e = $ II and III, requires that $\bar{\mathbf{K}}^e$ from Eqs 13.26a and 13.27a, with $\mathbf{T}$ from Eq. 13.31a, be substituted into Eq. 13.33a. The following matrix multiplication applies to the global co-ordinate system, i.e. the Cartesian co-ordinate frame (x, y) shown inset in Fig. 13.4 in which φ defines the orientation of each bar axis with respect to x:

$$
\mathbf{K}^e = \left(\frac{AE}{L}\right)_e \begin{bmatrix} \cos\varphi & -\sin\varphi & 0 & 0 \\ \sin\varphi & \cos\varphi & 0 & 0 \\ 0 & 0 & \cos\varphi & -\sin\varphi \\ 0 & 0 & \sin\varphi & \cos\varphi \end{bmatrix} \begin{bmatrix} 1 & 0 & -1 & 0 \\ 0 & 0 & 0 & 0 \\ -1 & 0 & 1 & 0 \\ 0 & 0 & 0 & 0 \end{bmatrix} \begin{bmatrix} \cos\varphi & \sin\varphi & 0 & 0 \\ -\sin\varphi & \cos\varphi & 0 & 0 \\ 0 & 0 & \cos\varphi & \sin\varphi \\ 0 & 0 & -\sin\varphi & \cos\varphi \end{bmatrix}
$$

$$= \left(\frac{AE}{L}\right)_e \begin{bmatrix} \cos^2\varphi & \cos\varphi\sin\varphi & -\cos^2\varphi & -\sin\varphi\cos\varphi \\ \sin\varphi\cos\varphi & \sin^2\varphi & -\sin\varphi\cos\varphi & -\sin^2\varphi \\ -\cos^2\varphi & -\cos\varphi\sin\varphi & \cos^2\varphi & \sin\varphi\cos\varphi \\ -\sin\varphi\cos\varphi & -\sin^2\varphi & \sin\varphi\cos\varphi & \sin^2\varphi \end{bmatrix} \qquad (13.33b)$$

Thus, the matrix in Eq. 13.33b defines the stiffness of an inclined bar in global co-ordinates. Notice that this matrix is square and symmetrical. For the bars I, II and III in Fig. 13.4, $\varphi = 0°$, $90°$ and $135°$, respectively. Setting $\varphi = 0°$ in Eq. 13.33b recovers Eq. 13.25b as would be expected for a bar aligned with x.

13.4.2 Frame's Global Stiffness Matrices

From here it will be taken that the frame's bars are of similar area and in the same material. Equations 13.33a,b will transform the local stiffness matrix to a global matrix for each bars within the frame. When applied to the horizontal bar I with $\varphi = 0°$ it is confirmed that the local stiffness matrix in Eq. 13.25a remains unaltered. However, when applied to the vertical bar II, where $\varphi = 90°$, Eq.13.33b gives,

$$\begin{bmatrix} f_{x1} \\ f_{y1} \\ f_{x3} \\ f_{y3} \end{bmatrix} = \left(\frac{AE}{L}\right)_{II} \begin{bmatrix} 0 & 0 & 0 & 0 \\ 0 & 1 & 0 & -1 \\ 0 & 0 & 0 & 0 \\ 0 & -1 & 0 & 1 \end{bmatrix} \begin{bmatrix} u_1 \\ v_1 \\ u_3 \\ v_3 \end{bmatrix} \qquad \mathbf{f}^{II} = \mathbf{K}^{II}\boldsymbol{\delta}^{II} \qquad (13.34a,b)$$

Comparing the local and global matrix multiplications from Eqs 13.26a and 13.34a shows what the transformation does in this case:

$$\bar{f}_{x1} = (AE/L)_{II}(\bar{u}_1 - \bar{u}_3) \quad \rightarrow \quad f_{y1} = (AE/L)_{II}(v_1 - v_3)$$

$$\bar{f}_{y1} = 0 \quad \rightarrow \quad f_{x1} = 0$$

$$\bar{f}_{x3} = (AE/L)_{II}(-\bar{u}_1 + \bar{u}_3) \quad \rightarrow \quad f_{y3} = (AE/L)_{II}(-v_1 + v_3)$$

$$\bar{f}_{y3} = 0 \quad \rightarrow \quad f_{x3} = 0$$

That is the bar's local co-ordinates $(\bar{x}, \bar{y})$, within which the force and displacement are defined, are interchanged with the frame's global co-ordinates (y, x). This is to be expected for the assembly of the frame's global stiffness matrix.

For the inclined bar III, setting $L_{III} = \sqrt{2}L$ and $\varphi = 135°$ in Eq. 13.33b, the global force-displacement relation for bar III becomes:

$$\begin{bmatrix} f_{x2} \\ f_{y2} \\ f_{x3} \\ f_{y3} \end{bmatrix} = \frac{AE}{2\sqrt{2}L} \begin{bmatrix} 1 & -1 & -1 & 1 \\ -1 & 1 & 1 & -1 \\ -1 & 1 & 1 & -1 \\ 1 & -1 & -1 & 1 \end{bmatrix} \begin{bmatrix} u_2 \\ v_2 \\ u_3 \\ v_3 \end{bmatrix} \qquad \mathbf{f}^{\mathrm{III}} = \mathbf{K}^{\mathrm{III}} \boldsymbol{\delta}^{\mathrm{III}} \qquad (13.35a,b)$$

It is now possible to assemble the overall stiffness matrix from Eqs 13.25a, 13.34a and 13.35a. As a learning exercise the three matrices are written in terms of the equations they represent. These provide the coefficients of *all* displacements u_1, v_1 ... etc. upon which *each* force f_{x1}, f_{y1} ... etc. depends as follows:

$$Lf_{x1}/AE = (1+0)u_1 + (0+0)v_1 + (-1)u_2 + (0)v_2 + (0)u_3 + (0)v_3$$

$$Lf_{y1}/AE = (0+0)u_1 + (0+1)v_1 + (0)u_2 + (0)v_2 + (0)u_3 + (-1)v_3$$

$$Lf_{x2}/AE = (-1)u_1 + (0)v_1 + (1+1/2\sqrt{2})u_2 + (0-1/2\sqrt{2})v_2 + (-1/2\sqrt{2})u_3 + (1/2\sqrt{2})v_3$$

$$Lf_{y2}/AE = (0)u_1 + (0)v_1 + (0-1/2\sqrt{2})u_2 + (0+1/2\sqrt{2})v_2 + (1/2\sqrt{2})u_3 + (-1/2\sqrt{2})v_3$$

$$Lf_{x3}/AE = (0)u_1 + (0)v_1 + (-1/2\sqrt{2})u_2 + (1/2\sqrt{2})v_2 + (0+1/2\sqrt{2})u_3 + (0-1/2\sqrt{2})v_3$$

$$Lf_{y3}/AE = (0)u_1 + (-1)v_1 + (1/2\sqrt{2})u_2 + (-1/2\sqrt{2})v_2 + (0-1/2\sqrt{2})u_3 + (1+1/2\sqrt{2})v_3$$

These coefficients become the components of the global stiffness matrix. When the known boundary conditions are applied in which u and v are both zero for nodes 1 and 3, this gives:

$$\begin{bmatrix} f_{x1} \\ f_{y1} \\ f_{x2} \\ f_{y2} \\ f_{x3} \\ f_{y3} \end{bmatrix} = \frac{AE}{L} \begin{bmatrix} 1 & 0 & -1 & 0 & 0 & 0 \\ 0 & 1 & 0 & 0 & 0 & -1 \\ -1 & 0 & 1+1/2\sqrt{2} & -1/2\sqrt{2} & -1/2\sqrt{2} & 1/2\sqrt{2} \\ 0 & 0 & -1/2\sqrt{2} & 1/2\sqrt{2} & 1/2\sqrt{2} & -1/2\sqrt{2} \\ 0 & 0 & -1/2\sqrt{2} & 1/2\sqrt{2} & 1/2\sqrt{2} & -1/2\sqrt{2} \\ 0 & -1 & 1/2\sqrt{2} & -1/2\sqrt{2} & -1/2\sqrt{2} & 1+1/2\sqrt{2} \end{bmatrix} \begin{bmatrix} u_1 = 0 \\ v_1 = 0 \\ u_2 \\ v_2 \\ u_3 = 0 \\ v_3 = 0 \end{bmatrix} \qquad (13.36a)$$

$$\mathbf{f} = \mathbf{K}\boldsymbol{\delta} \qquad (13.36b)$$

We see from Eqs 13.36a,b that the overall stiffness matrix $\mathbf{K}$ for this frame is symmetrical with a zero sum for the elements of its rows and columns, similar to the matrix for the spring assembly (see Eq. 13.23c). Expanding Eq. 13.36a gives the following simultaneous equations:

$$f_{x1} = - \frac{AE}{L}u_2, \quad f_{y1} = 0$$

$$f_{x2} = \frac{AE}{L}\left[\left(1 + \frac{1}{2\sqrt{2}}\right)u_2 - \frac{1}{2\sqrt{2}}v_2\right]$$

$$f_{y2} = \frac{AE}{L}\left[-\frac{1}{2\sqrt{2}}u_2 + \frac{1}{2\sqrt{2}}v_2\right] \qquad (13.36\text{c–h})$$

$$f_{x3} = \frac{AE}{L}\left[-\frac{1}{2\sqrt{2}}u_2 + \frac{1}{2\sqrt{2}}v_2\right]$$

$$f_{y3} = \frac{AE}{L}\left[\frac{1}{2\sqrt{2}}u_2 - \frac{1}{2\sqrt{2}}v_2\right]$$

Solving Eqs 13.36c–h provides the displacements at node 2:

$$u_2 = \frac{L}{AE}\left(f_{x2} + f_{y2}\right) \qquad (13.37\text{a})$$

$$v_2 = \frac{L}{AE}\left[(1 + 2\sqrt{2})f_{y2} + f_{x2}\right] \qquad (13.37\text{b})$$

Combining Eqs 13.36c–h with the force equilibrium conditions provides the reactions at the support points (nodes 1 and 3):

$$f_{x1} = -\left(f_{x2} + f_{x3}\right) = -\left(f_{x2} + f_{y2}\right)$$

$$f_{y1} = -\left(f_{y2} + f_{y3}\right) = 0$$

$$f_{x3} = f_{y2}, \quad f_{y3} = -f_{y2}$$

Finally, combining Eqs 13.24a,b with Eqs 13.37a,b provides the global bar forces p from the individual bar stiffness AE/L as:

$$p_{\mathrm{I}} = \left(\frac{AE}{L}\right)_{\mathrm{I}}(u_2 - u_1) = \left(\frac{AE}{L}\right)_{\mathrm{I}}u_2 = f_{x2} + f_{y2} \qquad (13.38\text{a})$$

$$p_{\mathrm{II}} = \left(\frac{AE}{L}\right)_{\mathrm{II}}(v_3 - v_1) = 0 \qquad (13.38\text{b})$$

$$p_{\mathrm{III}} = \left(\frac{AE}{L}\right)_{\mathrm{III}}(\bar{u}_3 - \bar{u}_2) \qquad (13.38\text{c})$$

In Eq. 13.38c, $\bar{u}_2$ and $\bar{u}_3$ are displacements in local co-ordinates, which appear in the equivalent matrix equation:

$$p_{\mathrm{III}} = \left(\frac{AE}{L}\right)_{\mathrm{III}} [1 \quad 0] \begin{bmatrix} (\bar{u}_3 - \bar{u}_2) \\ (\bar{v}_3 - \bar{v}_2) \end{bmatrix} \tag{13.39a}$$

The next step is to apply Eq. 13.30a to convert Eq. 13.39a from local to global displacements:

$$p_{\mathrm{III}} = \left(\frac{AE}{L}\right)_{\mathrm{III}} [1 \quad 0] \begin{bmatrix} \cos\varphi & \sin\varphi \\ -\sin\varphi & \cos\varphi \end{bmatrix} \begin{bmatrix} (u_3 - u_2) \\ (v_3 - v_2) \end{bmatrix} \tag{13.39b}$$

Setting $L_{\mathrm{III}} = \sqrt{2}L$ and $\varphi = -45°$ in Eq. 13.39b gives p_{III} as:

$$p_{\mathrm{III}} = \left(\frac{AE}{\sqrt{2}L}\right) [1 \quad 0] \begin{bmatrix} 1/\sqrt{2} & -1/\sqrt{2} \\ 1/\sqrt{2} & 1/\sqrt{2} \end{bmatrix} \begin{bmatrix} (u_3 - u_2) \\ (v_3 - v_2) \end{bmatrix} \tag{13.39c}$$

The multiplication in Eq. 13.39c gives

$$p_{\mathrm{III}} = \left(\frac{AE}{2L}\right) [(u_3 - u_2) - (v_3 - v_2)] \tag{13.39d}$$

Finally, setting $u_3 = v_3 = 0$ in Eq. 13.39d and substituting u_2 and v_2 from Eq. 13.37a,b, gives $p_{\mathrm{III}} = \sqrt{2}f_{y2}$. The direction of the bar forces follow the positive differences assumed between the nodal displacements, for example, as $u_2 - u_1$ is positive then p_1 acts as shown.

The reader may well ask what advantage this protracted procedure offers compared to bar force calculations using those direct equilibrium methods adopted in § 2.4 (p. 43). The answer is that the steps outlined above are programmed more readily, becoming most effective for frames with many bars. The user need only to specify the nodal co-ordinates and forces and to prescribe the boundary conditions. Both the bar forces and the nodal displacements appear in the numerical solution.

13.5 Torsion Element

In the formulation of the stiffness matrix for torsion the element is taken as a uniform circular shaft subjected to a torque t that varies with length z, i.e. $t = t(z)$. Thus with the end nodal torques t_1 and t_2, applied as shown in Fig. 13.6, the element nodes 1 and 2, twist by amounts θ_1 and θ_2 relative to the undeformed bar.

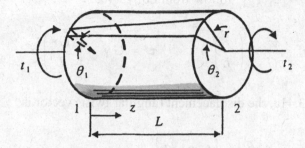

Figure 13.6 Bar element under pure torsion

The torsion Eq. 7.6 applies to the element's length L as:

$$\frac{t}{J} = \frac{G\theta}{L} = \frac{\tau}{r} \qquad (13.40a)$$

where $\theta = \theta_1 - \theta_2$ is the twist between the ends, $J = \pi d^4/32$ is the polar second moment of area, G is the rigidity modulus and τ is the shear stress at a radius r. Equation 13.40a provides the components of the torsional stiffness matrix as:

$$k = \frac{dt}{d\theta} = \frac{JG}{L} \qquad (13.40b)$$

With the torques t_1 and t_2 being the only actions applied at each node, the angular twists θ_1 and θ_2 are assumed to vary linearly with z:

$$\theta(z) = \alpha_1 + \alpha_2 z \qquad (13.41a)$$

from which the 'rate' of twist becomes constant:

$$\frac{d\theta}{dz} = \alpha_2 \qquad (13.41b)$$

It can be deduced that the element stiffness matrix is 2×2 since there are two nodes with a single degree of freedom at each node, similar to that derived previously for the bar element under tension. For a single bar element, $\mathbf{K}^e$ is found directly from Eqs 13.40b:

$$\begin{bmatrix} t_1 \\ t_2 \end{bmatrix} = \left(\frac{JG}{L} \right) \begin{bmatrix} 1 & -1 \\ -1 & 1 \end{bmatrix} \begin{bmatrix} \theta_1 \\ \theta_2 \end{bmatrix} \qquad \mathbf{f}^e = \mathbf{K}^e \boldsymbol{\delta}^e \qquad (13.42a,b)$$

where $\mathbf{f}^e = [\, t_1 \;\; t_2 \,]^{\mathrm{T}}$ and $\boldsymbol{\delta}^e = [\, \theta_1 \;\; \theta_2 \,]^{\mathrm{T}}$.

Once the nodal twists have been found the 'vectors' of shear strain $\boldsymbol{\gamma}^e = [\, \gamma_1 \; \gamma_2 \,]^{\mathrm{T}}$ and shear stress $\boldsymbol{\tau}^e = [\, \tau_1 \; \tau_2 \,]^{\mathrm{T}}$ follow from Eq. 13.40a:

$$\boldsymbol{\gamma}^e = \frac{\boldsymbol{\tau}^e}{G} = \left(\frac{r}{L} \right) \boldsymbol{\delta}^e, \qquad \boldsymbol{\tau}^e = G\boldsymbol{\gamma} = \left(\frac{Gr}{L} \right) \boldsymbol{\delta}^e \qquad (13.43\mathrm{a,b})$$

where from Eq. 13.41a, the displacement (angular twist) vector is:

$$\boldsymbol{\delta}^e = \begin{bmatrix} \theta_1 \\ \theta_2 \end{bmatrix} = \begin{bmatrix} 1 & 0 \\ 1 & L \end{bmatrix} \begin{bmatrix} \alpha_1 \\ \alpha_2 \end{bmatrix} \qquad\qquad (13.43\mathrm{c})$$

Again, $\mathbf{K}^e$ has been found by inspection. The reader should confirm that Eq. 13.42a also follows from applying either one of the energy principles outlined in § 13.2 (see Exercise 13.6).

13.5.1 Overall Stiffness Matrix

Equation 13.42a is used to form the overall stiffness matrix for the stepped shaft shown in Fig. 13.7. Here the lengths L and shear moduli G are taken to vary within each of the four cylindrical elements shown.

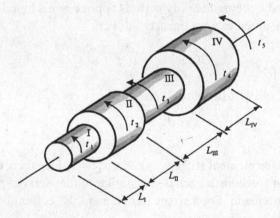

Figure 13.7 Stepped shaft under concentrated torques

Let elements ($e = $ I, II, III and IV) describe the four cylinders with their nodes 1, 2, 3, 4 and 5 co-incident with each junction. The torques $t_1, t_2 \ldots t_5$ are applied at these junctions and so the overall stiffness matrix is assembled immediately as:

$$\begin{bmatrix} t_1 \\ t_2 \\ t_3 \\ t_4 \\ t_5 \end{bmatrix} = \begin{bmatrix} (JG/L)_{\mathrm{I}} & (JG/L)_{\mathrm{I}} & 0 & 0 & 0 \\ -(JG/L)_{\mathrm{I}} & (JG/L)_{\mathrm{I}} + (JG/L)_{\mathrm{II}} & -(JG/L)_{\mathrm{II}} & 0 & 0 \\ 0 & -(JG/L)_{\mathrm{II}} & (JG/L)_{\mathrm{II}} + (JG/L)_{\mathrm{III}} & -(JG/L)_{\mathrm{III}} & 0 \\ 0 & 0 & -(JG/L)_{\mathrm{III}} & (JG/L)_{\mathrm{III}} + (JG/L)_{\mathrm{IV}} & -(JG/L)_{\mathrm{IV}} \\ 0 & 0 & 0 & -(JG/L)_{\mathrm{IV}} & (JG/L)_{\mathrm{IV}} \end{bmatrix} \begin{bmatrix} \theta_1 \\ \theta_2 \\ \theta_3 \\ \theta_4 \\ \theta_5 \end{bmatrix} \quad (13.44a)$$

The form of the banded stiffness matrix, that appears in Eq. 13.44a, is particularly suitable for the inversion:

$$\mathbf{f} = \mathbf{K}\boldsymbol{\delta} \quad \therefore \quad \boldsymbol{\delta} = \mathbf{K}^{-1}\mathbf{f} \qquad (13.44b,c)$$

The solution for $\boldsymbol{\delta}$ follows a Gauss-Seidal numerical iterative procedure, this being simplified further when, more usually G is constant.

13.5.2 Combined Torsion and Tension

The separate stiffness matrices, derived above for axial tension and torsion must be combined where a bar is subjected to a combination of each. Moreover, there are many examples where torsion is combined with an axial compression. The corresponding element in Fig. 13.8, carries axial forces f_{x1} and f_{x2}, tensile or compressive, combined with torques t_1 and t_2 at its nodes.

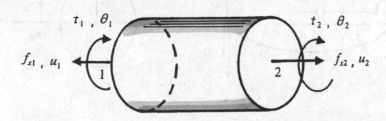

Figure 13.8 Bar element under combined tension–torsion

Here the solution requires a combination of the bar element stiffness matrices appropriate to torsion and axial tension. With two degrees of freedom at each of the two nodes, it follows that $\mathbf{K}^e$ is 4×4. The stiffness matrix is found from superposition of Eqs 13.21b and 13.42b:

$$
\begin{bmatrix} f_{x1} \\ t_1 \\ f_{x2} \\ t_2 \end{bmatrix} = \begin{bmatrix} (AE/L) & 0 & -(AE/L) & 0 \\ 0 & (JG/L) & 0 & -(JG/L) \\ -(AE/L) & 0 & (AE/L) & 0 \\ 0 & -(JG/L) & 0 & (JG/L) \end{bmatrix} \begin{bmatrix} u_1 \\ \theta_1 \\ u_2 \\ \theta_2 \end{bmatrix} \qquad \mathbf{f}^e = \mathbf{K}^e\,\boldsymbol{\delta}^e
$$

(13.45a,b)

In the symbolic matrix notation of Eq. 13.45a, the element stiffness matrix $\mathbf{K}^e$ connects the following combined force and displacement vectors:

$$
\mathbf{f}^e = \begin{bmatrix} \mathbf{f}_1 & \mathbf{f}_2 \end{bmatrix}^T = \begin{bmatrix} f_{x1} & t_1 & f_{x2} & t_2 \end{bmatrix}^T
$$

(13.45cb)

$$
\boldsymbol{\delta}^e = \begin{bmatrix} \boldsymbol{\delta}_1 & \boldsymbol{\delta}_2 \end{bmatrix}^T = \begin{bmatrix} u_1 & \theta_1 & u_2 & \theta_2 \end{bmatrix}^T
$$

(13.45d)

A similar combination applies to the components of $\mathbf{K}^e$ when forces f_x and f_y act together with a torque at each node. For this, it is necessary to combine with 13.42b the (4×4) stiffness matrix $\mathbf{K}^e$ in Eq. 13.25a, that applies to the global, x - aligned bar I in the frame of Fig. 13.4, (see Exercise 13.8).

13.6 Beam Element

The beam element shown in Fig. 13.9a will provide the displacements, strains and stresses under bending. At each of the element's nodes, 1 and 2, a shear force q and a moment m act to produce a deflection v and a rotation θ.

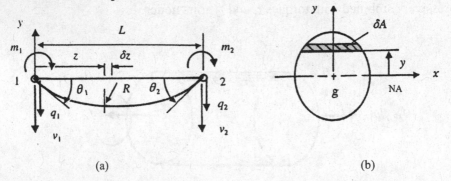

(a)   (b)

Figure 13.9 Beam element showing length and cross section

The directions of q, m, v and θ are aligned with those given in Fig. 13.9a and are positive when the sense of each is as shown. The column vectors of nodal forces and displacements are, respectively:

$$
\mathbf{f} = \begin{bmatrix} q & m \end{bmatrix}^T \quad \text{and} \quad \boldsymbol{\delta} = \begin{bmatrix} v & \theta \end{bmatrix}^T
$$

The axes x and y pass through the centroid g of the arbitrary cross section given in Fig. 13.9b. Direction z coincides with the longitudinal centroidal axis for the unloaded beam. This axis deflects in the loaded beam to a curve with a radius R at position z in Fig. 13.9a. It was shown in § 5.1, (see p. 149) that, despite its curvature R, the centroidal axis remains unstrained, i.e. it is the *neutral axis* (NA).

13.6.1 Displacements

In Chapter 6 (see Example 6.1, p. 195) it can be seen that, following a second integration of the flexure equation $d^2v/dz^2 = m(z)$, the displacement v equation for a beam under concentrated forces and moments becomes cubic in z. Here too the dependence of v upon z is expressed in a cubic polynomial function:

$$v = \alpha_1 + \alpha_2 z + \alpha_3 z^2 + \alpha_4 z^3 \qquad (13.46a)$$

where α depends upon the loading (i.e. the coefficients in Examples 6.1–6.3). The slope and curvature of the deflected beam follow as:

$$\theta = \frac{dv}{dz} = \alpha_2 + 2\alpha_3 z + 3\alpha_4 z^2 \qquad (13.46b)$$

$$\frac{1}{R} = \frac{d^2v}{dz^2} = 2\alpha_3 + 6\alpha_4 z \qquad (13.46c)$$

For any position z in the length, the beam's slope and deflection constitute the beam's displacements vector $\delta = [v \ \theta]^T$, which follows from Eqs 13.46a,b as:

$$\begin{bmatrix} v \\ \theta \end{bmatrix} = \begin{bmatrix} 1 & z & z^2 & z^3 \\ 0 & 1 & 2z & 3z^2 \end{bmatrix} \begin{bmatrix} \alpha_1 \\ \alpha_2 \\ \alpha_3 \\ \alpha_4 \end{bmatrix} \qquad\qquad \delta = A\,\alpha \qquad (13.47a,b)$$

Setting $z = 0$ and $z = L$ in Eq. 13.47a, provides the beam element's two nodal displacement vectors:

$$\delta_1 = [v_1 \ \theta_1]^T = [\alpha_1 \ \alpha_2]^T$$

$$\delta_2 = [v_2 \ \theta_2]^T = [\alpha_1 + L\alpha_2 + L^2\alpha_3 + L^3\alpha_4; \ \ \alpha_2 + 2L\alpha_3 + 3L^2\alpha_4]^T$$

showing that each node has two degrees of freedom. These vectors are combined in the matrix form:

$$\begin{bmatrix} \boldsymbol{\delta}_1 \\ \boldsymbol{\delta}_2 \end{bmatrix} = \begin{bmatrix} v_1 \\ \theta_1 \\ v_2 \\ \theta_2 \end{bmatrix} = \begin{bmatrix} 1 & 0 & 0 & 0 \\ 0 & 1 & 0 & 0 \\ 1 & L & L^2 & L^3 \\ 0 & 1 & 2L & 3L^2 \end{bmatrix} \begin{bmatrix} \alpha_1 \\ \alpha_2 \\ \alpha_3 \\ \alpha_4 \end{bmatrix} \qquad \boldsymbol{\delta}^e = \mathbf{A}^e \boldsymbol{\alpha} \qquad (13.48a,b)$$

where, in Eq. 13.48b, $e = $ I, II, III, etc., refer to the beam element number. Equations 13.47b and 13.48b express $\boldsymbol{\delta}$ in terms of $\boldsymbol{\delta}^e$

$$\boldsymbol{\delta} = \mathbf{A} \left(\mathbf{A}^e \right)^{-1} \boldsymbol{\delta}^e \qquad (13.49)$$

where $\left(\mathbf{A}^e \right)^{-1}$ is the inverse of the matrix $\mathbf{A}^e$, the former being derived from the component displacement equations, taken from Eq. 13.48a:

$$v_1 = \alpha_1$$
$$\theta_1 = \alpha_2$$
$$v_2 = \alpha_1 + L\alpha_2 + L^2\alpha_3 + L^3\alpha_4 \qquad (13.50a\text{-}d)$$
$$\theta_2 = \alpha_2 + 2L\alpha_3 + 3L^2\alpha_4$$

Solving the simultaneous Eqs 13.50a–d for the coefficients α_1–α_4:

$$\alpha_1 = v_1$$

$$\alpha_2 = \theta_1$$

$$\alpha_3 = \frac{3}{L^2} \left(-v_1 + v_2 \right) - \frac{1}{L} \left(2\theta_1 + \theta_2 \right) \qquad (13.51a\text{-}d)$$

$$\alpha_4 = \frac{2}{L^3} \left(v_1 - v_2 \right) + \frac{1}{L^2} \left(\theta_1 + \theta_2 \right)$$

Then, writing Eqs 13.51a–d in matrix form reveals the components of the 4×4 matrix $(\mathbf{A}^e)^{-1}$:

$$\begin{bmatrix} \alpha_1 \\ \alpha_2 \\ \alpha_3 \\ \alpha_4 \end{bmatrix} = \begin{bmatrix} 1 & 0 & 0 & 0 \\ 0 & 1 & 0 & 0 \\ -3/L^2 & -2/L & 3/L^2 & -1/L \\ 2/L^3 & 1/L^2 & -2/L^3 & 1/L^2 \end{bmatrix} \begin{bmatrix} v_1 \\ \theta_1 \\ v_2 \\ \theta_2 \end{bmatrix} \qquad \boldsymbol{\alpha} = \left(\mathbf{A}^e \right)^{-1} \boldsymbol{\delta}^e \qquad (13.52a,b)$$

13.6.2 Strains

Equation 5.1a (see p. 150) gives the bending strain in a fibre lying at a distance y from the neutral axis (NA in Fig. 13.9b). Substituting $1/R$ from Eq. 13.46c:

$$\varepsilon = \frac{y}{R} = y(2\alpha_3 + 6\alpha_4 z) \tag{13.53a}$$

The general matrix form of Eq. 13.53a is

$$\varepsilon = y\begin{bmatrix} 0 & 0 & 2 & 6z \end{bmatrix}\begin{bmatrix} \alpha_1 \\ \alpha_2 \\ \alpha_3 \\ \alpha_4 \end{bmatrix} \qquad \varepsilon = y\mathbf{C}\boldsymbol{\alpha} \tag{13.53b,c}$$

Eliminating $\boldsymbol{\alpha}$ between Eqs 13.52b and 13.53c:

$$\varepsilon = y\,\mathbf{C}\left(\mathbf{A}^e\right)^{-1}\boldsymbol{\delta}^e \tag{13.54a}$$

In full, Eq. 13.54a becomes:

$$\varepsilon = y\begin{bmatrix} 0 & 0 & 2 & 6z \end{bmatrix}\begin{bmatrix} 1 & 0 & 0 & 0 \\ 0 & 1 & 0 & 0 \\ -3/L^2 & -2/L & 3/L^2 & -1/L \\ 2/L^3 & 1/L^2 & -2/L^3 & 1/L^2 \end{bmatrix}\begin{bmatrix} v_1 \\ \theta_1 \\ v_2 \\ \theta_2 \end{bmatrix} \tag{13.54b}$$

Matrix multiplication from Eq. 13.54b gives the bending strain at the position y in the cross-section and position z in the length of the beam element in Figs 13.9a,b:

$$\varepsilon = y\left[\left(-\frac{6}{L^2} + \frac{12z}{L^3}\right) \left(-\frac{4}{L} + \frac{6z}{L^2}\right) \left(\frac{6}{L^2} - \frac{12z}{L^3}\right) \left(-\frac{2}{L} + \frac{6z}{L^2}\right)\right]\begin{bmatrix} v_1 \\ \theta_1 \\ v_2 \\ \theta_2 \end{bmatrix} \qquad \varepsilon = y\mathbf{B}\boldsymbol{\delta}^e \tag{13.55a,b}$$

where $\mathbf{B} = \mathbf{C}\left(\mathbf{A}^e\right)^{-1}$. The component strains, ε_1 and ε_2 for the element's nodal strain vector $\boldsymbol{\varepsilon}^e = \begin{bmatrix} \varepsilon_1 & \varepsilon_2 \end{bmatrix}^{\mathrm{T}}$, are found from putting $z = 0$ and $z = L$, respectively, in Eq. 13.55a. This gives:

$$
\begin{bmatrix} \varepsilon_1 \\ \varepsilon_2 \end{bmatrix} = y \begin{bmatrix} -6/L^2 & -4/L & 6/L^2 & -2/L \\ 6/L^2 & 2/L & -6/L^2 & 4/L \end{bmatrix} \begin{bmatrix} v_1 \\ \theta_1 \\ v_2 \\ \theta_2 \end{bmatrix} \qquad \boldsymbol{\varepsilon}^e = y\mathbf{B}^e\boldsymbol{\delta}^e \qquad (13.56\text{a,b})
$$

13.6.3 Stresses

The modulus of elasticity E connects the uniaxial bending stress and strain simply as $\sigma = E\varepsilon$. Thus, the stress at any position z in the length is found from multiplying Eq. 13.55b by E:

$$
\sigma = Ey\mathbf{B}\,\boldsymbol{\delta}^e \qquad (13.57\text{a})
$$

Similarly, using $\boldsymbol{\sigma}^e = E\boldsymbol{\varepsilon}^e$, the element's nodal stress vector $\boldsymbol{\sigma}^e = [\sigma_1 \ \ \sigma_2]^T$ is found from its nodal strain vector $\boldsymbol{\varepsilon}^e = [\varepsilon_1 \ \ \varepsilon_2]^T$. That is, replacing $\boldsymbol{\varepsilon}^e$ with $\boldsymbol{\sigma}^e/E$ in Eq. 13.56b gives:

$$
\boldsymbol{\sigma}^e = Ey\mathbf{B}^e\boldsymbol{\delta}^e \qquad (13.57\text{b})
$$

13.6.4 Element Stiffness Matrix

The PVD (Eq. 13.5a) is used to determine the beam element's stiffness matrix $\mathbf{K}^e$. Substituting the general Eqs 13.55b and 13.57a for stress and strain into Eq. 13.5b:

$$
\left(\boldsymbol{\delta}^{ev}\right)^T \mathbf{f}^e = \int_V \left(y\mathbf{B}\boldsymbol{\delta}^{ev}\right)^T \left(Ey\mathbf{B}\boldsymbol{\delta}^e\right) \mathrm{d}V \qquad (13.58\text{a})
$$

Removing $\boldsymbol{\delta}^e$ from the integral in Eq. 13.58a:

$$
\left(\boldsymbol{\delta}^{ev}\right)^T \mathbf{f}^e = \left(\boldsymbol{\delta}^{ev}\right)^T \left(\int_V \mathbf{B}^T E\mathbf{B}y^2\mathrm{d}V\right) \times \boldsymbol{\delta}^e \qquad (13.58\text{b})
$$

Cancelling $\left(\boldsymbol{\delta}^{ev}\right)^T$ in Eq. 13.58b and substituting $\mathrm{d}V = \mathrm{d}A \times \mathrm{d}z$, leads to:

$$
\mathbf{f}^e = \left[\int_z \mathbf{B}^T E\mathbf{B}\left(\int_A y^2\mathrm{d}A\right)\mathrm{d}z\right] \times \boldsymbol{\delta}^e \qquad (13.58\text{c})
$$

Now $I = \int_A y^2\mathrm{d}A$, is the second moment of area of the arbitrary beam section in Fig. 13.9b. Comparing Eq. 13.58c with $\mathbf{f}^e = \mathbf{K}^e\boldsymbol{\delta}^e$, defines the element stiffness matrix:

$$
\mathbf{K}^e = \int_z \mathbf{B}^T EI\mathbf{B}\,\mathrm{d}z \qquad (13.59\text{a})
$$

The flexural rigidity product EI may be removed from the integration when E and I do not vary with z. Substituting $\mathbf{B}$ and $\mathbf{B}^T$ from Eq. 13.55a into 13.59a:

$$\mathbf{K}^e = EI \int_0^L \begin{bmatrix} (-6/L^2 + 12z/L^3) \\ (-4/L + 6z/L^2) \\ (6/L^2 - 12z/L^3) \\ (-2/L^2 + 6z/L^2) \end{bmatrix} \begin{bmatrix} (-6/L^2 + 12z/L^3) & (-4/L + 6z/L^2) & (6/L^2 - 12z/L^3) & (-2/L^2 + 6z/L^2) \end{bmatrix} dz$$

Multiplying rows into columns and then integrating, gives the 16 components of the beam element's stiffness matrix:

$$\mathbf{K}^e = \frac{EI}{L^3} \begin{bmatrix} 12 & 6L & -12 & 6L \\ 6L & 4L^2 & -6L & 2L^2 \\ -12 & -6L & 12 & -6L \\ -6L & 2L^2 & -6L & 4L^2 \end{bmatrix} \tag{13.59b}$$

The resulting 4×4 symmetrical matrix $\mathbf{K}^e$ is expected from the two degrees of freedom that apply to each node.

Example 13.1. Apply Eq. 13.59b to the single-element cantilever beam shown in Fig. 13.10a. Determine the free-end slope and deflection and the fixed-end bending stress. Compare with the standard expression given by the flexure equation and the theory of bending.

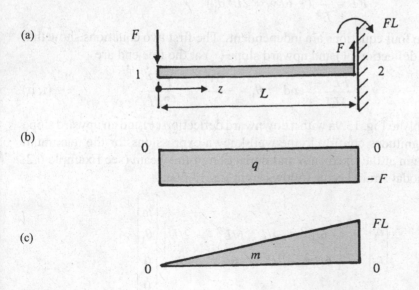

Figure 13.10 Single beam-element cantilever

Firstly, the force q and moment m at each node are established from the respective shear force and bending moment diagrams (see Figs 13.10b,c). These are drawn to our earlier convention in which hogging moments are positive and sagging moments are negative. Here, the diagrams conform to Fig. 13.9a when $q_1 = +F$ and $m_1 = 0$ for $z = 0$ at the free end (node 1). At the fixed end (node 2) the actions exerted by the support on the beam are are $q_2 = -F$ and $m_2 = +FL$ for $z = L$. Also, it is known that the deflection and slope for node 2 are zero. Hence, the stiffness matrix in Eq. 13.59b will connect the vector of force, $\mathbf{f} = [\; q_1 \;\; m_1 \;\; q_2 \;\; m_2 \;]^{\mathrm{T}}$ to the vector of displacement, $\boldsymbol{\delta} = [\; v_1 \;\; \theta_1 \;\; v_2 \;\; \theta_2 \;]^{\mathrm{T}}$, as follows

$$
\begin{bmatrix} F \\ 0 \\ -F \\ FL \end{bmatrix} = \frac{EI}{L^3} \begin{bmatrix} 12 & 6L & -12 & 6L \\ 6L & 4L^2 & -6L & 2L^2 \\ -12 & -6L & 12 & -6L \\ -6L & 2L^2 & -6L & 4L^2 \end{bmatrix} \begin{bmatrix} v_1 \\ \theta_1 \\ 0 \\ 0 \end{bmatrix}
$$

Expanding this matrix gives the simultaneous equations:

$$
F = \frac{EI}{L^3}(12v_1 + 6L\theta_1)
$$

$$
0 = 6Lv_1 + 4L^2\theta_1
$$

$$
-F = \frac{EI}{L^3}(-12v_1 - 6L\theta_1)
$$

$$
FL = \frac{EI}{L^3}(-6Lv_1 + 2L^2\theta_1)
$$

Only two of the four equations are independent. The first two equations show that the downward deflection (+) and upward slope (−) at the free end are

$$
v_1 = \frac{FL^3}{3EI} \quad \text{and} \quad \theta_1 = -\frac{3v_1}{2L} = -\frac{FL^2}{2EI} \tag{i, ii}
$$

The signs comply to Fig. 13.9a with a downward deflection (+) and an upward slope (−). Their magnitudes confirm to the well-known expressions for the maximum slope (with origin at the fixed end) and deflection of this beam (see Example 6.2, p. 197). The nodal strain vector follows from Eq. 13.56a:

$$
\begin{bmatrix} \varepsilon_1 \\ \varepsilon_2 \end{bmatrix} = y \begin{bmatrix} -6/L^2 & -4/L & 6/L^2 & -2/L \\ 6/L^2 & 2/L & -6/L^2 & 4/L \end{bmatrix} \begin{bmatrix} v_1 \\ \theta_1 \\ 0 \\ 0 \end{bmatrix}
$$

Multiplying out gives the strains within the cantilever's cross section at each node:

$$\varepsilon_1 = - \frac{2y}{L} \left(\frac{3v_1}{L} + 2\theta_1 \right) \qquad\qquad\qquad (iii)$$

$$\varepsilon_2 = \frac{2y}{L} \left(\frac{3v_1}{L} + \theta_1 \right) \qquad\qquad\qquad (iv)$$

where y is the position in the depth (see Fig. 13.9b). Substituting θ_1 and v_1 from Eqs i and ii into Eqs iii and iv shows:

$$\varepsilon_1 = - \frac{2y}{L} \left(\frac{3v_1}{L} - \frac{3v_1}{L} \right) = 0$$

$$\varepsilon_2 = \frac{2y}{L} \left(\frac{3v_1}{L} - \frac{3v_1}{2L} \right) = \frac{3yv_1}{L^2}$$

$$= \frac{3y}{L^2} \left(\frac{FL^3}{3EI} \right) = \frac{(FL)y}{EI} \qquad\qquad\qquad (v)$$

and from Eq. 13.57b the single element's nodal stresses are simply:

$$\sigma_1 = E\varepsilon_1 = 0, \qquad \sigma_2 = E\varepsilon_2 = \frac{(FL)y}{I}$$

which agree with the simple bending theory stresses from Eq. 5.4 (p. 150): $\sigma = My/I$, where $M = FL$ at the fixing.

13.6.5 Overall Stiffness Matrix

Where the beam is sub-divided into a number of elements then Eq. 13.59b may be applied only to each individual element. The stiffness for the whole structure requires an assembly between these individual stiffness matrices. There are a number of ways to assemble **K** depending upon the structure, as the following examples will show. The purpose of the assembly is to enable a solution to the displacements by inverting **K**, from Eqs 13.44b,c, in the most efficient manner computationally. Such efficiency can be achieved through the use of sub-matrices and reduced stiffness matrices. Where possible these aim to produce a banded matrix having many zero off-diagonal components to assist with a rapid inversion.

(a) Propped Cantilever

Here it will be shown how to assemble **K** by the direct stiffness method and then reduce this with the known slope and deflection (boundary conditions) that apply to the prop and the encastré fixing.

Example 13.2 The propped cantilever shown in Fig. 13.11 has a uniform cross section in a single material for which $EI = 10 \times 10^3$ kNm^2. Evaluate the individual stiffness matrices from the sub-division of the cantilever into the two elements I and II shown. Assemble the overall stiffness matrix for the structure then reduce it from applying its boundary conditions. Use the reduced matrix to find the slope at the prop and the slope and deflection at the free end.

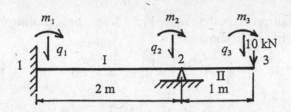

Figure 13.11 Propped cantilever

With the node numbering given Eq. 13.59b is applied to provide each individual element stiffness matrix with components, in units of kN/m:

$$\mathbf{K}^{I} = 10 \times 10^3 \begin{array}{c} \begin{array}{cccc} v_1 & \theta_1 & v_2 & \theta_2 \end{array} \\ \begin{bmatrix} 3/2 & 3/2 & -3/2 & 3/2 \\ 3/2 & 2 & -3/2 & 1 \\ -3/2 & -3/2 & 3/2 & -3/2 \\ 3/2 & 1 & -3/2 & 2 \end{bmatrix} \end{array}$$

$$\mathbf{K}^{II} = 10 \times 10^3 \begin{array}{c} \begin{array}{cccc} v_2 & \theta_2 & v_3 & \theta_3 \end{array} \\ \begin{bmatrix} 12 & 6 & -12 & 6 \\ 6 & 4 & -6 & 2 \\ -12 & -6 & 12 & -6 \\ 6 & 2 & -6 & 4 \end{bmatrix} \end{array}$$

In assembling the *direct stiffness matrix* the three nodes, each with two degrees of freedom, yields a 6×6 overall stiffness matrix $\mathbf{K}$ connecting the nodal force and displacement vectors as $\mathbf{f} = \mathbf{K}\delta$ where

$$\mathbf{f} = \begin{bmatrix} q_1 & m_1 & q_2 & m_2 & q_3 & m_3 \end{bmatrix}^T \text{ and } \delta = \begin{bmatrix} v_1 & \theta_1 & v_2 & \theta_2 & v_3 & \theta_3 \end{bmatrix}^T$$

and the stiffness matrix is

$$\mathbf{K} = 10 \times 10^3 \begin{bmatrix} 3/2 & 3/2 & -3/2 & 3/2 & 0 & 0 \\ 3/2 & 2 & -3/2 & 1 & 0 & 0 \\ -3/2 & -3/2 & (3/2+12) & (-3/2+6) & -12 & 6 \\ 3/2 & 1 & (-3/2+6) & (2+4) & -6 & 2 \\ 0 & 0 & -12 & -6 & 12 & -6 \\ 0 & 0 & 6 & 2 & -6 & 4 \end{bmatrix}$$

The slopes at the prop (node 2) together with the slope and deflection at the free end (node 3) appear in a *reduced stiffness matrix* between the load and displacement vectors, restricted to these nodes. Note that the net moment at node 2 is zero when the hogging and sagging moments of 10 kNm oppose in the adjacent spans, i.e. there is no external moment applied at node 2 and nor is there a moment applied to node 3. Also, with no deflection at node 2, the force vector reduces to:

$$\mathbf{f} = \begin{bmatrix} m_2 & q_3 & m_3 \end{bmatrix}^T = \begin{bmatrix} 0 & 10 & 0 \end{bmatrix}^T \text{ for } \boldsymbol{\delta} = \begin{bmatrix} \theta_2 & v_3 & \theta_3 \end{bmatrix}^T$$

That is

$$\begin{bmatrix} 0 \\ 10 \\ 0 \end{bmatrix} = 10 \times 10^3 \begin{bmatrix} 6 & -6 & 2 \\ -6 & 12 & -6 \\ 2 & -6 & 4 \end{bmatrix} \begin{bmatrix} \theta_2 \\ v_3 \\ \theta_3 \end{bmatrix}$$

from which the solution to the displacement vector components follows:

$$\theta_2 = 0.5 \times 10^{-3} \text{ rad}, \quad v_3 = 0.833 \times 10^{-3} \text{ m and } \theta_3 = 1 \times 10^{-3} \text{ rad}$$

(b) Continuous Beams

A beam resting on more than two supports is statically indeterminate. This means that the shear forces and bending moments cannot be found from applying the force equilibrium equations alone. An additional compatibility condition is required in which the displacements on either side of a node point are matched when sub-dividing a beam. For example, consider the beam resting upon three supports shown in Fig. 13.12. Four elements I, II, III and IV, are chosen so that their five nodes coincide with concentrated forces and reactions. Shear forces $q_1, q_2 .. q_5$ and moments m_1, m_2 .. m_5, act at each of the five nodes in the positive directions indicated. Positive nodal displacements v_1, v_2 .. v_5 and rotations θ_1, θ_2 ... θ_5, follow a similar sense to positive nodal 'forces' q and m respectively.

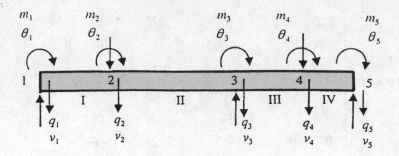

Figure 13.12 Continuous beam on three supports

The nodal force and displacement vectors are related through the following four element stiffness matrices (e = I, II, III and IV):

$$
\begin{bmatrix} q_1 \\ m_1 \\ q_2 \\ m_2 \end{bmatrix} = \begin{bmatrix} k_{11}^{I} & k_{12}^{I} & k_{13}^{I} & k_{14}^{I} \\ k_{21}^{I} & k_{22}^{I} & k_{23}^{I} & k_{24}^{I} \\ k_{31}^{I} & k_{32}^{I} & k_{33}^{I} & k_{34}^{I} \\ k_{41}^{I} & k_{42}^{I} & k_{43}^{I} & k_{44}^{I} \end{bmatrix} \begin{bmatrix} v_1 \\ \theta_1 \\ v_2 \\ \theta_2 \end{bmatrix} \qquad \mathbf{f}^{I} = \mathbf{K}^{I}\boldsymbol{\delta}^{I} \qquad (13.60\text{a,b})
$$

$$
\begin{bmatrix} q_2 \\ m_2 \\ q_3 \\ m_3 \end{bmatrix} = \begin{bmatrix} k_{11}^{II} & k_{12}^{II} & k_{13}^{II} & k_{14}^{II} \\ k_{21}^{II} & k_{22}^{II} & k_{23}^{II} & k_{24}^{II} \\ k_{31}^{II} & k_{32}^{II} & k_{33}^{II} & k_{34}^{II} \\ k_{41}^{II} & k_{42}^{II} & k_{43}^{II} & k_{44}^{II} \end{bmatrix} \begin{bmatrix} v_2 \\ \theta_2 \\ v_3 \\ \theta_3 \end{bmatrix} \qquad \mathbf{f}^{II} = \mathbf{K}^{II}\boldsymbol{\delta}^{II} \qquad (13.61\text{a,b})
$$

$$
\begin{bmatrix} q_3 \\ m_3 \\ q_4 \\ m_4 \end{bmatrix} = \begin{bmatrix} k_{11}^{III} & k_{12}^{III} & k_{13}^{III} & k_{14}^{III} \\ k_{21}^{III} & k_{22}^{III} & k_{23}^{III} & k_{24}^{III} \\ k_{31}^{III} & k_{32}^{III} & k_{33}^{III} & k_{34}^{III} \\ k_{41}^{III} & k_{42}^{III} & k_{43}^{III} & k_{44}^{III} \end{bmatrix} \begin{bmatrix} v_3 \\ \theta_3 \\ v_4 \\ \theta_4 \end{bmatrix} \qquad \mathbf{f}^{III} = \mathbf{K}^{III}\boldsymbol{\delta}^{III} \qquad (13.62\text{a,b})
$$

$$
\begin{bmatrix} q_4 \\ m_4 \\ q_5 \\ m_5 \end{bmatrix} = \begin{bmatrix} k_{11}^{IV} & k_{12}^{IV} & k_{13}^{IV} & k_{14}^{IV} \\ k_{21}^{IV} & k_{22}^{IV} & k_{23}^{IV} & k_{24}^{IV} \\ k_{31}^{IV} & k_{32}^{IV} & k_{33}^{IV} & k_{34}^{IV} \\ k_{41}^{IV} & k_{42}^{IV} & k_{43}^{IV} & k_{44}^{IV} \end{bmatrix} \begin{bmatrix} v_4 \\ \theta_4 \\ v_5 \\ \theta_5 \end{bmatrix} \qquad \mathbf{f}^{IV} = \mathbf{K}^{IV}\boldsymbol{\delta}^{IV} \qquad (13.63\text{a,b})
$$

Defining the overall force and displacement vectors as, respectively:

$$\mathbf{f} = [\, q_1 \; m_1 \; q_2 \; m_2 \; q_3 \; m_3 \; q_4 \; m_4 \; q_4 \; m_5 \,]^{\mathrm{T}} \qquad (13.64a)$$

$$\boldsymbol{\delta} = [\, v_1 \; \theta_1 \; v_2 \; \theta_2 \; v_3 \; \theta_3 \; v_4 \; \theta_4 \; v_5 \; \theta_5 \,]^{\mathrm{T}} \qquad (13.64b)$$

then $\mathbf{f}$ and $\boldsymbol{\delta}$ in Eqs 13.64a,b are connected by $\mathbf{f} = \mathbf{K}\boldsymbol{\delta}$ where $\mathbf{K}$ is an overall stiffness matrix. The latter is assembled from the individual element stiffness matrices in Eqs 13.60a, 13.61a, 13.62a and 13.63a, as follows:

$$\mathbf{K} = \begin{bmatrix}
k_{11}^{\mathrm{I}} & k_{12}^{\mathrm{I}} & k_{13}^{\mathrm{I}} & k_{14}^{\mathrm{I}} & 0 & 0 & 0 & 0 & 0 & 0 \\
k_{21}^{\mathrm{I}} & k_{22}^{\mathrm{I}} & k_{23}^{\mathrm{I}} & k_{24}^{\mathrm{I}} & 0 & 0 & 0 & 0 & 0 & 0 \\
k_{31}^{\mathrm{I}} & k_{32}^{\mathrm{I}} & k_{33}^{\mathrm{I}}+k_{11}^{\mathrm{II}} & k_{34}^{\mathrm{I}}+k_{12}^{\mathrm{II}} & k_{13}^{\mathrm{II}} & k_{14}^{\mathrm{II}} & 0 & 0 & 0 & 0 \\
k_{41}^{\mathrm{I}} & k_{42}^{\mathrm{I}} & k_{43}^{\mathrm{I}}+k_{21}^{\mathrm{II}} & k_{44}^{\mathrm{I}}+k_{22}^{\mathrm{II}} & k_{23}^{\mathrm{II}} & k_{24}^{\mathrm{II}} & 0 & 0 & 0 & 0 \\
0 & 0 & k_{31}^{\mathrm{II}} & k_{32}^{\mathrm{II}} & k_{33}^{\mathrm{II}}+k_{11}^{\mathrm{III}} & k_{34}^{\mathrm{II}}+k_{12}^{\mathrm{III}} & k_{13}^{\mathrm{III}} & k_{14}^{\mathrm{III}} & 0 & 0 \\
0 & 0 & k_{41}^{\mathrm{II}} & k_{42}^{\mathrm{II}} & k_{43}^{\mathrm{II}}+k_{21}^{\mathrm{III}} & k_{44}^{\mathrm{II}}+k_{22}^{\mathrm{III}} & k_{23}^{\mathrm{III}} & k_{24}^{\mathrm{III}} & 0 & 0 \\
0 & 0 & 0 & 0 & k_{31}^{\mathrm{III}} & k_{32}^{\mathrm{III}} & k_{33}^{\mathrm{III}}+k_{11}^{\mathrm{IV}} & k_{34}^{\mathrm{III}}+k_{12}^{\mathrm{IV}} & k_{13}^{\mathrm{IV}} & k_{14}^{\mathrm{IV}} \\
0 & 0 & 0 & 0 & k_{41}^{\mathrm{III}} & k_{42}^{\mathrm{III}} & k_{43}^{\mathrm{III}}+k_{21}^{\mathrm{IV}} & k_{44}^{\mathrm{III}}+k_{22}^{\mathrm{IV}} & k_{23}^{\mathrm{IV}} & k_{24}^{\mathrm{IV}} \\
0 & 0 & 0 & 0 & 0 & 0 & k_{31}^{\mathrm{IV}} & k_{32}^{\mathrm{IV}} & k_{33}^{\mathrm{IV}} & k_{34}^{\mathrm{IV}} \\
0 & 0 & 0 & 0 & 0 & 0 & k_{41}^{\mathrm{IV}} & k_{42}^{\mathrm{IV}} & k_{43}^{\mathrm{IV}} & k_{44}^{\mathrm{IV}}
\end{bmatrix} \qquad (12.65a)$$

This stiffness matrix is banded and symmetrical about its leading diagonal. However, unlike the overall stiffness matrices for the spring and frame assemblies (Eqs 13.23c and 13.36a, respectively), the rows and columns in Eq. 13.65a do not sum to zero. The assembly of $\mathbf{K}$ may be simplified by extracting 2×2 nodal submatrices within each $\mathbf{K}^e$ given in Eqs 13.60a–13.63a as follows:

$$\begin{bmatrix} \mathbf{f}_1 \\ \mathbf{f}_2 \\ \mathbf{f}_3 \\ \mathbf{f}_4 \\ \mathbf{f}_5 \end{bmatrix} = \begin{bmatrix}
\mathbf{S}_{11}^{\mathrm{I}} & \mathbf{S}_{12}^{\mathrm{I}} & 0 & 0 & 0 \\
\mathbf{S}_{21}^{\mathrm{I}} & \mathbf{S}_{22}^{\mathrm{I}}+\mathbf{S}_{22}^{\mathrm{II}} & \mathbf{S}_{23}^{\mathrm{II}} & 0 & 0 \\
0 & \mathbf{S}_{32}^{\mathrm{II}} & \mathbf{S}_{33}^{\mathrm{II}}+\mathbf{S}_{33}^{\mathrm{III}} & \mathbf{S}_{34}^{\mathrm{III}} & 0 \\
0 & 0 & \mathbf{S}_{43}^{\mathrm{III}} & \mathbf{S}_{44}^{\mathrm{III}}+\mathbf{S}_{44}^{\mathrm{IV}} & \mathbf{S}_{45}^{\mathrm{IV}} \\
0 & 0 & 0 & \mathbf{S}_{54}^{\mathrm{IV}} & \mathbf{S}_{55}^{\mathrm{IV}}
\end{bmatrix} \begin{bmatrix} \boldsymbol{\delta}_1 \\ \boldsymbol{\delta}_2 \\ \boldsymbol{\delta}_3 \\ \boldsymbol{\delta}_4 \\ \boldsymbol{\delta}_5 \end{bmatrix} \qquad (13.65b)$$

where the sub-stiffness matrix $\mathbf{S}_{ij}^e$ connects the nodal force vector $\mathbf{f}_i = [\, q \;\; m \,]^T$, to the nodal displacement vector $\boldsymbol{\delta}_j = [v \;\; \theta\,]^T$. When Eq. 13.65a is simplified in the manner of Eq. 13.65b it contains, for example, the node 2 stiffness equations:

$$\mathbf{f}_2 = \mathbf{S}_{21}^I \boldsymbol{\delta}_1 + \left(\mathbf{S}_{22}^I + \mathbf{S}_{22}^{II}\right)\boldsymbol{\delta}_2 + \mathbf{S}_{23}^{II}\boldsymbol{\delta}_3 \tag{13.66a}$$

In its expanded form Eq. 13.66a represents:

$$\begin{bmatrix} q_2 \\ m_2 \end{bmatrix} = \begin{bmatrix} k_{31}^I & k_{32}^I \\ k_{41}^I & k_{42}^I \end{bmatrix}\begin{bmatrix} v_1 \\ \theta_1 \end{bmatrix} + \left(\begin{bmatrix} k_{33}^I & k_{34}^I \\ k_{43}^I & k_{44}^I \end{bmatrix} + \begin{bmatrix} k_{11}^{II} & k_{12}^{II} \\ k_{21}^{II} & k_{22}^{II} \end{bmatrix} \right)\begin{bmatrix} v_2 \\ \theta_2 \end{bmatrix} + \begin{bmatrix} k_{13}^{II} & k_{14}^{II} \\ k_{23}^{II} & k_{24}^{II} \end{bmatrix}\begin{bmatrix} v_3 \\ \theta_3 \end{bmatrix} \tag{13.66b}$$

Expanding Eq. 13.66b contains the two force-displacement relations for node 2 that would follow directly from the overall stiffness matrix in Eq. 13.65a,

$$q_2 = k_{31}^I v_1 + k_{32}^I \theta_1 + (k_{33}^I + k_{11}^{II})v_2 + (k_{34}^I + k_{12}^{II})\theta_2 + k_{13}^{II}v_3 + k_{14}^{II}\theta_3 \tag{13.66c}$$

$$m_2 = k_{41}^I v_1 + k_{42}^I \theta_1 + (k_{43}^I + k_{21}^{II})v_2 + (k_{44}^I + k_{22}^{II})\theta_2 + k_{23}^{II}v_3 + k_{24}^{II}\theta_3 \tag{13.66d}$$

That is, Eqs 13.66c,d are confirmed from direct matrix multiplication within $\mathbf{f} = \mathbf{K}\boldsymbol{\delta}$ where $\mathbf{K}$ is given by Eq. 13.65a. Displacements v_2 and θ_2, being common to adjoining elements I and II, ensure that equilibrium and compatibility are satisfied. Similar conditions are satisfied for other nodes within the strict pattern of assembly connecting the individual elements.

Example 13.3 Derive the stiffness matrix for the continuous beam loaded as shown in Fig. 13.13. Take the element nodes 1, 2 .. 5 to coincide with the points of applied forces and supporting reactions. The second moment of area for the first bay is twice that of the second bay.

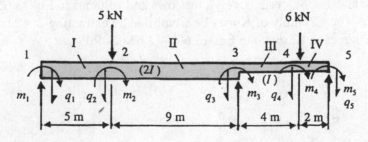

Figure 13.13 Beam with varying section on three supports

The sub-division chosen for the beam allows for the calculation of each individual stiffness from Eq. 13.59b. These are (in units of <u>kN</u> and <u>m</u>):

Element I:

$$k_{11}^{I} = \frac{12E(2I)}{L^3} = \frac{12E \times (2I)}{5^3} = 0.192\,EI \left(= k_{33}^{I} = -k_{13}^{I} = -k_{31}^{I} \right)$$

$$k_{22}^{I} = \frac{4E(2I)}{L} = \frac{4E \times (2I)}{5} = 1.6\,EI \left(= k_{44}^{I} \right)$$

$$k_{12}^{I} = \frac{6E(2I)}{L^2} = \frac{6E \times (2I)}{5^2} = 0.48\,EI \left(= k_{14}^{I} = k_{21}^{I} = -k_{23}^{I} = -k_{32}^{I} = -k_{34}^{I} = k_{41}^{I} = -k_{43}^{I} \right)$$

$$k_{24}^{I} = \frac{2E \times (2I)}{L} = \frac{2E \times (2I)}{5} = 0.80\,EI \left(= k_{42}^{I} \right)$$

Element II:

$$k_{11}^{II} = \frac{12E(2I)}{L^3} = \frac{12E \times (2I)}{9^3} = 0.033\,EI \left(= k_{33}^{II} = -k_{13}^{II} = -k_{31}^{II} \right)$$

$$k_{22}^{II} = \frac{4E(2I)}{L} = \frac{4E \times (2I)}{9} = 0.899\,EI \left(= k_{44}^{II} \right)$$

$$k_{12}^{II} = \frac{6E(2I)}{L^2} = \frac{6E \times (2I)}{9^2} = 0.148\,EI \left(= k_{14}^{II} = k_{21}^{II} = -k_{23}^{II} = -k_{32}^{II} = -k_{34}^{II} = k_{41}^{II} = -k_{43}^{II} \right)$$

$$k_{24}^{II} = \frac{2E \times (2I)}{L} = \frac{2E \times (2I)}{9} = 0.444\,EI \left(= k_{42}^{II} \right)$$

Element III:

$$k_{11}^{III} = \frac{12EI}{L^3} = \frac{12EI}{4^3} = 0.1875\,EI \left(= k_{33}^{III} = -k_{13}^{III} = -k_{31}^{III} \right)$$

$$k_{22}^{III} = \frac{4EI}{L} = \frac{4EI}{4} = EI \left(= k_{44}^{III} \right)$$

$$k_{12}^{III} = \frac{6EI}{L^2} = \frac{6EI}{4^2} = 0.375\,EI \left(= k_{14}^{III} = k_{21}^{III} = -k_{23}^{III} = -k_{32}^{III} = -k_{34}^{III} = k_{41}^{III} = -k_{43}^{III} \right)$$

$$k_{24}^{III} = \frac{2EI}{L} = \frac{2EI}{4} = 0.5\,EI \left(= k_{42}^{III} \right)$$

Element IV:

$$k_{11}^{IV} = \frac{12EI}{L^3} = \frac{12EI}{2^3} = 1.5\,EI \left(= k_{33}^{IV} = -k_{13}^{III} = -k_{31}^{III} \right)$$

$$k_{22}^{IV} = \frac{4EI}{L} = \frac{4EI}{2} = 2\,EI \left(= k_{44}^{IV} \right)$$

$$k_{12}^{IV} = \frac{6EI}{L^2} = \frac{6EI}{2^2} = 1.5\,EI \left(= k_{14}^{IV} = k_{21}^{IV} = -k_{23}^{IV} = -k_{32}^{IV} = -k_{34}^{IV} = k_{41}^{IV} = -k_{43}^{IV} \right)$$

$$k_{24}^{IV} = \frac{2EI}{L} = \frac{2EI}{2} = EI \left(= k_{42}^{IV} \right)$$

Substituting k_{ij}^{e} into Eq. 13.65a, we find the overall **K** matrix:

$$\begin{bmatrix} q_1 \\ m_1 \\ q_2=5 \\ m_2 \\ q_3 \\ m_3 \\ q_4=6 \\ m_4 \\ q_5 \\ m_5=0 \end{bmatrix} = EI \begin{bmatrix} 0.192 & 0.480 & -0.192 & 0.480 & 0 & 0 & 0 & 0 & 0 & 0 \\ 0.480 & 1.600 & -0.480 & 0.800 & 0 & 0 & 0 & 0 & 0 & 0 \\ -0.192 & -0.480 & 0.225 & -0.332 & -0.033 & 0.148 & 0 & 0 & 0 & 0 \\ 0.480 & 0.800 & -0.322 & 2.489 & -0.148 & 0.444 & 0 & 0 & 0 & 0 \\ 0 & 0 & -0.033 & -0.148 & 0.221 & 0.227 & -0.188 & 0.375 & 0 & 0 \\ 0 & 0 & 0.148 & 0.444 & 0.227 & 1.889 & -0.375 & 0.500 & 0 & 0 \\ 0 & 0 & 0 & 0 & -0.188 & -0.375 & 1.688 & 1.125 & -1.500 & 1.500 \\ 0 & 0 & 0 & 0 & 0.375 & 0.500 & 1.125 & 3.000 & -1.500 & 1.000 \\ 0 & 0 & 0 & 0 & 0 & 0 & -1.500 & -1.500 & 1.500 & -1.500 \\ 0 & 0 & 0 & 0 & 0 & 0 & 1.500 & 1.000 & -1.500 & 2.000 \end{bmatrix} \begin{bmatrix} v_1 \\ \theta_1 \\ v_2 \\ \theta_2 \\ v_3=0 \\ \theta_3 \\ v_4 \\ \theta_4 \\ v_5=0 \\ \theta_5 \end{bmatrix}$$

in which known values of q, m and v have been inserted. The ten equations within the matrix are taken with three further equations expressing the static force and moment equilibrium conditions:

$$\sum F_y = q_1 + q_3 + q_5 + 11 = 0; \quad \sum M_2 = m_2 - 5q_1 = 0; \quad \sum M_4 = m_4 + 2q_5 = 0$$

It is seen then that there are, in total, thirteen equations in thirteen unknowns. The solution to these equations will provide the slopes, θ, and deflections, v, at the five nodal points.

(c) Continuous Beam with Spring Support

Where one or more support are compression springs the overall stiffness of the structure must combine the stiffness of each beam element (Eq. 13.59b) with the stiffness matrix for each spring element (Eq. 13.21b). For example, consider the structure shown in Fig. 13.14 in which four supports at beam nodes 2, 3, 4 and 6 react to the three external loads R, S and T.

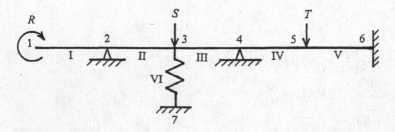

Figure 13.14 Continuous beam with spring support

With the sub-division into the six elements I, II, III . . . VI shown, a zero slope applies to the encastré fixing with no deflection here and also at the two knife-edges and spring fixing. These apply to their nodes as: $\theta_6 = 0$ and $v_2 = v_4 = v_6 = v_7 = 0$.

In the assembly of the overall stiffness matrix we may account for these boundary conditions through the following series of reduced stiffness matrices for each of the six elements:

$$
\mathbf{K}^{I} =
\begin{array}{c}
\begin{array}{cccc} v_1 & \theta_1 & v_2 & \theta_2 \end{array} \\
\begin{bmatrix}
k_{11}^{I} & k_{12}^{I} & - & k_{14}^{I} \\
k_{21}^{I} & k_{22}^{I} & - & k_{24}^{I} \\
- & - & - & - \\
k_{41}^{I} & k_{42}^{I} & - & k_{44}^{I}
\end{bmatrix}
\end{array}
\qquad
\mathbf{K}^{II} =
\begin{array}{c}
\begin{array}{cccc} v_2 & \theta_2 & v_3 & \theta_3 \end{array} \\
\begin{bmatrix}
- & - & - & - \\
- & k_{22}^{II} & k_{23}^{II} & k_{24}^{II} \\
- & k_{32}^{II} & k_{33}^{II} & k_{34}^{II} \\
- & k_{42}^{II} & k_{43}^{II} & k_{44}^{II}
\end{bmatrix}
\end{array}
\qquad (13.67a,b)
$$

$$
\mathbf{K}^{III} =
\begin{array}{c}
\begin{array}{cccc} v_3 & \theta_3 & v_4 & \theta_4 \end{array} \\
\begin{bmatrix}
k_{11}^{III} & k_{12}^{III} & - & k_{14}^{III} \\
k_{21}^{III} & k_{22}^{III} & - & k_{24}^{III} \\
- & - & - & - \\
k_{41}^{III} & k_{42}^{III} & - & k_{44}^{III}
\end{bmatrix}
\end{array}
\qquad
\mathbf{K}^{IV} =
\begin{array}{c}
\begin{array}{cccc} v_4 & \theta_4 & v_5 & \theta_5 \end{array} \\
\begin{bmatrix}
- & - & - & - \\
- & k_{22}^{IV} & k_{23}^{IV} & k_{24}^{IV} \\
- & k_{32}^{IV} & k_{33}^{IV} & k_{34}^{IV} \\
- & k_{42}^{IV} & k_{43}^{IV} & k_{44}^{IV}
\end{bmatrix}
\end{array}
\qquad (13.67c,d)
$$

$$
\mathbf{K}^{V} =
\begin{array}{c}
\begin{array}{cccc} v_5 & \theta_5 & v_6 & \theta_6 \end{array} \\
\begin{bmatrix}
k_{11}^{V} & k_{12}^{V} & - & - \\
k_{21}^{V} & k_{22}^{V} & - & - \\
- & - & - & - \\
- & - & - & -
\end{bmatrix}
\end{array}
\qquad
\mathbf{K}^{VI} =
\begin{array}{c}
\begin{array}{cc} v_3 & v_7 \end{array} \\
\begin{bmatrix}
k_0 & - \\
- & -
\end{bmatrix}
\end{array}
\qquad (13.67e,f)
$$

The spring influences the lateral stiffness of the beam reacting the force S at node 3 within the product of its stiffness k_0 and the vertical displacement at node 3. The spring does not contribute the rotational stiffness at this node. When node 3 is taken with other nodal supports where no rotation or displacement occurs it follows that the structure has eight non-zero degrees of freedom. That is, the nodal displacement vector, when reduced to its non-zero components becomes:

$$
\boldsymbol{\delta} = \begin{bmatrix} v_1 & \theta_1 & \theta_2 & v_3 & \theta_3 & \theta_4 & v_5 & \theta_5 \end{bmatrix}^T
\qquad (13.68a)
$$

An overall, reduced, stiffness matrix $\mathbf{K}$ connects these eight non-zero displacements to the vector of external forces referred to the respective element's nodes:

$$\mathbf{f} = \begin{bmatrix} q_1 & m_1 & m_2 & q_3 & m_3 & m_4 & q_5 & m_5 \end{bmatrix} = \begin{bmatrix} 0 & R & 0 & S & 0 & 0 & T & 0 \end{bmatrix}^{\mathrm{T}} \qquad (13.68b)$$

Thus from the individual reductions to each stiffness matrix $\mathbf{K}^e$ (e = I, II, ... VI), given above in Eqs 13.67a–f, the reduced stiffness matrix for the whole structure is assembled between the force and displacement vectors in Eqs 13.68a,b as follows

$$
\mathbf{K} =
\begin{array}{cccccccc}
v_1 & \theta_1 & \theta_2 & v_3 & \theta_3 & \theta_4 & v_5 & \theta_5
\end{array}
$$

$$
\mathbf{K} =
\begin{bmatrix}
k_{11}^{\mathrm{I}} & k_{12}^{\mathrm{I}} & k_{14}^{\mathrm{I}} & 0 & 0 & 0 & 0 & 0 \\
k_{21}^{\mathrm{I}} & k_{22}^{\mathrm{I}} & k_{24}^{\mathrm{I}} & 0 & 0 & 0 & 0 & 0 \\
k_{41}^{\mathrm{I}} & k_{42}^{\mathrm{I}} & (k_{14}^{\mathrm{I}} + k_{22}^{\mathrm{II}}) & k_{23}^{\mathrm{II}} & k_{24}^{\mathrm{II}} & 0 & 0 & 0 \\
0 & 0 & k_{32}^{\mathrm{II}} & (k_{33}^{\mathrm{II}} + k_{11}^{\mathrm{III}} + k_0) & (k_{34}^{\mathrm{II}} + k_{12}^{\mathrm{III}}) & k_{14}^{\mathrm{III}} & 0 & 0 \\
0 & 0 & k_{42}^{\mathrm{II}} & (k_{43}^{\mathrm{II}} + k_{21}^{\mathrm{III}}) & (k_{43}^{\mathrm{II}} + k_{21}^{\mathrm{III}}) & k_{24}^{\mathrm{III}} & 0 & 0 \\
0 & 0 & 0 & k_{41}^{\mathrm{III}} & k_{42}^{\mathrm{III}} & (k_{22}^{\mathrm{III}} + k_{22}^{\mathrm{IV}}) & k_{23}^{\mathrm{IV}} & k_{24}^{\mathrm{IV}} \\
0 & 0 & 0 & 0 & 0 & k_{32}^{\mathrm{IV}} & (k_{33}^{\mathrm{IV}} + k_{11}^{\mathrm{V}}) & (k_{34}^{\mathrm{IV}} + k_{12}^{\mathrm{V}}) \\
0 & 0 & 0 & 0 & 0 & k_{42}^{\mathrm{IV}} & (k_{43}^{\mathrm{IV}} + k_{21}^{\mathrm{V}}) & (k_{44}^{\mathrm{IV}} + k_{22}^{\mathrm{V}})
\end{bmatrix}
$$

$$(13.69)$$

The banded, symmetrical matrix in Eq. 13.69 provided by this reduction procedure allows a more efficient programming to the solution for those displacements in Eq. 13.68a. Note here that for each element e = I, II, III etc, the non-zero matrix components $k_{ij}^{\,e}$ appearing at the position given by row i and column j within each $\mathbf{K}^e$ (see Eqs 13.67a–e) are to be calculated from the standard 4×4 beam element stiffness matrix in Eq. 13.59b.

13.7 Plane Triangular Element

Many instances arise where two-dimensional elements are required to discretise the shape of a load-bearing component, as with plates and shells under in-plane biaxial stress. Both triangular and rectangular elements may be used. Their basic forms have 3 and 4 nodes but more sophisticated triangular and rectangular plane elements are available having additional nodes taken at the mid-positions of their sides. Only the simplest three-noded, plane triangular element under the nodal forces shown in

Fig. 13.15a, is examined here. This is sufficient to illustrate the technique to be followed when forming its individual stiffness matrix and its assembly with the stiffness matrices of other similar elements having interconnected nodes.

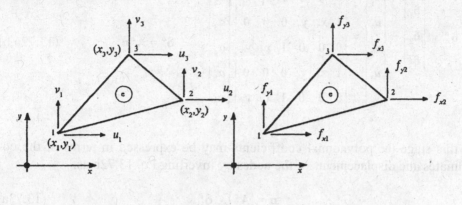

(a)   (b)

Figure 13.15 Nodal forces and displacements for a plane triangular element

13.7.1 Displacements

The analysis begins with assumed displacement functions in the global, Cartesian co-ordinates (x, y) shown in Fig. 13.15b:

$$u = \alpha_1 + \alpha_2 x + \alpha_3 y \tag{13.70a}$$

$$v = \alpha_4 + \alpha_5 x + \alpha_6 y \tag{13.70b}$$

Writing the general displacement vector as: $\boldsymbol{\delta} = [u \ v]^T$, Eqs 13.70a,b appear in their equivalent matrix forms:

$$\boldsymbol{\delta} = \begin{bmatrix} u \\ v \end{bmatrix} = \begin{bmatrix} 1 & x & y & 0 & 0 & 0 \\ 0 & 0 & 0 & 1 & x & y \end{bmatrix} \begin{bmatrix} \alpha_1 \\ \alpha_2 \\ \alpha_3 \\ \alpha_4 \\ \alpha_5 \\ \alpha_6 \end{bmatrix} \qquad \boldsymbol{\delta} = \mathbf{A}\boldsymbol{\alpha} \tag{13.71a,b}$$

In particular, when the co-ordinates for each node are substituted into Eq. 13.71a there follows the matrix form for the element's nodal point displacement vector $\boldsymbol{\delta}^e = [\boldsymbol{\delta}_1 \ \boldsymbol{\delta}_2 \ \boldsymbol{\delta}_3]^T$:

$$
\mathbf{\delta}^e = \begin{bmatrix} \mathbf{\delta}_1 \\ \mathbf{\delta}_2 \\ \mathbf{\delta}_3 \end{bmatrix} = \begin{bmatrix} u_1 \\ v_1 \\ u_2 \\ v_2 \\ u_3 \\ v_3 \end{bmatrix} = \begin{bmatrix} 1 & x_1 & y_1 & 0 & 0 & 0 \\ 0 & 0 & 0 & 1 & x_1 & y_1 \\ 1 & x_2 & y_2 & 0 & 0 & 0 \\ 0 & 0 & 0 & 1 & x_2 & y_2 \\ 1 & x_3 & y_3 & 0 & 0 & 0 \\ 0 & 0 & 0 & 1 & x_3 & y_3 \end{bmatrix} \begin{bmatrix} \alpha_1 \\ \alpha_2 \\ \alpha_3 \\ \alpha_4 \\ \alpha_5 \\ \alpha_6 \end{bmatrix} \qquad \mathbf{\delta}^e = \mathbf{A}^e \mathbf{\alpha} \qquad (13.72\text{a,b})
$$

At this stage the polynomial coefficients may be expressed in terms of the co-ordinates and displacements of the nodes by inverting Eq. 13.72b as:

$$
\mathbf{\alpha} = \left(\mathbf{A}^e \right)^{-1} \mathbf{\delta}^e \qquad (13.73\text{a})
$$

Equation 13.73a employs the inverted matrix $\left(\mathbf{A}^e \right)^{-1}$ whose components appear in the solution to the coefficients $\alpha_1, \alpha_2 .. \alpha_6$ from the six simultaneous equations of Eq. 13.72a. In full, $\left(\mathbf{A}^e \right)^{-1}$ is defined as

$$
\left(\mathbf{A}^e \right)^{-1} = \frac{1}{2\Delta} \begin{bmatrix} x_2 y_3 - x_3 y_2 & 0 & x_3 y_1 - x_1 y_3 & 0 & x_1 y_2 - x_2 y_1 & 0 \\ y_2 - y_3 & 0 & y_3 - y_1 & 0 & y_1 - y_2 & 0 \\ x_3 - x_2 & 0 & x_1 - x_3 & 0 & x_2 - x_1 & 0 \\ 0 & x_2 y_3 - x_3 y_2 & 0 & x_3 y_1 - x_1 y_3 & 0 & x_1 y_2 - x_2 y_1 \\ 0 & y_2 - y_3 & 0 & y_3 - y_1 & 0 & y_1 - y_2 \\ 0 & x_3 - x_2 & 0 & x_1 - x_3 & 0 & x_2 - x_1 \end{bmatrix} \qquad (13.73\text{b})
$$

where Δ is the area of the element in terms of its nodal co-ordinates:

$$
\Delta = \frac{1}{2} \left[(x_1 - x_2)(y_2 - y_3) - (x_2 - x_3)(y_1 - y_2) \right] \qquad (13.73\text{c})
$$

Combining Eq. 13.71b and Eq. 13.73a provides the element's internal displacements $\mathbf{\delta} = [u \ v]^T$ in terms of its nodal displacement vector $\mathbf{\delta}^e$:

$$
\mathbf{\delta} = \mathbf{A} \left(\mathbf{A}^e \right)^{-1} \mathbf{\delta}^e = \mathbf{N} \mathbf{\delta}^e \qquad (13.74\text{a})
$$

in which the matrix $\mathbf{N}$ arises from the product

$$
\mathbf{N} = \mathbf{A} \left(\mathbf{A}^e \right)^{-1} \qquad (13.74\text{b})
$$

this defining this element's *shape function* matrix, which takes the 2×6 matrix form for connecting $\mathbf{\delta}$ to $\mathbf{\delta}^e$ directly in Eq. 13.74a:

$$\delta = \begin{bmatrix} u \\ v \end{bmatrix} = \begin{bmatrix} N_1 & 0 & N_2 & 0 & N_3 & 0 \\ 0 & N_1 & 0 & N_2 & 0 & N_3 \end{bmatrix} \begin{bmatrix} u_1 \\ v_1 \\ u_2 \\ v_2 \\ u_3 \\ v_3 \end{bmatrix}$$

(13.74c)

Alternatively, the displacements appear in the indicial notation as:

$$u = N_k u_k$$

(13.75a)

$$v = N_k v_k$$

(13.75b)

in which a summation is implied for $k = 1$, 2 and 3. The non-zero elements N_k of matrix **N** may be rearranged in a column to show their dependence upon each node's co-ordinates (x, y) that define the element's shape (see Exercise 13.13):

$$\begin{bmatrix} N_1 \\ N_2 \\ N_3 \end{bmatrix} = \frac{1}{2\Delta} \begin{bmatrix} x_2 y_3 - x_3 y_2 & y_2 - y_3 & x_3 - x_2 \\ x_3 y_1 - x_1 y_3 & y_3 - y_1 & x_1 - x_3 \\ x_1 y_2 - x_2 y_1 & y_1 - y_2 & x_2 - x_1 \end{bmatrix} \begin{bmatrix} 1 \\ x \\ y \end{bmatrix}$$

(13.75c)

13.7.2 Plane Strains

The three strains ε_x, ε_y and γ_{xy} within the x, y plane follow from applying the three partial derivatives in Eqs 11.27a–c to the two displacement functions $u = u(x, y)$ and $v = v(x, y)$, assumed in Eqs 13.70a,b (see p. 463). This gives compatible strains:

$$\varepsilon_x = \frac{\partial u}{\partial x} = \alpha_2$$

(13.76a)

$$\varepsilon_y = \frac{\partial v}{\partial y} = \alpha_6$$

(13.76b)

$$\gamma_{xy} = \frac{\partial u}{\partial y} + \frac{\partial v}{\partial x} = \alpha_3 + \alpha_5$$

(13.76c)

Equations 13.76a–c have an equivalent matrix form that describes the uniform (constant) strain throughout the triangular element:

$$
\begin{bmatrix} \varepsilon_x \\ \varepsilon_y \\ \gamma_{xy} \end{bmatrix} = \begin{bmatrix} 0 & 1 & 0 & 0 & 0 & 0 \\ 0 & 0 & 0 & 0 & 0 & 1 \\ 0 & 0 & 1 & 0 & 1 & 0 \end{bmatrix} \begin{bmatrix} \alpha_1 \\ \alpha_2 \\ \alpha_3 \\ \alpha_4 \\ \alpha_5 \\ \alpha_6 \end{bmatrix} \qquad \varepsilon = \mathbf{C}\,\alpha \qquad\qquad (13.77a,b)
$$

and in terms of the nodal displacements these strains follow from Eqs 13.73a and 13.77b:

$$
\varepsilon = \mathbf{C}\left(\mathbf{A}^e\right)^{-1}\delta^e = \mathbf{B}\,\delta^e \qquad\qquad (13.78a)
$$

where the matrix $\mathbf{B} = \mathbf{C}\left(\mathbf{A}^e\right)^{-1}$ follows from the multiplication between the matrices $\mathbf{C}$ and $\left(\mathbf{A}^e\right)^{-1}$ from Eqs 13.77a and 13.73b, respectively:

$$
\mathbf{B} = \frac{1}{2\Delta}\begin{bmatrix} y_2 - y_3 & 0 & y_3 - y_1 & 0 & y_1 - y_2 & 0 \\ 0 & x_3 - x_2 & 0 & x_1 - x_3 & 0 & x_2 - x_1 \\ x_3 - x_2 & y_2 - y_3 & x_1 - x_3 & y_3 - y_1 & x_2 - x_1 & y_1 - y_2 \end{bmatrix} \qquad (13.78b)
$$

Equations 13.78a,b show how each element has associated with it a constant strain dependent only upon the nodal displacement vector $\delta^e = [u_1 \; v_1 \; u_2 \; v_2 \; u_3 \; v_3]^T$. Compared to other plane elements, one having a constant strain property is the exception. Indeed, had a higher-order polynomial been assumed for $u = u(x, y)$ and $v = v(x, y)$ in Eq. 13.70a,b then at this stage x and y would have appeared in $\mathbf{B}$. It follows that the constant strain element is suitable to use where strain gradients are known to be gradual. For example, to match the linear strain distribution through the depth of a beam the proportional change in strain is replicated across the centroids of adjacent elements by allowing the constant element strains to vary linearly between neighbouring elements.

13.7.3 Constitutive Relationships

In extending Hooke's law to the multi-axial stress-strain relations in § 4.4 (p. 120) two elastic constants appeared: E and v. When stress is to be connected to strain in two dimensions, E and v define an elasticity matrix $\mathbf{D}$ which connects the vectors of stress and strain, $\sigma = \mathbf{D}\varepsilon$, whilst accounting for the precise nature of the plane problem. For example, in its description of thin plates and vessels, with a thickness equal to a single element thickness, the problem may be formulated under a *general plane stress state* for which:

$$\varepsilon_x = \frac{1}{E}(\sigma_x - v\sigma_y) \tag{13.79a}$$

$$\varepsilon_y = \frac{1}{E}(\sigma_y - v\sigma_x) \tag{13.79b}$$

$$\gamma_{xy} = \frac{\tau_{xy}}{G} = \frac{2}{E}(1 + v)\tau_{xy} \tag{13.79c}$$

Equations 13.79a–c appear in their matrix representations

$$\begin{bmatrix} \varepsilon_x \\ \varepsilon_y \\ \gamma_{xy} \end{bmatrix} = \begin{bmatrix} 1/E & -v/E & 0 \\ -v/E & 1/E & 0 \\ 0 & 0 & 2(1+v)/E \end{bmatrix} \begin{bmatrix} \sigma_x \\ \sigma_y \\ \tau_{xy} \end{bmatrix} \qquad \boldsymbol{\varepsilon} = \mathbf{P}_\sigma \boldsymbol{\sigma} \tag{13.80a,b}$$

On the other hand, a two-dimensional loading imposed upon the section of a thick plate or a long bar defines a *general plane strain state*. Here a through-thickness stress σ_z arises from the geometric constraint an increased z-dimension imposes, giving either a constant value to the through-thickness strain ε_z for a thick plate or a zero value for the z-axis of a long bar. The constitutive equations are, generally:

$$\varepsilon_x = \frac{1}{E}[\sigma_x - v(\sigma_y + \sigma_z)] \tag{13.81a}$$

$$\varepsilon_y = \frac{1}{E}[\sigma_y - v(\sigma_x + \sigma_z)] \tag{13.81b}$$

$$\varepsilon_z = \frac{1}{E}[\sigma_z - v(\sigma_x + \sigma_y)] = \text{constant} \tag{13.81c}$$

$$\gamma_{xy} = \frac{\tau_{xy}}{G} = \frac{2}{E}(1 + v)\tau_{xy} \tag{13.81d}$$

In particular, when $\varepsilon_z = 0$ then, consequently $\sigma_z = v(\sigma_x + \sigma_y)$, so that the matrix representation of Eqs 13.81a–d remains two-dimensional:

$$\begin{bmatrix} \varepsilon_x \\ \varepsilon_y \\ \gamma_{xy} \end{bmatrix} = \begin{bmatrix} \dfrac{1-v^2}{E} & -\dfrac{v(1+v)}{E} & 0 \\ -\dfrac{v(1+v)}{E} & \dfrac{1-v^2}{E} & 0 \\ 0 & 0 & \dfrac{2(1+v)}{E} \end{bmatrix} \begin{bmatrix} \sigma_x \\ \sigma_y \\ \tau_{xy} \end{bmatrix} \qquad \boldsymbol{\varepsilon} = \mathbf{P}_\varepsilon \boldsymbol{\sigma} \tag{13.82a,b}$$

Inverting the flexibility matrices $\mathbf{P}$ for each plane condition is required develop the stiffness matrix. The plane stress and strain matrices $\mathbf{D}_\sigma$ and $\mathbf{D}_\varepsilon$ follow from the respective solutions to Eqs 13.79a–c and 13.81a–d in a common inverse form: $\boldsymbol{\sigma} = \mathbf{P}^{-1}\boldsymbol{\varepsilon} = \mathbf{D}\boldsymbol{\varepsilon}$ where, specifically:

$$
\begin{bmatrix} \sigma_x \\ \sigma_y \\ \tau_{xy} \end{bmatrix} = \frac{E}{1-v^2}\begin{bmatrix} 1 & v & 0 \\ v & 1 & 0 \\ 0 & 0 & \tfrac{1}{2}(1-v) \end{bmatrix}\begin{bmatrix} \varepsilon_x \\ \varepsilon_y \\ \gamma_{xy} \end{bmatrix} \qquad \boldsymbol{\sigma} = \mathbf{D}_\sigma\boldsymbol{\varepsilon} \qquad (13.83\text{a,b})
$$

$$
\begin{bmatrix} \sigma_x \\ \sigma_y \\ \tau_{xy} \end{bmatrix} = \frac{E(1-v)}{(1+v)(1-2v)}\begin{bmatrix} 1 & \dfrac{v}{1-v} & 0 \\ \dfrac{v}{1-v} & 1 & 0 \\ 0 & 0 & \dfrac{1-2v}{2(1-v)} \end{bmatrix}\begin{bmatrix} \varepsilon_x \\ \varepsilon_y \\ \gamma_{xy} \end{bmatrix}
$$

$$
\boldsymbol{\sigma} = \mathbf{D}_\varepsilon\boldsymbol{\varepsilon} \qquad (13.84\text{a,b})
$$

The analysis now proceeds with a general matrix symbol $\mathbf{D}$, thereby allowing it to take either form, $\mathbf{D}_\sigma$ or $\mathbf{D}_\varepsilon$, given in Eqs 13.83a and 13.84a, respectively. Hence the stress vector is defined from the strain vector as

$$
\boldsymbol{\sigma} = \mathbf{P}^{-1}\boldsymbol{\varepsilon} = \mathbf{D}\boldsymbol{\varepsilon} = \mathbf{D}\mathbf{B}\boldsymbol{\delta}^e \qquad (13.84\text{c})
$$

Note that with the three constant strain components being independent of position within the triangular element so too are the three stress components constant. The strain and stress does, however, vary from element to element and normally when showing their distributions they are referred to the centroid of each triangle.

13.7.4 Element Stiffness Matrix

An element stiffness matrix $\mathbf{K}^e$ of dimension 6×6 connects the nodal forces and displacements between Figs 13.13a,b when each is expressed as a vector:

$$
\mathbf{f}^e = [f_{x1}\ f_{y1}\ f_{x2}\ f_{y2}\ f_{x3}\ f_{y3}]^{\mathrm{T}} \qquad (13.85\text{a})
$$

$$
\boldsymbol{\delta}^e = [u_1\ v_1\ u_2\ v_2\ u_3\ v_3]^{\mathrm{T}} \qquad (13.85\text{b})
$$

Thus, in the symbolic matrix notation the relationship between Eqs 13.85a,b and, more usefully, the inverse relationship appear as:

$$
\mathbf{f}^e = \mathbf{K}^e\boldsymbol{\delta}^e \qquad \therefore \quad \boldsymbol{\delta}^e = (\mathbf{K}^e)^{-1}\mathbf{f}^e \qquad (13.86\text{a,b})
$$

in which $\mathbf{K}^e$ will now be established. The two methods outlined in § 13.2 (p. 544) may be applied. The *virtual work* Eq. 13.5a requires substitutions for virtual displacement/strain and real force/stress. Taking these from Eqs 13.78a and 13.84c:

$$(\delta^{e\mathrm{v}})^{\mathrm{T}}\mathbf{f}^e = \int_V (\mathbf{B}\delta^{e\mathrm{v}})^{\mathrm{T}}(\mathbf{D}\mathbf{B}\delta^e)\,\mathrm{d}V \qquad (13.87\mathrm{a})$$

and as $\mathbf{B}$ and $\mathbf{D}$ are independent of x and y, Eq. 13.87a is simplified:

$$(\delta^{e\mathrm{v}})^{\mathrm{T}}\mathbf{f}^e = (\delta^{e\mathrm{v}})^{\mathrm{T}}\left(\mathbf{B}^{\mathrm{T}}\mathbf{D}\mathbf{B}\delta^e\right)\int_V \mathrm{d}V \qquad (13.87\mathrm{b})$$

Cancelling the virtual displacement in Eq. 13.87b, then comparing with Eq. 13.86a, shows how the stiffness matrix is formed:

$$\mathbf{K}^e = \mathbf{B}^{\mathrm{T}}\mathbf{D}\mathbf{B}\,V \qquad (13.87\mathrm{c})$$

in which V is the element's volume. The stiffness matrix $\mathbf{K}^e$ in Eq. 13.87c may be confirmed from applying the stationary potential energy Eq. 13.10b. Substituting for the stress and strain matrices from Eqs 13.78a and 13.84c gives:

$$\frac{1}{2}\int_V \left[(\mathbf{D}\varepsilon)^{\mathrm{T}}\frac{\partial}{\partial\delta^e}(\mathbf{B}\delta^e) + \varepsilon^{\mathrm{T}}\frac{\partial}{\partial\delta^e}(\mathbf{D}\mathbf{B}\delta^e)\right]\mathrm{d}V - \mathbf{f}^{e^{\mathrm{T}}} = 0$$

$$\frac{1}{2}\int_V \left(\varepsilon^{\mathrm{T}}\mathbf{D}^{\mathrm{T}}\mathbf{B} + \varepsilon^{\mathrm{T}}\mathbf{D}\mathbf{B}\right)\mathrm{d}V - \mathbf{f}^{e^{\mathrm{T}}} = 0 \qquad (13.88\mathrm{a})$$

Equations 13.83a and 13.84a show that $\mathbf{D}^{\mathrm{T}} = \mathbf{D}$ under either plane condition. Therefore, it follows from Eq. 13.88a that:

$$\int_V \varepsilon^{\mathrm{T}}\mathbf{D}\mathbf{B}\,\mathrm{d}V - \mathbf{f}^{e^{\mathrm{T}}} = 0$$

$$\int_V (\mathbf{B}\,\delta^e)^{\mathrm{T}}\mathbf{D}\mathbf{B}\,\mathrm{d}V - \mathbf{f}^{e^{\mathrm{T}}} = 0$$

$$\mathbf{f}^{e^{\mathrm{T}}} = \delta^{e^{\mathrm{T}}}\left(\mathbf{B}^{\mathrm{T}}\mathbf{D}\mathbf{B}\right)\int_V \mathrm{d}V \qquad (13.88\mathrm{b})$$

The final operation requires the transpose of both sides of Eq. 13.88b:

$$\mathbf{f}^e = \left[\delta^{e^{\mathrm{T}}}\left(\mathbf{B}^{\mathrm{T}}\mathbf{D}\mathbf{B}\right)\right]^{\mathrm{T}}V = \left(\mathbf{B}^{\mathrm{T}}\mathbf{D}\mathbf{B}\right)^{\mathrm{T}}\delta^e V$$

$$= (\mathbf{D}\mathbf{B})^{\mathrm{T}}\mathbf{B}\delta^e V = \mathbf{B}^{\mathrm{T}}\mathbf{D}^{\mathrm{T}}\mathbf{B}\delta^e V = \mathbf{B}^{\mathrm{T}}\mathbf{D}\mathbf{B}\delta^e V \qquad (13.88\mathrm{c})$$

Again, comparing Eq. 13.88c with Eq. 13.86a confirms that $\mathbf{K}^e = \mathbf{B}^{\mathrm{T}}\mathbf{D}\mathbf{B}V$.

13.7.5 Overall Stiffness Matrix

In Fig. 13.16 a thin square plate is loaded in the x- and y-directions at its four corners as shown. It will now be shown how the overall stiffness matrix is assembled from the individual stiffness matrices of the two adjoining triangular elements shown. Despite this minimal discretisation a similar principle of assembly would be used had a multiple connection of smaller triangular elements been used. In taking the smaller number it is easier to see how the assembly procedure can be automated with skilful programming.

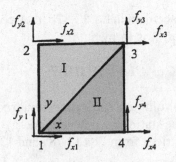

Figure 13.16 Two-element plate with nodal forces

Thus the overall stiffness matrix is assembled from the individual stiffness matrices for the plate's two triangular elements I and II. The assembly procedure that follows works symbolically with the elements k_{ij} of each stiffness matrix $\mathbf{K}^I$ and $\mathbf{K}^{II}$ and with a clockwise, sequential node numbering.

Element I: $\mathbf{F}^I = \mathbf{K}^I \boldsymbol{\delta}^I$ represents:

$$
\begin{bmatrix} f_{x1} \\ f_{y1} \\ f_{x2} \\ f_{y2} \\ f_{x3} \\ f_{y3} \end{bmatrix} = \begin{bmatrix} k_{11}^I & k_{12}^I & k_{13}^I & k_{14}^I & k_{15}^I & k_{16}^I \\ k_{21}^I & k_{22}^I & k_{23}^I & k_{24}^I & k_{25}^I & k_{26}^I \\ k_{31}^I & k_{32}^I & k_{33}^I & k_{34}^I & k_{35}^I & k_{36}^I \\ k_{41}^I & k_{42}^I & k_{43}^I & k_{44}^I & k_{45}^I & k_{46}^I \\ k_{51}^I & k_{52}^I & k_{53}^I & k_{54}^I & k_{55}^I & k_{56}^I \\ k_{61}^I & k_{62}^I & k_{63}^I & k_{64}^I & k_{65}^I & k_{66}^I \end{bmatrix} \begin{bmatrix} u_1 \\ v_1 \\ u_2 \\ v_2 \\ u_3 \\ v_3 \end{bmatrix} \qquad (13.89a)
$$

Element II: $\mathbf{F}^{II} = \mathbf{K}^{II}\boldsymbol{\delta}^{II}$ represents:

$$
\begin{bmatrix} f_{x1} \\ f_{y1} \\ f_{x3} \\ f_{y3} \\ f_{x4} \\ f_{y4} \end{bmatrix} = \begin{bmatrix} k_{11}^{II} & k_{12}^{II} & k_{13}^{II} & k_{14}^{II} & k_{15}^{II} & k_{16}^{II} \\ k_{21}^{II} & k_{22}^{II} & k_{23}^{II} & k_{24}^{II} & k_{25}^{II} & k_{26}^{II} \\ k_{31}^{II} & k_{32}^{II} & k_{33}^{II} & k_{34}^{II} & k_{35}^{II} & k_{36}^{II} \\ k_{41}^{II} & k_{42}^{II} & k_{43}^{II} & k_{44}^{II} & k_{45}^{II} & k_{46}^{II} \\ k_{51}^{II} & k_{52}^{II} & k_{53}^{II} & k_{54}^{II} & k_{55}^{II} & k_{56}^{II} \\ k_{61}^{II} & k_{62}^{II} & k_{63}^{II} & k_{64}^{II} & k_{65}^{II} & k_{66}^{II} \end{bmatrix} \begin{bmatrix} u_1 \\ v_1 \\ u_3 \\ v_3 \\ u_4 \\ v_4 \end{bmatrix}
\tag{13.89b}
$$

Elements I and II: Note, firstly, that all stiffness matrices, both individual and overall, are square and symmetrical about the leading diagonal with the elements in Eqs 13.89a,b conforming to:

$$
k_{ij}^{I} = k_{ji}^{I}, \quad k_{kl}^{II} = k_{lk}^{II} \quad \text{and} \quad k_{ij}^{I} + k_{kl}^{II} = k_{ji}^{I} + k_{lk}^{II}
$$

The overall stiffness equation $\mathbf{F} = \mathbf{K}\boldsymbol{\delta}$ is assembled from Eqs 13.89a,b ensuring that nodes 1 and 3 are common to the forces and displacements within each element. This gives:

$$
\begin{bmatrix} f_{x1} \\ f_{y1} \\ f_{x2} \\ f_{y2} \\ f_{x3} \\ f_{y3} \\ f_{x4} \\ f_{y4} \end{bmatrix} = \begin{bmatrix} k_{11}^{I}+k_{11}^{II} & k_{12}^{I}+k_{12}^{II} & k_{13}^{I} & k_{14}^{I} & k_{15}^{I}+k_{13}^{II} & k_{16}^{I}+k_{14}^{II} & k_{15}^{II} & k_{16}^{II} \\ k_{21}^{I}+k_{21}^{II} & k_{22}^{I}+k_{22}^{II} & k_{23}^{I} & k_{24}^{I} & k_{25}^{I}+k_{23}^{II} & k_{26}^{I}+k_{24}^{II} & k_{25}^{II} & k_{26}^{II} \\ k_{31}^{I} & k_{32}^{I} & k_{33}^{I} & k_{34}^{I} & k_{35}^{I} & k_{36}^{I} & 0 & 0 \\ k_{41}^{I} & k_{42}^{I} & k_{43}^{I} & k_{44}^{I} & k_{45}^{I} & k_{46}^{I} & 0 & 0 \\ k_{51}^{I}+k_{31}^{II} & k_{52}^{I}+k_{32}^{II} & k_{53}^{I} & k_{54}^{I} & k_{55}^{I}+k_{33}^{II} & k_{56}^{I}+k_{34}^{II} & k_{35}^{II} & k_{36}^{II} \\ k_{61}^{I}+k_{41}^{II} & k_{62}^{I}+k_{42}^{II} & k_{63}^{I} & k_{64}^{I} & k_{65}^{I}+k_{43}^{II} & k_{66}^{I}+k_{44}^{II} & k_{45}^{II} & k_{46}^{II} \\ k_{51}^{II} & k_{52}^{II} & 0 & 0 & k_{53}^{II} & k_{54}^{II} & k_{55}^{II} & k_{56}^{II} \\ k_{61}^{II} & k_{62}^{II} & 0 & 0 & k_{63}^{II} & k_{64}^{II} & k_{65}^{II} & k_{66}^{II} \end{bmatrix} \begin{bmatrix} u_1 \\ v_1 \\ u_2 \\ v_2 \\ u_3 \\ v_3 \\ u_4 \\ v_4 \end{bmatrix}
\tag{13.89c}
$$

Given the symmetry in $\mathbf{K}$ its individual elements k_{ij} need only be calculated from the upper half of the matrix product $\mathbf{B}^{T}\mathbf{D}\mathbf{B}$ applied to each element in Eq. 13.87c.

The following example applies Eq. 13.89c to find the displacements at the four corners of a plate in equilibrium with its applied forces. It will be seen from the solution why the two-element model is an inadequate split.

Example 13.4 A 5 mm thick, 2 m square, aluminium plate is loaded with inclined forces of 0.5 kN and 1.0 kN at its nodes 2 and 3. These loads are reacted by a hinged support at node 1 and a vertical cable anchor at node 4 (see Fig. 13.17). Determine the displacements at each corner node taking $E = 70$ GPa and $v = 1/3$. Comment upon the accuracy of the solution.

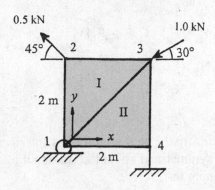

Figure 13.17 Thin plate with supports for inclined loading

Firstly, the corner forces are found from applying the equations of static equilibrium. Taking x, y forces and clockwise moments to be positive:

$$\sum f_x = 0: \quad f_{x1} - 0.5\cos 45° - 1.0\cos 30° = 0, \quad \rightarrow \quad f_{x1} = 1.2196 \text{ kN}$$

$$\sum M_1 = -2 \times 0.5\cos 45° - 2 \times 1.0\cos 30° + 2 \times 1.0\sin 30° - 2f_{y4} = 0, \quad \rightarrow f_{y4} = -0.7196 \text{ kN}$$

$$\sum f_y = f_{y1} + 0.5\sin 45° - 1.0\sin 30° + f_{y4} = 0, \quad \rightarrow \quad f_{y1} = 0.8661 \text{ kN}$$

Resolving the applied forces at nodes 2 and 3 gives the remaining nodal force components:

$$f_{x2} = -0.3536 \text{ kN}, \; f_{y2} = 0.3536 \text{ kN}, \; f_{x3} = -0.866 \text{ kN}, \; f_{y3} = -0.5 \text{ kN and } f_{x4} = 0$$

With the CW node numbering of element I we must interchange 2 and 3 for the **B** matrix in Eq. 13.78b to apply. Also, element II requires the replacement of its node numbers 2 with 4 for Eq. 13.78b to apply. Substituting the co-ordinates of each element's node in Fig. 13.17:

$$\mathbf{B}^{\text{I}} = \frac{1}{4}\begin{bmatrix} 0 & 0 & 2 & 0 & -2 & 0 \\ 0 & -2 & 0 & 0 & 0 & 2 \\ -2 & 0 & 0 & 2 & 2 & -2 \end{bmatrix}, \quad \mathbf{B}^{\text{II}} = \frac{1}{4}\begin{bmatrix} -2 & 0 & 2 & 0 & 0 & 0 \\ 0 & 0 & 0 & -2 & 0 & 2 \\ 0 & -2 & -2 & 2 & 2 & 0 \end{bmatrix}$$

Now for a thin plate the plane stress elasticity matrix $\mathbf{D}_\sigma$ in Eq. 13.83a applies to each element. With the volume of each element being:

$$V = \Delta \times t = \left(\frac{1}{2} \times 2 \times 2 \right) \times (5 \times 10^{-3}) = 10^{-2} \ \mathrm{m}^3$$

and with the coefficients for matrices $\mathbf{B}$ and $\mathbf{D}$ as, respectively

$$2\Delta = 2 \times \frac{1}{2} \times 2 \times 2 = 4 \ \mathrm{m}^2 \quad \text{and} \quad \frac{E}{1 - v^2} = \frac{70}{1 - (1/3)^2} = 78.75 \ \mathrm{GN/m}^2,$$

the stiffness for element I follows from Eq. 13.87c as $\mathbf{K}^I = \mathbf{B}^T \mathbf{D} \mathbf{B} V$:

$$\mathbf{K}^I = \frac{78.55 \times 10^{-2}}{4^2}
\begin{bmatrix}
0 & 0 & -2 \\
0 & -2 & 0 \\
2 & 0 & 0 \\
0 & 0 & 2 \\
-2 & 0 & 2 \\
0 & 2 & -2
\end{bmatrix}
\begin{bmatrix}
1 & 1/3 & 0 \\
1/3 & 1 & 0 \\
0 & 0 & 1/3
\end{bmatrix}
\begin{bmatrix}
0 & 0 & 2 & 0 & -2 & 0 \\
0 & -2 & 0 & 0 & 0 & 2 \\
-2 & 0 & 0 & 2 & 2 & -2
\end{bmatrix}$$

in which the units of $\mathbf{B}^T \mathbf{D} \mathbf{B} V$ are: $\dfrac{1}{\mathrm{m}} \times \dfrac{\mathrm{GN}}{\mathrm{m}^2} \times \dfrac{1}{\mathrm{m}} \times \mathrm{m}^3 = \dfrac{\mathrm{GN}}{\mathrm{m}}.$

$$\mathbf{K}^I = 4.922 \times 10^{-2}
\begin{bmatrix}
4/3 & 0 & 0 & -4/3 & -4/3 & 4/3 \\
0 & 4 & -4/3 & 0 & 4/3 & -4 \\
0 & -4/3 & 4 & 0 & -4 & 4/3 \\
-4/3 & 0 & 0 & 4/3 & 4/3 & -4/3 \\
-4/3 & 4/3 & -4 & 4/3 & 16/3 & -8/3 \\
4/3 & -4 & 4/3 & -4/3 & -8/3 & 16/3
\end{bmatrix} \quad \text{(i)}$$

Similarly, the stiffness for element II follows from $\mathbf{K}^{II} = \mathbf{B}^T \mathbf{D} \mathbf{B} V$. With the volume being scalar it may be combined with the constant coefficients for each matrix so that the three matrices multiply in the correct order as follows:

$$
\mathbf{K}^{II} = \frac{78.55 \times 10^{-2}}{4^2}
\begin{bmatrix}
-2 & 0 & 0 \\
0 & 0 & -2 \\
2 & 0 & -2 \\
0 & -2 & 2 \\
0 & 0 & 2 \\
0 & 2 & 0
\end{bmatrix}
\begin{bmatrix}
1 & 1/3 & 0 \\
1/3 & 1 & 0 \\
0 & 0 & 1/3
\end{bmatrix}
\begin{bmatrix}
-2 & 0 & 2 & 0 & 0 & 0 \\
0 & 0 & 0 & -2 & 0 & 2 \\
0 & -2 & -2 & 2 & 2 & 0
\end{bmatrix}
$$

in which the matrix multiplication leads to the components of the stiffness matrix $\mathbf{K}^{II}$, again in units of GN/m ($\equiv$ kN/mm):

$$
\mathbf{K}^{II} = 4.922 \times 10^{-2}
\begin{bmatrix}
4 & 0 & -4 & 4/3 & 0 & -4/3 \\
0 & 4/3 & 4/3 & -4/3 & -4/3 & 0 \\
-4 & 4/3 & 16/3 & -8/3 & -4/3 & 4/3 \\
4/3 & -4/3 & -8/3 & 16/3 & 4/3 & -4 \\
0 & -4/3 & -4/3 & 4/3 & 4/3 & 0 \\
-4/3 & 0 & 4/3 & -4 & 0 & 4
\end{bmatrix}
\tag{ii}
$$

Assembling the two matrices in Eqs i and ii the manner of Eq. 13.89c gives the following form of $\mathbf{F} = \mathbf{K\delta}$, here in the equivalent units of kN and mm. The nodal forces, and known displacements $u_1 = v_1 = 0$ have been inserted:

$$
\begin{bmatrix}
1.2196 \\
0.8661 \\
-0.3536 \\
0.3536 \\
-0.8660 \\
-0.5 \\
0 \\
-0.7196
\end{bmatrix}
= 4.922 \times 10^{-2}
\begin{bmatrix}
16/3 & 0 & 0 & -4/3 & -16/3 & 8/3 & 0 & -4/3 \\
0 & 16/3 & -4/3 & 0 & 8/3 & -16/3 & -4/3 & 0 \\
0 & -4/3 & 4 & 0 & -4 & 4/3 & 0 & 0 \\
-4/3 & 0 & 0 & 4/3 & 4/3 & -4/3 & 0 & 0 \\
-16/3 & 8/3 & -4 & 4/3 & 32/3 & -16/3 & -4/3 & 4/3 \\
8/3 & -16/3 & 4/3 & -4/3 & -16/3 & 32/3 & 4/3 & -4 \\
0 & -4/3 & 0 & 0 & -4/3 & 4/3 & 4/3 & 0 \\
-4/3 & 0 & 0 & 0 & 4/3 & -4 & 0 & 4
\end{bmatrix}
\begin{bmatrix}
0 \\
0 \\
u_2 \\
v_2 \\
u_3 \\
v_3 \\
u_4 \\
v_4
\end{bmatrix}
$$

which represents eight simultaneous equations, now in units of N and mm:

$24.78 = (0)u_1 + (0)v_1 + (0)u_2 - (4/3)v_2 - (16/3)u_3 + (8/3)v_3 + (0)u_4 - (4/3)v_4$

$17.6 = (0)u_1 + (0)v_1 - (4/3)u_2 + (0)v_2 + (8/3)u_3 - (16/3)v_3 - (4/3)u_4 + (0)v_4$

$-7.18 = (0)u_1 + (0)v_1 + (4)u_2 + (0)v_2 - (4)u_3 + (4/3)v_3 + (0)u_4 + (0)v_4$

$7.18 = (0)u_1 + (0)v_1 + (0)u_2 + (4/3)v_2 + (4/3)u_3 - (4/3)v_3 + (0)u_4 + (0)v_4$

$-17.6 = (0)u_1 + (0)v_1 - (4)u_2 + (4/3)v_2 + (32/3)u_3 - (16/3)v_3 - (4/3)u_4 + (4/3)v_4$

$-10.16 = (0)u_1 + (0)v_1 + (4/3)u_2 - (4/3)v_2 - (16/3)u_3 + (32/3)v_3 + (4/3)u_4 - (4)v_4$

$0 = (0)u_1 + (0)v_1 + (0)u_2 + (0)v_2 - (4/3)u_3 + (4/3)v_3 + (4/3)u_4 + (0)v_4$

$-14.62 = (0)u_1 + (0)v_1 + (0)u_2 + (0)v_2 + (4/3)u_3 - (4)v_3 + (0)u_4 + (4)v_4$

Writing from the first and second equations:

$$v_2 = -4u_3 + 2v_3 - v_4 - 18.585$$

$$u_2 = 2u_3 - 4v_3 - u_4 - 13.2$$

and substituting these into the remaining equations allows for the solution to the nodal displacements by Gaussian elimination. These are (in mm):

$$u_1 = 0, \ v_1 = 0; \ u_2 = -8.094, \ v_2 = 8.728,$$

$$u_3 = -7.618, \ v_3 = -4.277; \ u_4 = -3.238, \ v_4 = -5.391$$

in which the signs follow the x, y co-ordinates. The two, plane triangular finite elements used here are often referred to as constant strain triangles. They may appear too restrictive but they do allow the displacements within them to vary linearly with x and y according to Eqs 13.70a,b. Hence the solution is tenable but a greater accuracy would be expected from a finer mesh. Here with the regions of constant strain being confined to smaller triangular areas they permit the strain distribution throughout a volume to be matched more closely as the strain varies between the interconnected triangles.

Exercises

13.1 Figure 13.18 shows two bars I and II each with similar area and elastic modulus in a pin-jointed frame construction. The global co-ordinates for the three nodes are 1 (0, 0), 2 (3 , 0) and 3 (0, 4). Determine the local and global stiffness matrices for each bar and assemble these matrices to form the overall stiffness matrix for the frame when a vertical force F is applied at node 2.

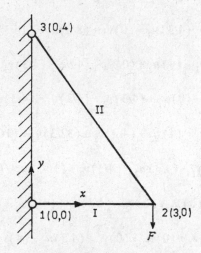

Figure 13.18

Ans to **K**:

$$\mathbf{K} = EA \begin{bmatrix} \dfrac{1}{3} & 0 & -\dfrac{1}{3} & 0 & 0 & 0 \\[2mm] 0 & 0 & 0 & 0 & 0 & 0 \\[2mm] -\dfrac{1}{3} & 0 & \dfrac{1}{3}+\dfrac{9}{125} & -\dfrac{12}{125} & \dfrac{9}{125} & \dfrac{12}{125} \\[2mm] 0 & 0 & -\dfrac{12}{125} & \dfrac{16}{125} & \dfrac{12}{125} & -\dfrac{16}{125} \\[2mm] 0 & 0 & \dfrac{9}{125} & \dfrac{12}{125} & \dfrac{9}{125} & -\dfrac{12}{125} \\[2mm] 0 & 0 & \dfrac{12}{125} & -\dfrac{16}{125} & -\dfrac{12}{125} & \dfrac{16}{125} \end{bmatrix}$$

13.2 The two bars shown in Fig. 13.19 support a vertical force F at their pinned junction each with a wall attachment at their opposite ends. The bars have the same cross-sectional area A and are each made from the same material with a Young's modulus E.

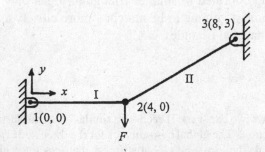

Figure 13.19

The length and orientation of each bar is defined by the global co-ordinates (x, y) for each node as follows 1 (origin 0, 0): 2 (4, 0) and 3 (8, 3). Determine the stiffness matrix for each bar element in terms of its local and global co-ordinates. Assemble each individual global stiffness matrix into an overall stiffness matrix **K** for the frame. (Ans to **K**:)

$$
\mathbf{K} = EA
\begin{bmatrix}
\frac{1}{4} & 0 & -\frac{1}{4} & 0 & 0 & 0 \\[2mm]
0 & 0 & 0 & 0 & 0 & 0 \\[2mm]
-\frac{1}{4} & 0 & \frac{1}{4}+\frac{16}{125} & \frac{12}{125} & -\frac{16}{125} & -\frac{12}{125} \\[2mm]
0 & 0 & \frac{12}{125} & \frac{9}{125} & -\frac{12}{125} & -\frac{9}{125} \\[2mm]
0 & 0 & -\frac{16}{125} & -\frac{12}{125} & \frac{16}{125} & \frac{12}{125} \\[2mm]
0 & 0 & -\frac{12}{125} & -\frac{9}{125} & \frac{12}{125} & \frac{9}{125}
\end{bmatrix}
$$

13.3 Assemble the overall stiffness matrix for the pin-jointed frame in Fig. 13.20. Determine the seven forces in bars I . . VII, given their areas and lengths are constant with included angles of 60°. (Ans in kg: $p_I = 1.05$, $p_{II} = -0.55$, $p_{III} = 0.35$, $p_{IV} = -0.65$, $p_V = -2.4$, $p_{VI} = 5.8$ and $p_{VII} = 4.1$)

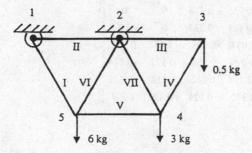

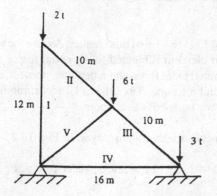

6 kg   3 kg   **Figure 13.20**

13.4 Determine the bar loading for the frame shown in Fig. 13.21 given that all bar areas are constant. The inner triangles are each isosceles and the outer triangle is right-angled. (Ans in tonne: $p_I = -2$, $p_{II} = 0$, $p_{III} = -5$, $p_{IV} = 4$ and $p_V = -5$)

Figure 13.21

13.5 Taking the origin of global co-ordinates (x, y) at note 1, determine the stiffness matrix for each of the three bars I, II and III of the cantilever frame shown in Fig. 13.22. Use both local and global co-ordinates for each bar given that the global co-ordinate for their nodes (in m) are: 1 (0, 0), 2 (12, 6) and 3(0, 10). Write down the overall stiffness matrix for the frame in global co-ordinates. The steel bars are each 10 mm diameter with $E = 200$ GPa.

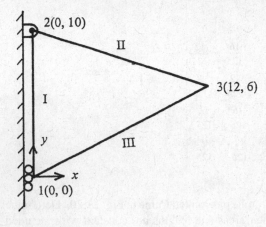

Figure 13.22

(Ans: **K** in MN/m)

$$\frac{\mathbf{K}}{100} = \begin{bmatrix} 0.936 & 0.468 & -0.936 & -0.468 & 0 & 0 \\ 0.468 & 1.804 & -0.468 & -0.234 & 0 & -1.57 \\ -0.936 & -0.468 & 2.054 & 0.096 & -1.118 & 0.372 \\ -0.468 & -0.234 & 0.096 & 0.358 & 0.372 & -0.124 \\ 0 & 0 & -1.118 & 0.372 & 1.118 & -0.372 \\ 0 & -1.57 & 0.372 & -0.124 & -0.372 & 1.694 \end{bmatrix}$$

13.6 Use the principals of virtual work and stationary potential energy to confirm the $\mathbf{K}^e$ matrix appearing in Eq. 13.42a, for a solid cylindrical shaft element under torsion.

13.7 Find the overall stiffness matrix when the stepped shaft under torsion in Fig. 13.7 suffers additional nodal displacements $u_1, u_2 \ldots u_5$, due to the application of axial nodal forces, f_1, $f_2 \ldots f_5$.

13.8 Using the stiffness matrices derived in § 13.3–§ 13.6 of this chapter, deduce the stiffness matrix in the case of a uniform circular bar element subjected to: (i) nodal forces f_x and f_y combined with an axial torque t, (ii) a combined axial force and a bending moment, and (iii) a combined axial force, bending moment and a torque. Take the axial force, moment and torque in (ii) and (iii) to vary between the two nodes of the element.

13.9 Determine the nodal forces and moments for the beam shown in Fig. 13.23 using a single element for each of its two bays.
(Ans: bending moments, tm : -14.953, -2.902, 15.127, -2.958; shear forces, t : -2.991, 2.009, 6.009, - 4.521, 1.479)

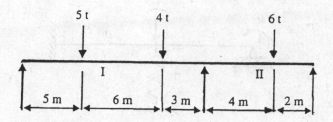

Figure 13.23

13.10 Assuming that the distributed loading upon the beam in Fig. 13.24 can be replaced with equal concentrated forces applied above each of its four supports, determine the forces and moments at nodes 1, 2, 3 and 4. (Ans: moments: $m_1 = m_4 = 0$, $m_2 = m_3 = 14.4$ kNm; shear forces: $f_1 = f_4 = -4.8$ kN, $f_2 = f_3 = -13.2$ kN)

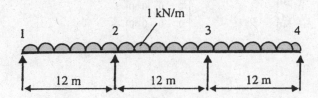

Figure 13.24

13.11 Write down the slope and deflection at node 1 and the moment and shear force at node 3 for the beam shown in Fig. 13.25. Show how these appear within the overall stiffness matrix with the beam's division into three elements I, II and II, as shown.

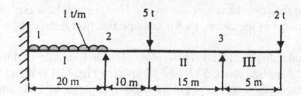

Figure 13.25

13.12 Sub-divide the beam shown in Fig.13.26 into suitable beam elements and express the individual stiffness of those elements in terms of E, I, L and k_o. Connect the load vector for the external loads to the vector for their nodal point displacements with the structure's overall, reduced stiffness matrix, i.e. $f_r = K_r \delta_r$.

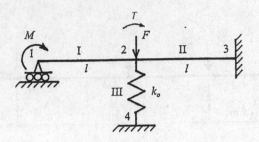

Figure 13.26

(Ans to stiffness matrices):

$$
\mathbf{K}^{\mathrm{I}} =
\begin{array}{c}
\begin{array}{cccc} v_1 & \theta_1 & v_2 & \theta_2 \end{array} \\
\begin{bmatrix}
0 & 0 & 0 & 0 \\
0 & k_{22}^{\mathrm{I}} & k_{23}^{\mathrm{I}} & k_{24}^{\mathrm{I}} \\
0 & k_{32}^{\mathrm{I}} & k_{33}^{\mathrm{I}} & k_{34}^{\mathrm{I}} \\
0 & k_{42}^{\mathrm{I}} & k_{43}^{\mathrm{I}} & k_{44}^{\mathrm{I}}
\end{bmatrix}
\end{array}
\qquad
\mathbf{K}^{\mathrm{II}} =
\begin{array}{c}
\begin{array}{cccc} v_2 & \theta_2 & v_3 & \theta_3 \end{array} \\
\begin{bmatrix}
k_{11}^{\mathrm{II}} & k_{12}^{\mathrm{II}} & 0 & 0 \\
k_{21}^{\mathrm{II}} & k_{22}^{\mathrm{II}} & 0 & 0 \\
0 & 0 & 0 & 0 \\
0 & 0 & 0 & 0
\end{bmatrix}
\end{array}
\qquad
\mathbf{K}^{\mathrm{III}} =
\begin{array}{c}
\begin{array}{cc} v_2 & v_4 \end{array} \\
\begin{bmatrix}
k_0 & 0 \\
0 & 0
\end{bmatrix}
\end{array}
$$

$$
\begin{bmatrix}
M \\
F \\
T
\end{bmatrix}
=
\begin{bmatrix}
k_{22}^{\mathrm{I}} & k_{23}^{\mathrm{I}} & k_{24}^{\mathrm{I}} \\
k_{32}^{\mathrm{I}} & k_{33}^{\mathrm{I}}+k_{11}^{\mathrm{II}}+k_0 & k_{34}^{\mathrm{I}}+k_{12}^{\mathrm{II}} \\
k_{42}^{\mathrm{I}} & k_{43}^{\mathrm{I}}+k_{21}^{\mathrm{II}} & k_{44}^{\mathrm{I}}+k_{22}^{\mathrm{II}}
\end{bmatrix}
\begin{bmatrix}
\theta_1 \\
v_2 \\
\theta_2
\end{bmatrix}
\qquad \rightarrow \qquad \mathbf{f}_r = \mathbf{K}_r\,\boldsymbol{\delta}_r
$$

13.13 Expand Eq. 13.74b fully for **A** and $(\mathbf{A}^e)^{-1}$ given by Eq. 13.71a and Eq. 13.73b respectively. Hence show that Eq. 13.75c applies to each node of a plane triangular element where x and y are the co-ordinates of that node. Hence show how Eq. 13.75c provides the non-zero components of the element's 6×6 shape function matrix **N**.

13.14 Find the nodal displacements for the plate given in Example 13.4 when, in its support of the two inclined applied loads, 5 kN and 10 kN (see Fig. 13.17), there is again a hinge at node 1 and a cable anchor at node 4 but inclined at an angle of 135° anticlockwise to the positive x-direction.

INDEX

Printed in the United States
By Bookmasters